FUNDAMENTALS OF PROGRAMMABLE LOGIC CONTROLLERS, SENSORS, AND COMMUNICATIONS

Third Edition

Jon Stenerson

PEARSON
Prentice Hall

Upper Saddle River, New Jersey
Columbus, Ohio

Library of Congress Cataloging-in-Publication Data

Stenerson, Jon.
Fundamentals of programmable logic controllers, sensors, and communications / Jon Stenerson.—3rd ed.
p. cm.
ISBN 0-13-061890-X (alk. paper)
1. Programmable controllers. 2. Telecommunication systems. I. Title.

TJ223.P76S74 2005
629.8'9—dc22 2003062448

Editor in Chief: Stephen Helba
Editor: Charles E. Stewart, Jr.
Assistant Editor: Mayda Bosco
Production Editor: Kevin Happell
Design Coordinator: Diane Ernsberger
Cover Designer: Kristina D. Holmes
Production Manager: Matthew Ottenweller
Marketing Manager: Ben Leonard

This book was set in 10/12 Times Roman by Carlisle Communications, Ltd. It was printed and bound by R.R. Donelley & Sons Company. The cover was printed by Phoenix Color Corp.

Pearson Education Ltd.
Pearson Education Singapore Pte. Ltd.
Pearson Education Canada, Ltd.
Pearson Education—Japan
Pearson Education Australia Pty. Limited
Pearson Education North Asia Ltd.
Pearson Educación de Mexico, S.A. de C.V.
Pearson Education Malaysia Pte. Ltd.

PEARSON
Prentice Hall

10 9 8 7 6 5 4 3 2 1
ISBN 0-13-061890-X

In memory of Tim Sevener of Control Technology Corporation

Brief Contents

Contents

Preface

First, thank you for choosing this textbook.

I am excited about this edition. This text includes a wonderful software package that can be used to learn how to program PLCs. (This software is included courtesy of Mr. Bill Simpson of TheLearningPit.com.) The software includes simulators so you can program and then watch the results. It is based on Rockwell Automation PLC programming. Several programming problems are included that utilize the software and simulators. The software enables the reader to learn programming on a home computer.

I am also excited about three new chapters on process control and instrumentation. For years I searched for easy-to-understand, practical material on process control and instrumentation. I developed the process control material to meet that need. I found that students had trouble learning and understanding PLC addressing, so I completely rewrote the addressing material to make it easy to learn and apply. I also found that many students have difficulty wiring PLC modules, so I completely revised the chapter on PLC wiring to make it easy to understand and apply as well. I added a chapter on robotics. I added material on safety to the lockout/tagout chapter. The communications chapters and other chapters have been updated and improved.

Each chapter begins with a generic approach to the topic and is explained clearly through the use of common, easy-to-understand, generic examples. Examples are then shown for specific controllers. The brands covered are AutomationDirect, GE Fanuc, Gould Modicon, Omron, and Rockwell Automation. There are many illustrations and practical examples. Each chapter has objectives and questions. Many chapters have programming exercises, and some also have programming exercises and assignments that utilize the included programming and simulation software. After completing the reading and exercises, the reader can easily learn any new industrial controller.

The text also covers new and emerging technologies, such as IEC 61131-3 programming, industrial automation controllers, ControlLogix, embedded controllers, supervisory control and data acquisition, fuzzy logic, and new programming techniques such as step, stage, and state logic programming. The book is not intended to replace the technical manual for any specific controller. The book does explain the common instructions, however, so that the reader can efficiently learn the use of new instructions from any technical manual.

- Chapter 1 examines safety and lockout/tagout. It covers safe work practices and lockout/tagout practices.
- Chapters 2–4 provide the basic foundation for the use of PLCs. Chapter 2 focuses on the history and fundamentals of the PLC; Chapter 3 covers number systems; Chapter 4 covers contacts, coils, and the fundamentals of programming.
- Chapter 5 focuses on Rockwell Automation addressing. Addressing can be a confusing topic and can inhibit progress in learning to program PLCs. This chapter takes a practical and understandable approach to the topic. Addressing for other brands of PLCs are covered in other chapters. (ControlLogix addressing is covered in Appendix B.)
- Chapter 6 examines I/O modules and wiring and covers digital and analog modules. The chapter also takes a step-by-step approach to wiring I/O modules. Practical examples make wiring understandable.
- Chapter 7 covers the programming and use of timers and counters for various brands of PLCs.
- Chapter 8 examines math instructions. Common arithmetic instructions, including add, subtract, multiply, divide, and compare, are covered for various PLCs. More complex and versatile instructions are also covered.
- Chapter 9 explains special instructions such as copy, move, block transfer, PID, time stamp, diagnostic, sequencers, shift registers, and many others. Examples of the use of these instructions help the reader understand their use.
- Chapter 10 covers advanced programming. Sequential logic, shift registers, step logic, stage logic, fuzzy logic, and state logic are all covered.
- Chapter 11 looks at industrial sensors and their wiring. The chapter focuses on types and uses of sensors. Sensors covered include photo, inductive, capacitive, encoders, resolvers, ultrasonic, and thermocouples. The wiring and practical application of sensors are stressed.
- Chapter 12 is an introduction to IEC 61131-3 programming methods. IEC 61131-3 is the standard that covers programming languages for industrial controllers. It is becoming much more common in PLCs and other controllers. ControlLogix controllers can utilize IEC 61131 programming languages.

- Chapter 13 covers supervisory control and data acquisition (SCADA) software. The fundamentals of SCADA are covered, and then a practical SCADA application is developed.
- Chapter 14 examines plant communication. Device communications are sure to increase in importance as companies integrate their enterprises. This chapter provides a foundation for the integration of plant floor devices.
- Chapter 15 looks at industrial networks. Industrial networks will radically change the way industry automates and integrates, and thus will change the demands on technicians and maintenance personnel.
- Chapter 16 considers PC-based control. PCs are gaining acceptance as industrial controllers. This chapter looks at some of the various approaches taken by software developers for PC-based control. PC-based hardware is also covered briefly. ControlLogix is examined.
- Chapter 17 begins the coverage of process control. This chapter examines open and closed-loop control. Proportional, integral, and derivative (PID) is also examined. Tuning of PID systems is covered.
- Chapter 18 focuses on the topic of instrumentation. Various measurement input and control output devices are covered in practical detail.
- Chapter 19 covers process control systems. Batch and continuous processes are illustrated with two practical examples of systems: beer brewing and whiskey distilling.
- Chapter 20 examines embedded controllers. Examples of simple applications and programming are included.
- Chapter 21 covers the fundamentals of robotics.
- Chapter 22 looks at industrial automation controllers. Industrial automation controllers are a new breed of controller. They are aimed at special applications such as motion control. These controllers are powerful and can easily be used to develop machine and motion control applications.
- Chapter 23 focuses on the installation and troubleshooting of PLC systems. The chapter begins with a discussion of cabinets, wiring, grounding, and noise. The chapter then covers troubleshooting. The ability to troubleshoot effectively is vital for technicians. This chapter provides the groundwork for proper installation and troubleshooting of integrated systems.
- Appendix A shows some common I/O device symbols. Appendix B covers ControlLogix addressing and project organization. A glossary and index conclude the text to help the reader define and locate terms and concepts.

Acknowledgments

This text would not have been possible without the help I received from corporations, companies, and individuals. I would especially like to thank the following companies for their assistance:

- AutomationDirect Inc.
- Control Technology Corporation
- ifm efector
- InTouch Corporation
- Omron Electronics, Inc.
- Rockwell Automation, Inc.
- Z-World Inc.

I would also like to thank the reviewers of this edition: Professor Ed Margraff, Marion Technical College; and Professor Larry B. Masten, Texas Tech University.

chapter 1

Safety and Lockout/Tagout

Safety should be the number-one concern in the workplace because of the numerous dangers: machinery, electrical hazards, and so forth. Stay on guard against the dangers to protect yourself and others. In this chapter we examine safe practices and lockout/tagout procedures.

OBJECTIVES

Upon completion of this chapter, you will be able to:

1. Define an accident and why accidents happen.
2. Describe how accidents can be prevented.
3. Describe lockout/tagout and how it is used in industry.
4. Develop a lockout/tagout procedure.

SAFETY

Be mindful of your safety and the safety of others as you work.

What Is an Accident?

An accident is an unexpected action that results in injury to people, machines, or tools. Note that an accident is always *unexpected.* We never think accidents will happen to us, but they can happen to anyone. An accident can easily cause you to lose a limb, your sight, or even your life. Accidents cost enterprises huge amounts of money.

Most accidents are minor, such as minor cuts, but all are important. We should look for ways to prevent even minor accidents from reoccurring because they can be worse than the first time. Every company has procedures for reporting even minor accidents, and all accidents should be reported. Companies often report accidents on incident report forms.

Causes of Accidents

Accidents can be prevented. The following list shows possible causes of accidents:

1. Carelessness.
2. People who ignore safety rules or bypass safety equipment.
3. The use of improper tooling or defective tools, or the improper use of tools.
4. Unsafe working practices such as lifting heavy objects incorrectly.
5. Horseplay (playing around on the job).

Common sources of accidents include the following:

1. Machines with moving or rotating parts.
2. Electrical machinery.
3. Equipment that uses high-pressure fluids.
4. Chemicals.
5. Sharp objects on machines or tools.

Remember that every piece of equipment is dangerous!

Accident Prevention

Here are a few ways to prevent accidents:

1. *Design safety into the equipment.* Machinery should be designed to prevent accidents. Guards and interlocks on machines can prevent injury from moving parts and electrical shock. There are lockouts on electrical panels so that a worker can lock the power out to repair the equipment.

Never remove guards or safety interlocks from machines while working on them.

2. *Use proper clothing and eye and hearing protection.* You must have self-discipline. Safety glasses are required in most workplaces. If you are in a noisy environment, wear ear protection also. Never wear gloves, loose clothing, or jewelry around moving machinery.
3. *Follow warning signs.* Warning signs indicate high voltage and other dangers. A yellow/black line means Danger—Do Not Cross This Line.
4. *Follow safety procedures closely.* Know the company policies regarding safety in all areas.

Safe Use of Lab Equipment and Hand Tools

It is important to follow safe procedures and learn good habits that will keep you safe on the job. Following is an important list of safety guidelines:

1. Wear proper safety equipment.
2. Wear proper eye protection.
3. Know how to use the equipment. Do not operate equipment you do not understand.
4. Do not hurry.
5. Do not fool around.
6. Keep the work area clean.
7. Keep the floor dry.
8. Always beware of electricity.
9. Use adequate light.
10. Know first aid.
11. Handle tools carefully.
12. Keep tools sharp. A dull tool requires more pressure and is more likely to cause injury.
13. Use the right tool.
14. Beware of fire or fumes. If you think you smell fire, stop immediately and investigate.
15. Use heavy extension cords. Thin extension cords can overheat, melting the insulation to expose the wires and cause a fire.

Overview of Lockout/Tagout

On October 30, 1989, the Lockout/Tagout Standard, 29 CFR 1910.147, went into effect. It was released by the Department of Labor. The standard is titled, "The Control of Hazardous Energy Sources (Lockout/Tagout)." The standard was intended to reduce the number of deaths and injuries related to servicing and maintaining machines and equipment. Deaths and tens of thousands of lost work days are attributable to maintenance and servicing activities each year.

The lockout/tagout standard covers the servicing and maintenance of machines and equipment in which the unexpected start-up or energizing of the machines or equipment, or the release of stored energy, could cause injury to employees. The standard is intended to cover energy sources such as electrical, mechanical, hydraulic, chemical, nuclear, and thermal. The standard establishes only minimum guides for the control of such hazardous energy; normal production operations, cords and plugs under exclusive control, and hot tap operations are not covered by the standard.

Hot tap operations are those involving transmission and distribution systems for substances such as gas, steam, water, or petroleum products. A hot tap is a procedure used in repair maintenance and in service activities that involve welding on a piece of equipment, such as on pipelines, vessels, or tanks under pressure, to install connections or appurtenances. Hot tap procedures are commonly used to replace or add sections of pipeline without the interruption of service for air, gas, water, steam, and petrochemical distribution systems. The standard does not apply to hot taps when they are performed on pressurized pipelines, provided that the employer demonstrates that continuity of service is essential, that shutdown of the system is impractical, that documented procedures are followed, and that special equipment is used to provide proven, effective protection for employees.

Employers are required to establish a program consisting of an energy control (lockout/tagout) procedure and employee training to ensure that before any employee performs any servicing or maintenance on a machine or equipment, where the unexpected energizing, start-up, or release of stored energy could occur and cause injury, the machine or equipment shall be isolated and rendered inoperative. The employer is also required to conduct annual inspections of the energy control procedure to ensure that the procedures and the requirements of this standard are being followed.

The standard defines an *energy source* as any source of electrical, mechanical, hydraulic, pneumatic, chemical, thermal, or other energy. Machinery or equipment is considered to be energized if it is connected to an energy source or contains residual or stored energy. Stored energy can be found in pneumatic and hydraulic systems, springs, capacitors, and even gravity.

Servicing and/or maintenance includes activities such as constructing, installing, setting up, adjusting, inspecting, modifying, and maintaining and/or servicing machines or equipment. These activities include lubricating, cleaning, or unjamming machines or equipment, and making adjustments or tool changes, during which time the employee may be exposed to the unexpected energizing or start-up of the equipment or release of hazardous energy.

Normal production operations are excluded from lockout/tagout restrictions. Normal production operation is the utilization of a machine or equipment to perform its intended production function. Any work performed to prepare a machine or equipment to perform its normal production operation is called setup.

If an employee is working on cord and plug–connected electrical equipment—for which exposure to unexpected energizing or start-up of the equipment is controlled by the unplugging of the equipment from the energy source, and the plug is under the exclusive control of the employee performing the servicing or maintenance—this activity is also excluded from requirements of the standard.

Only authorized employees may lock out machines or equipment. An authorized employee is one who has been trained and has the authority to lock or tag out machines or equipment to perform the servicing or maintenance.

An energy isolating device is a mechanical device that physically prevents the transmission or release of energy. Energy isolating devices include the following: manually operated electrical circuit breakers; disconnect switches (Figure 1–1); manually operated switches by which the conductors of a circuit can be disconnected from all ungrounded supply conductors and, in addition, no pole can be operated independently; line valves (Figure 1–2); and locks or similar devices used to block or isolate energy. Push buttons, selector switches, and other control circuit type devices are not energy isolating devices.

An energy isolating device is capable of being locked out if it has a hasp or other means of attachment to which, or through which, a lock can be affixed, or it has a built-in locking mechanism. Other energy isolating devices are capable of being locked out, if lockout can be achieved without the need to dismantle, rebuild, or replace the energy isolating device or permanently alter its energy control capability. Since January 2, 1990, industry has required that new machines or equipment with energy isolating devices be designed to accept a lockout device.

Lockout

Lockout is the placement of a lockout device on an energy isolating device, in accordance with an established procedure, to ensure that the device cannot be operated until the lockout device is removed.

A lockout device utilizes a positive means (such as a lock) to hold an energy isolating device in the safe position and prevent the energizing of a machine or equipment. A lock may be either key or combination type. If an energy isolating device is incapable of being locked out, then the employer's energy control program shall utilize a tagout system.

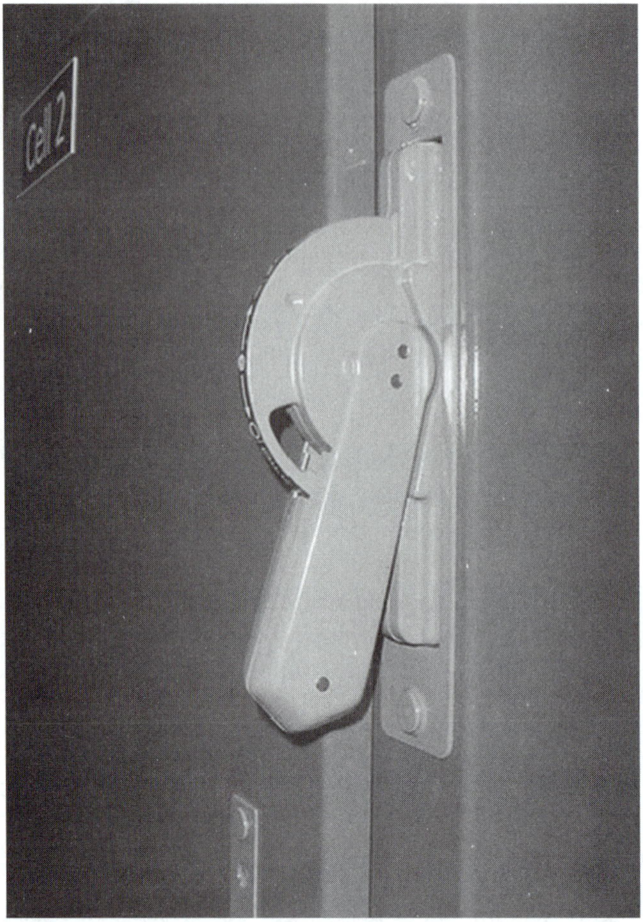

Figure 1–1 A typical electrical disconnect. In this case, it is not locked out.

Notification of Employees Affected employees must be notified by the employer or authorized employee of the application and removal of lockout devices or tagout devices. Notification shall be given before the controls are applied, and after they are removed from the machine or equipment. Affected employees are defined as employees whose job requires them to operate or use a machine or equipment on which servicing or maintenance is being performed under lockout or tagout, or whose job requires them to work in an area in which such servicing or maintenance is being performed.

Figure 1–2 A typical pneumatic disconnect.

Tagout

Tagout is the placement of a tagout device on an energy isolating device, in accordance with an established procedure, to indicate that the energy isolating device and the equipment being controlled may not be operated until the tagout device is removed. Tagout shall be performed only by the authorized employees who are performing the servicing or maintenance.

A tagout device is a prominent warning device, such as a tag and a means of attachment, that can be securely fastened to an energy isolating device in accordance with an established procedure. This procedure indicates that the energy isolating device and the equipment being controlled may not be operated until the tagout device is removed from each energy isolating device by the employee who applied the device. When the authorized employee who applied the lockout or tagout device is not available to remove it, that device may be removed under the direction of the employer, provided that specific procedures and training for such removal have been developed, documented, and incorporated into the energy control program.

Tagout devices, where used, must be affixed to indicate clearly that the operation or movement of energy isolating devices from the "safe" or "off" position is prohibited. Where tagout devices are used with energy isolating devices designed with the capability of being locked, the tag attachment must be fastened at the same point at which the lock would have been attached. Where a tag cannot be affixed directly to the energy isolating device, the tag must be located as close as safely possible to the device, in a position that will be immediately obvious to anyone attempting to operate the device.

TRAINING

Training must be provided by the employer to ensure that the purpose and function of the energy control program are understood by employees and that the knowledge and skills required for the safe application, usage, and removal of energy controls are attained by employees. Employees should be trained to do the following:

Recognize hazardous energy sources.

Know the type and magnitude of the energy available in the workplace.

Understand the methods and means for energy isolation and control.

Understand the purpose and use of the lockout/tagout procedures.

All other employees whose work operations are or may be in an area where lockout/tagout procedures may be used shall be instructed about the procedure and about the prohibition against attempting to restart or reenergize machines or equipment which are locked or tagged out.

When tagout procedures are used, employees must be taught about the following limitations of tags:

Tags are really just warning devices and do not provide physical restraint on equipment.

When a tag is attached it is not to be removed without authorization of personnel responsible for it, and it is never to be bypassed, ignored, or otherwise defeated.

Tags must be legible. All authorized employees, affected employees, and others who work in the area must understand the tags.

Tags may create a false sense of security. Everyone must understand the purpose of them.

Retraining

Retraining shall be provided for all authorized and affected employees when there is a change in job assignments; a change in machines, equipment, or processes that present a new hazard; or a change in the energy control procedures.

Additional retraining shall also be conducted when a periodic inspection reveals, or when the employer has reason to believe, that there are deviations from or inadequacies in the employee's knowledge or use of the energy control procedures.

The retraining shall reestablish employee proficiency and introduce new or revised control methods and procedures, as necessary.

The employer shall certify that employee training has been accomplished and is being kept up to date. The certification shall contain each employee's name and dates of training.

Requirements for Lockout/Tagout Devices

Lockout and tagout devices must be singularly identified, must be the only device(s) used for controlling energy, and must not be used for other purposes. The devices must be durable, which means they must be capable of withstanding the environment to which they are exposed for the maximum period of time that exposure is expected. Tagout devices must be constructed and printed so that exposure to weather conditions or wet and damp locations will not cause the tag to deteriorate or the message on the tag to become illegible (Figure 1–3). Tags must not deteriorate when used in corrosive environments such as areas where acid and alkali chemicals are handled and stored.

Lockout and tagout devices must be standardized within the facility according to either color, shape, or size. Print and format must also be standardized for tagout devices.

Lockout devices must be substantial enough to prevent removal without the use of excessive force or unusual techniques, such as with the use of bolt cutters or other metal-cutting tools.

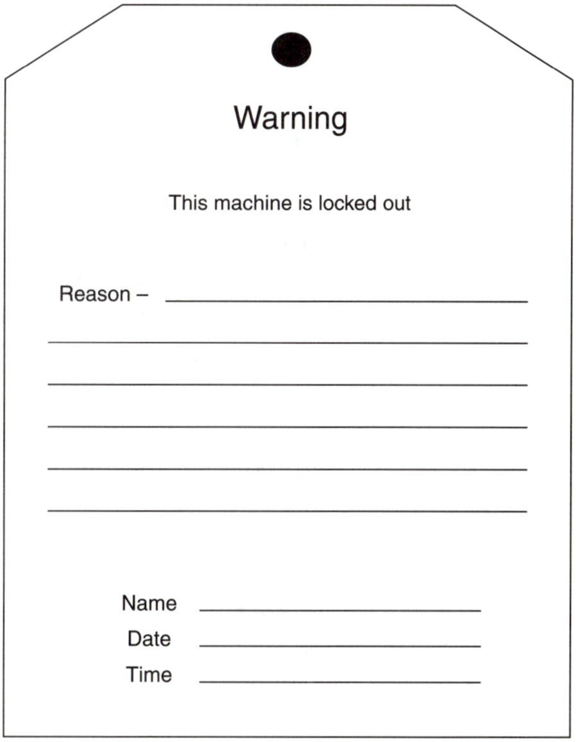

Figure 1–3 A typical tagout tag.

Tagout devices, including their means of attachment, shall be substantial enough to prevent inadvertent or accidental removal. Tagout devices must be attached with a nonreusable type of attachment. It must be attachable by hand, self-locking, and nonreleasable with a minimum unlocking strength of at least 50 pounds. It should have the general design and basic characteristics of being at least equivalent to a one-piece, all environment–tolerant nylon cable tie.

Lockout and tagout devices must identify the employee who applied them.

Tagout devices must clearly warn against hazardous conditions if the machine or equipment is energized, such as Do Not Start, Do Not Open, Do Not Close, Do Not Energize, or Do Not Operate.

Application of Control

The established procedures for the application of lockout or tagout shall cover the following elements and actions and shall be done in the following sequence:

1. Notify all affected employees that a lockout or tagout system will be used. They must also understand the reason for the lockout. Before an authorized or affected employee turns off a machine or equipment, the authorized employee must understand the types and magnitudes of the energy, the hazards of the energy to be controlled, and the method or means to control the energy for the machine or equipment being serviced or maintained.
2. The machine or equipment shall be turned off or shut down using the procedures established for the machine or equipment. An orderly shutdown must be utilized to avoid any additional or increased hazards to employees as a result of the equipment stoppage.
3. All energy isolating devices needed to control the energy to the machine or equipment shall be physically located and operated so that they isolate the machine or equipment from the energy sources.
4. Lockout or tagout devices shall be affixed to each energy isolating device by authorized employees. Lockout devices, where used, shall be affixed in a manner that will hold the energy isolating devices in a safe or off position.
5. Following the application of lockout or tagout devices to energy isolating devices, stored energy must be dissipated or restrained by methods such as repositioning, blocking, and bleeding down. If there is a possibility of reaccumulation of stored energy to a hazardous level, verification of isolation shall be continued until the servicing or maintenance is complete, or until the possibility of such accumulation of energy no longer exists.

6. Prior to working on machines or equipment that has been locked out or tagged out, the authorized employee shall verify that the machine or equipment has actually been isolated and de-energized. This is done by operating the push button or other normal operating controls to ensure that the equipment will not operate.

Caution: Be sure to return the operating controls to the neutral or off position after the test.

The machine is now locked or tagged out. Before lockout or tagout devices are removed and energy is restored to the machine or equipment, authorized employees take action to ensure the following:

- The work area shall be inspected to ensure that nonessential items have been removed and that machine or equipment components are operationally intact.
- The work area shall be checked to ensure that all employees have been safely positioned or removed.
- Before lockout or tagout devices are removed and before machines or equipment are energized, affected employees shall be notified that the lockout or tagout devices have been removed.

Each lockout or tagout device must be removed from each energy isolating device by the employee who applied the device. The only exception to this stipulation is when that person is unavailable to remove it. The device may then be removed under the direction of the employer, provided that specific procedures and training for such removal have been developed, documented, and incorporated into the employer's energy control program. The employer must demonstrate that the specific procedure includes at least the following elements:

- The employer must verify that the authorized employee who applied the device is not at the facility.
- All reasonable efforts must be made to contact the authorized employee to inform him or her that the lockout or tagout device has been removed.
- The authorized employee must be made aware that the lockout/tagout device was removed before he or she resumes work at that facility.

Testing of Machines, Equipment, or Components

In certain situations, the lockout or tagout devices may be temporarily removed from the energy isolating device and the machine or equipment energized to test or position the machine, equipment, or component. In this case, the following sequence of actions must be taken:

1. Clear the machine or equipment of tools and materials.
2. Remove employees from the machine or equipment area.
3. Remove the lockout or tagout devices as specified in the standard.
4. Energize and proceed with testing or positioning.
5. De-energize all systems and reapply energy control measures in accordance with the standard to continue the servicing and/or maintenance.

Outside Personnel Working in the Plant

When outside servicing personnel (such as contractors) are to be engaged in activities covered by the lockout/tagout standard, the onsite employer and the outside employer must inform each other of the respective lockout or tagout procedures. The onsite employer must ensure that all employees understand and comply with the restrictions and prohibitions of the outside employer's energy control program.

Group Lockout or Tagout

When servicing and/or maintenance is performed by a group of people, they must use a procedure to protect them to the same degree as would a personal lockout or tagout procedure. The lockout/tagout standard specifies requirements for group procedures. Primary responsibility is vested in an authorized employee for a set number of employees. These employees work under the protection of a group lockout or tagout device (Figure 1–4). The group lockout device ensures that no individual can start up or energize the machine or equipment. All lockout or tagout devices must be removed to reenergize the machine or equipment. The authorized employee who is responsible for the group must ascertain the exposure status of individual group members with regard to the lockout or tagout of the machine or

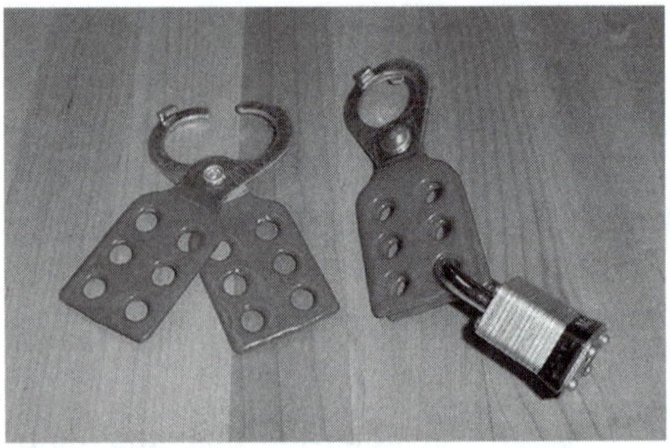

Figure 1–4 A hasp allows multiple personnel to lock out machines or equipment.

equipment. When more than one crew, craft, or department is involved, overall job-associated lockout or tagout control responsibility is assigned to an authorized employee. This employee is designated to coordinate affected workforces and ensure continuity of protection. Each authorized employee must affix a personal lockout or tagout device to the group lockout device, group lock-box, hasp (Figure 1–4), or comparable mechanism when beginning work, and shall remove those devices when stopping work on the machine or equipment being serviced or maintained.

Shift or Personnel Changes

Specific procedures must be utilized during shift or personnel changes to ensure the continuity of lockout or tagout protection, and therefore the orderly transfer of lockout or tagout device protection between shifts of employees, to minimize exposure to hazards from the unexpected energizing or start-up of the machine or equipment, or the release of stored energy.

Sample Lockout Procedure

Below is a sample lockout procedure. Tagout procedures may be used when the energy isolating devices are not lockable, provided the employer complies with the provisions of the standard that require additional training and more rigorous periodic inspections. When tagout is used and the energy isolating devices are lockable, the employer must provide full employee protection and additional training as well as more rigorous periodic inspections. When more complex systems are involved, additional comprehensive procedures may need to be developed, documented, and utilized.

Lockout Procedure for Machine 37

Note: This would normally name the machine when multiple procedures exist. If only one exists, it would normally be the company name.

Purpose

This procedure establishes the minimum requirements for the lockout of energy isolating devices when maintenance or servicing is done on machine 37. This procedure must be used to ensure that the machine is stopped, isolated from all potentially hazardous energy sources, and locked out before employees perform any servicing or maintenance during which time the unexpected energizing or start-up of the machine or equipment or the release of stored energy could cause injury.

Employee Compliance

All employees, including authorized employees, are required to comply with the restrictions and limitations imposed on them during the use of this lockout procedure. All employees, upon observing a machine or piece of equipment locked out to perform servicing or maintenance, shall not attempt to start, energize, or use that machine or equipment.

(A company may want to list disciplinary actions taken in the event of an employee violating the procedure.)

Lockout Sequence

1. Notify all affected employees that servicing or maintenance is required on the machine and that the machine must be shut down and locked out to perform the servicing or maintenance.

   The procedure should list the names and/or job titles of affected employees and how to notify them.

2. The authorized employee must refer to the company procedure to identify the type and magnitude of the energy that the machine utilizes, must understand the hazards of the energy, and must know the methods to control the energy.

   The types and magnitudes of energy, their hazards, and the methods to control the energy should be detailed here.

3. If the machine or equipment is operating, shut it down by the normal stopping procedure (e.g., depress the stop button, open switch, close valve).

   The types and locations of machine or equipment operating controls should be detailed here.

4. De-activate the energy isolating devices so that the machine or equipment is isolated from the energy sources.

   The types and locations of energy isolating devices should be detailed here.

5. Lock out the energy isolating devices with assigned individual locks.

6. Stored or residual energy (such as that in capacitors; springs; elevated machine members; rotating flywheels; hydraulic systems; and air, gas, steam, or water pressure) must be dissipated or restrained by methods such as grounding, repositioning, blocking, or bleeding down.

   The types of stored energy, as well as methods to dissipate or restrain the stored energy, should be detailed here.

7. Ensure that the equipment is disconnected from the energy sources by first checking that no personnel are exposed; then verify the isolation of the equipment by operating the push button or other normal operating controls or by testing to make certain the equipment will not operate. *Caution:* Return operating controls to the neutral or off position after you verify the isolation of the equipment.

   The method of verifying the isolation of the equipment should be detailed here.

8. The machine or equipment is now locked out.

Returning the Machine or Equipment to Service

When the servicing or maintenance is complete and the machine or equipment is ready to return to normal operating condition, the following steps are taken:

1. Check the machine or equipment and the immediate area around the machine to ensure that nonessential items have been removed and that the machine or equipment components are operationally intact. Check the work area to ensure that all employees have been safely positioned or removed from the area.
2. After all tools have been removed from the machine or equipment, and guards have been reinstalled and employees are in the clear, remove all lockout or tagout devices. Verify that the controls are in neutral and reenergize the machine or equipment. Note that the removal of some forms of blocking may require reenergization of the machine before safe removal. Notify affected employees that the servicing or maintenance is complete and the machine or equipment is ready for use.

Sample Lockout/Tagout Checklist

Notification

I have notified all affected employees that a lockout is required and the reason for the lockout.

Date _____ Time _____ Signature ______________________

Shutdown

I understand the reason the equipment is to be shut down following normal procedures.

Date _____ Time _____ Signature ______________________

Disconnection of Energy Sources

I operated the switches, valves, and other energy isolating devices so that each energy source has been disconnected or isolated from the machinery or equipment. I have dissipated or restrained all stored energy such as springs, elevated machine members, capacitors, rotating flywheels, and pneumatic and hydraulic systems.

Date _____ Time _____ Signature ______________________

Lockout

I have locked out the energy isolating devices using my assigned individual locks.

Date _____ Time _____ Signature ______________________

Safety Check

After ensuring that no personnel are exposed to hazards, I have operated the start button and other normal operation controls to ensure that all energy sources have been disconnected and that the equipment will not operate.

Date _____ Time _____ Signature ______________________

The Machine Is Now Locked Out.

QUESTIONS

1. What is an accident?
2. Explain three ways to prevent accidents.
3. Explain at least five precautions you can take to be safe in the lab.
4. List and explain the sources of energy that are typically found in an industrial environment.
5. Who is an affected employee?
6. Who is an authorized employee?
7. Define the term *lockout.*
8. Define the term *tagout.*
9. Describe the typical steps in a lockout/tagout procedure.
10. Write a lockout/tagout procedure for a cell that contains electrical and pneumatic energy.

chapter 2

Overview of Programmable Logic Controllers

Within a short time, programmable logic controllers (PLCs) have become an integral and invaluable tool in industry. In this chapter we examine how and why PLCs have gained such wide application. We also take an overall look at the basics of a PLC.

OBJECTIVES

Upon completion of this chapter, you will be able to:

1. Explain some of the reasons why PLCs are replacing hardwired logic in industrial automation.
2. Explain terms such as *ladder logic, CPU, programmer, input devices,* and *output devices.*
3. Explain some of the features of a PLC that make it an easy tool for an electrician to use.
4. Explain how the PLC is protected from electrical noise.
5. Draw a block diagram of a PLC.
6. Explain the types of programming devices available.

PLC Components

The programmable logic controller (PLC) is really just an industrial computer in which the hardware and software have been specifically adapted to both the industrial environment and electrical technician. Figure 2–1 shows the functional components of a typical PLC. Note the similarity to a computer.

Central Processing Unit

The central processing unit (CPU) is the brain of the PLC (Figure 2–2). It contains one or more microprocessors to control the PLC. The CPU also handles the communication and interaction with other components of the system. The CPU contains the same type of microprocessor as found in a microcomputer, except that the program used with the PLC microprocessor is written to accommodate ladder logic instead of other programming languages. The CPU executes the operating system, manages memory, monitors inputs, evaluates the user logic (ladder diagram), and turns on the appropriate outputs.

The factory floor is a noisy environment where motors, motor starters, wiring, welding machines, and even fluorescent lights create electrical noise. PLCs are hardened to be noise immune.

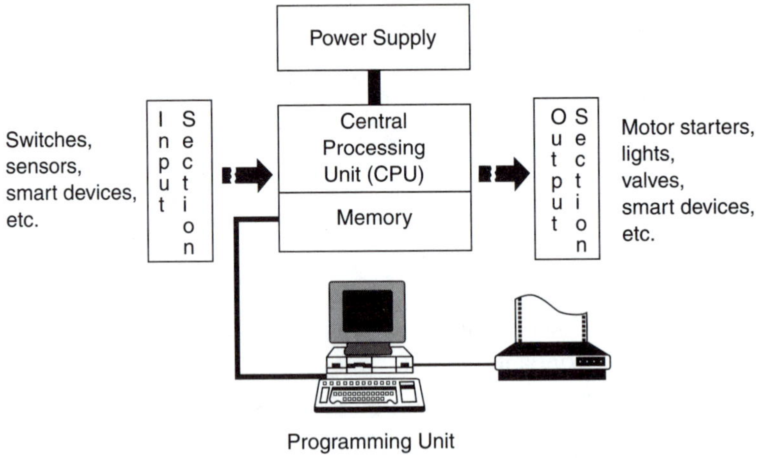

Figure 2–1 Block diagram of the typical components that comprise a PLC. Note particularly the input section, the output section, and the central processing unit (CPU). The broken arrows from the input section to the CPU and from the CPU to the output section represent protection that is necessary to isolate the CPU from the real-world inputs and outputs. The programming unit is used to write the control program (ladder logic) for the CPU. It is also used for documenting the programmer's logic and troubleshooting the system.

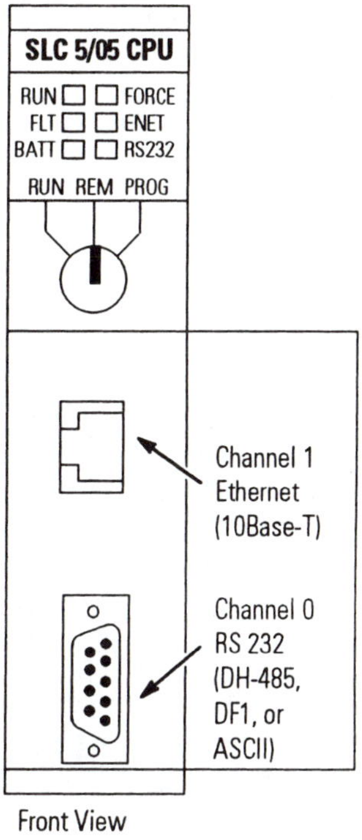

Figure 2–2 Rockwell PLC processor module (CPU). Note the many indicators for error checking and the keyswitch for switching modes between run and program mode. *(Courtesy Rockwell Automation Inc.)*

PLCs also have elaborate memory-checking routines to be sure that the PLC memory has not been corrupted by noise or other problems. Memory checking is undertaken for safety reasons. It helps ensure that the PLC will not execute if memory is corrupted.

CPU Modes

The keyswitch on Rockwell PLC CPUs is used to switch among program, run, and remote modes.

Program Mode In the program mode, all outputs are forced to the off condition regardless of their state in the logic. The program mode can be used to:

Develop and download/upload programs.

Set up input/output (I/O) forcing without them being enabled.

Transfer program files to and from backup memory modules.

Run Mode In the run mode, the PLC continuously scans and executes the ladder logic.

Remote Mode In the remote mode, a computer that is attached can control in which mode the processor is operating. The user can switch the SLC between the run and program modes from the computer without changing the key position.

CPU Models

5/01 The SLC 5/01 has program memory options of 1K or 4K instruction words. It can address up to 256 local I/Os. The 5/01 has DH485 peer-to-peer communications.

5/02 The SLC 5/02 has a more advanced instruction set than the 5/01. It has 4K of program memory. It can address up to 480 local I/Os that can be expanded via remote I/O or DeviceNet. Up to 4000 remote inputs and 4000 remote outputs can be addressed. The 5/02 has DH485 communications.

5/03 The SLC 5/03 has 12K words and 4K of additional data words; 960 local I/Os can be addressed. An additional 4000 inputs and 4000 outputs can be addressed remotely. The 5/03 has two communications ports: DH485 and RS232/DF1 or DH485 port. The 5/03 also has a real-time clock; online programming; indirect addressing; and additional math, block transfer, time stamping, scale instructions, and advanced diagnostics.

5/04 The 5/04 can have 12K, 28K, or 60K words of memory and an additional 4K of data words. The 5/04 has all the features and capabilities of the 5/03, as well as a Data Highway Plus (DH+) port and an RS232/DF1, DH485, or ASCII protocol. The 5/04 has a real-time clock and a math coprocessor, and it can be programmed online. Networking between 5/04s is accomplished by utilizing DH+. Networking to a microcomputer network can be done through either a Rockwell KTX card or ControlLogix.

5/05 The 5/05 has the features of the 5/04 plus Ethernet communications capability. The 5/05 can be networked utilizing Ethernet communications.

Memory

PLC memory can be of various types. Some of the PLC memory is used to hold operating system memory and some is used to hold user memory.

Operating System Memory Read-only memory (ROM) is used by the PLC for the operating system. The operating system is burned into ROM by the PLC manufacturer and controls functions such as the system software used to program the

PLC. The ladder logic that the programmer creates is a high-level language. A high-level language is a computer language that makes it easy for people to program. The system software must convert the electrician's ladder diagram (high-level language program) to instructions that the microprocessor can understand. ROM is not changed by the user. ROM is nonvolatile memory, which means that even if the electricity is shut off, the memory is retained.

User Memory The memory of a PLC is broken into blocks that have specific functions. Some sections of memory are used to store the status of inputs and outputs. These are normally called I/O image tables. The states of inputs and outputs are kept in I/O image tables. The real-world state of an input is stored as either a 1 or a 0 in a particular bit of memory. Each input or output has one corresponding bit in memory (Figures 2–3 and 2–4). Other portions of the memory are used to store the contents of variables of a user program. For example, a timer or counter value would be stored in this portion of memory. Memory is also reserved for processor work areas.

Random Access Memory Random access memory (RAM) is designed so that the user can read or write to the memory. RAM is commonly used for user

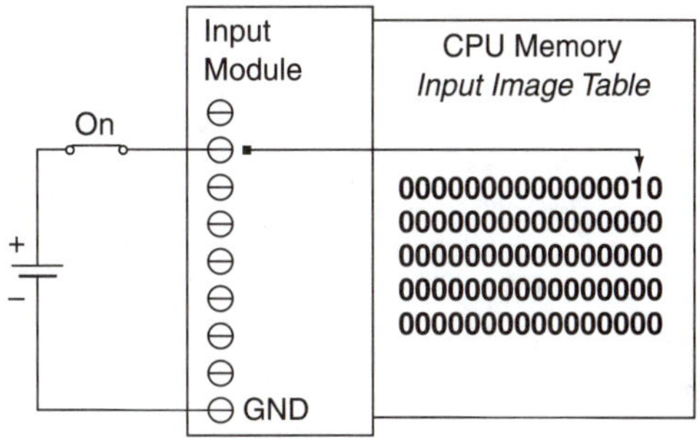

Figure 2–3 How the status of a real-world input becomes a 1 or a 0 in a word of memory. Each bit in the input image table represents the status of one real-world input.

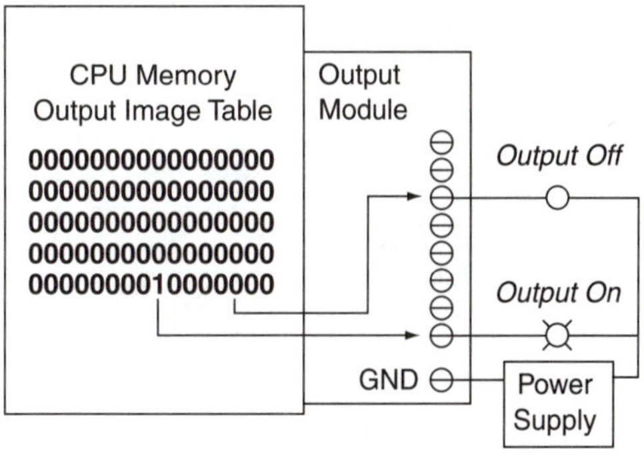

Figure 2–4 How a bit in memory controls one output. Using active-high logic, if the bit is a 1, the output will be on; if the bit is a 0, the output will be off.

memory. The user's program, timer/counter values, input/output status, and so forth, are stored in RAM.

RAM is volatile, which means that if the electricity is shut off, the data in memory are lost. This problem is solved by the use of a lithium battery. The battery takes over when the PLC is shut off. Most PLCs use CMOS-RAM technology for user memory. CMOS-RAM chips have very low current draw and can maintain memory with a lithium battery for a long time—2 to 5 years in many cases. A good preventive maintenance program should include a schedule to change batteries so serious losses can be avoided.

Figure 2–5 shows an example of battery replacement for a Rockwell SLC. Such processors have a capacitor that provides at least 30 minutes of battery backup while the battery is being changed. The data in RAM are not lost if the battery is replaced within 30 minutes. To replace the battery, the technician (1) removes the battery from the retaining clips, (2) inserts a new battery into the retaining clips, (3) plugs the battery connector into the socket, and (4) reinstalls the module into the

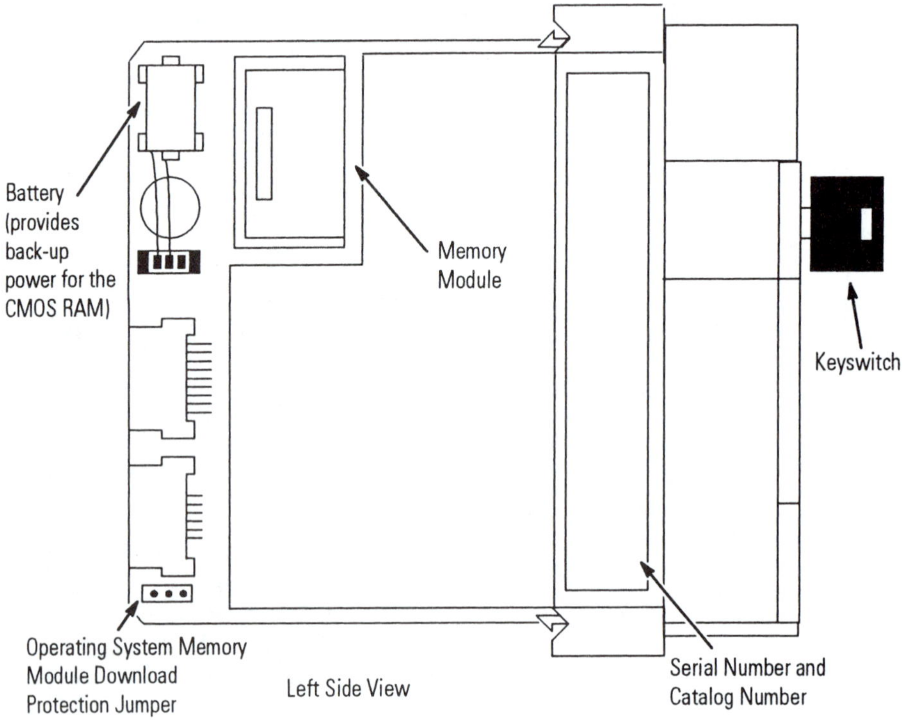

Figure 2–5 Battery replacement in a Rockwell SLC. *(Courtesy Rockwell Automation Inc.)*

rack. The battery in an SLC will last for approximately 2 years. The BATT LED on the front of the processor will light when the battery voltage falls below a threshold level.

Electrically Erasable Programmable Read-Only Memory Electrically erasable programmable read-only memory (EEPROM) can function in almost the same manner as RAM. The EEPROM can be erased electrically. It is nonvolatile memory, and so it does not require battery backup. Functionally, EEPROM is almost like a hard drive.

PLC Programming Devices

Computers are the main devices used to program PLCs. Handheld programmers are also available. Programming devices do not need to be attached to the PLC once the ladder is written because they are simply used to write the user program for the PLC. They also may be used to troubleshoot the PLC.

Handheld Programmers Handheld programmers (terminals) (Figure 2–6) must be attached to a PLC to be used. They are handy for troubleshooting because they

Figure 2–6 Handheld terminal.

can be carried easily to the manufacturing system and plugged into the PLC. Once plugged in, they can be used to monitor the status of inputs, outputs, variables, counters, timers, and so on. This feature eliminates the need to carry a large programming device to the factory floor. Handheld programmers can also be used to turn inputs and outputs on or off for troubleshooting. Turning I/O off or on by overriding the logic is called forcing. Handheld programmers are designed for the factory floor. They typically have membrane keypads that are immune to the contaminants in the factory environment. One disadvantage of these programmers is that they can show only a small amount of a ladder on the screen at one time.

Microcomputers The microcomputer is the most commonly used programming device. It can be used for offline programming and the storage of programs. One disk can hold many ladder diagrams. The microcomputer also can upload and download programs to a PLC and can be used to force inputs and outputs on and off.

This upload/download capability is vital for industry. Occasionally, PLC programs are modified on the factory floor to get a system running for a short time. It is vital that once the system has been repaired, the correct program is reloaded into the PLC. It is also useful to verify from time to time that the program in the PLC has not been modified, which helps avoid dangerous situations on the factory floor. Some automobile manufacturers have set up communications networks that regularly verify the programs in PLCs to ensure that they are correct.

The microcomputer also can be used to document the PLC program. Notes for technicians can be added and the ladder can be output to a printer for hardcopy so that the technicians can study the ladder diagram.

RSLogix 500 is an example of Windows-based microcomputer software for programming Rockwell Automation PLCs (Figure 2–7). RSLogix 500 is a powerful software package that allows offline and online programming. It allows ladder diagrams to be stored to a floppy disk, and then uploaded/downloaded to and from the PLC. The software allows monitoring of the ladder operation while it is executing, and the forcing of system I/Os on and off. Both features are extremely valuable for troubleshooting. RSLogix 500 also allows the programmer to document the ladder. This documentation is invaluable for understanding and troubleshooting ladder diagrams. The programmer can add notes, names of input or output devices, and comments that may be useful for troubleshooting and maintenance. The addition of notes and comments helps technicians to understand the ladder diagram, which allows them, not just the person who developed it, to troubleshoot the system. The notes and comments could even specify replacement part numbers, if so desired, to facilitate rapid repair of any problems due to faulty parts.

Previously, the person who developed the system had great job security because no one else could understand the operation. A properly documented ladder allows any technician to understand it.

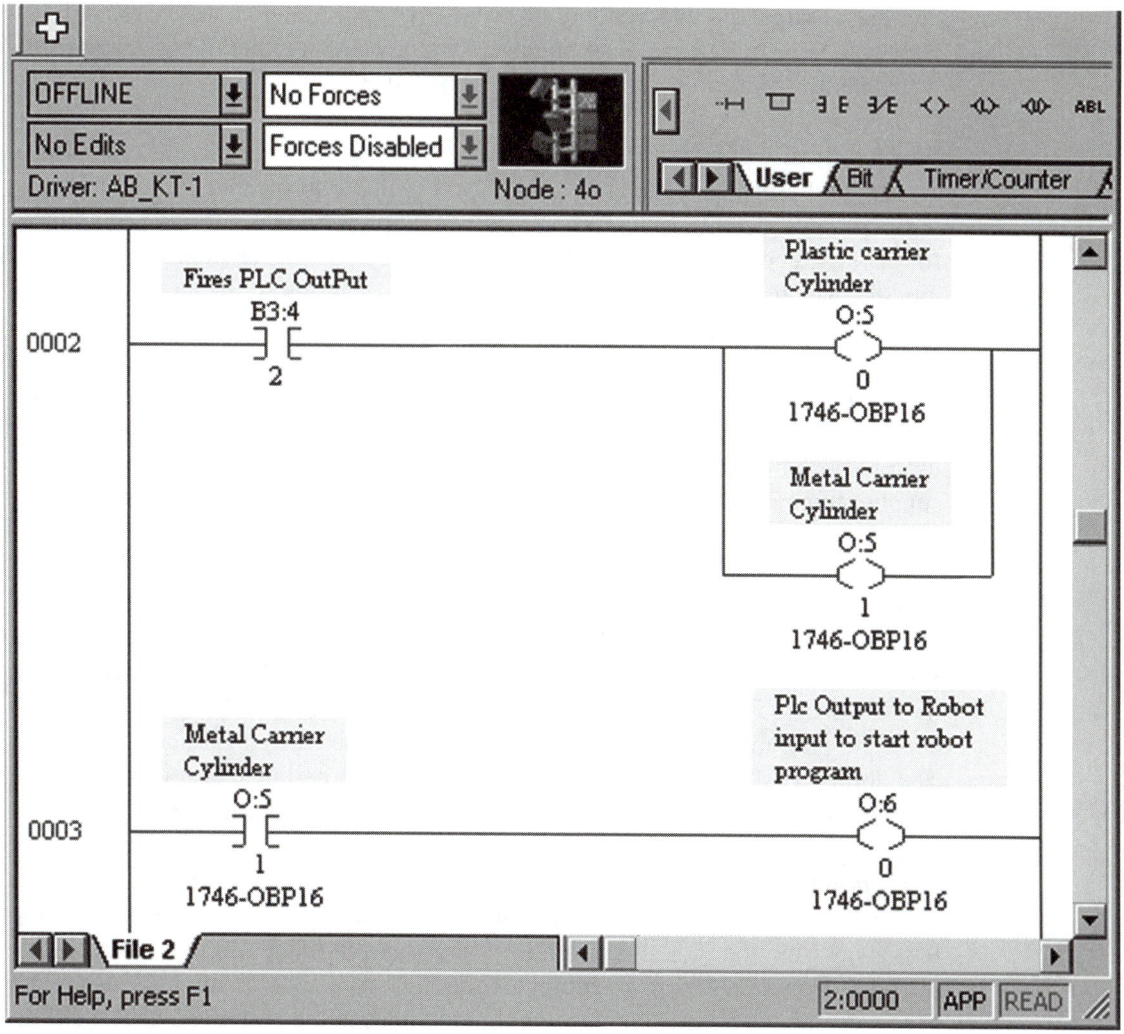

Figure 2–7 Typical PLC programming software. Note the comments and tag names to make the ladder logic more understandable. *(Courtesy Rockwell Automation Inc.)*

IEC 61131-3 Programming The International Electrotechnical Commission (IEC) has developed a standard for PLC programming. The latest IEC standard (IEC 61131-3) has attempted to merge PLC programming languages under one international standard. PLCs can now be programmed in function block diagrams, instruction lists, and C and structured text. The standard is accepted by an increasing number of suppliers and vendors of process control systems, safety-related systems, and industrial personal computers. An increasing number of application software vendors offer products based on IEC 61131-3.

Power Supply The power supply is used to supply power for the central processing unit. Most PLCs operate on 115 VAC, which means that the input voltage to the power supply is 115 VAC. The power supply provides various DC voltages for the PLC components and CPU. On some PLCs, the power supply is a separate module, which is usually the case when extra racks are used. Each rack must have its own power supply.

The user must determine how much current will be drawn by the I/O modules to ensure that the power supply provides adequate current. Different types of modules draw different amounts of current. *Note:* The PLC power supply is not typically used to power external inputs or outputs. The user must provide separate power supplies to power the inputs and outputs of the PLC. Some of the smaller PLCs do supply voltage to be used to power the inputs, however.

PLC Racks The PLC rack serves several functions. It is used to hold physically the CPU, power supply, and I/O modules. The rack also provides the electrical connections and communications among the modules, power supply, and CPU through the backplane. The modules are plugged into slots on the rack (Figure 2–8). This ability to plug modules in and out easily is one reason why PLCs are so popular. The ability to change modules quickly allows rapid maintenance and repair. Input/output numbering corresponds to a slot that holds a specific module.

Sometimes it is necessary to have more than one rack. Some midsize and large PLCs provide this feature. In certain applications, there are more I/O points than one rack can handle. There are also cases when it is desirable to locate some of the I/Os away from the PLC. For example, imagine a very large machine. Rather than run wires from every input and output to the PLC, an extra rack might be used. The I/O on one end of the machine is wired to modules in the remote rack. The I/O on the other end of the machine is wired to the main rack. The two racks communicate with one set of wiring rather than running all of the wiring from one end to the other. Many PLCs allow the use of multiple racks. When more than one rack is used it is necessary to identify the rack where the I/O is located.

Input Section The input portion of the PLC performs two vital tasks. It takes inputs from the outside world and in the process protects the CPU. Inputs can be almost any device. The input module converts the real-world logic level to the logic level required by the CPU. For example, a 250 VAC input module would convert a 250 VAC input to a low-level DC signal for the CPU.

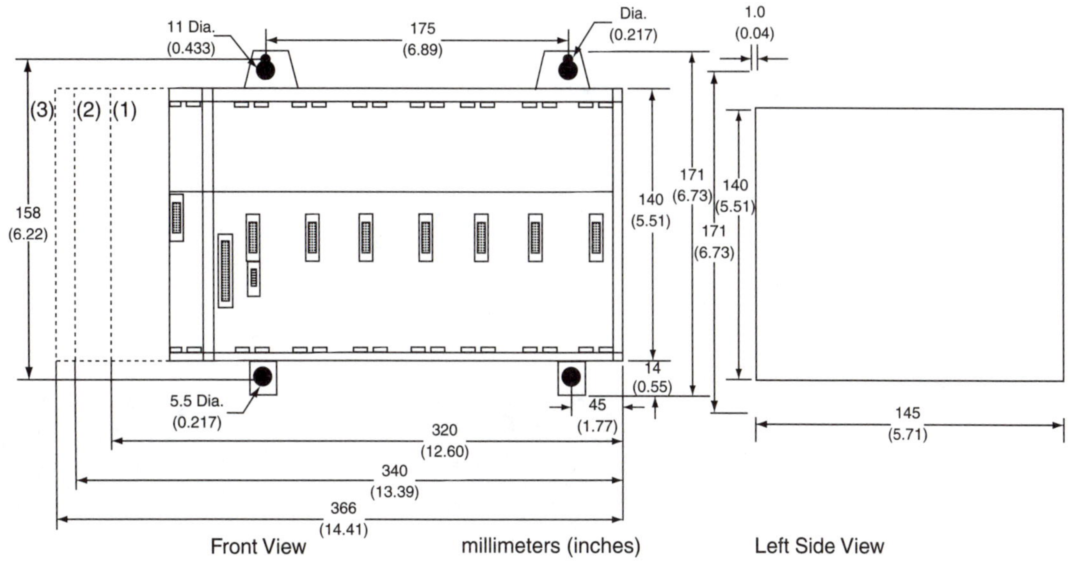

Figure 2–8 A seven-slot rack (modular chassis). *(Courtesy Rockwell Automation Inc.)*

Input Modules There are a wide variety of input modules available. Some are available for digital and analog input. Certain modules have 4, 8, 16, and 32 inputs available. Digital modules are available for AC, DC, and transistor-transistor logic (TTL) (5 volt) signals. Modules are available for positive or negative input voltages. Analog modules are available to measure voltage or current. Some are user configurable to do either. Specialty modules are available for inputs such as thermocouples, RTDs (temperature sensors), and high-speed signals such as counters or encoders. Combination I/O modules are also available.

Common input devices include switches and sensors. These are often called field devices. Field devices are gaining extensive capability, especially the ability to communicate over industrial communications networks. Other smart devices, such as robots, computers, and even PLCs, can act as inputs to the PLC.

The inputs are provided through the use of input modules. The user simply chooses input modules that will meet the needs of the application. These modules are installed in the PLC rack (Figure 2–9).

Optical Isolation The other task that the input section of a PLC performs is isolation. The PLC CPU must be protected from the outside world and at the same time be able to take input data from there. This is typically done by optical isolation, or

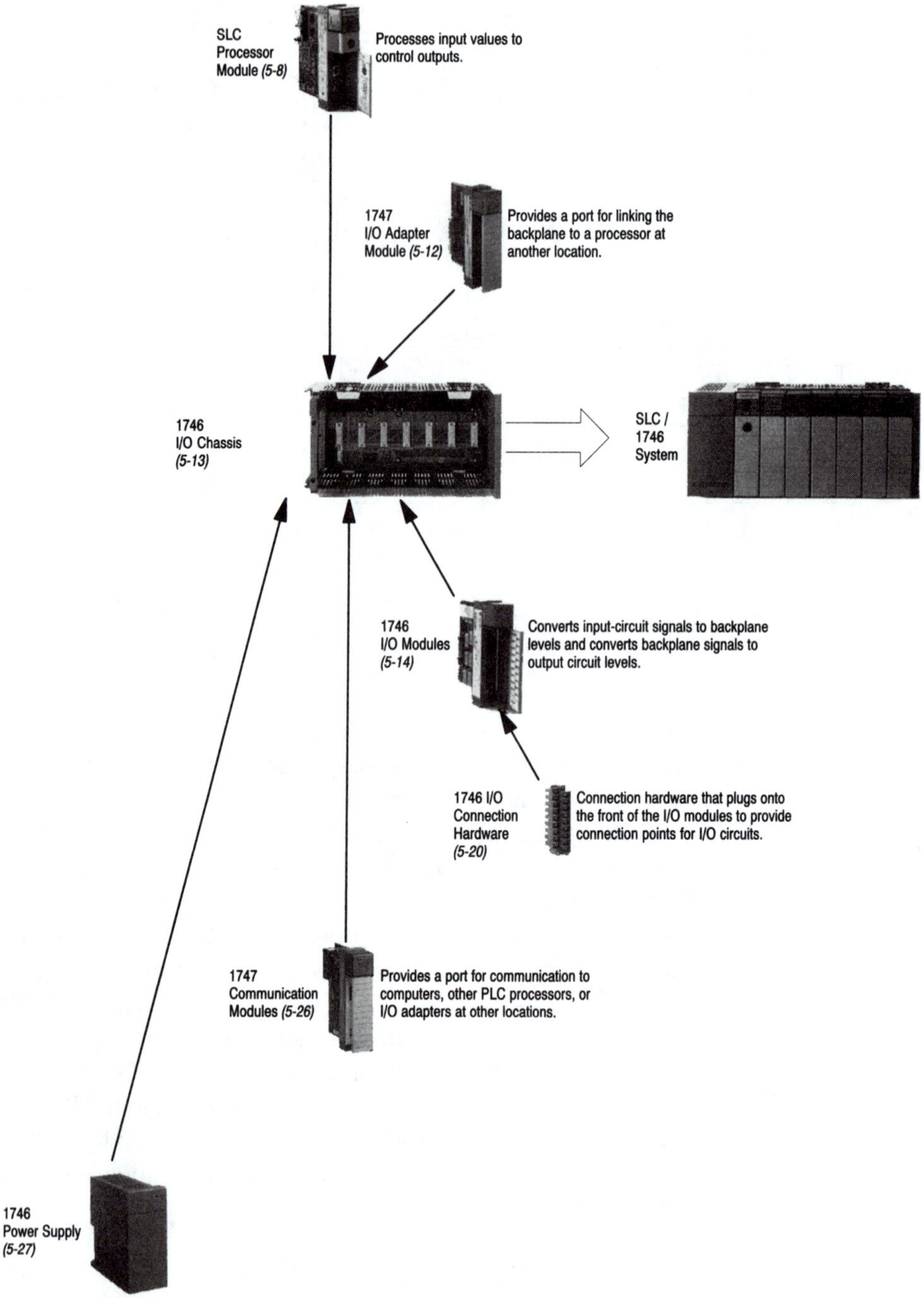
SLC
Processor
Module (5-8)
Processes input values to control outputs.
1747
I/O Adapter
Module (5-12)
Provides a port for linking the backplane to a processor at another location.
1746
I/O Chassis
(5-13)
SLC /
1746
System
1746
I/O Modules
(5-14)
Converts input-circuit signals to backplane levels and converts backplane signals to output circuit levels.
1746 I/O
Connection
Hardware
(5-20)
Connection hardware that plugs onto the front of the I/O modules to provide connection points for I/O circuits.
1747
Communication
Modules (5-26)
Provides a port for communication to computers, other PLC processors, or I/O adapters at other locations.
1746
Power Supply
(5-27)

Figure 2–9 Modules, racks, and a rack filled with modules. *(Courtesy Rockwell Automation Inc.)*

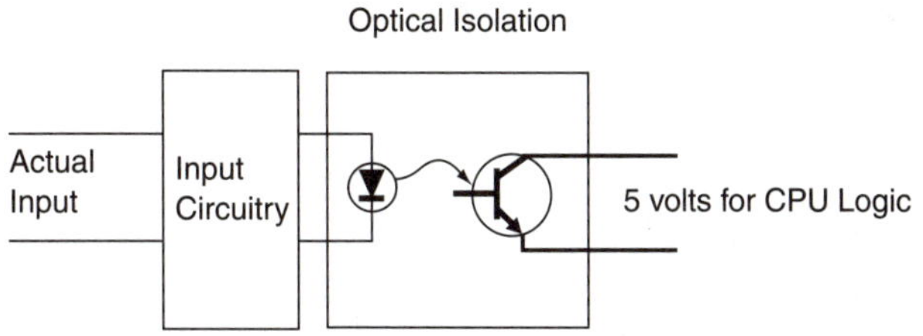

Figure 2–10 Typical optical isolation circuit. The arrow represents the fact that only light travels between the input circuitry and the CPU circuitry. There is no electrical connection.

opto-isolation (Figure 2–10), which means that there is no electrical connection between the outside world and the CPU. The two are separated optically (with light). The outside world supplies a signal that turns on a light in the input card. The light shines on a receiver and the receiver turns on. There is no electrical connection between the two.

The light separates the CPU from the outside world up to very high voltages. Even if there was a large surge of electricity, the CPU would be safe. (Of course, if the voltage is too large, the opto-isolator can fail and may cause a circuit failure.) Optical isolation is used for inputs and outputs.

Input modules provide the user with various troubleshooting aids. There are normally light emitting diodes (LEDs) for each input. If the input is on, the CPU should see the input as a high (or a 1). Input modules also provide circuits that debounce the input signal. Many input devices are mechanical and have contacts. When these devices close or open, unwanted "bounces" occur that close and open the contacts. Debounce circuits ensure that the CPU sees only debounced signals. The debounce circuit also helps eliminate the possibility of electrical noise from firing the inputs.

Inputs to Input Modules Sensors are commonly used as inputs to PLCs and can be purchased for a variety of purposes. They can sense part presence; count pieces; measure temperature, pressure, or size; and sense for proper packaging. There are also sensors that can sense any type of material. Inductive sensors can sense ferrous metal objects, capacitive sensors can sense almost any material, and optical sensors can detect any type of material. Other devices can also act as inputs to a PLC. Smart devices such as robots, computers, and vision systems often have the ability to send signals to a PLC's input modules (Figure 2–11). This can be used for handshaking during operation. A robot, for example, can send the PLC an input when it has finished a program.

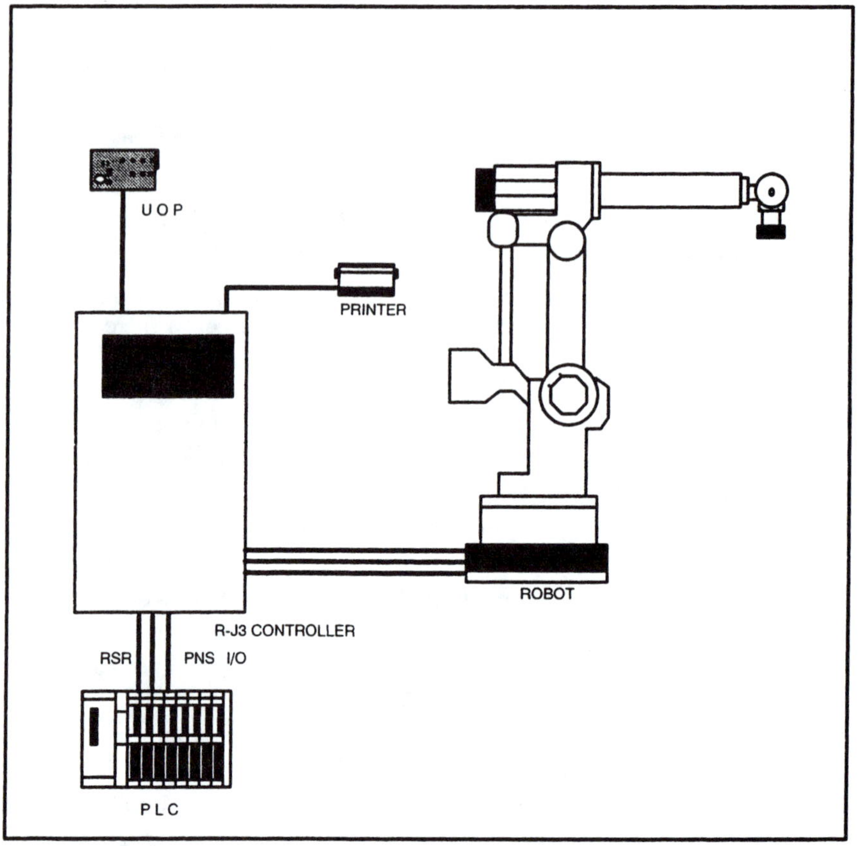

Figure 2–11 A "smart" device can also act as an input device to a PLC. The PLC can also output to the robot. Robots and some other devices typically have a few digital outputs and inputs available for this purpose. The use of these inputs and outputs allows for some basic handshaking between devices. Handshaking means that the devices give each other permission to perform tasks at some times during execution to ensure proper performance and safety. When devices communicate with a digital signal, it is called primitive communication. *(Courtesy Fanuc Robotics North America, Inc.)*

Output Section The output section of the PLC provides the connection to real-world output devices such as motor starters, lights, coils, and valves. These, too, are often called field devices. Field devices can be either input or output devices. Output modules can be purchased to handle DC or AC voltages. They can be used to output analog or digital signals. A digital output module acts like a switch. The output is either energized or deenergized. If the output is energized, then the output is turned on, like a switch.

Output modules can be purchased with various output configurations. Digital and analog output modules are available. Digital modules are available for DC and AC voltages. They are available as modules with eight, sixteen, and thirty-two outputs. Modules with more than eight outputs are sometimes called high-density modules. They are generally the same size as the eight-output modules but have many more components within the module. High-density modules therefore will not handle as much current for each output because of the size of the components and the heat generated by them.

The analog output module is used to output an analog signal. An example of this is a motor whose velocity we would like to control. An analog module puts out a voltage that corresponds to the desired speed. Analog modules are available for current or voltage. Combination modules are also available.

Current Ratings Module specifications will list an overall current rating and an output current rating. For example, the specification may give each output a current limit of 1 ampere (A). If there are eight outputs, we would assume that the output module overall rating would be 8A, but this is poor logic. The overall rating of the module current may be less than the total of the individual components. The overall rating might be 6A. The user must take this into consideration when planning the system. Normally, each of the eight devices would not pull their 1A at the same time. Figure 2–12 shows an example of I/O wiring.

Output Image Table The output image table is part of CPU memory (Figure 2–13). The user's logic determines whether an output should be on or off. The CPU evaluates the user's ladder logic. If it determines that an output should be on, it stores a 1 in the bit that corresponds to that output. The 1 in the output image table is used to turn on the actual output through an isolation circuit (Figure 2–14).

The outputs of small PLCs are often relays. This allows the user to mix and match output voltages or types. For example, some outputs could then be AC and some DC. Relay output modules are also available for some of the larger PLCs. The other choices are transistors for DC outputs and triacs for AC outputs. Many types of field devices can be connected to outputs. Figure 2–15 shows a few examples.

PLC Applications

Programmable logic controllers are used for various applications (Figures 2–16 and 2–17). They are used to replace hardwired logic in older machines, which can reduce the downtime and maintenance of older equipment. More important, PLCs can increase the speed and capability of older equipment. Retrofitting an older piece of

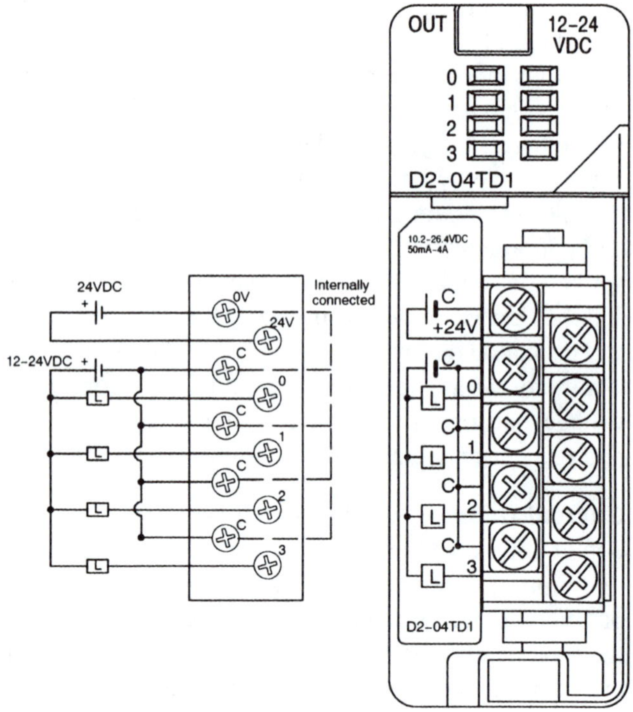

Figure 2–12 PLC I/O wiring. *(Courtesy Rockwell Automation Inc.)*

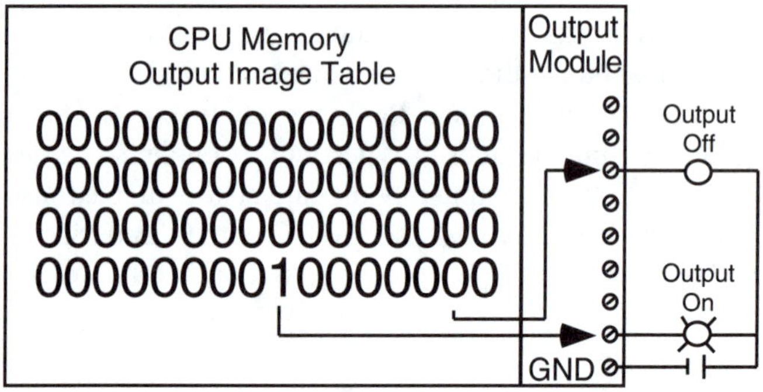

Figure 2–13 How a typical PLC handles outputs. The CPU memory contains a section called the output image table, which contains the desired states of all outputs. Using active-high logic, if there is a 1 in the bit that corresponds to the output, then the output is turned on; if there is a 0, then the output is turned off.

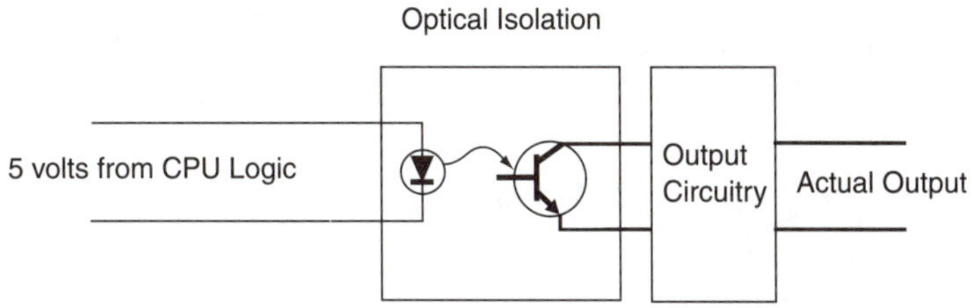

Figure 2–14 How PLC output isolation works. The CPU provides a 5-volt signal that turns on the LED. The light from the LED is used to fire the base of the output transistor. There is no electrical connection between the CPU and the outside world.

Figure 2–15 A few possible output devices including a contactor, AC motor, starter, and valve.

Figure 2–16 A PLC-controlled injection molding machine. *(Courtesy Rockwell Automation Inc.)*

equipment with a PLC for control is almost like getting a new machine. PLCs are being used to control processes such as chemical production, paper production, steel production, and food processing. In processes such as these, they are used to control temperature, pressure, mixture, and concentration. They are used to control position and velocity in many kinds of production processes. For example, they can be used to control complex automated storage and retrieval systems. They also can be used to control equipment such as robots and production machining equipment.

Many small companies have started up recently to produce special-purpose equipment that is normally controlled by PLCs. It is quite cost effective for these companies to use PLCs. Examples are conveyors and palletizing, packaging, pro-

Figure 2–17 Large flexographic printing press controlled by a PLC. Flexographic presses are used to produce various packaging and printed materials.

cessing, and material handling. Without PLC technology, many small equipment design companies might not exist.

PLCs are often used for motor-control applications. They can control simple stepper motor systems or complex AC and DC control of motor drives. PLCs are being used extensively in position and velocity control. A PLC can control position and velocity much more quickly and accurately than can mechanical devices such as gears and cams. An electronic system of control is not only faster, but also does not wear out and lose accuracy as do mechanical devices.

PLCs are used for almost any process imaginable. Certain companies own PLC-equipped railroad cars that regrind and true the rail track as they travel. PLCs have been used to ring the perfect sequences of bells in church bell towers, at the exact times during the day and week. PLCs are used in lumber mills to grade, size, and cut lumber for optimal output. The uses of PLCs are limited only by the imagination of the engineers and technicians who use them.

Questions

1. The PLC is programmed by technicians using:
   a. the C programming language.
   b. ladder logic.
   c. the language determined by the manufacturer.
   d. none of the above.
2. Changing relay-control-type circuits involves changing:
   a. the input circuit devices.
   b. the voltage levels of most I/Os.
   c. the circuit wiring.
   d. the input and output devices.
3. The most common programming device for PLCs is the:
   a. dumb terminal.
   b. dedicated programming terminal.
   c. handheld programmer.
   d. personal computer.
4. True or false: CPU stands for central processing unit.
5. Opto-isolation is:
   a. used to protect the CPU from real-world inputs.
   b. not used in PLCs, so isolation must be provided by the user.
   c. used to protect the CPU from real-world outputs.
   d. both a and c.
6. EEPROM is:
   a. electrically erasable memory.
   b. electrically programmable RAM.
   c. erased by exposing it to ultraviolet light.
   d. programmable.
   e. a, b, and d.
7. True or false: RAM typically holds the operating system.
8. Typical program storage devices for ladder diagrams include:
   a. computer disks.
   b. EEPROM.
   c. static RAM cards.
   d. all of the above.
   e. none of the above.

9. The IEC 61131-3 standard specifies characteristics for:
   a. PLC communications.
   b. EEPROM.
   c. memory.
   d. PLC programming languages.
   e. none of the above.
10. Input devices include the following:
   a. switches.
   b. sensors.
   c. other smart devices.
   d. all of the above.
   e. none of the above.
11. Troubleshooting a PLC system:
   a. requires special PLC diagnostic equipment.
   b. is much more difficult than for relay-type systems.
   c. is easier because of indicators such as I/O indicators on I/O modules.
   d. all of the above.
   e. none of the above.
12. Output modules can be purchased with which of the following output devices?
   a. Transistor outputs.
   b. Triac outputs.
   c. Relay outputs.
   d. All of the above.
   e. Both a and c.
13. Field devices would include the following:
   a. switches.
   b. sensors.
   c. valves.
   d. all of the above.
   e. none of the above.
14. True or false: If an output module's current rating is 1 A per output and there are eight outputs, the current rating for the module is 8 A. Explain your answer.
15. Describe how the status of real-world inputs are stored in PLC memory.
16. Describe how the status of real-world outputs are stored in PLC memory.

17. Define the term *debounce.* Why is it so important?
18. What is the difference between online and offline programming?
19. What does it mean to force I/O?

chapter

Overview of Number Systems 3

A knowledge of various numbering systems is essential to the use of PLCs. In addition to the decimal system, binary, octal, and hexadecimal systems are regularly used. An understanding of the systems will make the task of working with PLCs an easier one. In this chapter we examine each of these systems.

OBJECTIVES

Upon completion of this chapter, you will be able to:

1. Explain each of the numbering systems.
2. Explain the benefits of typical number systems and why each is used.
3. Convert from one number system to another.
4. Explain how input or output modules might be numbered using the octal or hexadecimal number system.
5. Use each number system properly.
6. Explain terms such as *most significant, least significant, nibble, byte,* and *word.*

Decimal System

A short review of the basics of the decimal system will help in a thorough understanding of the other number systems. Although calculators can do the tedious work of number conversion between systems, it is vital that the technician be comfortable with the number systems. Binary, octal, and hexadecimal systems are regularly used to identify items such as inputs/outputs and memory addresses. The technician who understands and uses these systems will have an easier time with PLCs.

The decimal number system uses ten digits: 0 through 9. The highest digit is 9 in the decimal system. Zero through 9 are the only digits allowed.

The first column in decimal can be used to count up to nine items. Another column must be added if the number is larger than 9. The second column can also use the digits 0 through 9. This column is weighted, however, as shown in Figure 3–1. The second column is used to tell the number of tens. For example, the number 23 represents two 10s and three 1s. By using one column we are able to count to 9. If we use two columns we can count to 99. The first column can hold up to 9. The second column can hold up to 90 (nine 10s), exactly 10 times as much as the first column. In fact, in the decimal system, each column is worth 10 times as much as the preceding column. The third column represents the number of hundreds (10 times 10). For example, 227 represents two 100s, two 10s, and seven 1s (Figure 3–2).

The decimal system is certainly the simplest and most familiar type of numbering system. The other systems are based on the same principles as the decimal system. It would certainly be easier if there were only one system, but the computer cannot "think" in decimal. The computer can only work with binary numbers. In fact, the other number systems are quite convenient for certain uses and actually simplify some tasks.

Decimal System					
100,000s	10,000s	1000s	100s	10s	1s

Figure 3–1 Weights of the decimal system.

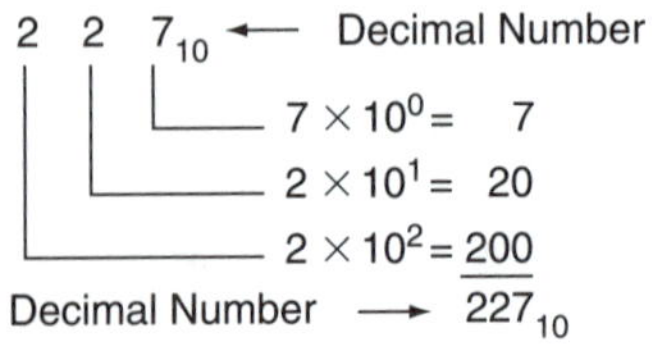

Figure 3–2 Relationship between the weights of each column and the decimal number 227.

Binary Numbering System

The binary numbering system is based on only two digits: 0 and 1. A computer is a digital device. It works with voltages, on or off. Computer memory is a series of 0s and 1s.

The binary system works like the decimal system. The first column holds the number of 1s (Figure 3–3). Because the only possible digits in the first column are 0 or 1, it should be clear that the first column can hold zero 1s or one 1. Thus, we can only count up to 1 using the first column in binary. The second column holds the number of 2s. There can be zero 2s or one 2. The binary number 10 would equal one 2 plus zero 1s. The number 10 in binary is 2 in the decimal system (one 2 + zero 1s). The binary number 11 would be 3 in decimal (one 2 + one 1) (see Figure 3–4). The third column is the number of 4s. Thus, binary 100 would be equal to decimal 4. The fourth column is the number of 8s, the fifth column is the number of 16s, the sixth column is the number of 32s, the seventh is the number of 64s, and the eighth column is the number of 128s. As you can see, each column's value is twice as large as that of the previous column.

The value of the column in binary can be found by raising 2 to the power represented by that column. For example, the third column's weight can be found by raising 2 to the second power ($2 \times 2 = 4$). Remember that the first column is column 0, the second is column 1, and so on. The weight of the fourth column is 8 ($2 \times 2 \times 2$) (Figure 3–5).

Binary is used extensively because it is the only numbering system usable by a computer. It is also quite useful when considering digital logic because a 1 can represent one state and a 0 the opposite state. For example, a light is either on or off. See Figure 3–6 for the appearance of a binary word.

Binary Coded Decimal System

Binary coded decimal (BCD) involves the blending of the binary and decimal systems. In BCD, 4 binary bits are used to represent a decimal digit. These 4 bits are used to represent the numbers 0 through 9. Thus, 0111 binary would be 7 decimal (Figure 3–7).

Binary System					
32s	16s	8s	4s	2s	1s

Figure 3–3 Weights of each column in the binary system. The first column is the number of 1s, the second is the number of 2s, and so on.

Figure 3–4 Comparison of the binary and decimal numbers from 0 through 15.

Binary				Decimal
8s	4s	2s	1s	
0	0	0	0	0
0	0	0	1	1
0	0	1	0	2
0	0	1	1	3
0	1	0	0	4
0	1	0	1	5
0	1	1	0	6
0	1	1	1	7
1	0	0	0	8
1	0	0	1	9
1	0	1	0	10
1	0	1	1	11
1	1	0	0	12
1	1	0	1	13
1	1	1	0	14
1	1	1	1	15

Figure 3–5 Relationship between binary and decimal. The binary number 11001101 is equal to 205 decimal.

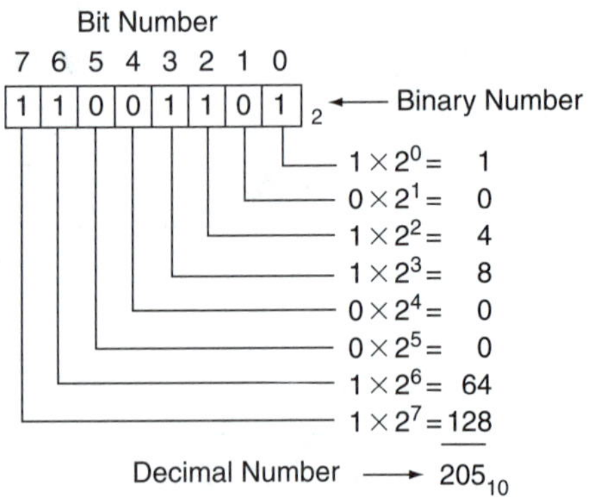

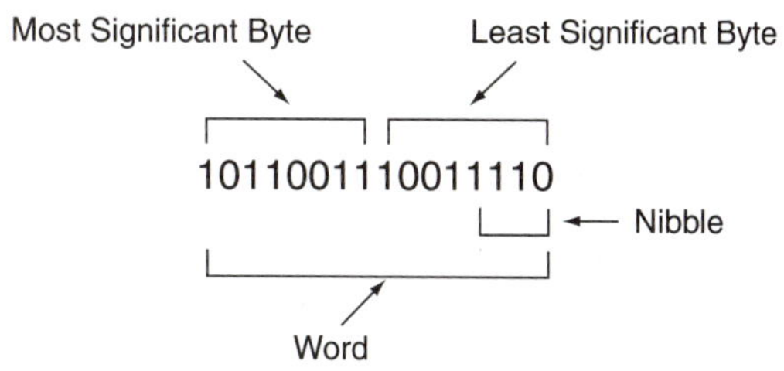

Figure 3–6 A 16-bit binary number. The bit on the right is the least significant bit. The bit on the left is the most significant bit. The next unit of grouping is the nibble. A nibble is 4 bits. The next grouping of a binary number is the byte. A byte is 8 bits. Note that the first eight digits are called the least significant byte and the last eight digits are called the most significant byte. The next grouping is called the word. The size of a word is dependent on the processor. A 16-bit processor has a 16-bit word. A 32-bit processor has a 32-bit word.

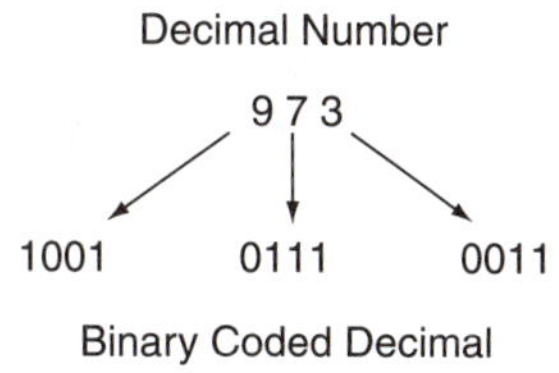

Figure 3–7 How the decimal number 973 would be represented in the binary coded decimal (BCD) system. Each decimal digit is represented by its four-digit binary number. *Caution:* BCD is not the same as binary. The decimal number 973 is 1001 0111 0011 in BCD and is 0011 1100 1101 in binary.

The difference in BCD is in the way numbers above decimal 9 are represented. For example, the decimal number 43 would be 0100 0011 in BCD. The first 4 bits (least significant bits) represent the decimal 3. The second 4 bits (most significant bits) represent the decimal 4. In BCD, the first 4 bits represent the number of 1s in a decimal number, the second 4 bits represent the number of 10s, the third 4 bits represent the number of 100s, and so on.

Use of the BCD format is popular for output from instruments. Measuring devices will typically output BCD values. Some input devices use the BCD system to output their value. Thumbwheels are one example. A person dials in a decimal digit between 0 and 9. The thumbwheel outputs 4 bits of data. The 4 bits are BCD. For example, if an operator were to dial in the number 8, then the output from the BCD thumbwheel would be 1000. PLCs can easily accept BCD input.

Octal System

The octal system is based on the same principles as the binary and decimal systems except that it is base 8. There are eight possible digits in the octal system: 0, 1, 2, 3, 4, 5, 6, and 7. The first column in an octal number is the number of 1s. The second column is the number of 8s, the third column is the number of 64s, the fourth column is the number of 512s, the fifth column is the number of 4096s, and so on.

Weights of the columns can be found by using the same method as that used for binary. The number 8 is simply raised to the power represented by that column. The first column (column 0) represents 8 to the zero power (1 by definition). Remember that the first column is column 0. The weight of the second column (column 1) is found by raising 8 to the first power (8×1). The weight of the third column (column 2) is found by raising 8 to the second power (8×8) (see Figure 3–8).

The actual digits in the octal number system are 1, 2, 3, 4, 5, 6, and 7. If we must count above 7, then we must use the next column. For example, let us count to 10 in octal: 1, 2, 3, 4, 5, 6, 7, 10, 11, and 12. The 12 represents one 8 and two 1s ($8 + 2 = 10$). The number 23 decimal would be 27 in octal. Two 8s and seven 1s is equal to 23. The number 3207 octal would be 1671 in decimal. Figure 3–9 shows how an octal number can be converted to a decimal number.

Some PLC manufacturers use octal to number input and output modules and also to number memory addresses. For example, assuming the use of input cards with eight inputs per card, the first eight inputs on the first card would be numbered 0, 1, 2, 3, 4, 5, 6, and 7. The next input card numbering would begin with octal 10, 11, 12, 13, 14, 15, 16, and 17 (Figure 3–10). This makes it easy to find the location of

Octal System					
32,768s	4,096s	512s	64s	8s	1s

Figure 3–8 Weights of the columns in the octal number system. The first column is the number of 1s, the second column is the number of 8s, and so on.

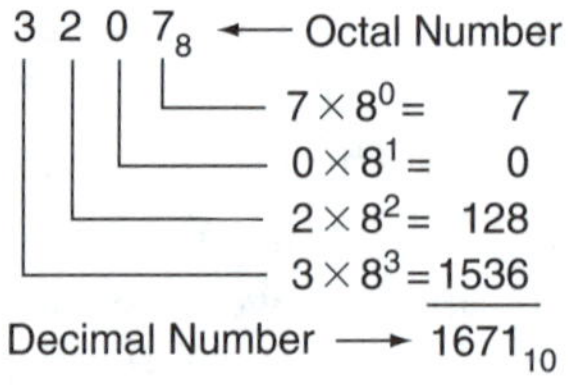

Figure 3–9 How the octal number 3207 is converted to the decimal number 1671.

Input Module 0	Input Module 1	Input Module 2	Input Module 3
I00	I10	I20	I30
I01	I11	I21	I31
I02	I12	I22	I32
I03	I13	I23	I33
I04	I14	I24	I34
I05	I15	I25	I35
I06	I16	I26	I36
I07	I17	I27	I37

Figure 3–10 I/O module addressing using the octal numbering system. The first input module is numbered 0. The first input on the module would be called input I00 (the first zero represents the first module, the second zero means that it is the first input, and the I stands for "input"). The eighth input on the first module would be called I07. The eighth input on the fourth module would be called input 37. (Remember that the first module is module 0; the fourth module is module 3.) The first output module would begin numbering as output module 0, and each output would be numbered just as the input modules were. For example, the fifth output on the second output module would be numbered output 14.

inputs or outputs. The least significant digit is used to specify the actual input/output number, and the most significant digit is used to specify the particular card where the input/output is located.

Octal is also used by some manufacturers for numbering memory. For example, Siemens Industrial Automation Inc. has 128 timers and counters available for the 405 series of PLCs. The timers are numbered 0 to 177 octal. This equates to 0 through 127 decimal.

HEXADECIMAL SYSTEM

The hexadecimal system normally causes the most trouble for people. Hexadecimal (or hex) is based on the same principles as the other numbering systems discussed thus far. Hex has sixteen possible digits, but with an unusual twist. It uses numbers and also the letters A to F. This can be a little confusing at first. In hexadecimal, we count 0, 1, 2, 3, 4, 5, 6, 7, 8, and 9. After the number 9, the counting changes: 10 becomes A, 11 is B, 12 is C, 13 is D, 14 is E, and 15 is F

Figure 3–11 Comparison of the hexadecimal and decimal systems.

Hexadecimal	Decimal
0	0
1	1
2	2
3	3
4	4
5	5
6	6
7	7
8	8
9	9
A	10
B	11
C	12
D	13
E	14
F	15
10	16
11	17
12	18
13	19
14	20

(Figure 3-11). The first column (column 0) in hex is the number of 1s (Figure 3–12). The second column is the number of 16s. The third column is the number of 256s, and so on.

Weights can be found in the same manner as in the binary system. In the hexadecimal system, 16 is raised to the power of the column. For example, the weight of the third column (column 2) is 16 to the second power ($16 \times 16 = 256$). Figure 3–13 shows how a hexadecimal number can be converted to a decimal number.

Hexadecimal System					
1,048,576s	65,536s	4,096s	256s	16s	1s

Figure 3–12 Weights of the columns in the hexadecimal system. The first column is the number of 1s, the second column is the number of 16s, and so on.

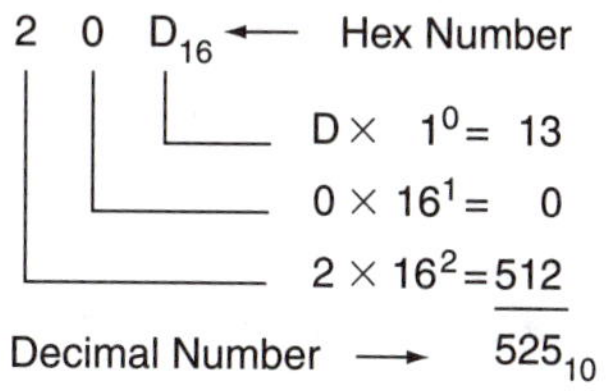

Figure 3–13 Weights of the hexadecimal number system converted to the decimal weighting system.

$$2 \quad 0 \quad D_{16} \longleftarrow \text{Hex Number}$$

$$D \times 1^0 = 13$$
$$0 \times 16^1 = 0$$
$$2 \times 16^2 = 512$$

Decimal Number $\longrightarrow$ 525_{10}

Hex 7 D 3 F

0111 1101 0011 1111

Binary 0111110100111111

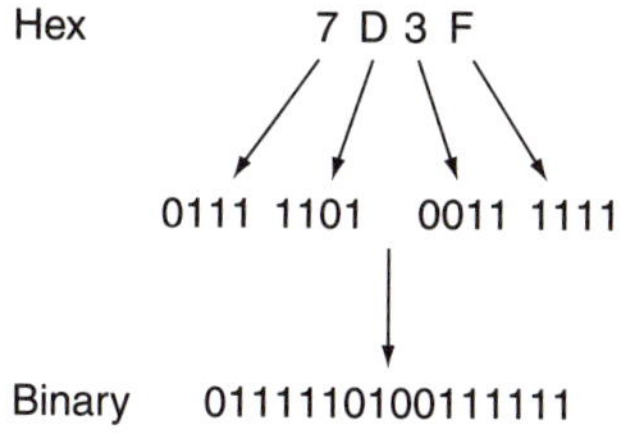

Figure 3–14 The conversion of a hexadecimal number to its binary equivalent. Each hex digit is simply converted to its four-digit binary. The result is a binary equivalent. In this case, hex 7D3F is equal to binary 0111110100111111. (Which number would you rather work with?)

Hexadecimal numbers are easier (less cumbersome) to work with than binary numbers. It is easy to convert between the two systems (Figures 3–14 and 3–15). Each hex digit is simply converted to its four-digit binary equivalent. The result is the binary equivalent of the whole hex number.

The same process works in reverse. Any binary number can be converted to its hex equivalent by breaking the binary number into four-digit pieces and converting each 4-bit piece to its hex equivalent. The resulting numbers are equal in value.

Figure 3–16 shows an overview of the four number systems.

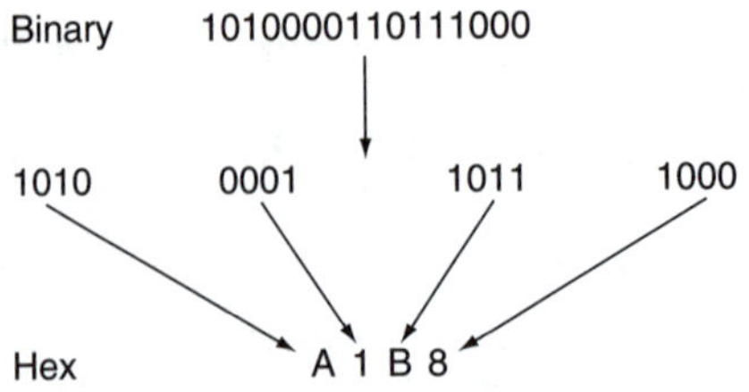

Figure 3–15 Conversion of a binary number to a hex number. The binary number is broken into 4-bit pieces (4 bits are called a nibble) and each 4-bit nibble is converted to its hex equivalent. The binary number 1010000110111000 is equal to hex A1B8. (Which would you prefer to work with?)

Hexadecimal	Decimal	Octal		Binary			
		8s	1s	8s	4s	2s	1s
0	0	0	0	0	0	0	0
1	1	0	1	0	0	0	1
2	2	0	2	0	0	1	0
3	3	0	3	0	0	1	1
4	4	0	4	0	1	0	0
5	5	0	5	0	1	0	1
6	6	0	6	0	1	1	0
7	7	0	7	0	1	1	1
8	8	1	0	1	0	0	0
9	9	1	1	1	0	0	1
A	10	1	2	1	0	1	0
B	11	1	3	1	0	1	1
C	12	1	4	1	1	0	0
D	13	1	5	1	1	0	1
E	14	1	6	1	1	1	0
F	15	1	7	1	1	1	1

Figure 3–16 A comparison of the four number systems for the numbers 0 to 15.

QUESTIONS

1. Complete the following table.

	Binary	Octal	Decimal	Hexadecimal
a.	101			5
b.		11		
c.			15	
d.				D
e.		16		
f.	1001011			
g.		47		
h.			73	

2. Number the following input/output modules using the octal method.

	Input Module 0	Input Module 1	Output Module 0	Input Module 2
a.				
b.				
c.				
d.				
e.				
f.				
g.				
h.				

3. Define each of the following:
   a. bit
   b. nibble
   c. byte
   d. word
4. Why is the binary number system used in computer systems?

5. Complete the following table.

	Binary	Hexadecimal
a.	1011001011111101	
b.		1A07
c.	100100000111	
d.		C17F
e.	0010001111000010	
f.		D91C
g.	0011010111100110	
h.		ECA9
i.	0101001011000101	

6. True or false: A word is sixteen bits. Explain your answer.
7. True or false: One K of memory is exactly 1000 bytes.
8. How can the weight of the fifth column of a binary number be calculated?
9. How can the weight of the fourth column of a hex number be calculated?
10. What is the BCD system normally used for?

chapter 4

Fundamentals of Programming

In this chapter we examine the basics of ladder logic programming, including terminology and common symbols. We also learn how to write basic ladder logic programs.

OBJECTIVES

Upon completion of this chapter, you will be able to:

1. Describe the basic process of writing ladder logic.
2. Define terms such as *contact, coil, rung, scan, normally open,* and *normally closed.*
3. Write ladder logic for simple applications.

LADDER LOGIC

Programmable controllers are primarily programmed in ladder logic, a symbolic representation of an electrical circuit. Symbols resemble those of the schematic symbols of electrical devices, which makes it easy for the plant electrician to learn how to use the PLC. An electrician who has never seen a PLC can understand a ladder diagram.

The main function of the PLC program is to control outputs based on the condition of inputs. The symbols used in ladder logic programming can be divided into two broad categories: contacts (inputs) and coils (outputs).

Contacts

Most inputs to a PLC are simple devices that are either on or off. These inputs are sensors and switches that detect part presence, empty or full, and so on. The two common symbols for contacts are shown in Figure 4–1.

Contacts can be thought of as switches. There are two basic kinds of switches, normally open and normally closed. A normally open switch will not pass current until pressed. A normally closed switch will allow current flow until it is pressed. Rockwell calls the normally open switch an examine if closed (XIC) instruction and the normally closed contact an examine if open (XIO) instruction. Think of a doorbell switch. Would you use a normally open switch or a normally closed switch for a doorbell? If you chose the normally closed switch, the bell would be on continuously until someone pushes the switch. Pushing the switch opens the contacts and stops current flow to the bell. The normally open switch is the necessary choice. If the normally open switch is used, the bell will not sound until someone pushes the button on the switch.

Sensors are used to detect the presence of physical objects or quantities. For example, one type of sensor might be used to sense when a box moves down a conveyor, and a different type might be used to measure a quantity such as heat. Most sensors are switchlike: they are on or off depending on what the sensor is sensing. Like switches, sensors that are either normally open or normally closed can be purchased.

Imagine, for example, a sensor that is designed to sense a metal part as the part passes the sensor. We could buy a normally open or a normally closed sensor for the application. If you wanted to notify the PLC every time a part passed the sensor, a normally open sensor might be chosen. The sensor would turn on only if a metal part passed in front of the sensor. The sensor would turn off again when the part was gone.

The PLC could then count the number of times the normally open switch turned on (closed) and would know how many parts had passed the sensor. Normally

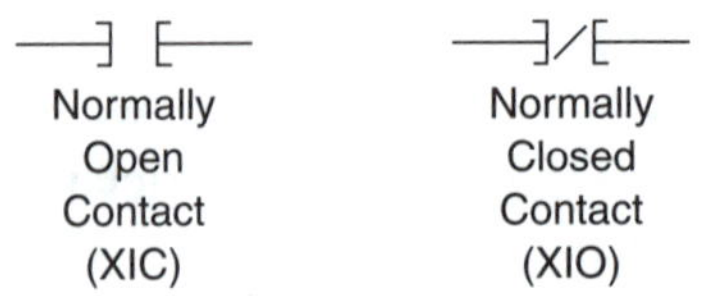

Figure 4–1 A normally open and a normally closed contact.

closed sensors and switches are often used when safety is a concern. These topics are examined later in the chapter.

Coils

Whereas contacts are input symbols, coils are output symbols. Outputs can take various forms, including motors, lights, pumps, counters, timers, and relays. A coil represents an output. The PLC examines the contacts (inputs) in the ladder and turns the coils (outputs) on or off depending on the condition of the inputs. The basic coil is shown in Figure 4–2. Coil symbols appear only on the right side of the rung. A specific coil number should appear only once at the right of a rung. The output that the coil represents, however, can be used many times on the left as contact(s).

Ladder Diagrams

The basic ladder diagram looks similar to a step ladder. Two uprights hold the rungs that make up the PLC ladder. The left and right uprights (sometimes called power rails) represent power. If we connect the left and right uprights, power can flow through the rung from the left upright to the right upright.

Consider the doorbell example again, with one input and one output. The ladder diagram for the PLC would be only one rung (Figure 4–3). The real-world switch would be connected to input number 0 of the PLC. The bell would be connected to output number 0 of the PLC (Figure 4–4). The uprights represent a DC voltage that will be used to power the doorbell. If the real-world doorbell switch is pressed, power can flow through the switch to the doorbell.

The PLC would then run the ladder. The PLC will monitor the input continuously and control the output, a process called scanning. The amount of time it takes for the PLC to check the states of inputs, evaluate the logic, and then update the I/O table each time is called scan time. Scan time varies among PLCs. Most applications do not require extreme speed, so any PLC is fast enough; even a slow PLC

Figure 4–2 Ladder logic symbol for a coil.

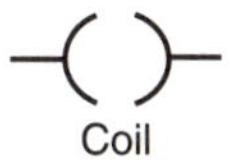

Figure 4–3 Simple conceptual view of a ladder diagram.

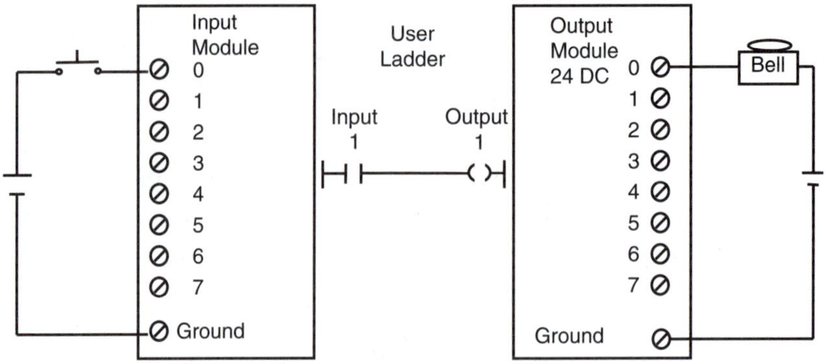

Figure 4–4 Conceptual view of a PLC system. The real-world inputs are attached to an input module (left side). Outputs are attached to an output module (right side). The center shows the logic that the CPU must evaluate. The CPU evaluates user logic by looking at the inputs and then turns on outputs based on the logic. In this case if input 0 (a normally open switch) is closed, then output 0 (the doorbell) will turn on.

Figure 4–5 An example of how a user's ladder logic is continually scanned.

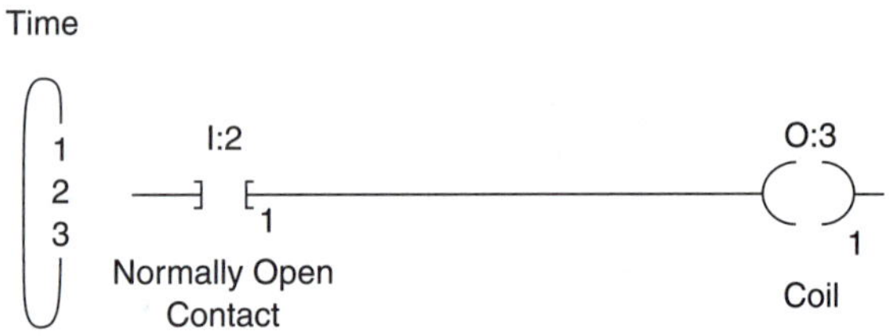

scan time would be in milliseconds. The longer or more complex the ladder logic, the more time it takes to scan.

The scan cycle is illustrated in Figure 4–5. Note that this one rung of logic represents our entire ladder. Each time the PLC scans the doorbell ladder, it checks the state of the input switch *before* it enters the ladder (time period 1). While in the ladder, the PLC then decides if it needs to change the state of any outputs (evaluation during time period 2). *After* the PLC finishes evaluating the logic (time period 2), it turns on or off any outputs based on the evaluation (time period 3). The PLC then returns to the top of the ladder, checks the inputs again, and repeats the entire process. The total of these three stages makes up scan time. We discuss the scan cycle more completely later in this chapter.

Normally Open Contacts

First we examine the normally open contact. A normally open contact does not pass power until the input associated with it is energized. A normally open contact was shown in Figure 4-5. Imagine the doorbell switch again. The actual switch is normally open. If we push the switch, it closes and passes power to sound the doorbell in the house. A normally open contact is similar to the normally open doorbell switch. In a ladder diagram this contact can be used to monitor a real-world switch. If the real-world switch is closed (energized), then the normally open contact in the ladder logic would pass power.

Normally Closed Contacts

The normally closed contact will pass power until it is activated. A normally closed contact in a ladder diagram will pass power while the real-world input associated with it is off.

A home security system is an example of normally closed logic. Assume that the security system was intended to monitor the two entrance doors to a house. One way to wire the house is to wire one normally open switch from each door to the alarm, like a doorbell switch (Figure 4–6). Then if a door opened, it would close the switch and the alarm would sound. This configuration would work, but certain problems exist. Assume that the switch fails. For example, a wire may be cut accidentally, or a connection may become loose, or a switch may break. The problem is that the homeowner would never know that the system was not working. An intruder could open the door, and the switch would not work and the alarm would not sound. Obviously, this is not a good way to design a system. The system should be set up so that the alarm will sound for an intruder and will also sound if a component fails. The homeowner surely wants to know if the system fails. It is far better for the alarm to sound when the system fails with no intruder than not to sound if the system fails with an intruder.

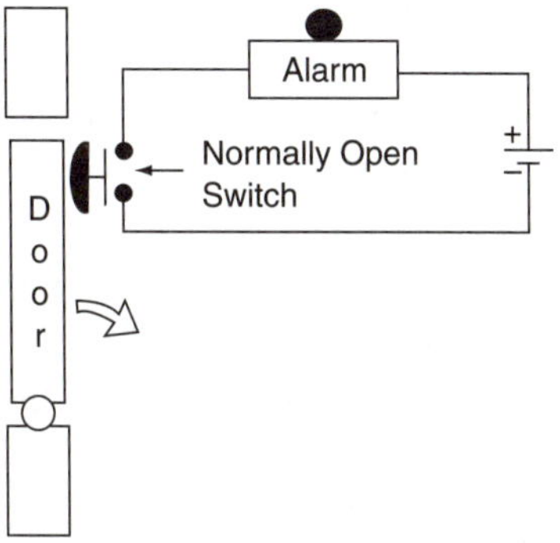

Figure 4–6 A conceptual diagram of the wrong way to construct a burglar alarm circuit. In this case, the homeowner would never know if the system failed. The correct method would be to use normally closed switches. The control system would then monitor the circuit continuously to see if the doors opened or a switch failed.

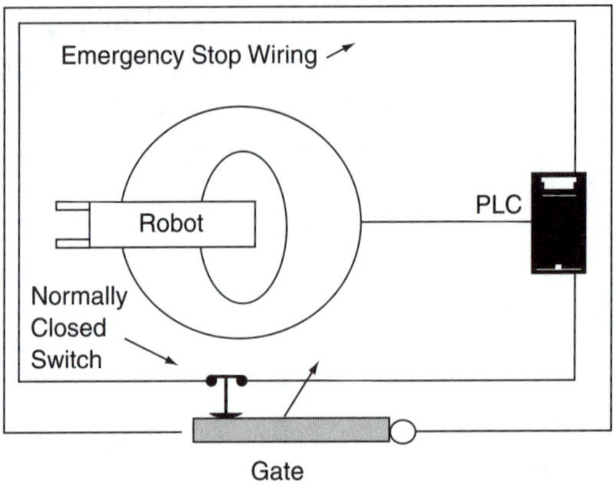

Figure 4–7 Robot cell application. Note the fence around the cell with one gate, the PLC used as a cell controller, and the safety switch to prevent anyone from entering the cell while the robot is running. If someone enters the cell, the PLC will sense that the switch opened and sound an alarm. In this case a normally closed switch is used. If the wiring or switch fails, the PLC will think someone entered the cell and sound an alarm. This system is called fail-safe.

Such considerations are just as important in an industrial setting where failure can cause an injury. The procedure of programming to ensure safety is called fail-safe. The programmer must carefully design the system and ladder logic so that if a failure occurs, people and processes are safe. As shown in Figure 4–7, if the gate is opened, it opens the normally closed switch. The PLC would see that the switch had opened and would sound an alarm immediately to protect whomever had entered the work cell. (In reality, we would sound an alarm and stop the robot to protect the intruder.)

The normally closed switch as used in ladder logic can be confusing. The normally closed contact in our ladder passes electricity if the input switch is off. The switch in the gate of the cell is a normally closed switch. (The switch in the cell normally allows electricity to flow.) If the gate switch is closed, the alarm will be off because the normally closed contact in the ladder logic will be open (Figure 4–8).

If someone opens the gate, the normally closed gate switch opens, stopping electrical flow. The PLC sees that there is no flow, the normally closed contact in the ladder allows electricity to flow, and the alarm is turned on (Figure 4–9).

What would happen if a tow motor was too close to the cell and cut the wire that connects the gate safety switch to the PLC? The alarm will sound because the wire being cut is similar to the gate opening the switch. Is it a good thing that the alarm sounds if the wire is cut? Yes. It warns the operator that something failed in the cell. The operator could then call maintenance and have the cell repaired. This is a fail-

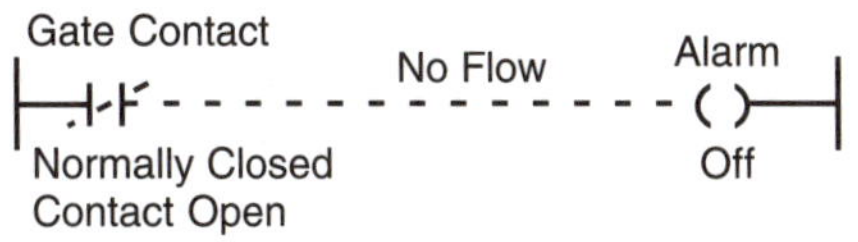

Figure 4–8 One rung of a ladder diagram. A normally closed contact is used in the ladder. If the switch associated with that contact is closed, it forces the normally closed contact open. No current flows to the output (the alarm). The alarm is off.

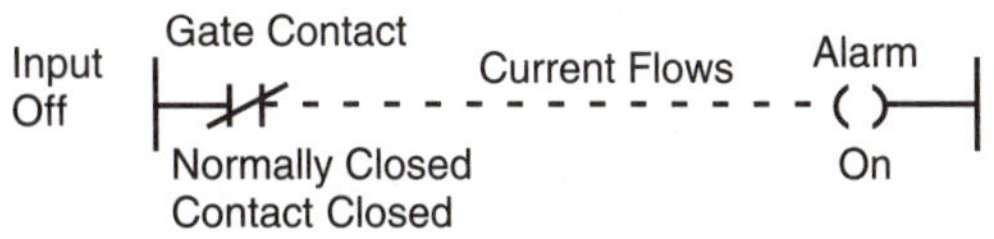

Figure 4–9 The gate switch is off. (Someone opened the gate and thus opened the switch.) The normally closed contact is true when the input is false, so the alarm sounds. The same thing would happen if a tow motor cut the wire that led to the safety sensor. The input would go low and the alarm would sound.

safe system. Something in the cell failed and the system was shut down by the PLC so that no one would be hurt. The same would be true if the gate safety switch were to fail. The alarm would sound. If the switch were opened (someone opened the gate to the cell), the PLC would see that there is no power at the input (Figure 4–9). The normally closed contact in the ladder logic is then closed, allowing electricity to flow. This causes the alarm to sound. Consider the rungs shown in Figure 4–10 and determine if the output coils are on or off. (The answers are listed below the caption.) Pay particular attention to the normally closed examples.

Transitional Contacts

Transitional contacts are a special type of contact. They are also called one-shot contacts. The symbol for this type of contact is shown in Figure 4–11. The one-shot rising (OSR) instruction is an input instruction that can trigger an event to occur one time, which explains why they are called one-shot contacts. When the rung conditions preceding the OSR instruction go from false to true, the OSR will be true for one scan.

There are many reasons to use the transitional contacts. They are often used to provide a pulse for timing, counting, or sequencing. They are also used when you want to perform an instruction only once (not every scan). For example, if an add instruction were used to add two numbers once, it would not be necessary to add them at every scan. A transitional contact would ensure that the instruction executes only on the desired transition.

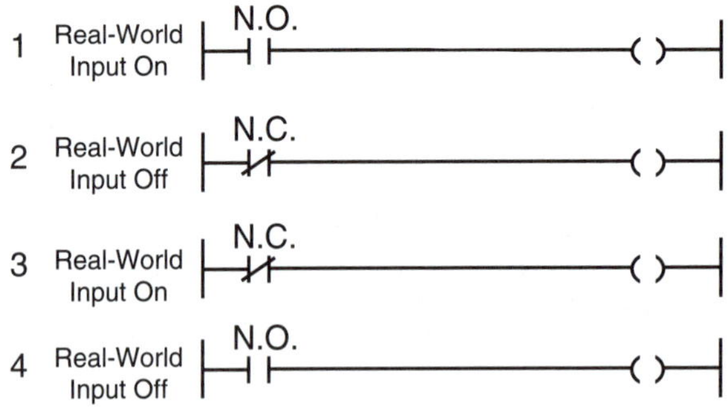

Figure 4–10 Ladder diagram exercise. N.O. equals "normally open"; N.C. equals "normally closed."

Answers to ladder logic in Figure 4–10:

1. The output in example 1 is on. The real-world input is on, so the normally open contact (or examine if closed—XIC) in the ladder would be energized and pass power to the output. In other words, the examine-on instruction is used to check if the real-world input is closed (has power). If so, the examine-on instruction passes power to the output coil in the ladder.
2. The output in example 2 is on. The real-world input is off. The normally closed contact (or examine if open—XIO) in the ladder is energized because the input is off. In other words, the examine-off instruction is used to check if the real-world input is open. If so, the XIO is energized and passes power to the output.
3. The output in example 3 is off. The real-world input is on, which forces the normally closed contact in the ladder open. In other words, the real-world input is on, so the XIO is de-energized and does not pass power. The output is off.
4. The output in example 4 is off. The real-world input is off, so the normally open contact remains open and the output is off. In other words, the real-world input is off, so the XIC is de-energized and does not pass power. The output is off.

Figure 4–11 A one-shot rising instruction.

Multiple Contacts

More than one contact can be put on the same rung. For example, think of a drill machine. The engineer wants the drill press to turn on only if there is a part present and the operator has one hand on each of the start switches (see Figures 4–12 and 4–13). This would ensure that the operator's hands could not be in the press while it is running.

Note that the switches were programmed as normally open contacts. They are all on the same rung (series). All will have to be on for the output to turn on. If there

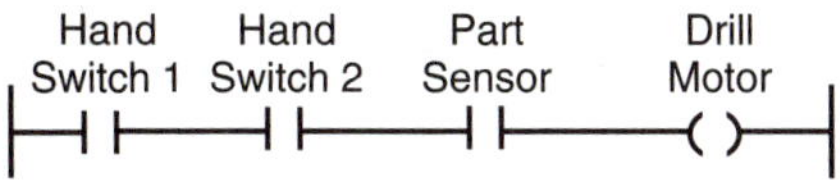

Figure 4–12 Series circuit. Hand switches 1 and 2 and the part sensor must be closed before the drill motor will be turned on, which ensures that there is a part in the machine and that the operator's hands are in a safe location.

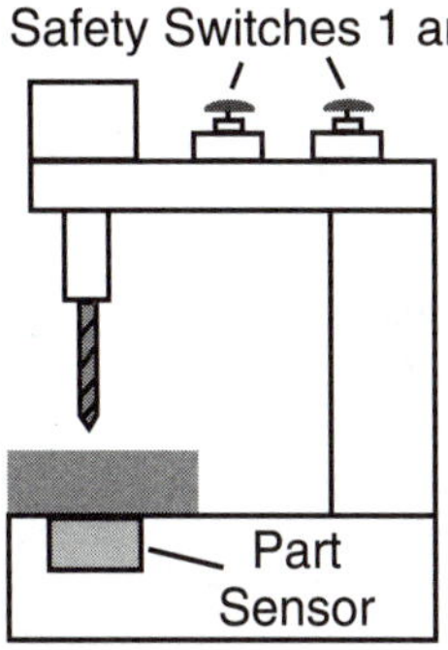

Figure 4–13 Simple drilling machine. There are two hand safety switches and one part sensor on the machine. Both hand switches and the part sensor must be true for the drill press to operate, which ensures that the operator's hands are not in the way of the drill. This is an AND condition. Switch 1 and switch 2 and the part sensor must be activated to make the machine operate. (The ladder for the PLC is shown in Figure 4–14.)

is a part present and the operator puts both hands on the start switches, the drill press will run. If the operator removes one hand for some reason, the press will stop. Contacts in a series such as this can be thought of as logical AND conditions. In this case, the part presence switch *and* the left-hand switch *and* the right-hand switch would have to be closed to run the drill press. Study the examples in Figure 4–14 and determine the status of the outputs.

Branching

You may often want to turn on an output for more than one condition. For example, in a house, the doorbell should sound if either the front or rear door button is pushed (the two conditions under which the bell should sound). The ladder is called a branch (Figure 4–15). As shown, two paths (or conditions) can turn on the doorbell. (This can also be called a logical OR condition.)

If the front door switch is closed, electricity can flow to the bell, or if the rear door switch is closed, electricity can flow through the bottom branch to the bell.

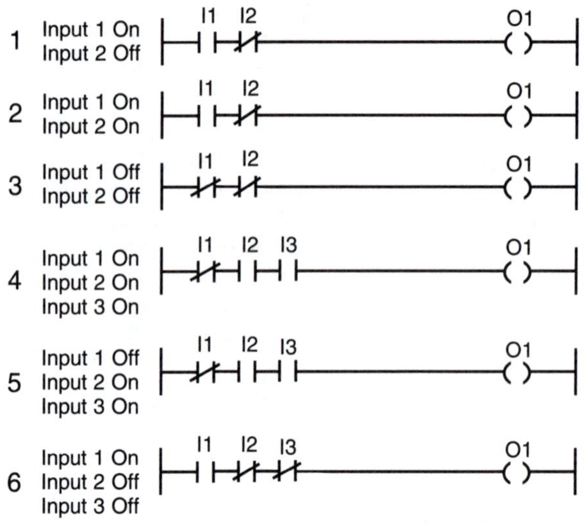

Figure 4–14 Ladder diagram exercise. I equals "input"; O equals "output."

Answers to ladder logic in Figure 4–14.

1. The output for rung 1 will be on. Input 1 is on, which energizes contact 1. Input 2 is off, so normally closed contact 2 is still energized. Both contacts are energized, so the output is on.
2. The output in rung 2 is off. Input 1 is on, which energizes normally open contact 1. Input 2 is on, which de-energizes normally closed contact 2. The output cannot be on because normally closed contact 2 is de-energized.
3. The input in rung 3 is on. Inputs 1 and 2 are off, so normally closed contacts 1 and 2 are energized.
4. The output in rung 4 is off. Input 1 is on, which de-energizes normally closed contact 1.
5. The output in rung 5 is on. Input 1 is off, so normally closed contact 1 is energized. Inputs 2 and 3 are on, which energizes normally open contacts 2 and 3.
6. The output in rung 6 is on. Input 1 is on, energizing normally open contact 1. Inputs 2 and 3 are off, which energizes normally closed contacts 2 and 3.

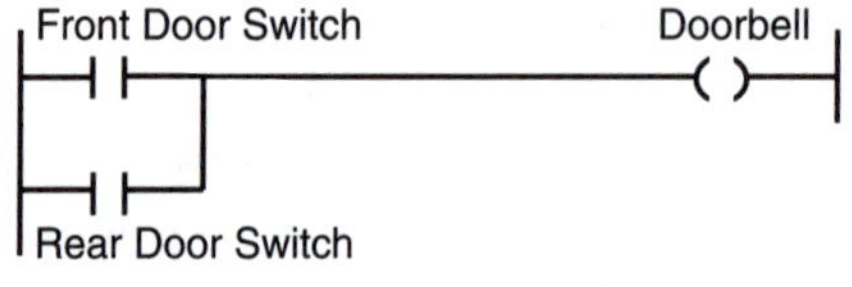

Figure 4–15 Parallel condition. If the front door switch is closed, the doorbell will sound; *or* if the rear door switch is closed, the doorbell will sound. These parallel conditions are also called OR conditions.

Branching can be thought of as an OR situation. One branch *or* another can control the output. ORs allow multiple conditions to control an output. This is very important in industrial control of systems. Think of a motor that is used to move the table of a machine. There are usually two switches to control table movement: a jog switch and a feed switch (Figure 4–16). Both switches are used to turn on the same motor—an OR condition. The jog switch *or* the feed switch can turn on the table feed motor. Evaluate the ladder logic shown in Figure 4–17 to determine the output states.

Start/Stop Circuit

Start/stop circuits are extremely common in industry. Machines will have a start button to begin a process and a stop button to shut off the system. Several important concepts can be learned from the simple logic of a start/stop circuit (Figure 4–18). Notice that the actual start switch is a normally open push button. When pressed, it closes the switch. When the button is released, the switch opens. The stop switch is a normally closed switch. When pressed, it opens.

Now examine the ladder. When the start switch is momentarily pressed, power passes through X000. Power also passes through X001 because the real-world stop switch is a normally closed switch. The output (Y1) is turned on. Note that Y1 is then also used as an input on the second line of logic. Output Y1 is on, so contact Y1 also closes. This process is called latching. The output latches itself on even if the start switch opens. Output Y1 will shut off only if the normally closed stop switch (X001) is pressed. If X001 opens, then Y1 is turned off. The system will require the start button to be pushed to restart the system. Note that the real-world stop switch is a normally closed switch, but that in the ladder, it is programmed normally open for safety reasons.

There are as many ways to program start/stop circuits (or ladder diagrams in general) as there are programmers. Figures 4–19 and 4–20 show examples of the wiring of start/stop circuits, where safety is always the main consideration.

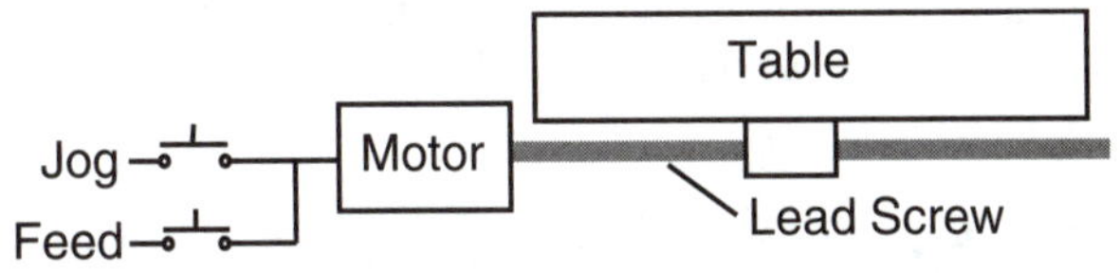

Figure 4–16 Conceptual drawing of a mill table. Note the two switches connected to the motor. These represent OR conditions. The jog switch *or* the feed switch can move the table.

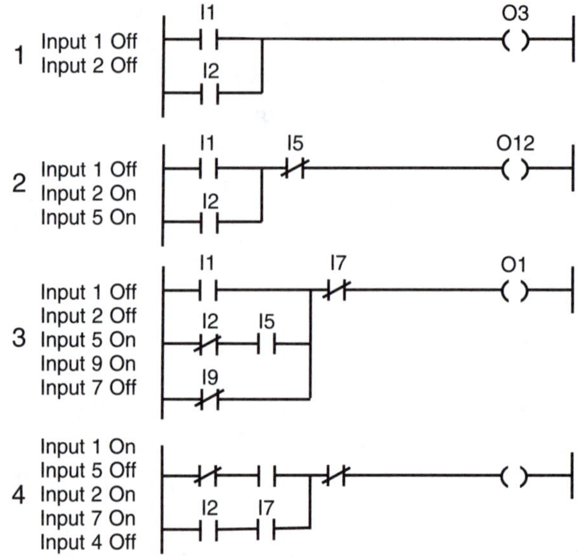

Figure 4–17 Ladder logic examples. I equals "input"; O equals "output."

Answers to ladder logic in Figure 4–17.

1. The output in example 1 would be on. Input 2 is off, so that normally closed contact 2 is energized, energizing the output.
2. The output in example 2 is off. Input 5 is on, which de-energizes normally closed contact 5, so the output cannot be on whether input 1 *or* input 2 is on. Note also that in these branching examples, we have combinations of ANDs and ORs. In English, this example would be input 1 *and* input 5 *or* input 2 *and* input 5 will turn on output 12.
3. The output in example 3 will be energized. Input 2 is off, which leaves normally closed contact input 2 energized *and* input 5 is on, which energizes normally open input 5 *and* normally closed input 7 is off, which energizes normally closed contact 7, which energizes output 1. In this ladder there are three OR conditions and combinations of ANDs.
4. In example 4, the output will be on. Input 5 is off, which energizes normally closed contact 5 *and* input 1 is on, which energizes normally open contact 1 *and* input 4 is off, which energizes normally closed contact 4 and input 4 is off, which energizes normally closed contact 4, energizing the output. Inputs 2 and 7 are also both on, energizing normally open contacts 2 *and* 7.

PLC Scanning and Scan Time

Now that you are familiar with some basic PLC instructions and programming, it is important to understand the way a PLC executes a ladder diagram. Most people would like to believe that a ladder provides a sequential process. We like to think of a ladder as first things first. We would like to believe that the first rung is evalu-

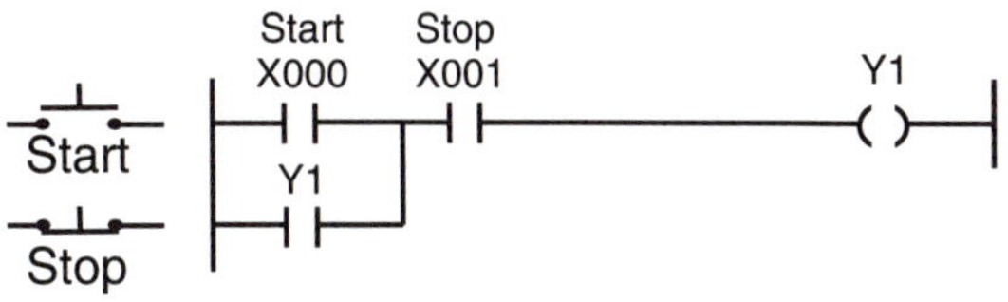

Figure 4–18 Ladder diagram and real-world switches for start/stop circuit.

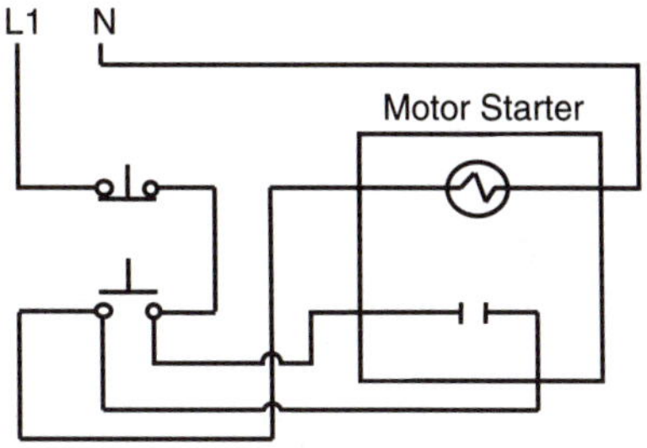

Figure 4–19 Start/stop circuit.

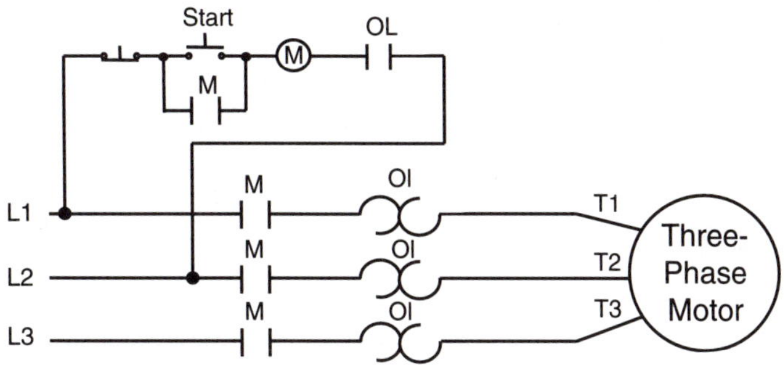

Figure 4–20 Start/stop circuit.

ated and acted on before the next, and so on. We would like to believe that the CPU looks at the first rung, goes out and checks the actual inputs for their present state, comes back, immediately turns on or off the actual output for that rung, and then evaluates the next rung. This is not exactly true, however, and misunderstanding the way the PLC scans a ladder can cause programming bugs.

Scan time can be divided into two components: I/O scan and program scan. When the PLC enters run mode, it first takes care of the I/O scan (Figure 4–21). The I/O scan can be divided into the output step and the input step. During these two steps, the CPU transfers data from the output image table to the output modules (output step) and then from the input modules to the input image table (input step).

The third step is logic evaluation. The CPU uses the conditions from the image table to evaluate the ladder logic. If a rung is true, then the CPU writes a 1 into the

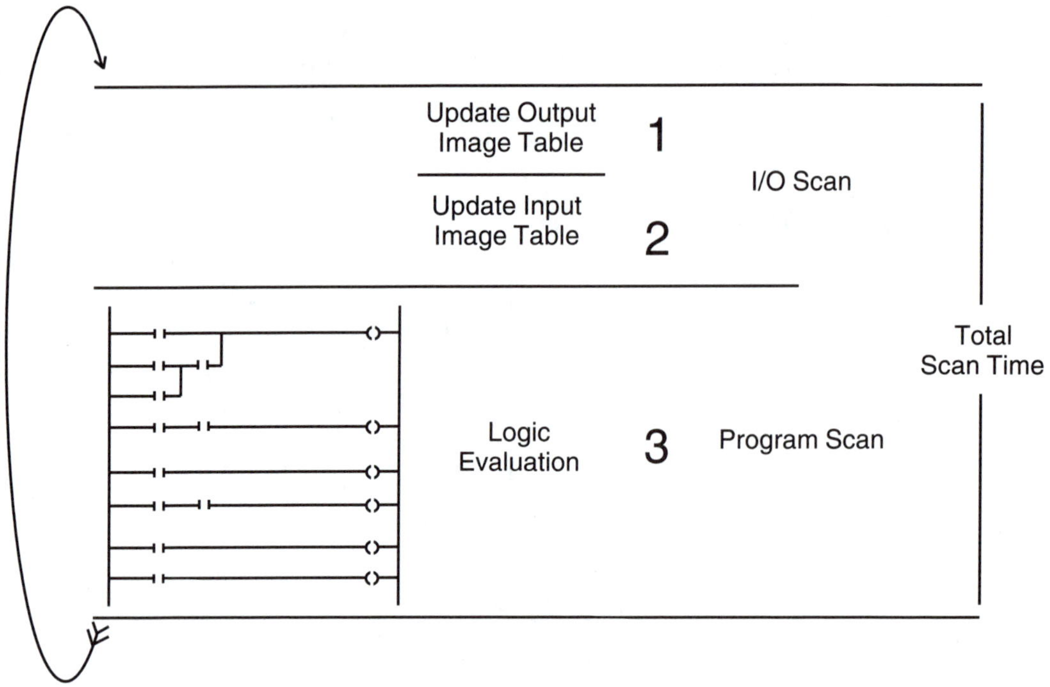

Figure 4–21 A generic example of PLC scanning.

corresponding bit in the output image table. If the rung is false, then the CPU writes a 0 into the corresponding bit in the output image table. Note that nothing concerning real-world I/O is occurring during the evaluation phase, which can be a point of confusion. The CPU is basing its decisions on the states of the inputs as they existed before the evaluation phase. We would like to believe that if an input condition changes while the CPU is in the evaluation phase, it would use the new state, but it cannot. (*Note:* It actually can use the new state if special instructions called immediate update contacts and coils are used. Most ladders will not utilize immediate instructions, however. The states of all inputs are frozen before the evaluation phase. The CPU does not turn outputs on and off during this phase either. This phase is only for evaluation and updating the output image table status.)

Once the CPU has evaluated the entire ladder, it performs the I/O scan again. During the I/O scan, the output states of real-world outputs are changed depending on the output image table. The real-world input states are then transferred again to the input image table.

Rockwell Automation Scan Order

Rockwell scanning occurs in the following order.

1. Input scan.
2. Program scan.
3. Output scan.
4. Service communications.
5. Overhead (timers, bits, integers, etc.).

The previous steps take only a few milliseconds (or less) to accomplish. This speed of PLCs creates problems when troubleshooting. Scan time is the sum of the times it takes to execute all of the individual instructions in the ladder. Simple contacts and coils take very little time. Complex math statements and other types of instructions take much more time. Even a long ladder diagram will normally execute in less than 50 milliseconds. There are considerable differences in the speeds of different brands and models of PLCs. Manufacturers normally will give scan time in terms of fractions of milliseconds per K of memory to provide a rough idea of the scan times of various brands.

QUESTIONS

1. What is a contact? A coil?
2. What is a transitional contact?
3. Describe the uses of transitional contacts.
4. Explain the term *normally open* (XIC—examine if closed).
5. Explain the term *normally closed* (XIO—examine if open).
6. What are some uses of normally open contacts?
7. Explain the terms *true* and *false* as they apply to contacts in ladder logic.
8. Design a ladder that shows series input (AND logic). Use X5, X6, AND NOT (normally closed contact) X9 for the inputs and use Y10 for the output.
9. Design a ladder that has parallel input (OR logic). Use X2 and X7 for the contacts.
10. Design a ladder that has three inputs and one output. The input logic should be: X1 AND NOT X2, OR X3. Use X1, X2, and X3 for the input numbers and Y1 for the output.
11. Design a three-input ladder that uses AND logic and OR logic. The input logic should be X1 OR X3, AND NOT X2. Use contacts X1, X2, and X3. Use Y12 for the output coil.

12. Design a ladder into which coil Y5 will latch itself. The input contact should be X1. The unlatch contact should be X2.
13. Draw a diagram of a PLC scan and thoroughly explain what occurs during such a scan.
14. Examine the following rungs and determine whether the output for each is on or off. The input conditions shown represent the states of real-world inputs.

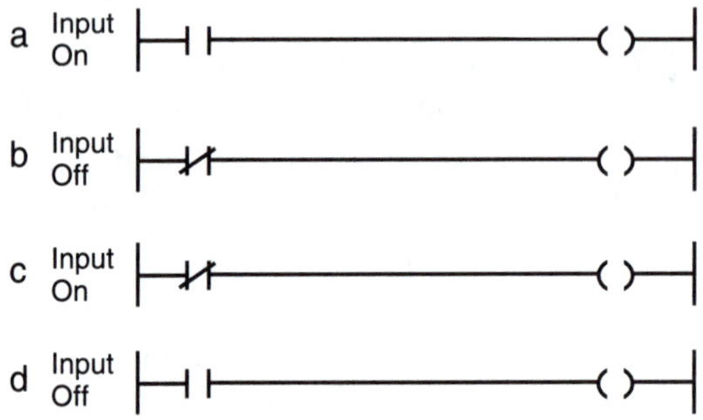

15. Examine the following rungs and determine whether the output for each is on or off. The input conditions shown represent the states of real-world inputs.

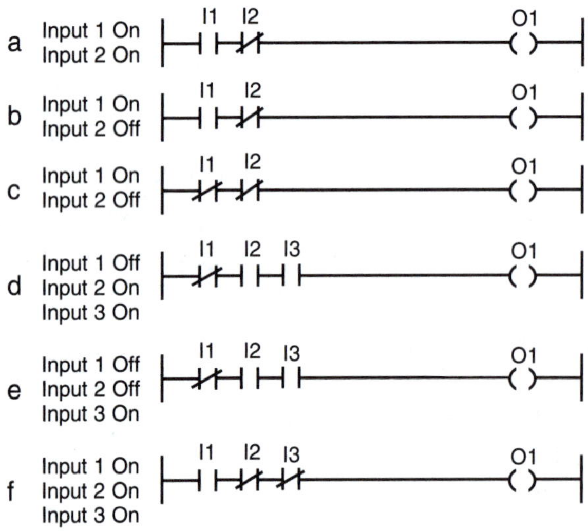

16. Examine the following rungs and determine whether the output for each is on or off. The input conditions shown represent the states of real-world inputs.

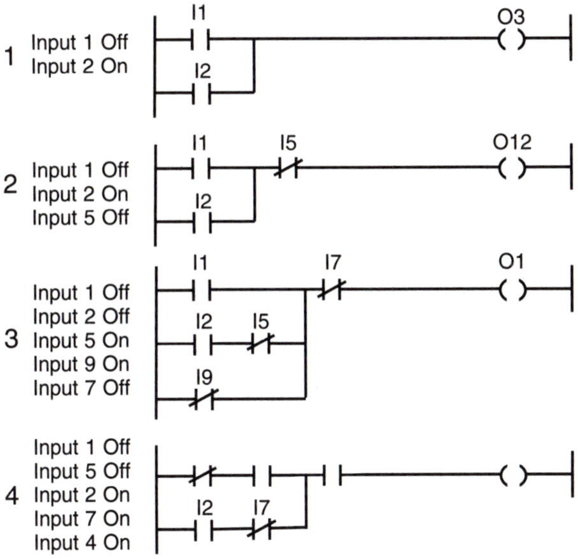

chapter 5

Rockwell Automation Addressing and Instructions

In this chapter we examine Rockwell Automation memory addressing and instructions. We also learn how to write basic ladder logic programs.

OBJECTIVES

Upon completion of this chapter, you will be able to:

1. Explain Rockwell Automation memory organization.
2. Explain Rockwell Automation addressing.
3. Define addresses for various file types.
4. Explain the use of various Rockwell Automation instructions.
5. Write programs that utilize Rockwell Automation instructions.

UNDERSTANDING ROCKWELL FILE ORGANIZATION AND ADDRESSING

Understanding file organization and addressing can be one of the more confusing topics when learning how to program a PLC, so it is vital that you study this material carefully. This approach will dramatically reduce the frustration you experience as you begin to write programs. First, Rockwell Automation divides its memory system into two types: program and data (Figure 5–1). These two types of memory are important to remember. Creating a mental image of the two separate

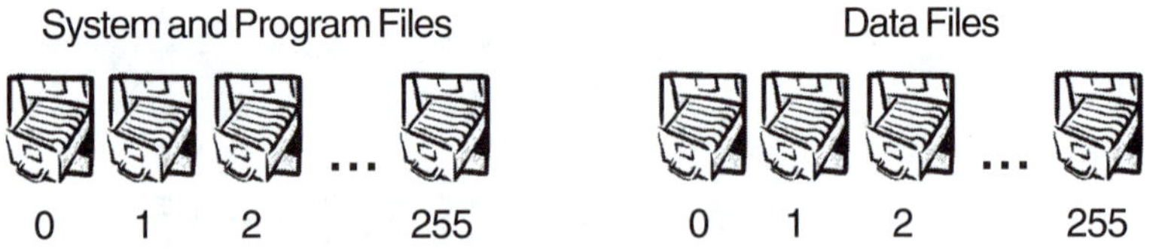

Figure 5–1 The two types of memory in a Rockwell Automation PLC.

memory areas may help. The program area of memory has 256 files; the data area also has 256 files.

To review, thus far you know there are two types of memory in a Rockwell Automation PLC: program and data. The program area of memory has 256 files (file 0 through file 255). The data area of memory has 256 files (file 0 through file 255). Let us now look at program memory.

Program File Memory

The 256 files in program memory (file 0 through file 255) contain controller information, the main ladder program, and any subroutine programs (Figure 5–2). The program files are as follows:

- File 0 contains various system-related information and user-programmed information such as processor type, I/O configuration, processor file name, and password.
- File 1 is reserved.
- File 2 contains the main ladder diagram.

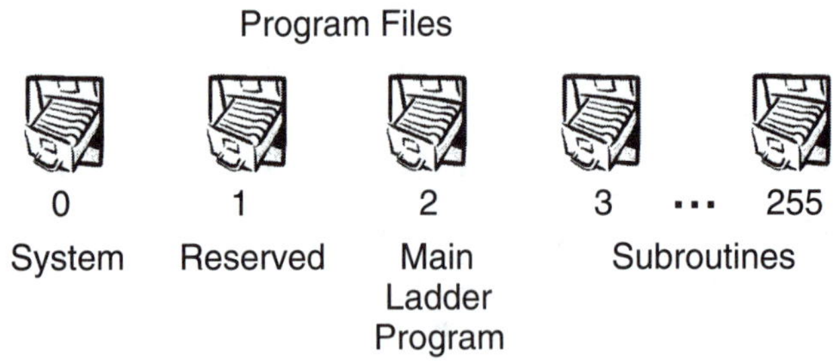

Figure 5–2 System and program files in a Rockwell Automation PLC.

- Files 3 through 255 are user created and accessed according to subroutine instructions residing in the main ladder program file. In other words, the user can break the ladder diagram into logical portions of the total application program. Each portion of the ladder diagram can then be accessed as needed from the main program in file 2. For example, the main ladder diagram in file 2 could contain the logic for the user to choose manual or automatic operation. If the user chooses automatic operation, then the logic in file 3 (or whatever file number the programmer desired) is executed. If the user chooses manual operation, then the logic in file 4 (or whatever file number the programmer desired) is executed.

Data Memory

The second type of memory that Rockwell PLCs have is data memory. As mentioned, data memory also has 256 files available. Data files are the files we need to write ladder logic. It may be helpful to imagine an office where information is stored in files. Imagine 256 files (Figure 5–1). Imagine that each file has a specific use.

Data files contain the status information associated with external I/O and all other instructions used in the main and subroutine ladder program files. In addition, these files store information concerning processor operation. You can also use the files to store "recipes" and look up tables if needed.

Data files are organized by the type of data they contain. The data file types are as follows:

- File 0 is used to store the status of the outputs of the PLC. If you needed to change the status of an output, you could put a 1 or 0 in the correct bit in file 0.
- File 1 is used to store the status of inputs (Figure 5–3). If you needed to check the status of inputs, you would look in file 1.
- File 2 is reserved for PLC status information, which could be helpful for troubleshooting and program operation.
- File 3 is used to store bit information. Bits can be helpful in logic programming. They can be used to store information about conditions or as contacts or coils for non–real-world I/O.
- File 4 is used for timer information.
- File 5 is reserved for counter information.
- File 6 is reserved for control. It is used when working with shift registers and sequencers.

File Number	Type	Use
0	Output	Stores the states of output terminals for the controller.
1	Input	Stores the states of input terminals for the controller.
2	Status	Stores the controller's operation information. This file can be useful for troubleshooting the controller and the program operation.
3	Bit	Can be used for internal relay bit storage.
4	Timer	Stores the accumulated value, preset value, and status bits for timers.
5	Counter	Stores the accumulated value, preset value, and status bits for counters.
6	Control	Stores the length, pointer position, and status bits for specific instructions such as sequencers and shift registers.
7	Integer	Can be used to store integer numbers or bit information.
8	Floating point	Can be used to store single-precision, nonextended 32-bit float numbers.
9–255	User defined	Can be used for any of the previously defined types by the user. Note that the whole file number must be used for the same type.

Figure 5–3 Data files and uses.

- File 7 is used to store integers (whole numbers).
- File 8 is used to store floating point numbers (decimal numbers). This covers the files reserved for special uses. There are still many files available.
- Files 9 through 255 are user configurable, meaning they are for any purpose a user would like. For example, if the user needs more room to store integers, the user could reserve any of the remaining files for more integers. Types may not be mixed files, however. The whole file must be reserved for integer use.

Figure 5–3 illustrates the files and their uses. Take some time to study them. It will help you to understand addressing.

Memory Addressing

I/O addressing is relatively straightforward. The first letter of the input address is called an identifier. Figure 5-4 shows the identifiers. In Figure 5–5 it is an input, so the identifier is I, indicating it will be an input. The second item in the address is a 1, which is the file number. The default number for input files is 1. The next item is a colon. This is a delimiter. A delimiter defines what will come next. If the delimiter is a colon, then the next item to follow the colon will always be the slot number of the module if we are naming inputs or outputs. Figure 5–6 shows an example of a PLC rack of modules. In our example in Figure 5–5, this input address is in slot 2. The next item is a slash. The slash is also a delimiter. The slash delimiter always means the next item will be a bit number. In this case it means that this will be input number 3. So, to review, this would be input 3 on the input module that is in slot 2.

Figure 5–4 Default file identifiers and numbers.

File Type	Identifier	File Number
Output	O	0
Input	I	1
Status	S	2
Bit	B	3
Timer	T	4
Counter	C	5
Control	R	6
Integer	N	7
Float	F	8

Figure 5–5 How address input 3 on the input module in slot 2 would be named.

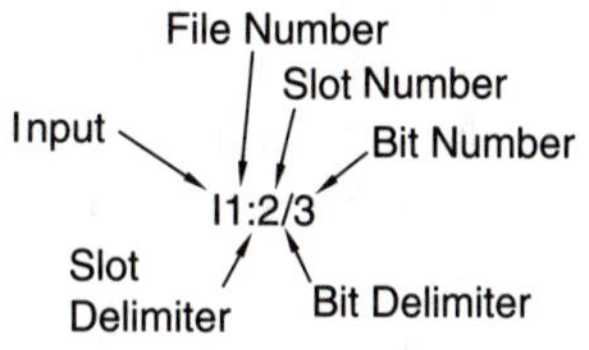

Figure 5–6 Note the input modules in slots 1, 2, and 3. The output modules are in slots 4, 5, and 6.

	Power Supply	CPU	INPUT	INPUT	INPUT	OUTPUT	OUTPUT	OUTPUT
Slot Number		0	1	2	3	4	5	6

Figure 5–7 Output addressing example.

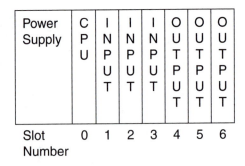

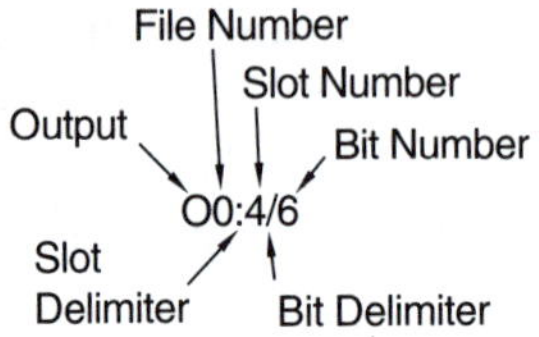

O0:5/12—Output 12 in slot 5, output file 0

O:6/7—Output 7 in slot 6, output file 0 (default file number 0)

I1:2/8—Input 8 in slot 2, input file 1

I:3/5—Input 5 in slot 3, input file 1 (default file number 1)

O0:4/12—Output 12 in slot 4, output file 0

O:5/1—Output 1 in slot 5, output file 0 (default file number 0)

I:1/9—Input 9 in slot 1, input file 1 (default file number 1)

Figure 5–8 Examples of I/O addressing.

Figure 5–7 shows an example of output addressing. The first item in the address is O, for output. The next item is the file number. It is the default file number for outputs (file 0). The next item is a delimiter. It is a colon. The colon delimiter means that the next item to follow will be the slot number. In this case it is the output module in slot 4. The next item is the delimiter. The slash means that the number that follows will be the bit number (output number). In this case it is output 6 on the output module in slot 4.

Figure 5–8 has several important examples of I/O addressing. Study them carefully.

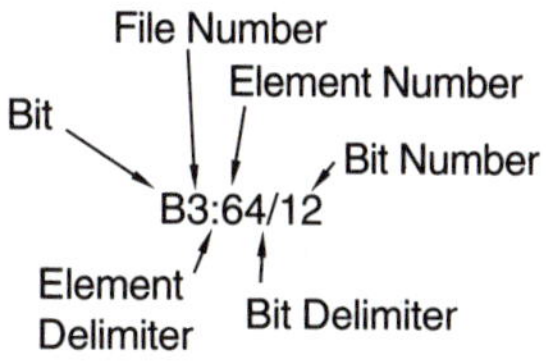

Figure 5–9 Bit element addressing example.

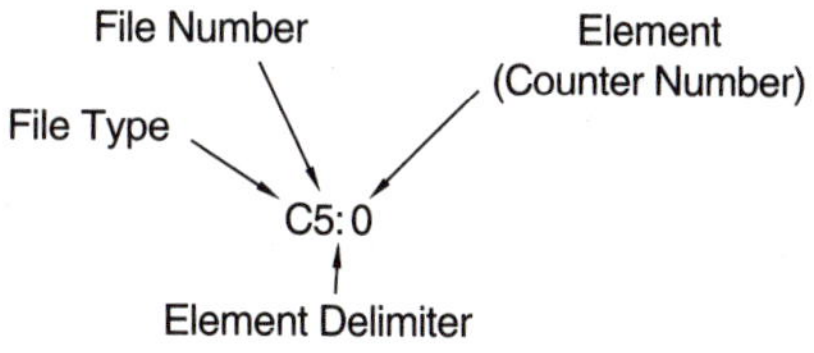

Figure 5–10 Example of a counter address.

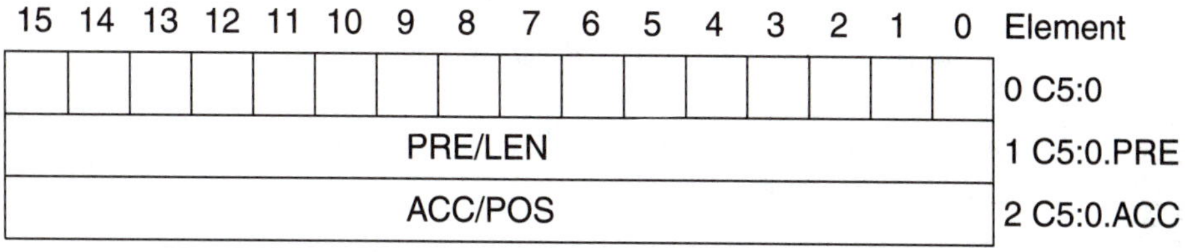

Figure 5–11 Example of counter addressing.

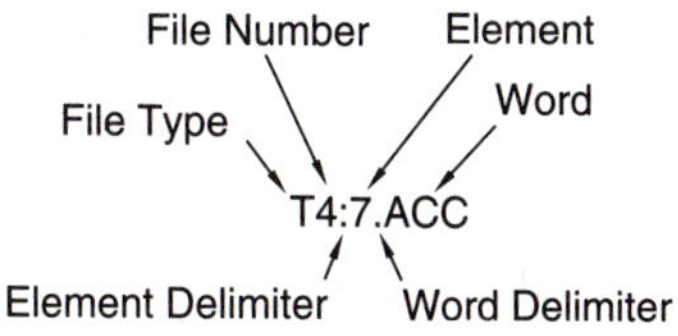

Figure 5–12 Timer element addressing example.

Figure 5–9 shows an example of bit element addressing. In this example the type identifier (B) specifies a bit type. The file number is 3. The element within the file is 64. The bit number is 12.

Elements for timers, counters, control, and ASCII files consist of three words. Figure 5–10 shows an example of counter addressing. In this case, the C stands for counter, 5 is the file number, and the counter is 0. Figure 5–11 shows that counters occupy three words of memory. C5:0.PRE would hold the preset value and C5:0.ACC would hold the accumulated value. Timers and counters will be covered in more detail in Chapter 7.

Figure 5–12 shows an element address of T4:7.ACC. The T identifier stands for a timer. The file number is 4. The colon is the delimiter. The timer number is 7 and

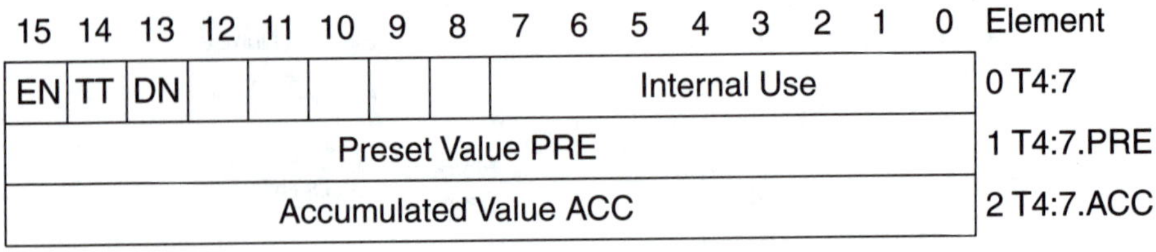
15 14 13 12 11 10 9 8 7 6 5 4 3 2 1 0 Element
EN
TT
DN
Internal Use
0 T4:7
Preset Value PRE
1 T4:7.PRE
Accumulated Value ACC
2 T4:7.ACC

Figure 5–13 Timer memory.

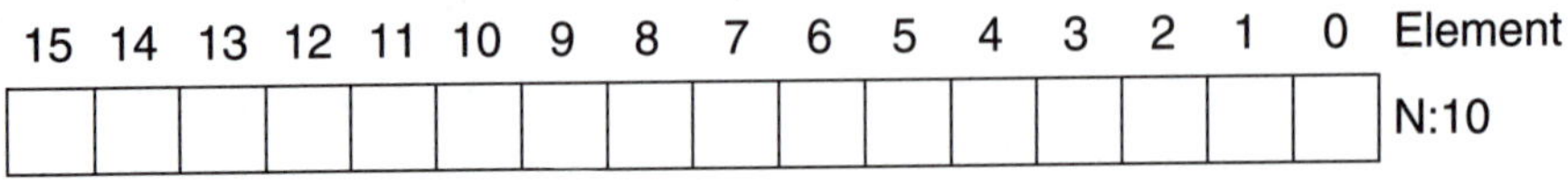
15 14 13 12 11 10 9 8 7 6 5 4 3 2 1 0 Element
N:10

Figure 5–14 Example of the addressing of integer elements.

Figure 5–15 Integer element addressing example.

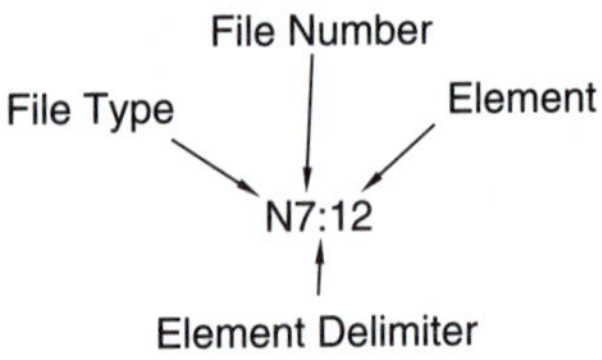

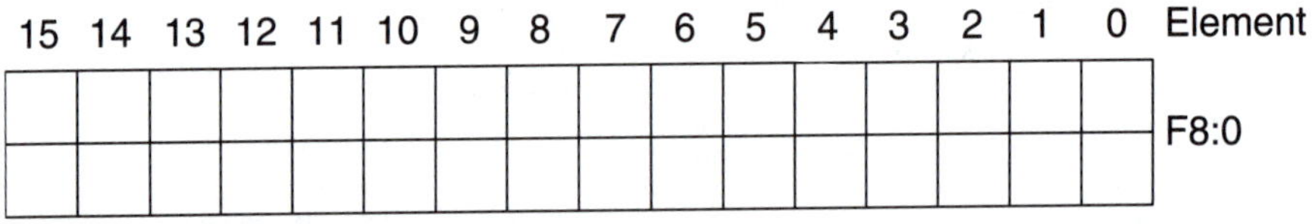
15 14 13 12 11 10 9 8 7 6 5 4 3 2 1 0 Element
F8:0

Figure 5–16 Example of the addressing of floating-point elements.

the ACC specifies the accumulated value word. Note that timers, like counters, utilize three words of memory (Figure 5–13). Timers will be covered in depth in Chapter 7.

Figures 5–14 and 5–15 show examples of integer element addressing. In Figure 5–15 the N stands for integer, 7 is the file number, and 12 identifies the element.

Floating-point files have two-word elements. Figure 5–16 shows an example of floating-point element addressing. Note that there are two words to hold the floating-point number. In this example, the F stands for floating point, 8 is the file number,

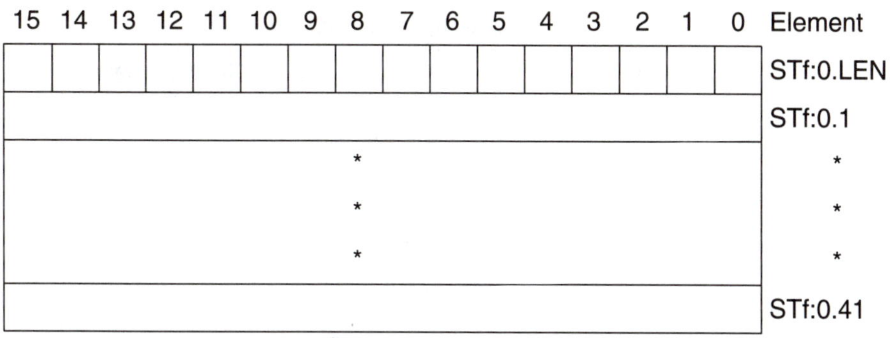

Figure 5–17 Example of the addressing of string file elements.

and 0 means the first two words in file 8. Remember that each floating-point number requires two words of storage.

String files have 42-word elements. Figure 5–17 shows an example. Note the use of the decimal point delimiter to specify the particular word in the string file.

User-Defined Files

Figure 5–18 shows the files that users can define. File numbers 9 through 255 are available for the user for any purpose. A whole file, however, must have the same use. For example, if the user wants to use file 10 for additional integers, file 10 can be used only for integers.

Rockwell Automation Contacts

Examine If Closed

The examine if closed instruction (XIC) is normally open. If a real-world input device is on, this type of instruction is true and passes power (Figure 5–19). If the input bit from the input image table associated with this instruction is a 1, then the instruction is true. If the bit in the input image table is a 0, then the instruction associated with this particular input bit is false.

Examine If Open

Rockwell Automation calls its normally closed contacts examine if open contacts. The examine if open instruction (XIO) can also be called a normally closed instruction. This instruction responds in the opposite fashion to the normally open instruction. If the bit associated with this instruction is a 0 (off), then the instruction

User-Defined Files		
File Type	**Identifier**	**File Number**
Bit	B	9–255
Timer	T	
Counter	C	
Control	R	
Integer	N	
Float	F	
String	St	
ASCII	A	

Figure 5–18 User-defined file numbers.

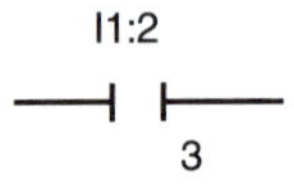

Figure 5–19 Examine if closed instruction (XIC). If the CPU sees an on condition at bit I1:2/3, this instruction is true. The numbering of the input instruction is as follows: This is an input in file 1. It is located on an input module in slot 2 and it is real-world input 3 of the input module. The numbering of inputs and outputs will be covered later.

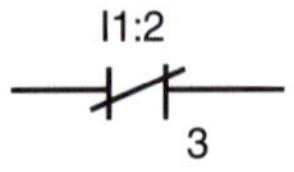

Figure 5–20 Examine if open instruction (normally closed).

is true and passes power. If the bit associated with the instruction is a 1 (true), then the instruction is false and does not allow power flow.

Figure 5–20 shows an examine if open instruction (XIO). The input is 3 on an input module in slot 2. If the bit associated with I1:2/3 is true (1), then the instruction is false (open) and does not allow power flow. If bit I0:2/3 is false (0), then the instruction is true (closed) and allows flow.

Special Contacts

There are many special-purpose contacts available to the programmer. The original PLCs had few available. Sharp programmers used normally open and normally closed contacts in ingenious ways to turn outputs on for one scan, to latch outputs on, and so on. PLC manufacturers added special contacts to their ladder programming languages to meet these needs. The programmer can now accomplish these special tasks with one contact instead of a few lines of logic.

Immediate Instructions

Immediate instructions are used when the input or output being controlled is highly time dependent. For example, for safety reasons we may have to update the status of a particular input every few milliseconds. If our ladder diagram is 10 milliseconds long, the scan time would be too slow, which can be dangerous. The use of immediate instructions allows inputs to be updated immediately as they are encountered in the ladder. The same is true of output coils.

Rockwell Automation SLC 500 Immediate Input with Mask The immediate input with mask (IIM) instruction is used to acquire the present state of one word of inputs (Figure 5–21). Normally, the CPU would have to finish all evaluation of the ladder logic and then update the output and input image tables. In this case, when the CPU encounters this instruction during ladder evaluation, it interrupts the scan. Data from a specified I/O slot are transferred through a mask to the input data file. This makes the data available to instructions following the IIM instruction. Thus, it gets the real-time states of the actual inputs at that time and puts them in that word of the input image table. The CPU then returns to evaluating the logic using the new states it acquired.

The IIM process is used only when time is a crucial factor. Normally, the few milliseconds a scan takes is fast enough for any job. There are cases, however, when scan time takes too long for some I/O updates. Motion control is one example. Updates on speed and position may be required every few milliseconds or less. We cannot safely wait for the scan to finish to update the I/O. In these cases immediate instructions are used.

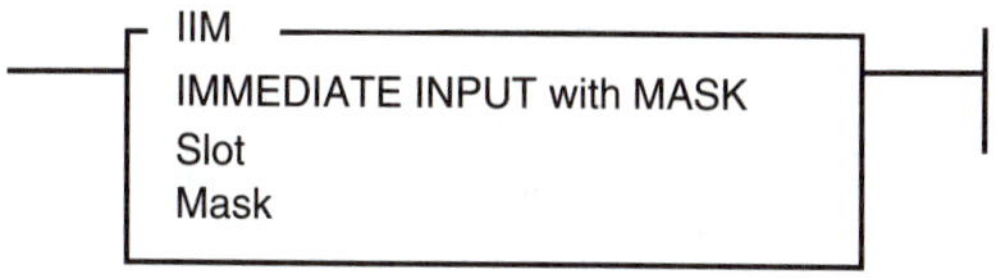

Figure 5–21 Instruction numbering. This is an immediate input with mask (IIM) instruction.

Figure 5–22 Immediate input instruction (IIM).

I:2	Inputs of slot 2, word 0
I:2.1	Inputs of slot 2, word 1
I:1	Inputs of slot 1, word 0

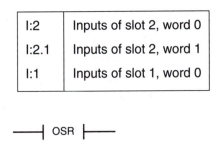

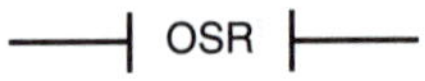

Figure 5–23 Rockwell Automation OSR instruction.

Figure 5–22 shows an IIM addressing format. When the CPU encounters the instruction during evaluation, it will immediately suspend what it is doing (evaluating) and will update the input image word associated with I/O slot 2, word 0.

Rockwell Automation One-Shot Rising Instruction The one-shot rising (OSR) instruction is a retentive input instruction that can trigger an event to happen one time (Figure 5–23). When the rung conditions that precede the OSR go from false to true, the OSR instruction will be true for one scan. After the one scan, the OSR instruction becomes false, even if the preceding rung conditions that precede it remain true. The OSR will become true again only if the rung conditions preceding it make a transition from false to true. Only one OSR can be used per rung.

Latching Instructions

Latches are used to lock in a condition. For example, if an input contact is on for only a short time, the output coil would be on for the same short time. If the programmer wanted to keep the output on even if the input goes low, a latch could be used. This can be done by using either the output coil to latch itself on (Figure 5–24) or a special latching coil (Figure 5–25). When a latching output is used, it will stay on until it is unlatched. Unlatching is done with a special unlatching coil. When activated, the latched coil of the same number is unlatched.

Rockwell Automation Coils

Output Energize

The output energize (OTE) instruction is the normal output instruction. The OTE instruction sets a bit in memory. If the logic in its rung is true, then the output bit will be set to a 1. If the logic of its rung is false, then the output bit is reset to a 0.

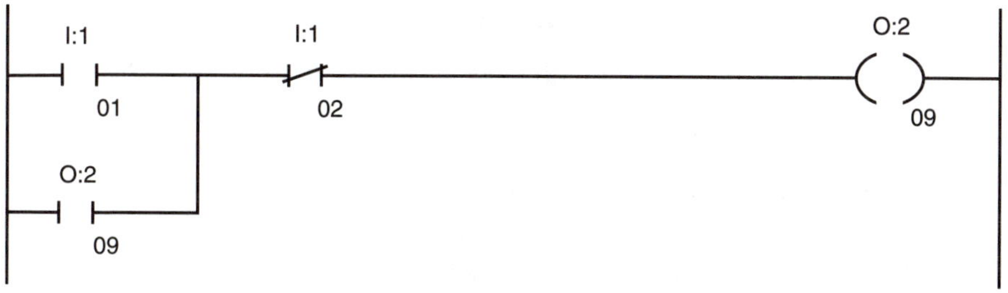

Figure 5–24 An example of latching an output on. If input 1:1/01 is true, coil O:2/09 will be energized. (Remember that contact I:1/02 is normally closed.) When coil O:2/09 energizes, it latches itself on by providing a parallel path around I:1/01. The only way to turn the latched coil off would be to energize normally closed contact I:1/02. This would open the rung and de-energize coil O:2/09.

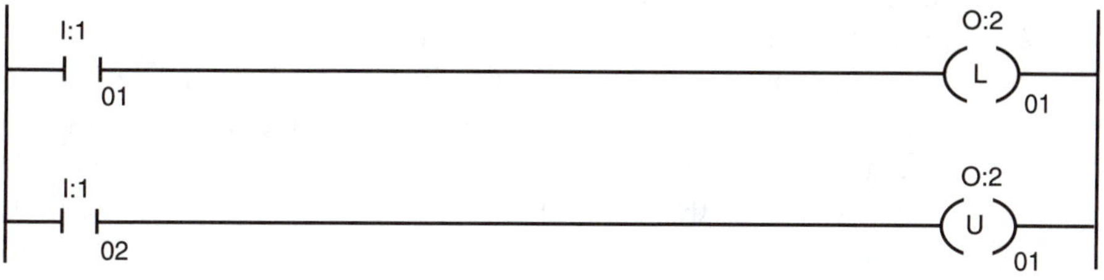

Figure 5–25 Use of a latching output. If input I:1/01 is true, output O:2/01 energizes. It will stay energized even if input I:1/01 becomes false. O:2/01 will remain energized until input I:1/02 becomes true and energizes the unlatch instruction (coil O:2/01). Note that the coil number of the latch is the same as that of the unlatch. This is an example of a Rockwell Automation latch and unlatch instruction.

Figure 5–26 Output energize (OTE) instruction.

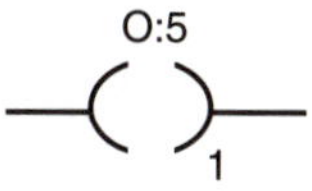

Figure 5–26 shows an OTE instruction. This particular example is real-world output 1 of the output module in slot 5. If the logic of the rung leading to this output instruction is true, then output bit O:5/1 will be set to a 1 (true). If the rung is false, then the output bit would be set to a 0 (false).

Rockwell Automation Immediate Outputs

The immediate output with mask (IOM) instruction is used to update output states immediately. In some applications the ladder scan time is longer than the needed update time for certain outputs. For example, it might cause a safety problem if an output were not turned on or off before an entire scan was complete. In these cases, or when performance requires immediate response, IOMs are used. When the CPU encounters an immediate instruction, it exits the scan and immediately transfers data to a specified I/O slot through a mask. The CPU will then resume evaluating the ladder logic. Figures 5–27 and 5–28 show examples of the addressing format for an IOM.

Rockwell Automation Output Latch Instruction

The output latch (OTL) instruction is a retentive instruction. If this input is turned on, it will stay on even if its input conditions become false. A retentive output can be turned off only by an unlatch instruction. Figure 5–29 shows an OTL. In this case if the rung conditions for this output coil are true, then the output bit will be set to a 1. It will remain a 1 even if the rung becomes false. The output will be latched on. Note that if the OTL is retentive and if the processor loses power, the actual output turns off, but when power is restored, the output is retentive and will turn on. This is also true in the case of switching from run to program mode. The actual output turns off, but the bit state of 1 is retained in memory. When the processor is switched to run again, retentive outputs will turn on again regardless of the rung conditions. Retentive instructions can help or hurt the programmer. Be care-

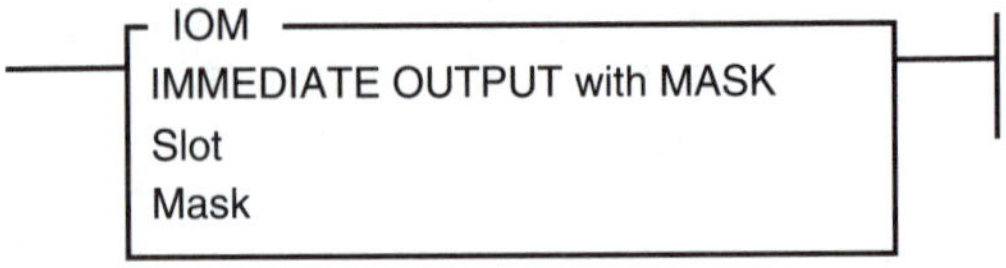

Figure 5–27 Format for an immediate output with mask (IOM) instruction.

Figure 5–28 Immediate output instruction I/O numbering.

O:2	Outputs of slot 2, word 0
O:2.1	Outputs of slot 2, word 1
O:1	Outputs of slot 1, word 0

Figure 5–29 Output latch (OTL) instruction.

ful from a safety standpoint when using retentive instructions. Use an unlatch instruction to turn a retentive output off.

Rockwell Automation Output Unlatch Instruction

The output unlatch (OTU) instruction is used to unlatch (change the state of) retentive output instructions. It is the only way to turn an OTL off. Figure 5–30 shows an OTU. If this instruction is true, it unlatches the retentive output coil of the same number.

PROGRAM FLOW INSTRUCTIONS

Many types of flow control instructions are available on PLCs. Flow control instructions can be used to control the sequence in which your program is executed. Flow instructions allow the programmer to change the order in which the CPU scans the ladder diagram. These instructions are typically used to minimize scan time and create a more efficient program. They can also be used to help troubleshoot ladder logic. They should be used with great care by the programmer. Serious consequences can occur if they are improperly used because their use causes portions of the ladder logic to be skipped.

Rockwell Automation Jump Instructions

Rockwell Automation has jump (JMP) and label (LBL) instructions available (Figure 5–31). These can be used to reduce program scan time by omitting a section of program until it is needed. It is possible to jump forward and backward in the ladder. The programmer must be careful not to jump backward an excessive amount of times. A counter, timer, logic, or the program scan register should be used to limit the amount of time spent looping inside a JMP/LBL instruction. If the rung containing the JMP instruction is true, the CPU skips to the rung containing the specified LBL and continues execution. You can jump to the same label from one or more JMP instructions.

Figure 5–30 Output unlatch (OTU) instruction.

Figure 5–31 Rockwell Automation jump (JMP) and label (LBL) instructions.

Rockwell Automation Jump to Subroutine Instructions

Rockwell Automation also has subroutine instructions available. The jump to subroutine (JSR), subroutine (SBR), and return (RET) are used for this purpose (Figures 5–32 and 5–33). Subroutines can be used to store recurring sections of logic that must be executed in several points in the program. A subroutine saves effort and memory because you program it only once. Subroutines can also be used for time-critical logic by using immediate I/O in them.

The SBR instruction must be the first instruction on the first rung in the program file that contains the subroutine. Subroutines can be nested up to eight deep. This allows the programmer to direct program flow from the main program to a subroutine and then to another subroutine, and so on.

The desired subroutine is identified by the file number entered in the JSR instruction. This instruction serves as the label or identifier for a program file as a regular subroutine file. The SBR instruction is always evaluated as true. The RET instruction is used to show the end of the subroutine file. It causes the CPU to return to the instruction following the previous JSR instruction.

Subroutines can be useful in breaking a control program into smaller logical sections. For example, the main file we use for our ladder logic is file 2. Many times all of the ladder logic is programmed in file 2. It makes more sense and is more understandable if the program is divided into logical modes of operation. For example, the programmer could make file 2 contain the logic needed for overall system operation; the main file (file 2) could contain logic to determine if the user wants to go into the automatic or manual mode of operation; it could also call another subroutine that monitors the system for safety. Let us use file 3 for the manual mode of operation, file 4 for the automatic mode of operation, and file 5 for the monitoring logic. We would use conditional logic to determine whether to call the manual subroutine or the automatic routine. If the user hit the manual key, the main would

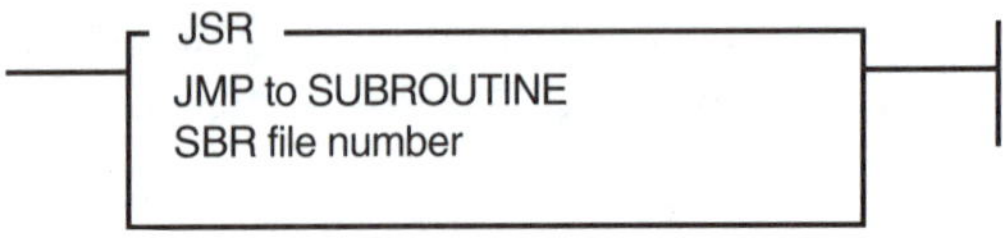

Figure 5–32 Rockwell Automation jump to subroutine (JSR) instruction.

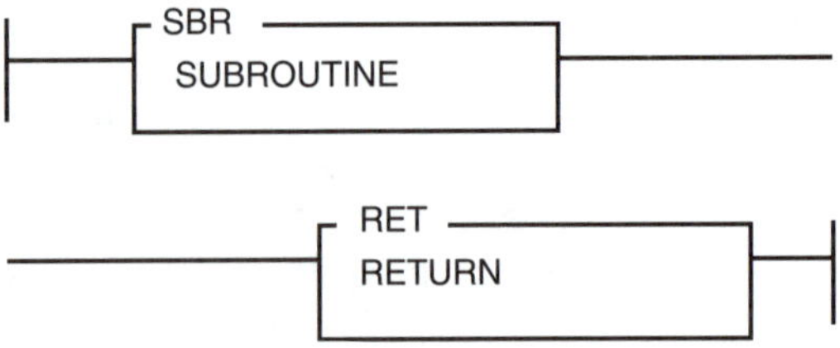

Figure 5–33 Rockwell Automation subroutine (SBR) and return (RET) instructions.

call the subroutine in file 3. If the user hits the auto key, the main would call the automatic subroutine in file 4. We would want the monitor subroutine to run continuously. The logic in the main can call the monitor logic or the other subroutines can call the monitor mode because subroutines can be nested and call other subroutines. It is much simpler to program the system when it can be divided into smaller logical sections, and it is much easier to understand and troubleshoot.

QUESTIONS

1. What is XIO? XIC?
2. What is IOM?
3. Write the address for the first input or output for each I/O module below.

Power Supply	CPU	INPUT	INPUT	INPUT	OUTPUT	OUTPUT	OUTPUT
Slot Number	0	1	2	3	4	5	6

4. Design a latching circuit using a Rockwell Automation latch and unlatch instruction. Use contact I:2/7 for the latch input, I:2/8 for the unlatch input, and O:5/3 for the coil.
5. Describe the purpose and use of subroutines.

Following are sample Rockwell Automation SLC addresses. Thoroughly explain each of them.

6. B3:16/12
7. O0:1/6
8. I1:1/3
9. O0:2/5
10. I1:1/2
11. T4:7.PRE
12. C5:0.ACC
13. F8:0
14. I1:1/2

15. T4:5.EN
16. T4:3.DN
17. O0:4/9
18. I1:2/4
19. N10:3
20. I:3.0/4
21. I:2/17
22. O:7/12
23. B3:3/5
24. S:42
25. N7:12

Programming Practice

RSLogix Relay Logic Instructions

These exercises utilize the LogixPro software included with this book. To install the software, simply insert the CD into your CD-ROM drive; it should self-install. Note that the software will operate only for a limited trial period. After that (or anytime before) you may register by paying a small fee to have the full version of the software.

Description of the Software

This is a complete, working version of LogixPro. In the trial or evaluation mode, only the I/O and Door simulations are fully available for user programming. In addition, until LogixPro has been registered, File Save and Printing functionality is disabled.

If you need help with the RSLogix addressing or instructions, you can go to two places. With your mouse, select Start, then Programs, then TheLearningPit, then LogixPro, and then you may choose the Instruction Set Reference, LogixPro, the readme file, or the student exercises. You can also log onto the Internet and go to the "LogixPro . . . Student Exercises and Documentation" page entry, which is listed on TheLearningPit.com home page. Also remember to try clicking on rungs, instructions, and so on, with the right mouse button to locate pop-up editing menus and so forth.

First Exercise

This exercise is designed to familiarize you with the operation of LogixPro and to walk you through the process of creating, editing, and testing simple PLC programs utilizing the relay logic instructions supported by RSLogix. This is a very important exercise because it will prepare you for programming the more complex applications in later chapters. It will also enable you to achieve basic competancy in the use of Rockwell Automation RSLogix software. Go to the Student Exercises as described and complete the Relay Logic—Introductory Lab.

Additional Exercise

1. Enter the circuit that you designed in question 4 and test it in the I/O simulator. Make sure you use appropriate addresses.
2. Go to the student exercises on the CD. Follow the instructions and program and test door simulation exercise 1. Document your program.
3. Complete student programming exercise 2 for the door simulation. Document your program.
4. Complete student programming exercise 3 for the door simulation. Document your program.

Extra Credit

Complete student programming exercise 4 for the door simulation. Document your program.

chapter

Input/Output Modules and Wiring 6

The PLC was originally designed for simple digital (on/off) control. Over the years, PLC manufacturers have added to its capabilities. Today, there are I/O cards available for almost any application imaginable. Because of this, wiring can seem very confusing at first. In this chapter we study the wiring of digital and analog devices.

OBJECTIVES

Upon completion of this chapter, you will be able to:

1. List at least two I/O cards that can be used for communication.
2. Describe at least five special-purpose I/O cards.
3. Define terms such as *resolution, high density, discrete,* and *TTL.*
4. Choose an appropriate I/O module for a given application.
5. Understand and be able to wire analog and discrete I/O modules.

I/O MODULES

PLCs were originally used to control one simple machine or process. Changes in U.S. manufacturing have required much more capability. The increasing speed of production and the demand for higher quality require closer control of industrial processes. PLC manufacturers have added modules to meet these new requirements.

Special modules have been developed to meet almost any imaginable need, including control processes. Temperature control is one example.

Industry is beginning to integrate its equipment so that data can be shared. Modules have been developed to allow the PLC to communicate with other devices such as computers, robots, and machines.

Velocity and position control modules have been developed to meet the needs of accurate high-speed machining. These modules also make it possible for entrepreneurs to start new businesses that design and produce special-purpose manufacturing devices such as packaging equipment, palletizing equipment, and various other production machinery.

These modules are also designed to be easy to use. They are intended to make it easier for the engineer to design an application. In the balance of this chapter, we examine many of the modules that are available.

Digital (Discrete) Modules

Digital modules are also called discrete modules because they are either on or off. A large percentage of manufacturing control can be accomplished through on/off control. Discrete control is easy and inexpensive to implement.

Digital Input Modules

Digital input modules accept either an on or off state from the real world. The module inputs are attached to devices such as switches or digital sensors. The modules must be able to buffer the CPU from the real world. Assume that the input is 250 VAC. The input module must change the 250 VAC level to a low-level DC logic level for the CPU. The modules must also optically isolate the real world from the CPU. Input modules usually have fuses for module protection.

Input modules typically have light emitting diodes (LEDs) for monitoring the inputs. There is one LED for every input. If the input is on, the LED is on. Some modules also have fault indicators. The fault LED turns on if there is a problem with the module. The LEDs on the modules are useful for troubleshooting.

Most modules also have plug-on wiring terminal strips. All wiring is connected to the terminal strip, which is plugged into the actual module. If there is a problem with a module, the entire strip is removed, a new module is inserted, and the terminal strip is plugged into the new module. Note that there is no rewiring. A module can be changed in minutes or less. This aspect is vital considering the huge cost of a system being down (unable to produce product).

Input modules usually need to be supplied with power. (Some of the small PLCs are the exception.) The power must be supplied to a common terminal on the module, through an input device, and back to a specific input on the module. Current

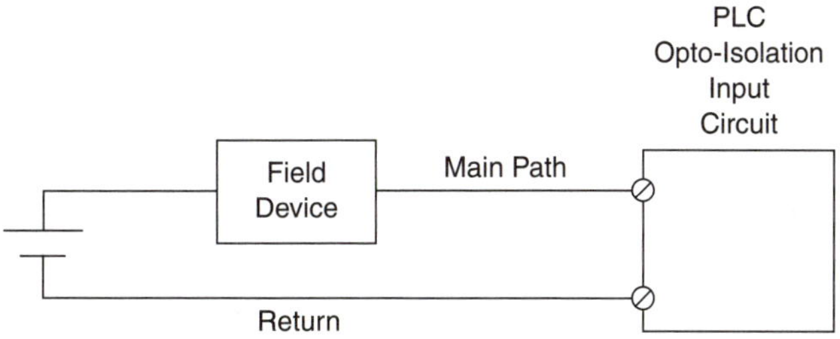

Figure 6–1 Typical I/O current path.

Figure 6–2 Typical shared return path (common).

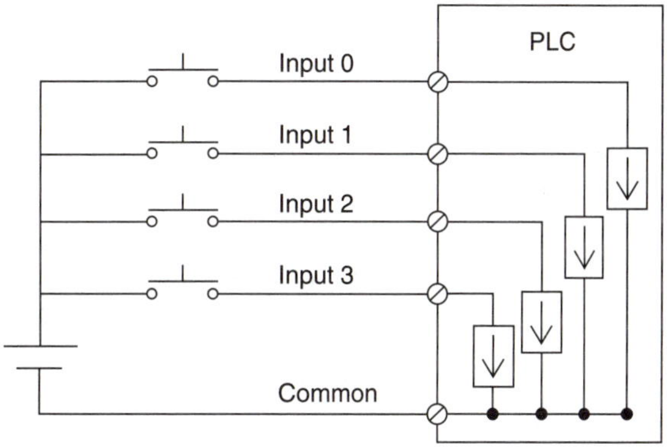

must enter at one terminal of the I/O module and exit at another terminal (Figure 6–1). The figure shows a power supply, field input device, the main path, I/O circuit, and the return path. Unfortunately this would require two terminals for every I/O point. Most I/O modules provide groups of I/O that share return paths (commons). Figure 6–2 shows the use of a common return path.

Wiring

One of the easier ways to figure out how to wire is to simplify the circuit and get it on paper first, where most people can better visualize the whole picture. Imagine that we have a system with many inputs to wire. Let us say we worry about one input first. If we can figure out the first one, the rest will be easy. Figure 6–3 shows an example of one input, a power supply, and wiring terminals for a sinking input module. Input and output circuits almost always come down to three categories: power, a device, and two terminals. Figure 6–4 shows how we would wire this simple circuit. It is easy to see the current path.

Figure 6–3 Sinking input module wiring example.

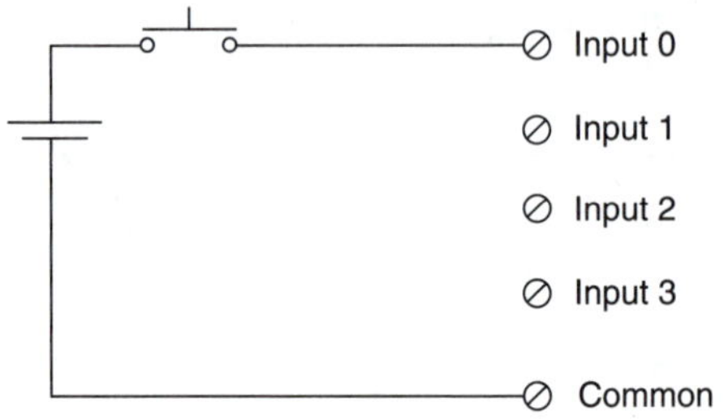

Figure 6–4 Wiring in a simple circuit.

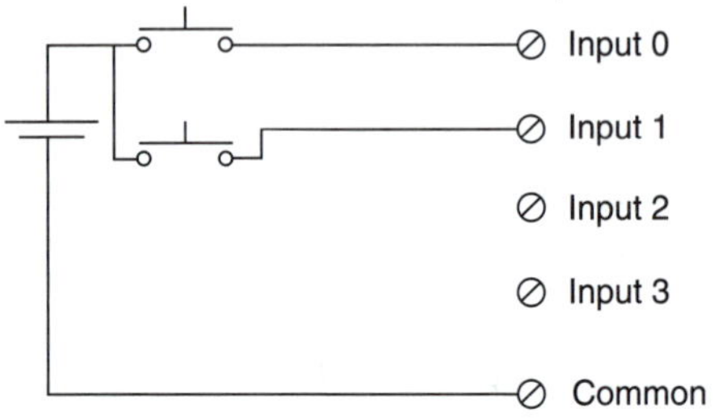

Figure 6–5 Additional input device.

Let's now add the second input. Figure 6–5 shows the addition of one more input device. This device must be wired the same as the first. Positive must be connected to one side of the switch, and the other side of the switch must be connected to the second input terminal.

Obviously this is a simple yet powerful technique. If you can figure out the wiring for the first device, the rest are easy. As stated, planning it on paper first will simplify the task of wiring.

Figure 6–6 shows a three-wire sensor that must be connected to an input. The sensor has a positive lead, a negative lead, and an output lead. The sensor is a sourcing sensor, so the output is positive. Figure 6–7 shows how this circuit was wired. The sensor had to be wired to + and − voltage to power the sensor. The + output was then connected to an input, and − from the supply was connected to the common on the sinking module.

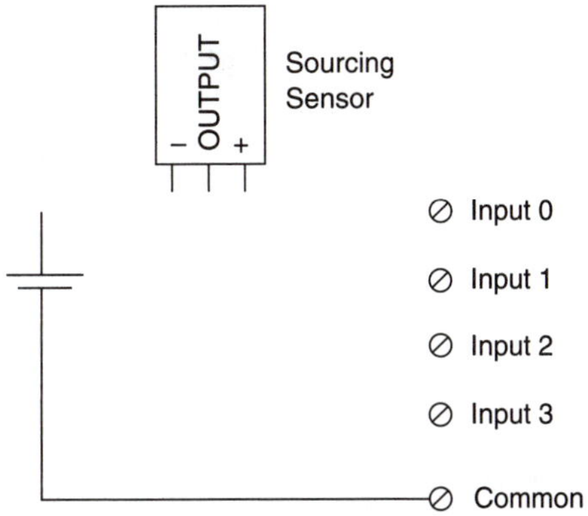

Figure 6–6 Three-wire sensor example.

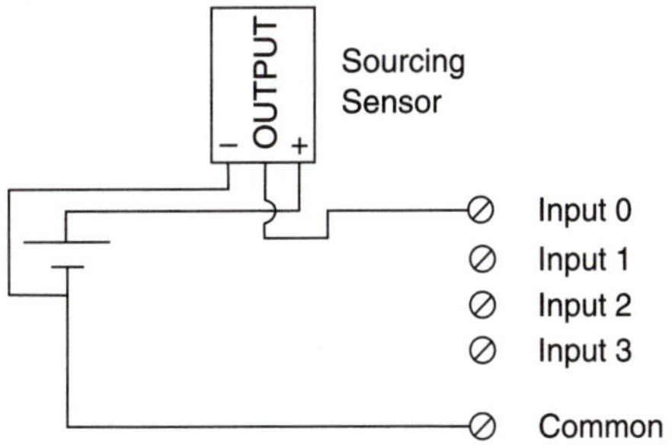

Figure 6–7 Complete wired circuit.

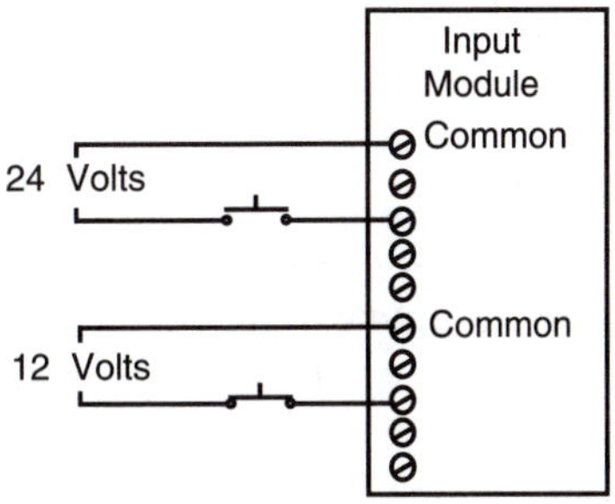

Figure 6–8 Input module with dual commons allows the user to mix input voltages.

Some modules provide multiple commons to allow the user to mix voltages on the same module (Figure 6–8). These commons can be jumpered together if desired (Figure 6–9).

Figure 6–10 shows an input module for a Rockwell Automation SLC. Note that the commons are connected internally for this module. Note also that the negative side of the DC voltage must be connected to the common. The positive voltage from the power supply is brought through the input devices and back to the input terminal. This is a sinking module.

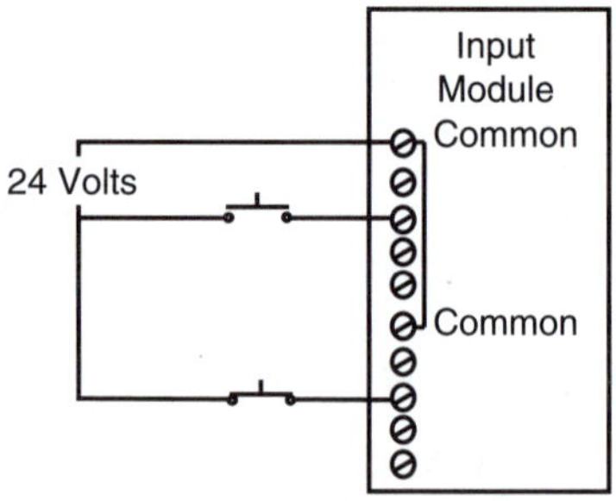

Figure 6–9 How dual commons can be wired together. All inputs would use the same voltage.

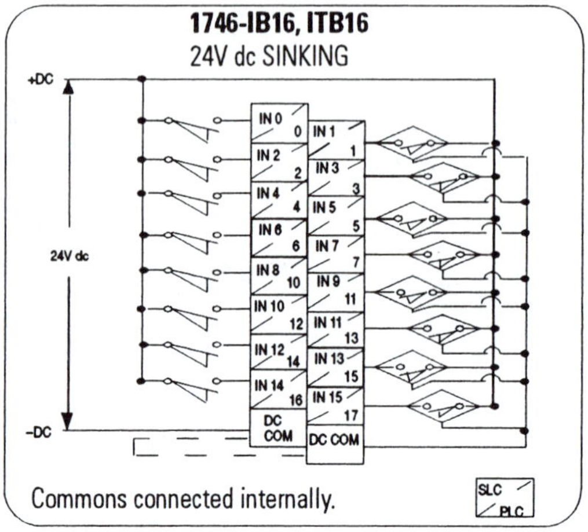

Figure 6–10 Example of SLC DC input module wiring. *(Courtesy Rockwell Automation Inc.)*

Input Wiring

Let's first look at what the module in Figure 6–10 requires for power before we consider the actual input wiring. There is no direct connection terminal on the module for + power from the power supply. There is a direct connection from the negative side of the power supply. The negative is connected to one of the commons. The dashed line shows us that the two commons are internally connected. Next let us look at the actual inputs. The inputs on the left are the equivalent of a switch or a two-wire sourcing sensor. Consider the top left switch. The left side of the switch is connected to the + side of the power supply. The other side of the switch is connected directly to the desired input terminal. As you can see, all of the switches' left sides were commoned to the + side of the power supply. The right side of each switch was connected to an individual input terminal. Imagine that the first switch closes. We have a complete path from the positive side of the supply through the switch to input terminal 0. The input would be on.

Next consider the right side of the input module. Three-wire sensors were used on this side. A three-wire sensor requires two wires for power and another for the output lead. Consider the first one. The negative power lead of the sensor (see the bottom of the sensor in Figure 6–10) was connected to the common (−) on the module. The positive from the power supply was connected to the positive lead on the sensor. The output lead of the sensor was connected to the input terminal. Last, note that all the negative leads on the three-wire sensors were connected to the − common on the module. All the positive leads were connected to the + side of the power supply. The output side of each three-wire sensor was connected to the individual input terminal.

When two-wire sensors are used, a small leakage current is always necessary for the operation of the sensor, but this is not normally a problem. In some cases, however, this leakage current is enough to trigger the input of the PLC module. In this case, a resistor can be added that will "bleed" the leakage current to ground (Figure 6–11). When a bleeder resistor is added, most of the current goes through it to common, which ensures that the PLC input turns on only when the sensor is really on.

Discrete Output Modules

Discrete output modules are used to turn real-world output devices either on or off. Discrete output modules can be used to control any two-state device. Output modules are available in AC and DC versions and in various voltage ranges and current capabilities. Output modules can be purchased with transistor output, triac output, or relay output. The transistor output would be used for DC outputs. There are various voltage ranges and current ranges available, as well as transistor-transistor logic (TTL) output. Triac outputs are used for AC devices.

The current specifications for a module are normally given as an overall module current and as individual output current. The specification may rate each output at 1 A. If it is an eight-output module, one might assume that the overall current limit

Figure 6–11 Use of a bleeder resistor.

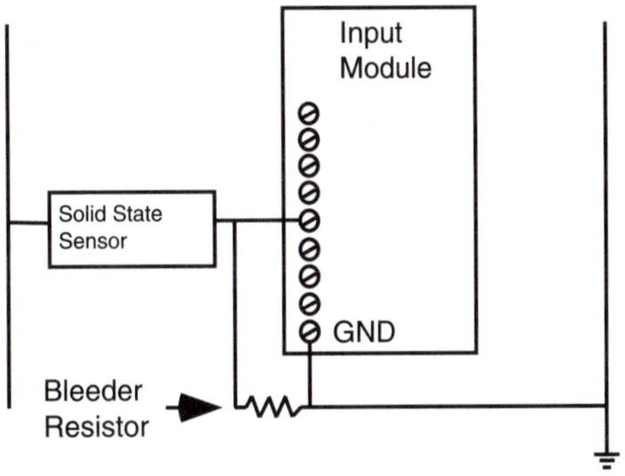

would be 8 A (8 × 1), but this is normally not the case. The overall current limit will probably be less than the sum of the individuals. For example, the overall current limit for the module might be 5 A. The user must be careful that the total current to be demanded does not exceed the total that the module can handle. Normally the user's outputs will not each draw the maximum current, nor will they normally all be on at the same time. The user should consider the worst case when choosing an appropriate output module.

Output modules are normally fused, and this fuse is normally intended to provide short-circuit protection for wiring only to external loads. If there is a short circuit on an output channel, it is likely that the output transistor, triac, or relay associated with that channel will be damaged. In that case, the module must be replaced or the output moved to a spare channel on the output module. The fuses are normally easily replaced. Check the technical manual for the PLC to find the exact procedure. Figure 6–12 shows the fuse location and jumper setting for a Rockwell SLC module. Note that by choosing the jumper location, the user can choose whether the processor faults or continues in the event of a fuse blowing.

Some output modules provide more than one common terminal, which allows the user to use different voltage ranges on the same card (Figure 6–13). These multiple commons can be tied together if the user desires. All outputs would be required to use the same voltage, however (Figure 6–14).

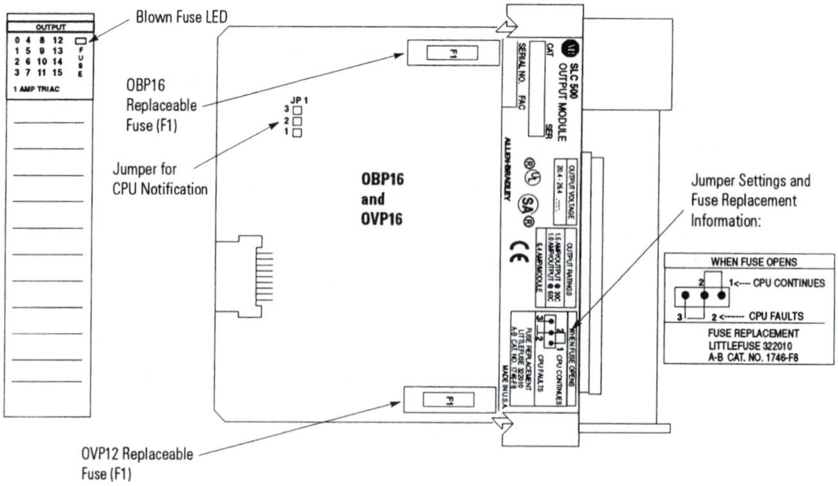

Figure 6–12 Fuse and jumper settings for a Rockwell SLC. Check the technical manual for your PLC to find the exact procedure. *(Courtesy Rockwell Automation Inc.)*

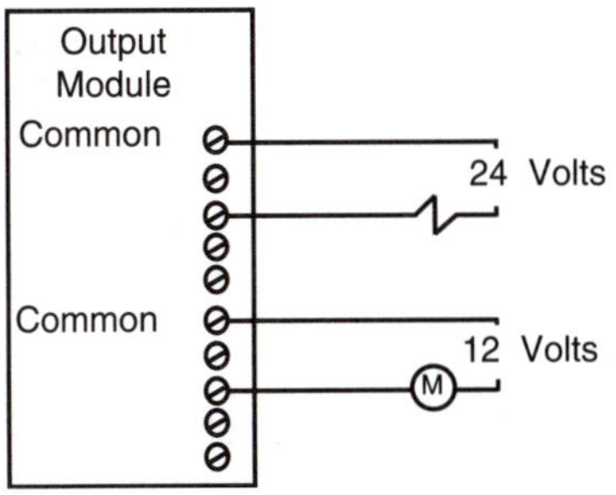

Figure 6–13 Use of a dual-common output module.

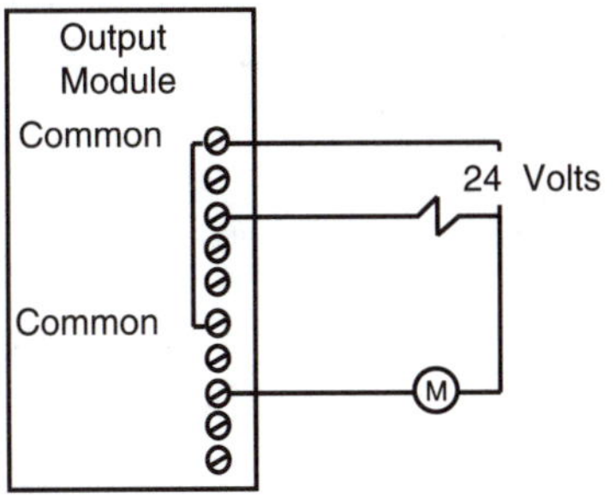

Figure 6–14 Use of a dual-common output module with the commons tied together. Note that the voltages must be the same if the commons are tied together.

Figure 6–15 Power supply, coil, and output terminals.

Output Wiring

Output wiring is similar in concept to input wiring. There are normally three types of components: power, a device, and output module terminals. Drawing a diagram before wiring can ease the task. Imagine a system that has many outputs. Let us first figure out one output circuit, then add more. Figure 6–15 shows an example of a power supply, a coil, and output terminals.

The wiring is shown completed in Figure 6–16. It is easy to see the current path. If the ladder logic turns on output 0, output terminal 0 will be internally connected to common and there will be a complete current path.

Next let's add another output (Figure 6–17). The output wiring should be the same for the output coil shown. One side of the coil must go to the output terminal and the other must be connected to the positive side of the power supply. Figure 6–18 shows the complete wiring diagram. It is difficult for most people to

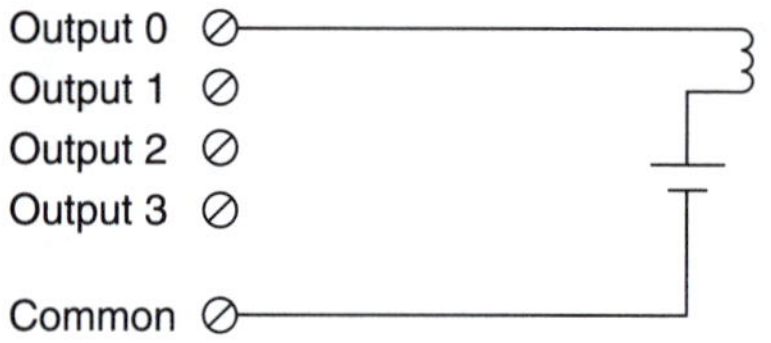

Figure 6–16 Complete wiring schematic.

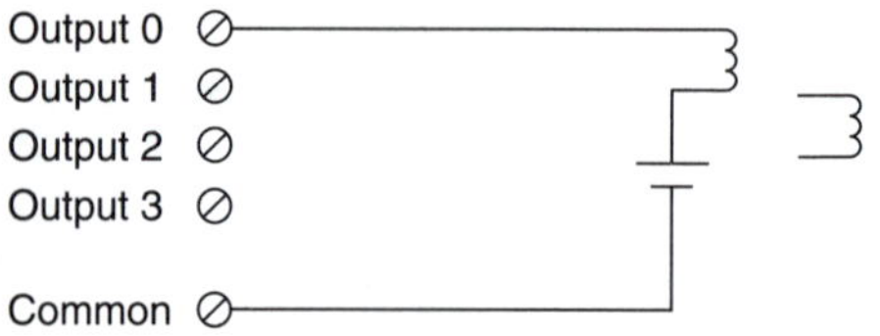

Figure 6–17 Additional output wiring example.

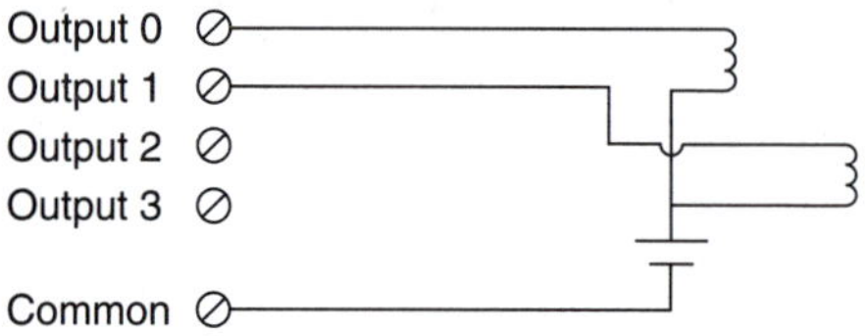

Figure 6–18 Complete wiring diagram.

envision a complete wiring diagram, especially when many devices are involved. A simple paper drawing can ease the task.

Figure 6–19 shows a DC output module for a Rockwell Automation SLC. Note also that the positive side of the DC voltage must be connected to the VDC terminal. The output devices are connected to an output terminal and then to the negative side of the power supply. The negative side of the supply must also be connected to the DC common.

Figure 6–20 shows a wiring diagram for a Rockwell AC triac output module. Note that L1 is connected to VAC 1 and VAC 2. L2 is connected through the actual output and then to the output module terminal.

Figure 6–21 shows the wiring of a Rockwell relay output module. First let us look at what the module requires for power before we consider the actual output wiring. There is a direct connection terminal on the top terminal of the module for +AC or DC power from the power supply. There is no direct connection for the – side of the power supply. Next let us look at the actual outputs. The outputs are the equivalent of coils. Consider the top left output. The left side of the output is connected to L2 or the −DC side of the power supply. The other side of the output is connected directly to the desired output terminal. As you can see, all of one side of the outputs are common to L2 or the −DC side of the power supply. The other side

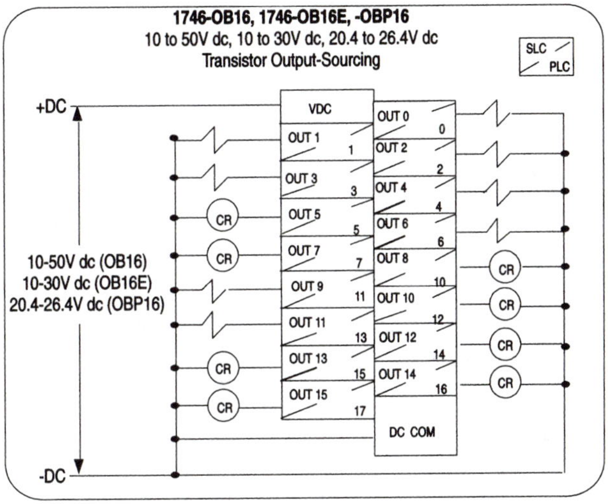

Figure 6–19 Example of SLC DC output module wiring. *(Courtesy Rockwell Automation Inc.)*

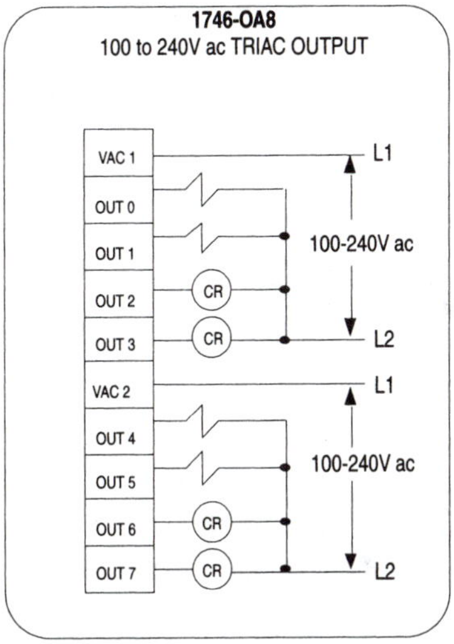

Figure 6–20 Example of SLC AC output module wiring. Note that these are triac outputs. *(Courtesy Rockwell Automation Inc.)*

of each output was connected to an individual output terminal. Imagine that the ladder logic turns on the first output. We have a complete path and the output would in turn be energized.

Note that the top and bottom half of the wiring diagrams are separate, which allows the user to use different voltages for outputs 1 through 7 and 10 through 17,

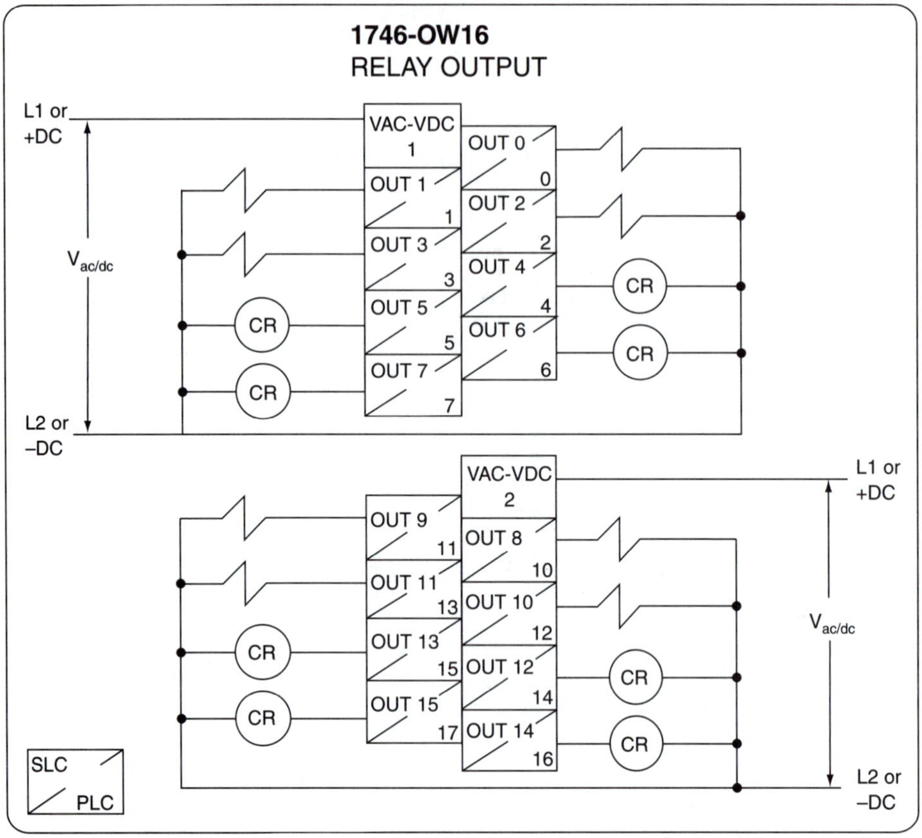

Figure 6–21 Wiring of a relay output module. *(Courtesy Rockwell Automation Inc.)*

or the user could use AC for half and DC for the other half. Relay outputs can be protected with a diode to prolong contact life (Figure 6–22).

Some applications require connecting a PLC output to the solid-state input of a device, usually to provide a low-level signal, not to power an actuator or coil. Figures 6–23 and 6–24 show how sinking and sourcing solid-state devices can be connected to a PLC sinking output. It is important to size the pull-up resistor properly. Figure 6–25 shows the formula for calculating the proper size resistor.

Solid-state modules can leak small unwanted current from output module outputs. If this is a problem, a bleeder resistor may be the solution. Figure 6–26 shows the use of a bleeder resistor to bleed off unwanted leakage through an output module.

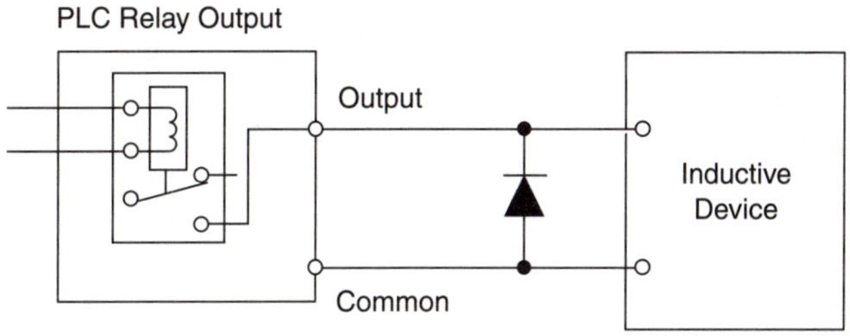

Figure 6–22 The use of a diode to help prolong contact life.

Figure 6–23 A PLC sinking output connected to a solid-state sourcing input on an output device.

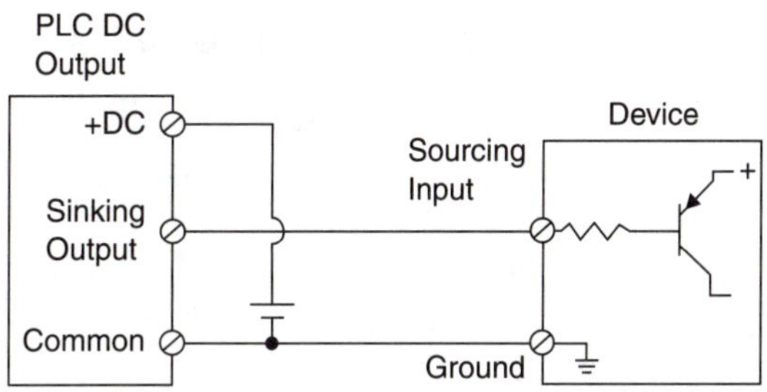

Figure 6–24 A PLC sinking output connected to a solid-state sinking input on an output device.

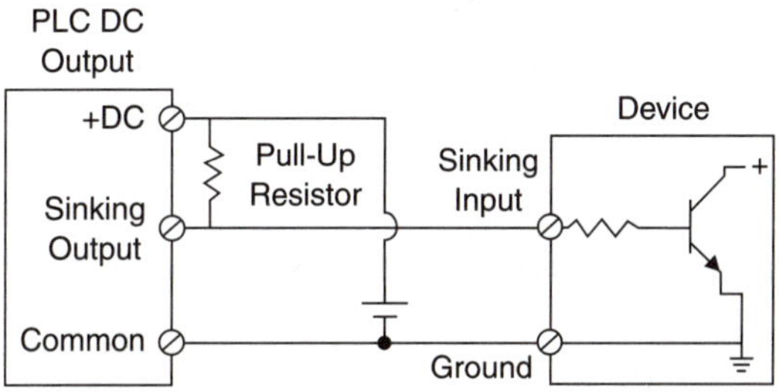

Figure 6–25 The formula for calculating the correct size of pull-up resistor.

$$I_{\text{INPUT}} = \frac{V_{\text{INPUT TURN-ON VOLTS}}}{R_{\text{INPUT}}}$$

$$P_{\text{PULL-UP}} = \frac{V_{\text{SUPPLY}}^2}{R_{\text{PULL-UP}}}$$

$$R_{\text{PULL-UP}} = \frac{V_{\text{SUPPLY}} - 0.7}{I_{\text{INPUT}}} - R_{\text{INPUT}}$$

Figure 6–26 Use of a bleeder resistor to bleed off unwanted leakage through an output module.

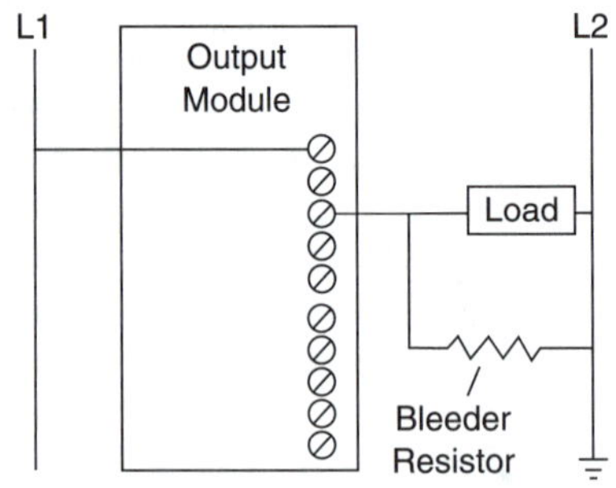

High-Density I/O Modules

High-density modules are digital I/O modules. A normal I/O module has eight inputs or outputs. A high-density module may have up to thirty-two inputs or outputs. The advantage is that there are a limited number of slots in a PLC rack. Each module uses a slot. With the high-density module, it is possible to install thirty-two inputs or outputs in one slot. The only disadvantage is that the high-density output modules typically cannot handle as much current per output.

Analog Modules

Computers (PLCs) are digital devices. They do not work with analog information. Analog data such as temperature must be converted to digital information before the computer can work with it.

Analog Input Modules

Cards have been developed to take analog information and convert it to digital information. These are called analog-to-digital (A/D) input cards. There are two basic types available: current sensing and voltage sensing. These cards will take the output from analog sensors (such as thermocouples) and change it to digital data for the PLC.

Voltage input modules are available in two types: unipolar and bipolar. *Unipolar modules* can take only one polarity for input. For example, if the application requires the card to measure only 0 V to +10 V, a unidirectional card will work. The bipolar module will take input of positive and negative polarity. For example, if the application produces a voltage between −10 V and +10 V, a bidirectional input card is required because the measured voltage could be negative or positive. Analog input modules are commonly available in 0 V to 10 V models for the unipolar and −10 V to +10 V for the bipolar module.

Analog models are also available to measure current. These typically measure from the smallest input value of 4 milliamperes (mA) to the largest input value of 20 mA.

Many analog modules can be configured by the user. Dip switches or jumpers are used to configure the module, to accommodate different voltages or current. Some manufacturers make modules that will accept voltage or current for input. The user simply wires to either the voltage or the current terminals, depending on the application.

Resolution in Analog Modules

Resolution can be thought of as how closely a quantity can be measured. Imagine a 1 foot ruler. If the only graduations on the ruler were inches, the resolution would be 1 inch. If the graduations were every 1/4 inch, the resolution would be 1/4 inch. The closest we would be able to measure any object would be 1/4 inch. That concept is the basis for the measure of an analog signal. The computer can only work with digital information. The A/D card changes the analog source into discrete steps.

Examine Figure 6–27(a). Ideally, the PLC would be able to read an exact temperature for every setting of the thermostat. Unfortunately, PLCs work only digitally. Consider Figure 6–27(b). The analog input card changes the analog voltage (temperature) into digital steps. In this example, the analog card changed the temperature

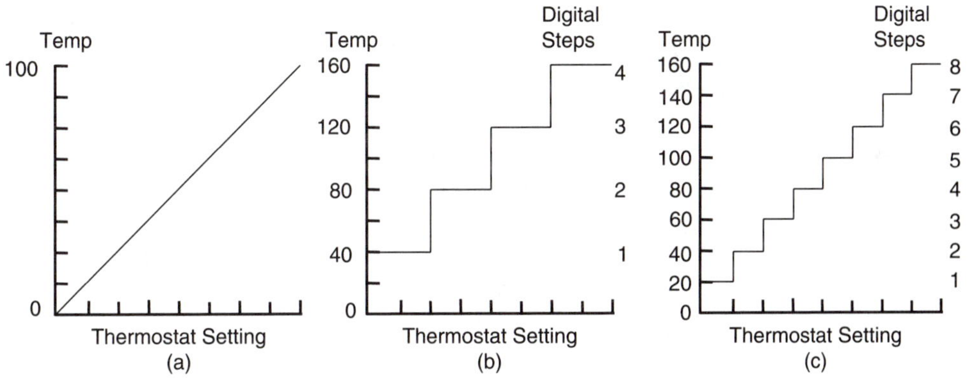

Figure 6–27 Graphs of temperature versus thermostat setting. Graph (a) represents a linear relationship of temperature versus setting. In reality, when analog control is used, the analog is a series of steps (resolution). Graph (b) shows how a four-step system would look. The resolution would be 40° per step. Graph (c) shows an eight-step system. The resolution is 20°.

from 40° to 160° in four steps. The PLC would read a number between 1 and 4 from the A/D card. A simple math statement in the ladder could change the number into a temperature. For example, assume that the temperature was 120°. The A/D card would output the number 3. The math statement in the PLC would take the number and multiply by 40 to get the temperature. In this case, it would be 3 × 40°, or 120°. If the PLC read 4 from the A/D card, the temperature would be 4 × 40°, or 160°. Assume now that the temperature is 97°. The A/D card would output the number 2. The PLC would read 2 and multiply by 40. The PLC would believe the temperature to be 80°. The closest the PLC can read the temperature is about 20° if four steps are used. (The temperature that the PLC calculates will always be in a range from 20° below the actual temperature to 20° above the measured temperature.) Each step is 40°. This is called *resolution.* The smallest temperature increment is 40°, the resolution would be 40.

Consider Figure 6–27(c). This A/D card has eight steps. The resolution would be twice as fine, or 20°. For a temperature of 67°, the A/D would output 3. The PLC would multiply 3 × 20° and assume the temperature to be 60°. The largest possible error would be approximately 10°. (The PLC calculated temperature would be within 10° below the actual temperature to 10° above the actual temperature.)

Industry requires very fine resolution. Typically, an industrial A/D card for a PLC would have 12-bit binary resolution, which means that there would be 4096 steps. In other words, the analog quantity to be measured would be broken into 4096 steps—very fine resolution—although there are cards available with even finer resolution. The typical A/D card is 12 bits (4096 steps) or 14 bits (16,384 steps).

Analog modules are available that can take between one and eight individual analog inputs.

Input modules have user-selectable dip switch settings to choose whether each input will be a current or voltage input.

Figure 6–28 shows the wiring for an analog input module. Note the shield around the signal wires, and that it is grounded only at one end. It is normally grounded to the chassis at the control end. Note that each analog source has two leads to the module inputs. None are connected together. This is called *differential wiring.* Note also that unused inputs are jumpered.

Input Resolution

Figure 6–29 shows the resolution for several analog input modules. Pay particular attention to the input ranges. Depending on which range the user selects, the second column of the table shows the range of decimal numbers that will appear in the input image table. The third column shows the number of significant bits that are

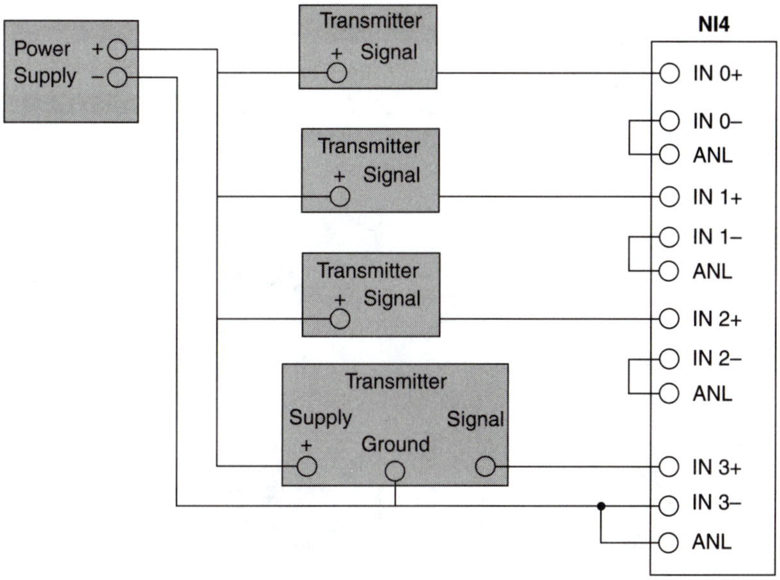

Figure 6–28 How single-ended (nondifferential) input connections are made. *(Courtesy Rockwell Automation Inc.)*

NI4, NIO4I, and NIO4V Input Range	Decimal Range (Input Image Table)	Number of Significant Bits	Nominal Resolution
±10 V dc – 1 LSB	−32,768 to + 32,767	16	305.176μV/LSB
0 to 10 V dc – 1 LSB	0 to 32,767	15	
0 to 5 V dc	0 to 16,384	14	
1 to 5 V dc	3,277 to 16,384	13.67	
±20 mA	±16,384	15	1.22070 μA/LSB
0 to 20 mA	0 to 16,384	14	
4 to 20 mA	3,277 to 16,384	13.67	

Figure 6–29 Resolution for various analog input modules. *(Courtesy Rockwell Automation Inc.)*

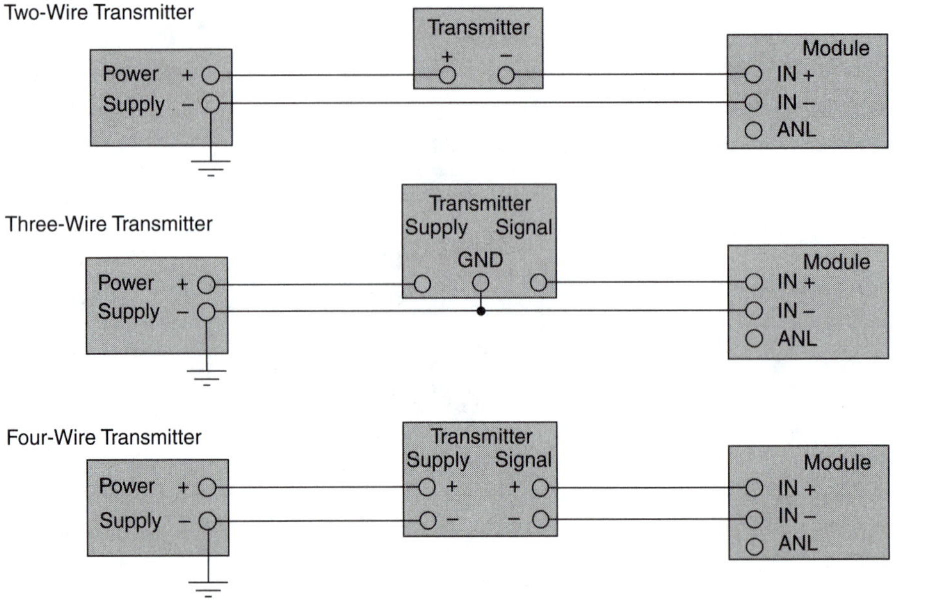

Figure 6–30 How two-, three-, and four-wire sensors are wired to an analog input module. *(Courtesy Rockwell Automation Inc.)*

used. The fourth column shows the resolution in microvolts or microamps per bit. Figure 6–30 shows how two-, three-, and four-wire sensors are wired to an analog input module.

Special-purpose A/D modules are also available. One example would be thermocouple modules. These are simply A/D modules that have been adapted to meet the needs of thermocouple input. These modules are available to make it easy to accept input from various types of thermocouples. Thermocouples output tiny voltages. To use the entire range of the module resolution, a thermocouple module amplifies the small output from the thermocouple so that the entire 12-bit resolution is used. The modules also provide cold junction compensation. Figure 6–31 shows how a Rockwell Automation thermocouple input module is wired.

Figure 6–32 shows a tank-filling application. A level sensor measures the level in the tank. If it is empty, the sensor outputs 0 V. If the tank is full, it outputs 10 V. The output is linear between 0 level (0 V) and full (10 V). The output from the sensor becomes an input to an analog input module.

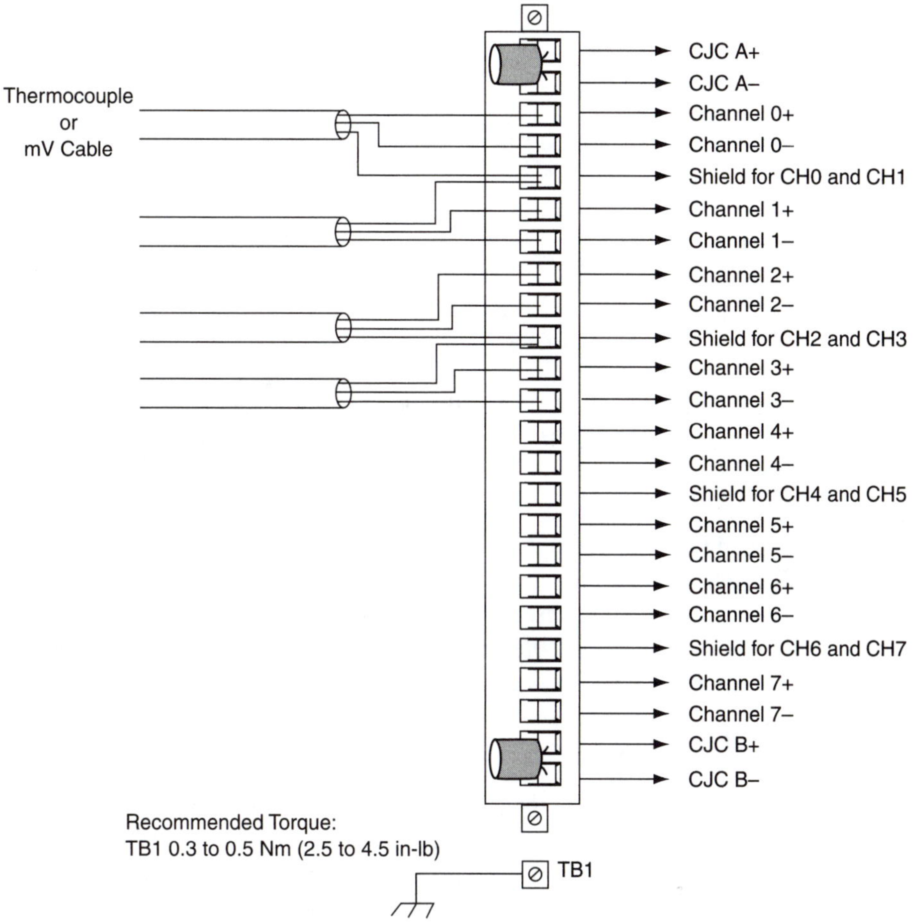

Figure 6–31 Wiring diagram for a Rockwell Automation thermocouple input module. *(Courtesy Rockwell Automation Inc.)*

Analog Output Modules

Analog output modules are also available. The PLC works in digital, so the PLC outputs a digital number (step) to the D/A converter module. The D/A converts the digital number from the PLC to an analog output. Analog output modules are available with voltage or current output. Typical outputs are 0 V to 10 V, −10 V to +10 V, and 4 mA to 20 mA.

Imagine a bakery. A temperature sensor (analog) in the oven could be connected to an A/D input module in the PLC. The PLC could read the voltage (steps) from

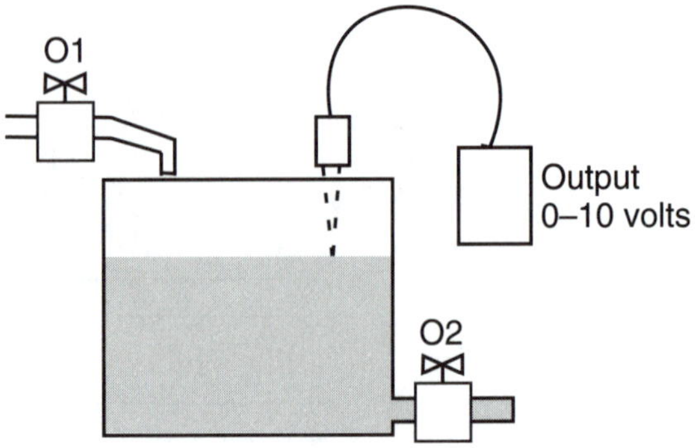

Figure 6–32 A tank-filling system.

the A/D card. The PLC would then know the temperature. The PLC can then send digital data to the D/A output module, which would control the heating element in the oven. This would create a fully integrated, closed-loop system to control the temperature in the oven.

Figure 6–33 shows the wiring for an analog output module. Note the shield around the signal wires. Note also that it is grounded only at one end—normally the output end. Unused outputs cannot be jumpered. Note that on this module the user can select whether to use an external 24 V supply or to use power from the backplane of the module.

Output Resolution

Figure 6–34 shows the resolution for several analog output modules. Pay particular attention to the output ranges for each module. The second column of the table shows the output range of current or voltage. The third column shows the range of decimal numbers that can be used in the output image table. The fourth column shows the number of significant bits that are used. The fifth column shows the resolution in microamps or millivolts per bit.

Rethink the tank system in Figure 6–32. An analog output would be used to control the variable inlet valve to the tank. A second output could be used to control the variable valve that controls the outflow from the tank.

Remote I/O Modules

Special modules are available for some PLCs that allow the I/O module to be positioned separately from the PLC. In some processes it is desirable (or necessary) to position the I/O at a different location. In some cases the machine or application is spread over a wide physical area. In these cases it may be desirable to position the I/O modules away from the PLC. Figure 6–35 shows an example of the use of an Allen-Bradley remote I/O adapter module.

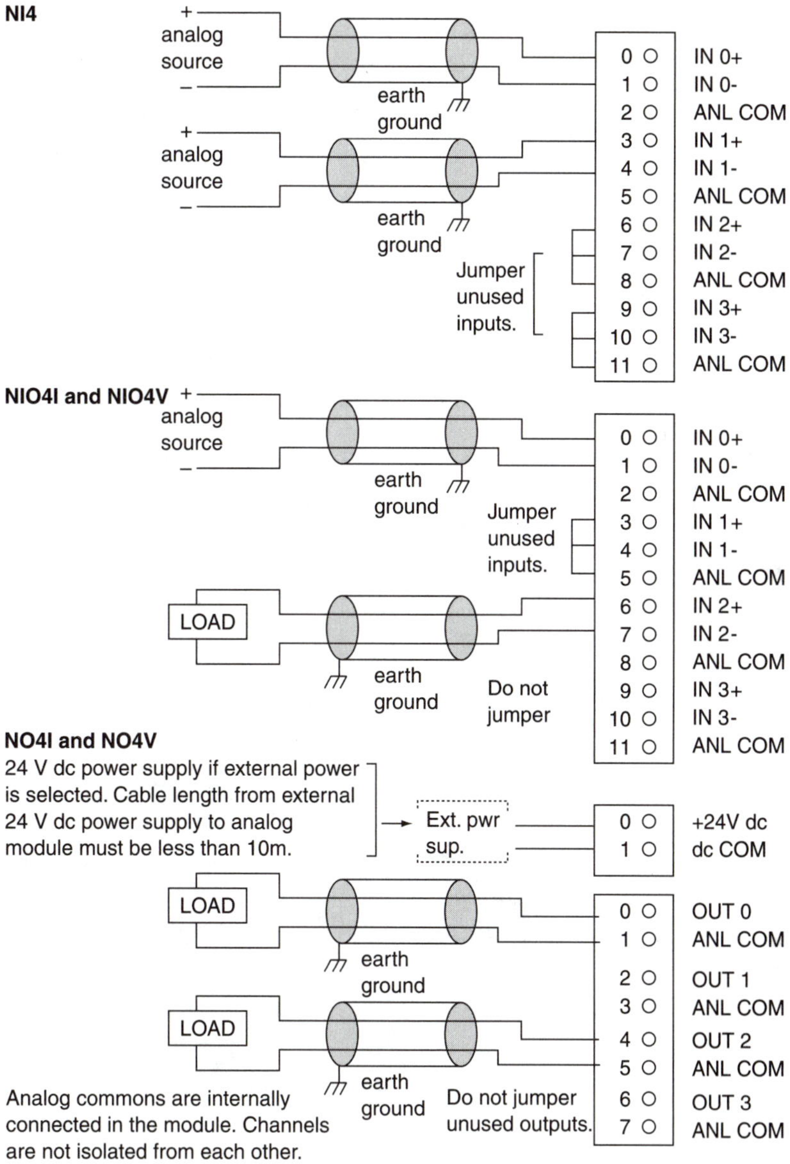

Figure 6–33 Wiring diagram for a Rockwell Automation combination analog input and output module. *(Courtesy Rockwell Automation Inc.)*

Module	Output Range	Decimal Range (Output Image Table)	Significant Bits	Resolution
FIO4I NIO4I NO4I	0 to 21 mA − 1 LSB	0 to 32,764	13 bits	2.56348 μA/LSB
	0 to 20 mA	0 to 31,208	12.92 bits	
	4 to 20 mA	6,242 to 31,208	12.6 bits	
FIO4V NIO4V NO4V	±10 V dc − 1 LSB	−32,768 to + 32,764	14 bits	1.22070 mV/LSB
	0 to 10 V dc − 1 LSB	0 to 32,764	13 bits	
	0 to 5 V dc	0 to 16,384	12 bits	
	1 to 5 V dc	3,277 to 16,384	11.67 bits	

Figure 6–34 Output resolution for various output modules. *(Courtesy Rockwell Automation Inc.)*

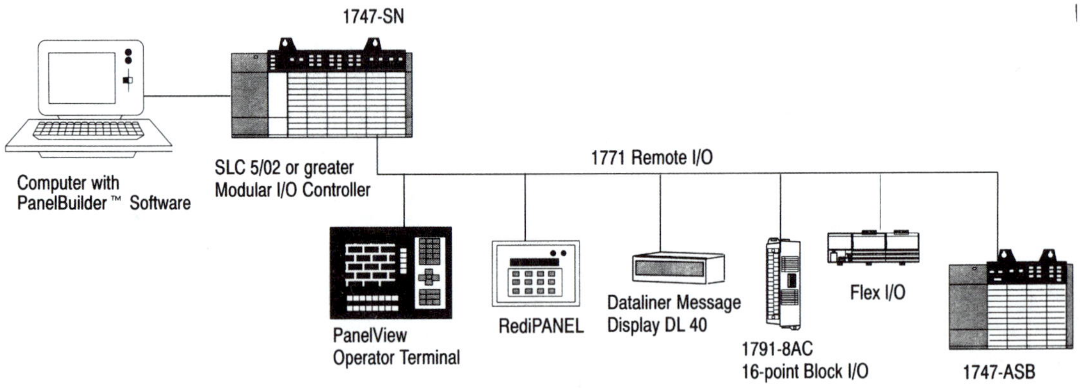

Figure 6–35 Example of the use of remote I/O modules. *(Courtesy Rockwell Automation Inc.)*

Twisted pair wiring is a common method for connecting the remote module to the processor. Two wires are twisted around each other and connected between the PLC and the remote I/O. Twisting reduces the possibility of electrical interference (noise) because any noise acts on both conductors equally. Twisted pair connections can transmit data thousands of feet.

OPERATOR I/O DEVICES

As systems become more integrated and automated, they become more complex. Operator information becomes crucial. Many devices are available for this information interchange.

Operator Terminals

Many PLC makers now offer their own operator terminals, from simple to highly complex. The simpler ones can display a short message. The more complex models can display graphics and text in color while taking operator input from touch screens (Figure 6–36), bar codes, keyboards, and so on. These display devices can cost from a couple hundred dollars to several thousand dollars. If we remember that the PLC has most of the valuable information about the processes it controls in its memory, we can see that the operator terminal can be a window into the memory of the PLC.

The greatest advances have been in the ease of use. Many PLC manufacturers have software available that runs on an IBM personal computer. The software essentially writes the application for the user. The user draws the screens and decides which variables from the PLC should be displayed. The user also decides what input is needed from the operator. When the screens are designed, they are downloaded to the display terminal.

These smart terminals can store hundreds of pages of displays in their memory. The PLC simply sends a message that tells the terminal which page and information to display. This helps reduce the load on the PLC. The memory of the display is used to hold the display data. The PLC only requests the correct display, and the terminal displays it.

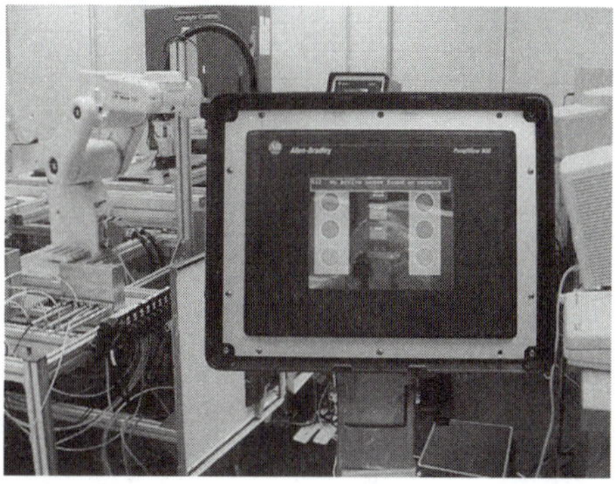

Figure 6–36 Operator display terminal. *(Courtesy Rockwell Automation Inc.)*

The PLC needs only to update the variables that may appear on the screen. The typical display would include graphics showing a portion of the process, variables showing times or counts, and any other information that might aid an operator in the use or maintenance of a system.

QUESTIONS

1. What voltages are typically available for I/O modules?
2. If an input module sensed an input from a load-powered sensor when it should not, what might be the possible problem, and what is a possible solution?
3. What types of output devices are available in output modules? List at least three.
4. Explain the purpose of A/D modules and how they function.
5. Explain the purpose of D/A modules and how they function.
6. What type of module can be used to communicate with a computer?
7. Explain the term *resolution.*
8. If a 16-bit input module is used to measure the level in a tank, and the tank can hold between 0 and 15 feet of fluid, what is the resolution in feet?
9. What are remote I/O modules?
10. List three reasons why operator I/O devices are becoming more prevalent.

chapter 7

Timers and Counters

Timers and counters are invaluable in PLC programming. Industry must count product, time sequences, and so on. In this chapter we will examine the types and programming of timers and counters. Timers and counters are similar in all PLCs. We will show examples of several of the leading brands.

OBJECTIVES

Upon completion of this chapter, you will be able to:

1. Describe the use of timers and counters in ladder logic.
2. Define terms such as *retentive, cascade, delay-on,* and *delay-off.*
3. Utilize timers and counters to develop applications.

TIMERS

Timing functions are important in PLC applications. Cycle times are critical in many processes. Timers are used to delay actions. They may be used to keep an output on for a specified time after an input turns off or to keep an output off for a specified time before it turns on.

Think of a garage light. It would be a nice feature if a person could touch the on switch and the light would immediately turn on, and then stay on for a given time (maybe 2 minutes). At the end of the time, the light would turn off. This would allow you to get into the house with the light on. In this example, the output (light) turned on instantly when the input (switch) turned on. The timer counted down the time (timed out) and turned the output (light) off. This is an example of a *delay-off timer.*

Consider Figure 7–1. When switch X1 is activated, the timer turns on and starts counting time intervals. The timer is used in the second rung as a contact. If the timer is on, the contact in rung 2 (timer 5) closes and turns on the light. When the timer times out (in this case, in 2 minutes), the contact in rung 2 opens and the light turns off. As you can see from this example, output coils can be used as contacts to control other outputs.

This type of timer is called a *delay-off timer.* The timer turns on instantly, counts time increments down, and then turns off (delay-off).

The other type of timer is called *delay-on.* When input X3 to the delay-on timer is activated, the timer begins counting time increments but remains off until the time has elapsed (see Figure 7–2). In this case, the switch is activated and the timer

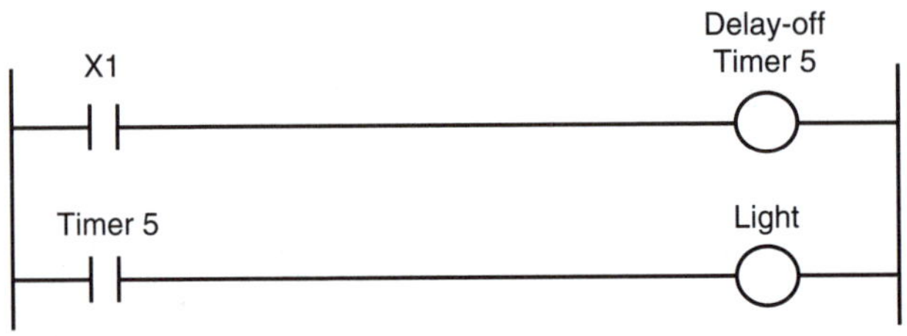

Figure 7–1 Delay-off timing circuit. If contact X1 closes, the delay-off timer immediately turns on, which turns on the light. When the timer reaches the programmed time it will turn off, which turns the light off also.

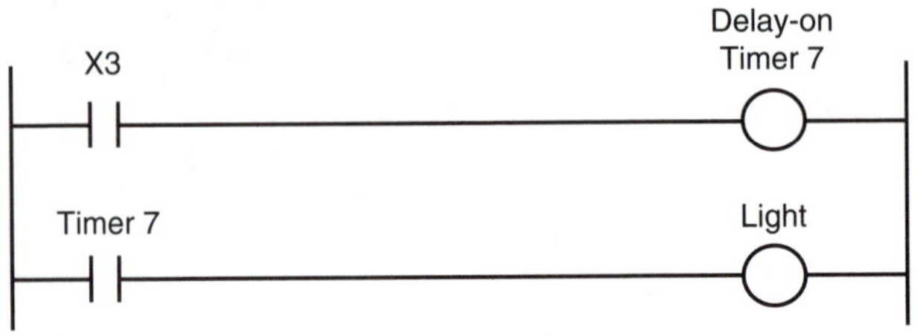

Figure 7–2 Delay-on timing circuit. In this example if contact X3 closes, the timer will begin timing. When the time reaches the programmed time, the timer will turn on, which will turn the light on.

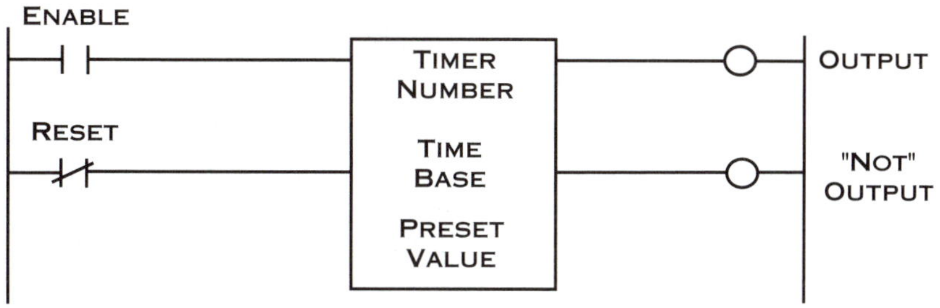

Figure 7–3 A typical block type timer.

starts to count, but it remains off until the total time has elapsed. Then the timer turns on, which closes the contact in rung 2 and turns on the light.

Many PLCs use block-style timers and counters (see Figure 7–3). No matter what the brand of PLC, there are many similarities in the way timers are programmed. Each timer will have a number to identify it. For some it will be as simple as T and a number: T7 (timer 7), for example. This chapter will utilize generic examples to illustrate some concepts about timers and counters but will then examine timers and counters for various manufacturers.

Every timer has a time base. Timers can typically be programmed with several different time bases: 1 second, 0.1 second, and 0.01 second are typical time bases. If a programmer entered .1 for the time base and 50 for the number of delay increments, the timer would have a 5-second delay (50×0.1 second = 5 seconds).

Timers must also have a preset value. The preset value is the number of time increments the timer must count before changing the state of the output. The actual time delay equals the preset value multiplied by the time base. Presets can be a constant value or a variable. If a variable is used, the timer uses the real-time value of the variable to calculate the delay. This allows delays to be changed depending on conditions during operation. An example is a system that produces two different products, each requiring a different time in the actual process. Product A requires a 10-second process time, so the ladder logic would assign 10 to the variable. When product B comes along, the ladder logic can change the value to that required by B. When a variable time is required, a variable number is entered into the timer block. Ladder logic can then be used to assign values to the variable.

Timers typically have one or two inputs. Every timer has one input that functions as a timer enable input. When this input is true (high), the timer will begin timing. Some timers have a second input. This is used to reset the timer's accumulated time to zero. If the reset line changes state, the timer clears the accumulated value. For example, the timer in Figure 7–3 requires a high for the timer to be active. If the reset line goes low, the timer clears the accumulated time to zero. Some timers have just the enable input and utilize a separate reset instruction to reset the accumulated time to zero.

Timers can be retentive or nonretentive. *Retentive timers* do not lose the accumulated time when the enable input line goes low. They retain the accumulated time until the line goes high again. They then add to the count when the input goes high again. *Nonretentive timers* lose the accumulated time every time the enable input goes low. If the enable input to the timer goes low, the timer count goes to zero. Retentive timers are sometimes called accumulating timers. They function like a stopwatch. Stopwatches can be started and stopped and retain their timed value. A reset button on a stopwatch resets the time to zero.

GE Fanuc Timers

First, we must examine GE Fanuc contacts, coils, and addressing to make GE Fanuc timers and counters understandable.

GE Fanuc Memory Organization and I/O Addressing

GE Fanuc uses the following letters to represent different types of contacts, coils and memory. Real-world, discrete inputs are identified as %I. %I0003 would be the name of real-world input 3 in a ladder diagram. Real-world outputs are identified as %Q. %Q0005 would be the name of real-world output 5. Discrete internal coils are identified as %M. See Figure 7–4 for a list of the types of discrete types that are available.

GE Fanuc uses the following letters to represent analog I/O data and register data. Real-world, analog inputs are identified as %AI. %AI0003 would be the name of real-world analog input 3 in a ladder diagram. Real-world outputs are identified as %AQ. %AQ0005 would be the name of real-world output 5. System registers are

LABEL	DESCRIPTION
%I	DISCRETE INPUT
%Q	DISCRETE OUTPUT
%M	DISCRETE INTERNAL COIL
%T	DISCRETE TEMPORARY COIL
%G	DISCRETE GLOBAL DATA
%S, %SA, %SB, %SC	DISCRETE SYSTEM STATUS REFERENCES

Figure 7–4 GE Fanuc programming labels for discrete-type data and I/O.

identified as %R. See Figure 7–5 for a list of the types of analog and register types that are available.

GE Fanuc Contacts

Several types of contacts are available (see Figure 7–6). The normally open and normally closed contacts function just as we have seen before. The continuation contact is a little different. The continuation coil passes power if the preceding contact is on (see Figure 7–7).

GE Fanuc Coils

Several types of coils are available (see Figure 7–3). There is, of course, the normally open coil that we are familiar with (-( )-). This coil is on if it receives power from the contacts to the left on the rung.

The set (-(S)-) coil is used to latch a coil on. If the preceding contacts on the rung are true, the set contact turns on. It stays on even if the contacts to its left become de-energized. The set coil can be turned off only if a reset (-(R)-) coil is used.

Figure 7–5 GE Fanuc label names for registers and analog-type data.

LABEL	DESCRIPTION
%R	SYSTEM REGISTER
%AI	ANALOG INPUTS
%AQ	ANALOG OUTPUTS

INSTRUCTION	FUNCTION	DESCRIPTION
<+>---	CONTINUATION CONTACT	THE CONTINUATION CONTACT PASSES POWER TO THE RIGHT IF THE PRECEDING CONTINUATION COIL IS SET ON.
-| |-	NORMALLY OPEN CONTACT	A NORMALLY OPEN CONTACT PASSES POWER IF THE ASSOCIATED REFERENCE IS IS ON.
-|/|-	NORMALLY CLOSED CONTACT	A NORMALLY CLOSED CONTACT PASSES POWER IF THE ASSOCIATED REFERENCE IS OFF.

Figure 7–6 GE Fanuc contacts.

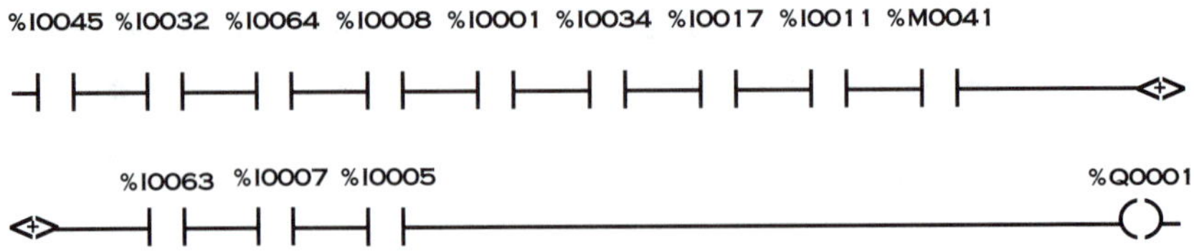

Figure 7–7 Sample GE Fanuc ladder diagram.

INSTRUCTION	FUNCTION	DESCRIPTION
ONDTR	ON-DELAY STOPWATCH TIMER	THIS TYPE OF TIMER ACCUMULATES TIME WHILE RECEIVING POWER. IT PASSES POWER IF THE CURRENT VALUE EXCEEDS THE PRESET VALUE. THE CURRENT VALUE IS RESET TO ZERO WHEN THE RESET (R) INPUT RECEIVES POWER.
OFTD	OFF-DELAY TIMER	THIS TIMER INCREMENTS WHILE POWER FLOW IS OFF AND RESETS TO ZERO WHEN POWER FLOW IS ON. TIME BASES AVAILABLE ARE TENTHS OF SECONDS (.1 DEFAULT), HUNDREDTHS OF SECONDS (.01), AND THOUSANDTHS OF SECONDS (.001). THE RANGE FOR THIS TIMER IS 0-32,767 TIME UNITS. THIS TIMER IS RETENTIVE ON POWER FAILURE. NOTE THAT NO AUTOMATIC INITIALIZATION OCCURS AT POWERUP.
TMR	ON-DELAY TIMER	THE CURRENT VALUE OF THE TIMER FUNCTION IS SET TO ZERO WHEN THE FUNCTION TRANSITIONS ON. THE FUNCTION ACCUMULATES TIME WHILE RECEIVING POWER AND PASSES POWER IF THE CURRENT VALUE IS GREATER THAN OR EQUAL TO A PRESET VALUE.

Figure 7–8 The types of GE Fanuc timers.

The negated coil (-(/)-) is a normally closed coil. If the coil receives power from the rung, it is off. If the coil does not receive power from the rung, it is on.

A retentive coil (-(M)-) is like a normally open coil. The difference is that it retains its state through a power failure and a stop-to-run transition. This feature is important if the state of a coil needs to be retained for a system to operate safely and correctly through a power failure or stop-to-run transition.

Two types of transition coils are available. The positive transition coil turns on for only one scan if it senses a low-to-high transition of the contacts on the rung. The negative transition coil turns on for one scan if it senses a high-to-low transition of the rung. These transitional coils can be used as one-shots.

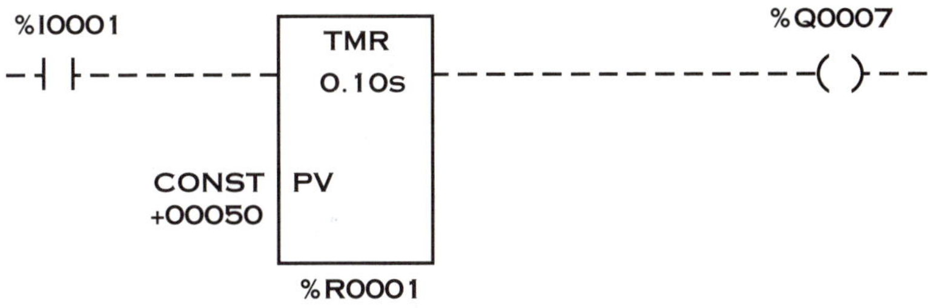

Figure 7–9 Use of a GE Fanuc timer.

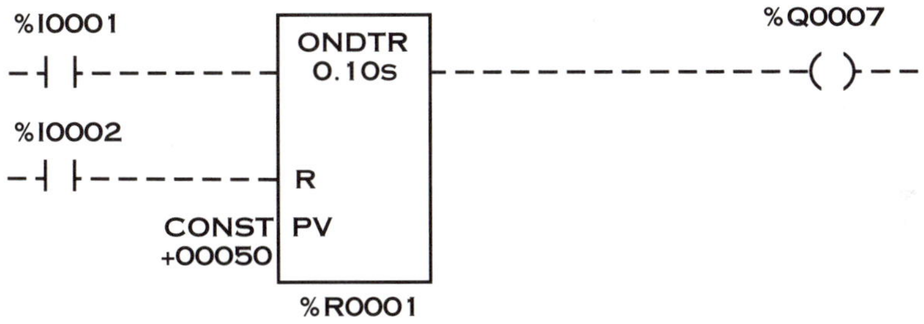

Figure 7–10 Use of an on-delay timer.

Figure 7–8 shows a typical GE Fanuc timer block with a typical enable input and reset input. The preset value for the timer must be entered. The time base can also be changed. The actual delay can be calculated by multiplying the preset value by the time base. The programmer can choose a register where the present value of the timer will be stored or it can be chosen automatically. There is one output from each timer.

Figure 7–9 shows a typical GE Fanuc timer block with a typical enable input and reset input. Note that the preset value in this case is a constant (a variable time can also be used through the use of a register). The value of the constant is 50. The time base is .1 second. This timer will time to 5 seconds (50 * .1 second). The current time will be held in register 1. If input 1 (%Q0001) becomes energized, the timer will begin to accumulate time. When the current value reaches 50 in register 1, the timer passes power to output coil 7 (%Q0007). Whenever the current value (register 1) is larger than the preset value, the timer will pass power to the coil. If the input coil (%I0001) becomes energized at any time, the timer current value is set to zero.

Figure 7–10 shows the use of a GE Fanuc on-delay timer (ONDTR). Note that the preset value in this example is a constant. The value of the constant is 50. The time base is .1 second. This timer will time to 5 seconds (50 * .1 second). The current time in the timer will be held in register 1. If input 1 (%I0001) becomes energized, the timer will begin to accumulate time.

When the value reaches 50 in register 1, the timer will pass power to output coil 7 (%Q0007). Anytime the current value (register 1) is larger than the preset value, the timer will pass power to the coil. If the input coil (%I0001) becomes de-energized at any time, the timer current value is retained. When the input (%I0001) becomes energized again, the timer begins to increment the time again. This is an accumulating timer. The timer's current value (accumulated time) is reset to zero only if the input to the reset input of the timer is energized.

Figure 7–11 shows the use of a GE Fanuc off-delay timer (OFDT). Note that the preset value in this example is a constant. The value of the constant is 50. The time base is 0.1 second. This timer will time to 5 seconds (50 * .1 second). This timer increments while power flow is off and resets to zero when power flow is on. This means that while the present value exceeds the preset value, the timer passes power. If the input (%I0001) is off, the timer increments the time. It continues to increment the time until the input is energized. When the input is energized, the timer is reset to zero.

Gould Modicon Timers

Figure 7–12 shows a typical Gould Modicon timer block with a typical enable input and reset input. This is a TMR-type timer. It is nonretentive. If the input goes low, the timer's accumulated time is reset to zero. The preset value for the timer must be entered. The time base must be chosen. The actual delay can be calculated by multiplying the preset value by the time base. The timer number is actually a storage register where the accumulated value of the timer will be stored. Two outputs are possible from the timer. The first output is one that will turn on when the timer reaches the timer preset value. The second output is the opposite. It is "nor-

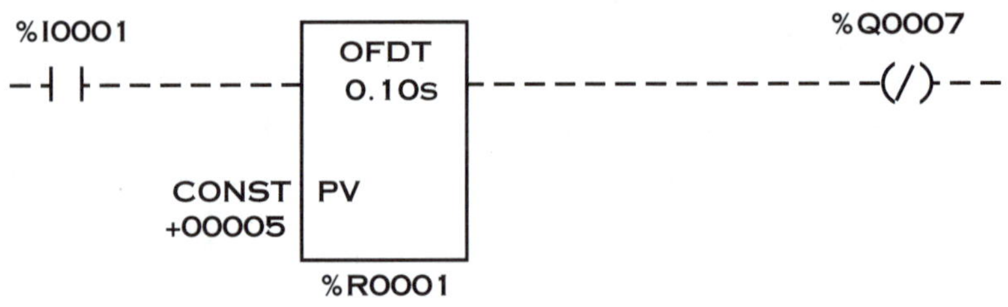

Figure 7–11 Example of the use of a GE Fanuc off-delay timer.

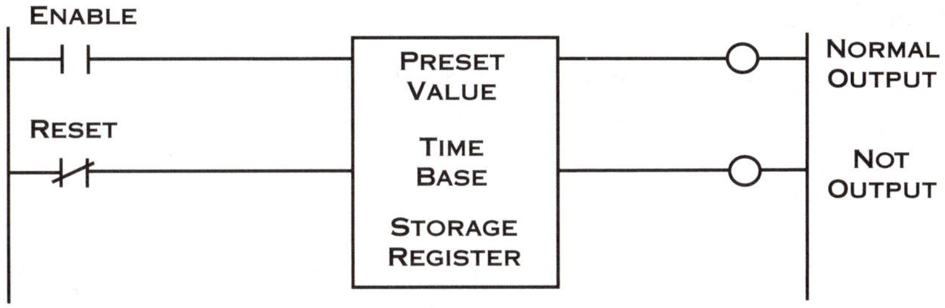

Figure 7–12 Gould Modicon timer block.

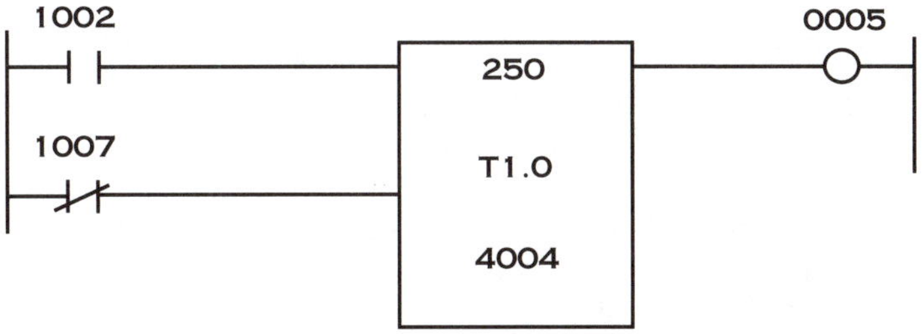

Figure 7–13 Gould Modicon timer example.

mally on" and it will turn off when the timer reaches the preset value. These two outputs make it very easy to program delay-on or delay-off logic.

Figure 7–13 shows the actual use of a Gould Modicon timer. Note the first number of the inputs, outputs, and storage register. The 1 in the inputs denotes that the number is an input. The first 0 in output 0005 shows that the number represents an output. The 4 of the number 4005 shows that the number is a storage register. The time base is 0.1 second. The preset is 250, so the value of the delay is 25 seconds (250 × 0.1 second = 25 seconds). It will be used as a delay-on timer because the top output on the timer block was used. Note that input 7 is programmed as a normally closed contact. The reset line must be high for the timer to time. If this line goes low, the timer is reset to a count of zero and is unable to time again until the reset line goes high. Input 2 is used as the enable line. The current value of the timer will be stored in register 4004. If the timer value reaches 25 seconds, output 5 will turn on and stay on. If input 2 goes low or input 7 goes low, output 5 will turn off.

Omron Timers

Omron C200H Memory Areas

Omron divides memory into areas. There are several types of areas, each with a specific use.

Holding Relay Area The holding relay (HR) area is used to store and manipulate numbers. It retains the values even when modes are changed or during a power failure. The address range for the HR area is 0000 to 9915. This memory area can be used by the programmer to store numbers that need to be retained.

Temporary Relay Area The temporary relay (TR) area is used for storing data at program branching points.

Auxiliary Relay Area The auxiliary relay (AR) area is used for internal data storage and manipulating data. A portion of this data area is reserved for system functions. The AR bits that can be written to by the user range from 0700 to 2215.

Link Relay Area The link relay (LR) area is used for communications to other processors. If it is not needed for communications, it can be used for internal data storage and data manipulation. The address range for the LR area is 0000 to 6315.

Timer/Counter Area The timer/counter (TC) area is used to store timer and counter data. The timer counter area ranges from 000 to 511. Note that a number can be used only once. The same number cannot be used for a timer and a counter. For example, if the programmer uses the numbers 0 to 5 for timers, they cannot be used for counters.

Data Memory Area The data memory (DM) area is used for internal storage and manipulation of data. It must be accessed in 16-bit channel units. If a multiplication sign is used before the DM (*DM), it means that indirect addressing is being used. Data memory ranges from addresses 0000 to 1999. The user can write only to addresses 0000 to 0999.

I/O and Internal Relay Area The internal relay (IR) area is used to store the status of inputs and outputs. Any of the bits that are not assigned to actual I/O can be used as work bits by the programmer. Channels 000 to 029 are allocated for I/O. The remaining addresses up to 24615 are for the work area. The IR area is addressed in bit or channel units. The addresses are accessed in channels or channel/bit combinations. Channel/bit addresses are 5 bits long. The two least significant digits are the bit within the channel. For example, address 01007 would be channel 10, bit (or terminal) 7.

Special Relay Area The special relay (SR) area can be used to monitor the PLC's operation. It can also be used to generate clock pulses and to signal errors. For example, bit 25400 is a 1-minute clock pulse. Bit 25500 is a 0.1-second clock pulse. These special bits can be used by the programmer in ladder logic. Many bits are provided to signal errors that could occur. The programmer can use these in ladder logic to indicate problems.

Program Memory The user memory (UM) area is where the user's program instructions are stored. Memory is available in various sizes (RAM and ROM). These data areas are used simply by using the prefix (such as HR) followed by the appropriate address within that data area.

There are two types of timers: TIM and TIMH. The TIM timer measures in increments of 0.1 second. It is capable of timing from 0 to 999.9 seconds with an accuracy of ±0.1 second. The high-speed timer (TIMH) measures in increments of 0.01 second. Both types are decrementing-style, delay-on timers. They require a timer number and a set value (SV). When the set value (SV) has elapsed, the timer output turns on. Timer counter numbers refer to an actual address in memory. Numbers must not be duplicated. You cannot use the same number for a timer and a counter. Timers/counters are numbered 0 to 511.

Figure 7–14 shows the use of a TIM timer in a ladder diagram. When the timer decrements to zero, the timer output turns on. Output 00110 will be on until the timer becomes true. Omron numbers their outputs and inputs based on a five-digit scheme. The three most significant digits represent the channel (slot) number. The two least significant digits represent the bit number on the module. So output 00110 would be output number 10 on the output module in channel (slot) 1. Note that a normally closed timer 000 has been used as the contact for output coil 00110. This means that as long as the timer is false, output 001100 will be energized. When the timer becomes true, output 00109 will turn on. This, of course,

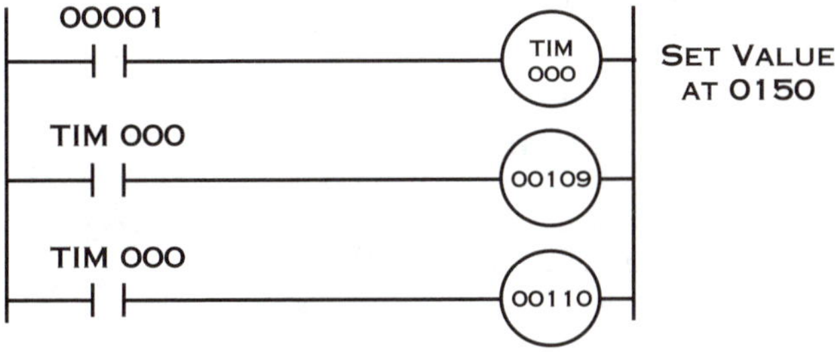

Figure 7–14 Omron TIM timer used as a normally open contact and a normally closed contact.

also means that output 00110 will de-energize. The timer will reset if input 00001 becomes false. A power failure would also cause the timer to reset.

This problem can be solved if a retentive timer is used. A counter is used to make a retentive timer. Remember that there is very little difference between a timer and a counter. Both count increments or events. If a counter is used to count time increments, it becomes a timer.

Figure 7–15 shows the use of a counter to make a retentive timer. A special bit is used to create the 1-second pulse. Bit 25502 is a special bit that Omron provides. It is a one-second clock pulse. If input 00001 is present and the one-second pulse bit makes twenty transitions, the counter will be true. Even if power is lost, the present value of the count is retained. This means that we now have a retentive timer.

The high-speed timer (TIMH) times in 0.01-second increments. The TIMH timer has a range of 0.00 to 99.99 seconds. Scan time can affect TIMH timers numbered 48 to 511. Timers 48 to 511 may be inaccurate if the scan time exceeds 10 milliseconds. If the scan time is more than 10 milliseconds, use timer numbers below 48.

Figure 7–16 shows the use of a TIMH timer. It operates just like the TIM timer except that the time increment is smaller. Remember that the TIM and TIMH timers are not retentive; if the enable input is lost, the present value is lost.

AutomationDirect Timers

Figure 7–17 shows an AutomationDirect nonretentive timer. Two values are required for the timer block. The first value is the timer number. The second is the preset value. If X0 becomes true (high), the timer will begin to time. If the time equals the preset, the timer contact turns on. The timer contact can then be used in another rung. In this case if the timer reaches its preset, output Y1 will turn on until X0 goes low again.

Figure 7–18 shows an AutomationDirect accumulating timer. Accumulating timers can time up to a maximum of 99,999,999. Note that the TMRA in the block

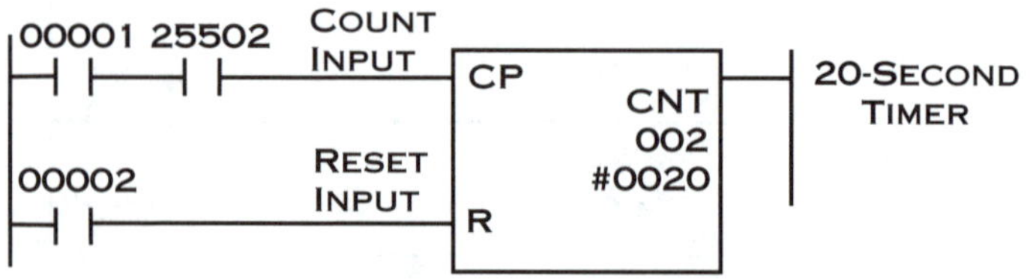

Figure 7–15 Use of a counter to create a retentive timer. This will preserve the present value of the timer in the event of a power interruption. Note that contact 25502 is a special internal bit. Bit 25502 is a 1-second timing pulse that can be used by the programmer.

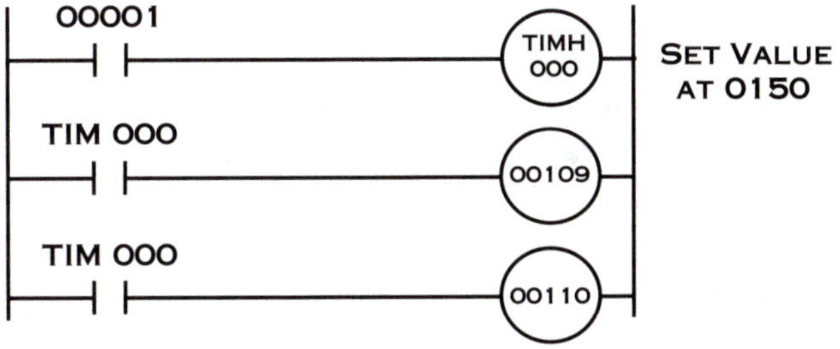

Figure 7–16 Omron high-speed timer (TIMH). Output 00110 will be on until the timer times out because a normally closed contact was used. When the timer times out, output 00109 will be energized.

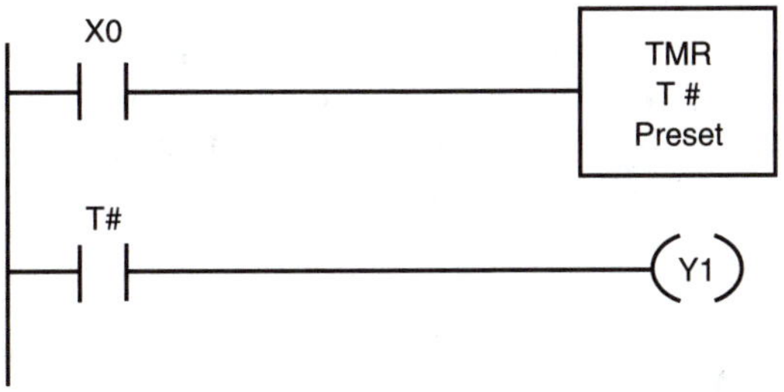

Figure 7–17 AutomationDirect timer.

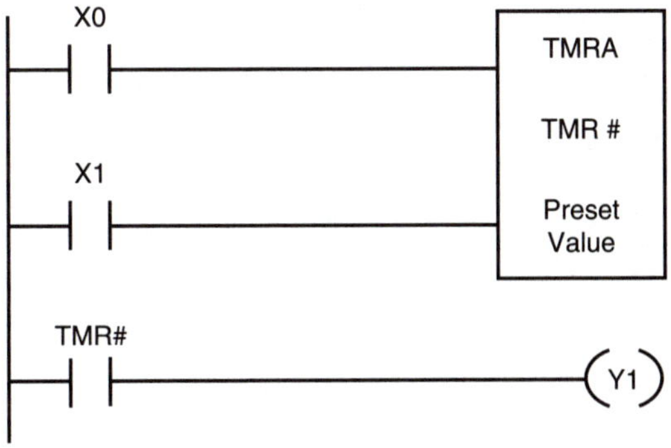

Figure 7–18 AutomationDirect accumulating timer example.

means “accumulating timer.” The AutomationDirect timer requires a low at the reset line to activate the timer. If the reset line goes high, the timer resets. In this case if X0 goes high, the timer will begin to time. If the timer count reaches the preset value, the timer contact will turn on. This would turn on output Y1. If X0 goes low before the count reaches the preset time, the timer will hold the present time and will not reset the time count to zero. When X0 goes high again, the timing will continue. An accumulating timer will never reset to zero unless the reset line goes high.

Figure 7–19 shows the use of another AutomationDirect. Note that this is a fast accumulating timer (TMRAF). It is really a retentive timer with a time base of 0.01 second. The timer number is 5. The enable input is X0. The reset input is X1. T5 controls output Y1. (If the timer preset has been reached, the timer is on and output Y1 is on.) The preset in this case is a variable (V1400). The value for the preset value will equal whatever is the value of V1400. Other ladder logic would establish the value of V1400.

The accumulated value of timers is kept in variables. There are 128 (0 to 177 octal) timers available. Their accumulated value is stored in variables V00000 to V00177. If we wanted to monitor the accumulated value of timer 5, we would examine variable V00005. These accumulated values can be used in ladder logic, just as any other variable could be.

Rockwell Automation Timers

Timer On-Delay

The timer on-delay instruction is used to turn an output on after a timer has been on for a preset time interval. The timer on-delay (TON) begins accumulating time when the rung becomes true and continues until one of the following conditions is met: the accumulated value is equal to the preset value, the rung goes false, a reset timer instruction resets the timer, or the associated SFC step becomes inactive.

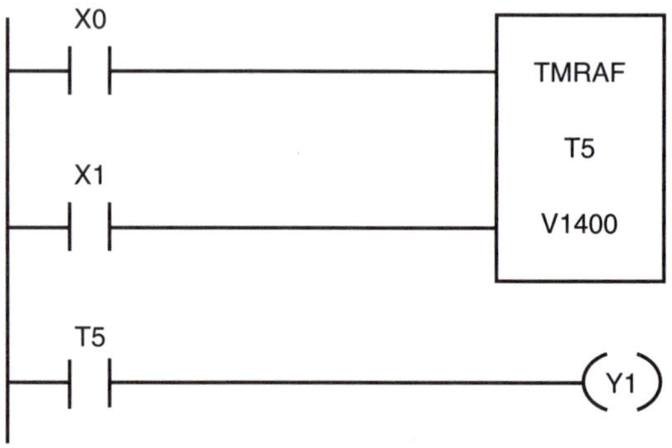

Figure 7–19 An AutomationDirect fast accumulating timer (TMRAF). The time increment for the fast timer is 0.01 second.

Figure 7–20 Format for a timer.

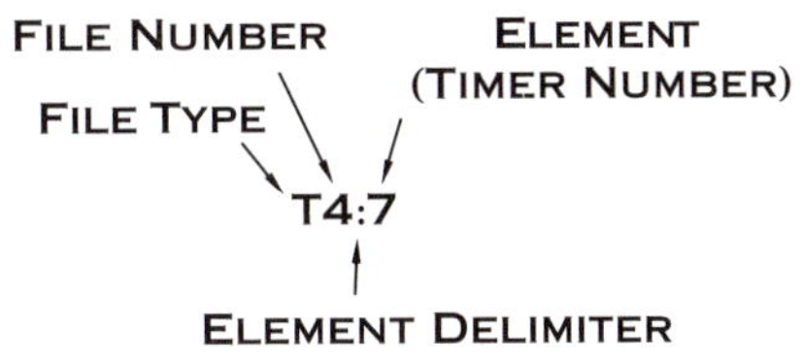

Figure 7–20 shows the numbering system for timers. The T stands for timer, the 4 is the file number of the timer, and 7 is the actual timer number. When entering the name of the timer into a timer instruction, enter T4: and the number you want to use for the timer. For example, T4:0 would be timer 0.

Status Bit Use Timer status bits can be used in ladder logic. Several bits are available for use (see Figure 7–21). Consider timer T4:0. The timer enable bit (EN) is set immediately when the rung goes true. It stays set until the rung goes false or a reset instruction resets the timer. The .EN bit indicates that the timer is enabled. The .EN bit from any timer can be used for logic. For example, T4:0.EN could be used as a contact in a ladder.

The timer timing bit (TT) can also be used. The TT bit is set when the rung goes true. It remains true until the rung goes false or the DN bit is set (accumulated value equals preset value). For example, T4:0/TT can be used as a contact in a ladder.

The timer done bit (DN) is set until the accumulated value is equal to the preset value, and the DN remains set until the rung goes false and a reset instruction resets the timer. When the DN bit is set, the timing operation is complete. For example, T4:0/DN can be used as a contact in the ladder. The preset (.PRE) is also available to the ladder. For example, T4:0.PRE accesses the preset value of T4:0. Note that the PRE value is an integer.

Accumulated Value Use The accumulated value can also be used by the programmer. The accumulated value (.ACC) is acquired in the same manner as the status bits and preset. For example, T4:0.ACC accesses the accumulated value of timer T4:0.

Time Bases Three time bases are available: 1.0-, 0.01-, or 0.001-second intervals (see Figure 7–22). The potential time ranges are also shown. If a longer time is needed, timers can be cascaded (discussed later in this chapter). Figure 7–23 shows how timers are handled in memory. Three bits are used in the first storage location for this timer to store the present status of the timer bits (EN, TT, and DN). The preset value (PRE) is stored in the second 16 bits of this timer storage. The third 16 bits hold the accumulated value of the timer.

Condition	Result
If the rung is true	.EN bit remains set (1)
	.TT bit remains set (1)
	.ACC value is cleared and begins counting up
If the rung is false	.EN bit is reset
	.TT bit is reset
	.DN bit is reset
	.ACC value is cleared and begins counting up

Figure 7–21 Use of special timer bits.

Time Base	Potential Time Range
1 Second	To 32,767 time-base intervals (up to 9.1 hours)
.01 Second (10 ms)	To 32,767 time-base intervals (up to 5.46 minutes)
.001 Second (1 ms)	To 32,767 time-base intervals (up to .546 minute)

Figure 7–22 Time bases available.

15	14	13	12	11	10	9	8	7	6	5	4	3	2	1	0	Element
EN	TT	DN						Internal Use								0 T4:0
Preset Value PRE																1 T4:0.PRE
Accumulated Value ACC																2 T4:0.ACC

Figure 7–23 Use of control words for timers.

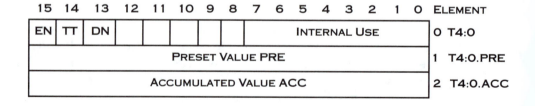

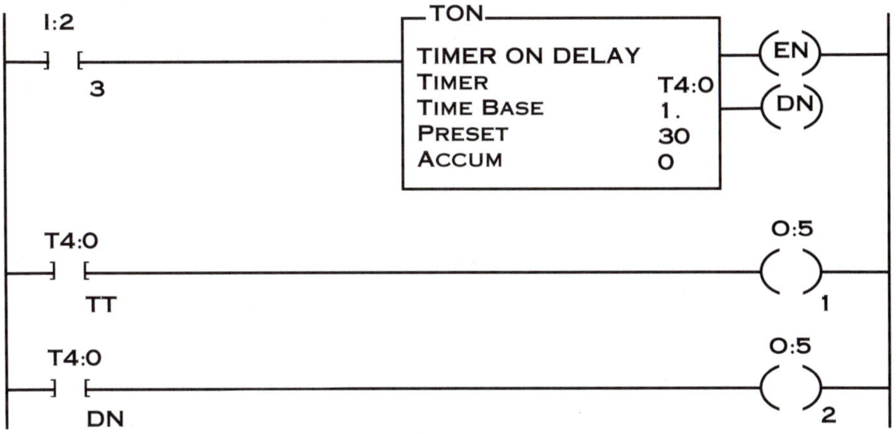

Figure 7–24 Use of a TON timer in a ladder logic program.

Figure 7–24 shows an example of the use of a TON timer in a ladder. When input I:2/3 is true, the timer begins to increment the accumulated value of TON timer 4:0 in 1-second intervals. The timer timing bit (TT) for timer 4:0 is used in the second rung to turn on output O:5/1, while the timer is timing (ACC < PRE). The timer done bit (DN) of timer 4:0 is used in rung 3 to turn on output O:5/2 when the timer is done timing (.ACC = .PRE). The preset for this timer is 30, which means that the timer will have to accumulate thirty 1-second intervals to time out. Note that this is not a retentive timer. If input I:2/3 goes low, the accumulated value is reset to zero.

Figure 7–25 shows an example of the timer done bit (T4:0/DN) used to reset the timer to zero accumulated time every time the timer accumulated time reaches the preset. This would turn T4:0/DN on for one scan cycle. This feature is useful when you need something to happen at regular intervals in a ladder diagram. For example, maybe you need to use a special instruction every 30 seconds. (Note that a special reset instruction can also do this. It will be covered later in this chapter.)

Timer Off-Delay

The timer off-delay instruction is used (TOF) to turn an output on or off after the rung has been off for a desired time. The TOF instruction starts to accumulate time when the rung becomes false. It will continue to accumulate time until the accumulated value equals the preset value or the rung becomes true. The timer enable bit (EN bit 15) is set when the rung becomes true (see Figure 7–26). It is reset when the rung become false. The timer timing bit (TT bit 14) is set when

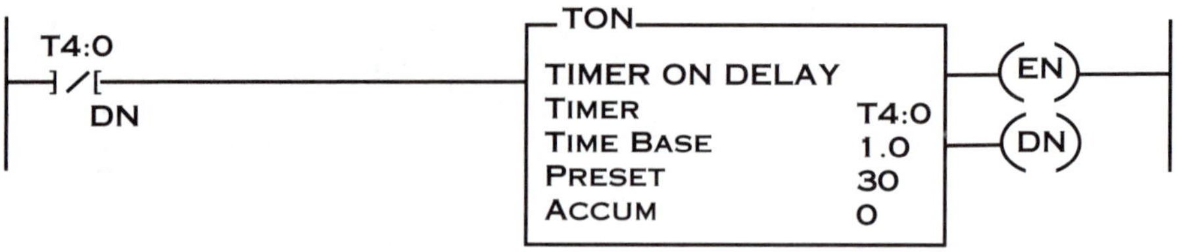

Figure 7–25 Use of a timer's DN bit to reset the counter's accumulated time to 0 everytime it reaches 30, the reset value. The timer DN bit would be true for one scan every thirty seconds.

BIT	SET WHEN	REMAINS SET TILL
TIMER DONE BIT (BIT 13 OR DN)	RUNG CONDITIONS ARE TRUE	RUNG CONDITIONS GO FALSE AND THE ACCUMULATED VALUE IS GREATER THAN OR EQUAL TO THE PRESET VALUE
TIMER TIMING BIT (BIT 14 OR TT)	RUNG CONDITIONS ARE FALSE AND THE ACCUMULATED VALUE IS LESS THAN THE PRESET VALUE	RUNG CONDITIONS GO TRUE OR WHEN THE DONE BIT IS RESET
TIMER ENABLE BIT (BIT 15 OR EN)	RUNG CONDITIONS ARE TRUE	RUNG CONDITIONS GO FALSE

Figure 7–26 Use of TOF bits.

the rung becomes false and ACC < PRE. The TT bit is reset when the rung becomes false, or the DN bit is reset (ACC = PRE), or a reset instruction resets the timer.

The done bit (DN bit 13) is reset when the accumulated value (ACC) is equal to the preset (PRE) value. The DN bit is set when the rung becomes true.

Figure 7–27 shows the use of a TOF timer in a ladder diagram. Input I:2/3 is used to enable the timer. When input I:2/3 is false, the accumulated value is incremented as long as the input stays false and ACC ≤ PRE. The timer timing bit for timer T4:0 (T4:0/TT) is used to turn on output O:05/1 while the timer is timing (ACC < PRE). The done bit (DN) for timer 4:0 (T4:0/DN) is used to turn on output O:5/2 when the timer has completed the timing (ACC = PRE).

The TOF timer can be confusing so the following is a simple, nonindustrial example. Think of a garage light. The garage owner wants to be able to touch a mo-

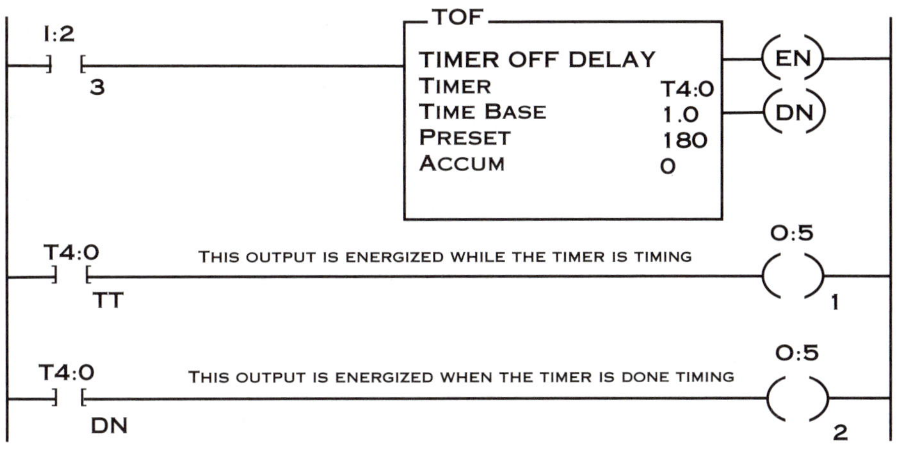

Figure 7–27 Use of a TOF timer in a ladder logic diagram.

mentary on switch so the light turns on immediately and then stays on for a given time (maybe 2 minutes). At the end of the time, the light turns off. This feature would allow time to get into the house with the light on. In this example, the output (light) turned on instantly when the input (switch) was momentarily turned on and off. The timer counted down the time (timed out) and turned the output (light) off. This is an example of a *timer off-delay.*

Consider Figure 7–28. When the momentary switch is pushed and then released, the timer starts counting time intervals. The timer done bit (T4:0/DN) is used in the second rung as a contact. As soon as the momentary switch has made the high-to-low transition, the time begins timing and turns the DN bit on. If the timer DN bit is energized, the light is energized. When the timer times out (in this case, in 2 minutes), the contact (T4:0/DN) in rung 2 opens and the light turns off.

Retentive Timer On

The retentive timer on instruction (RTO) is used to turn an output on after a set time period (see Figure 7–29). The RTO timer is an accumulating timer. It retains the accumulated value even if the rung goes false. The only way to zero the accumulated value is to use a reset (RES) instruction in another rung with the same address as the RTO you wish to reset. The RTO retains the accumulated count even if power is lost, or you switch modes, the rung becomes false, or the associated SFC becomes inactive. Remember, the only way to zero the accumulated value is to use a

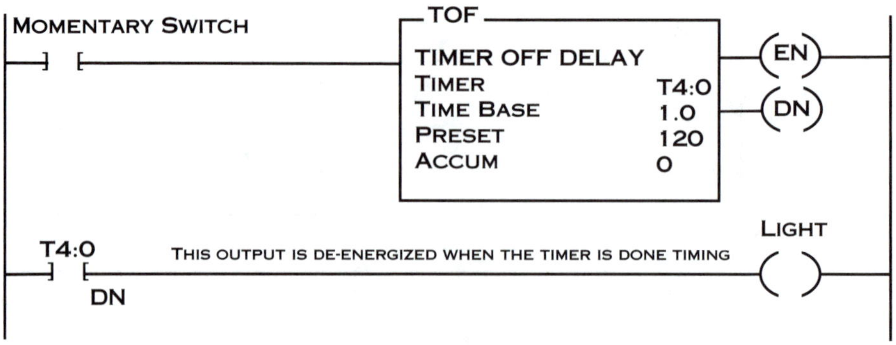

Figure 7–28 Delay-off timing circuit. If the momentary switch closes, delay-off timer T4:0 immediately begins timing when the momentary switch makes the transition from high to low. The done bit (T4:0/DN) is immediately energized, which turns on the light. When the timer reaches 2 minutes (120 seconds), it will de-energize the DN bit, which also turns the light off.

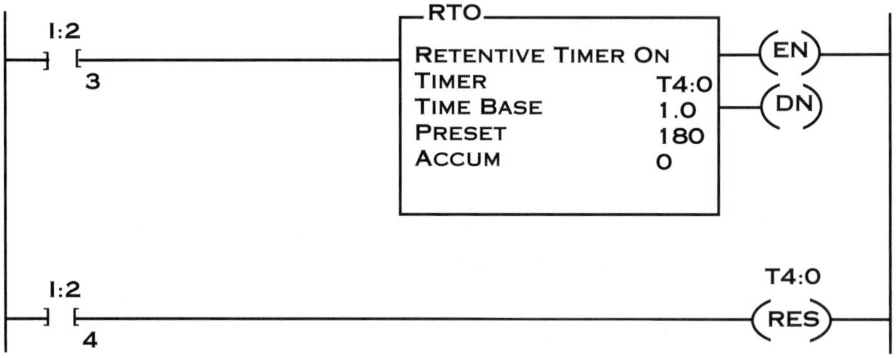

Figure 7–29 Use of an RTO timer.

reset instruction. The reset (RES) instruction must have the same address as the timer you want to reset. Note that the RES instruction can also be used for other types of timers and counters.

The status bits can be used as contacts in a ladder diagram (see Figure 7–30). The timer enable (EN) bit is set when the rung becomes true. When the EN bit is a 1, it indicates that the timer is timing. It remains set until the rung becomes false or a reset instruction zeros the accumulated value.

Bit	Set When	Remains Set Till
Timer Done Bit (Bit 13 or DN)	Accumulated Value Is Equal to or Greater Than the Preset Value	The Appropriate RES Instruction Is Enabled
Timer Timing Bit (Bit 14 or TT)	Rung Conditions Are True and the Accumulated Value Is Less Than the Preset Value	Rung Conditions Go False or When the Done Bit Is Set
Timer Enable Bit (Bit 15 or EN)	Rung Conditions Are True	Rung Conditions Go False

Figure 7–30 Use of RTO bits.

The timer timing (TT) bit is set when the rung becomes true. It remains set until the accumulated value equals the preset value or a reset instruction resets the timer. When the TT bit is a 1, it indicates that the timer is timing.

The TT bit is reset when the rung becomes false or when the done (DN) bit is set. The timer done (DN) bit is set when the timer's accumulated value is equal to the preset value. When the DN bit is set, it indicates that the timing is complete. The DN is reset with the reset instruction.

Cascading Timers

Applications sometimes require longer time delays than one timer can accomplish. Multiple timers can then be used to achieve a longer delay than would otherwise be possible. One timer acts as the input to another. When the first timer times out, it becomes the input to start the second timer timing. This is called *cascading*. Figure 7–31 shows an example of cascading timers.

Counters

Counting is important in industrial applications. Often, the product must be counted so that another action can take place. For example, if twenty-four cans go into a case, the twenty-fourth can should be sensed by the PLC and the case should be sealed. Counters are required in almost all applications.

Several types of counters are available, including up counters, down counters, and up/down counters. The choice of which to use depends on the task to be done. For example, if we are counting the finished product leaving a machine, we might use an up counter. If we are tracking how many parts are left, we might use a down

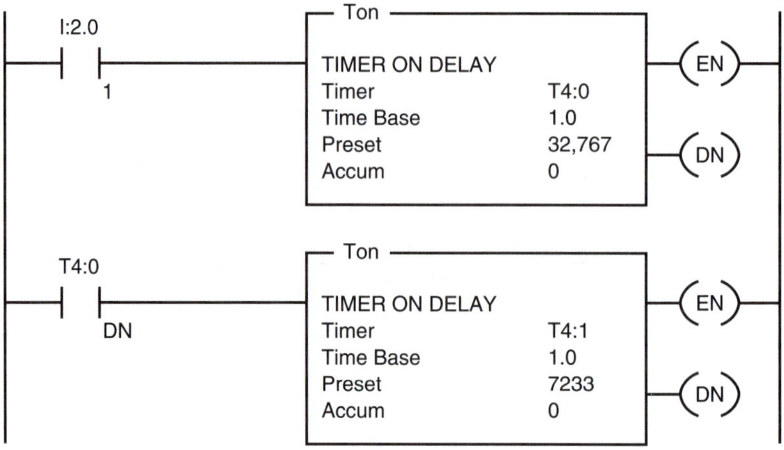

Figure 7–31 Two timers used to extend the time delay. The first timer output, T4:0/DN, acts as the input to start the second timer. When input I:2.0/1 becomes true, timer 1 begins to count to 32,767 seconds. (The limit for these timers is 32,767 seconds.) When it reaches 32,767, output T4:0/DN turns on. This energizes timer T4:1. Timer T4:1 then times to 7233 seconds and then turns T4:1/DN. The delay was 40,000 seconds.

counter. If we are using a PLC to monitor an automated storage system, we might use an up/down counter to track how many are coming and how many are leaving to establish the total number in stock.

Counters normally use a low-to-high transition from an input to trigger the counting action. Figure 7–32 shows a generic counter example. Counters have a reset input or a separate reset instruction to clear the accumulated count. Counters are very similar to timers. Timers count the number of time increments; counters count the number of low-to-high transitions on the input line.

Note that the counter in Figure 7–32 is edge-sensitive triggered. The rising, or leading, edge triggers the counter. X000 is used to count the pulses. Everytime there is an off-to-on transition on X000, the counter adds 1 to its count. When the accumulated count equals the preset value, the counter turns on, which turns on output Y20. X001 is used as a reset/enable. If contact X001 is closed, the counter is returned to zero. The counter is active (enabled or ready to count) only if X000 is off (open). The example just described is an up counter.

Down counters cause a count to decrease by 1 everytime there is a pulse. Some brands of PLCs also have up/down counters available. An up/down counter has one input that causes it to increment the count and another that causes it to decrement the count. Ladder diagram statements can utilize these counts for comparing and/or

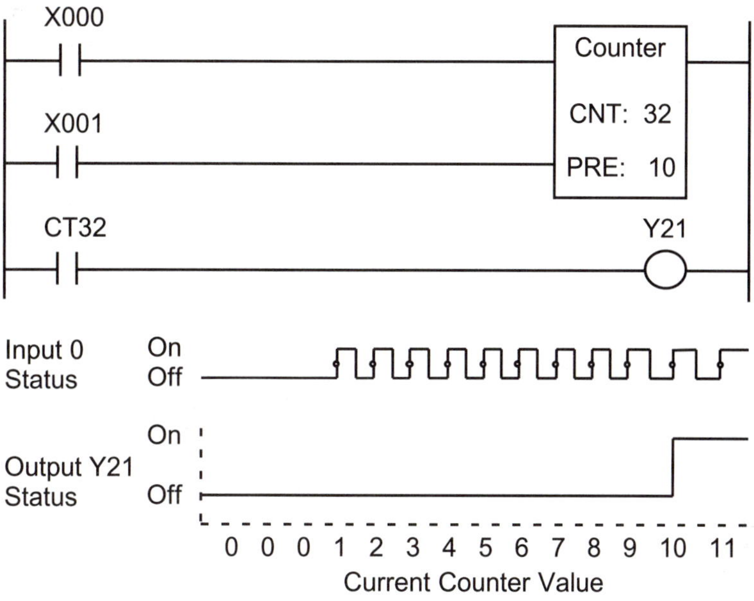

Figure 7–32 How a typical counter works. When ten or more low-to-high transitions of input X000 have been made, counter CT32 is energized, which energizes output Y21.

decision making. Counts can also be compared to constants or variables to control outputs.

GE Fanuc Counters

Figure 7–33 shows two types of GE Fanuc counters. The up counter increments the count by 1 each time the counter receives transitional power. The down counter decrements the count everytime the counter receives transitional power.

The GE Fanuc up counter (UPCTR) is shown in Figure 7–34. This counter preset value is set to a constant value of 5. The current value of the count is held in register 7 (%R0007). For every transition of the enable input (%I0001), the counter's current value is incremented by one count. When the counter current value is equal to or greater than the preset value of 5, the counter will pass power and, in this example, turn coil 5 (%Q0005) on. The counter continues to increment the current value with every power transition on the enable input. The counter's current value will be reset to zero only if the reset input (%I0002) is energized.

The GE Fanuc down counter (DNCTR) is shown in Figure 7–35. This counter preset value is set to a constant value of 5. The present value of the count is held in

Instruction	Function	Description
upctr	Up Counter	This type of counter increments by 1 each time the function receives transitional power. If the current value in the timer is greater than or equal to a preset value, the function passes power. The R input is used to reset the counter to zero.
dnctr	Down Counter	This counter counts down every time the function receives transitional power. If the current value of the counter is zero, the function passes power. The R input is used to set the current value to equal the preset value.

Figure 7–33 GE Fanuc counters.

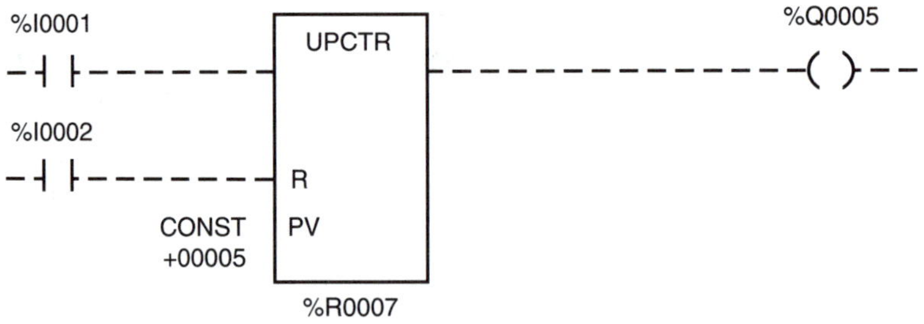

Figure 7–34 Use of a GE Fanuc up counter.

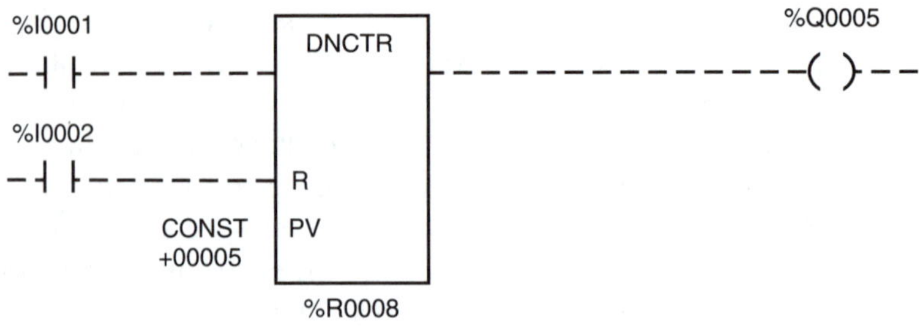

Figure 7–35 Use of a GE Fanuc down counter.

register 8 (%R0008). For every transition of the enable input (%I0001), the counter's current value is decremented from the preset value. When the current value is equal to zero, the counter will pass power. In this case output 5 (%Q0005) is turned on. The counter's current value will be reset to the preset value if the reset input (%I0002) is energized.

Gould Modicon Counters

Figure 7–36 shows an example of a Gould Modicon counter. The counter has two inputs: an enable and a reset. The reset line must be high for the counter to count. Every time the enable input makes a transition from low to high, the counter increments. The preset value for this counter was set to 6. When the count equals 6, the outputs change state. The current count is kept in data register 4005. The top output is off unless the actual count is equal to or greater than the preset. The bottom "NOT" output is on unless the count in 4005 is equal to or greater than the value of the preset.

Figure 7–37 shows the use of a Gould Modicon counter. Every time input 1001 makes a low-to-high transition, the counter increments the count in data register 4005. When the count reaches 6 or more, output 0005 will turn on. Output 0005 will stay on until the counter is reset by the reset line going low.

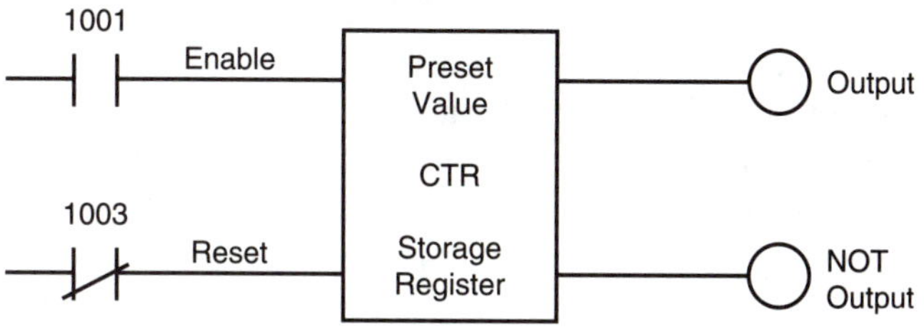

Figure 7–36 Gould Modicon counter. Note that two outputs are available.

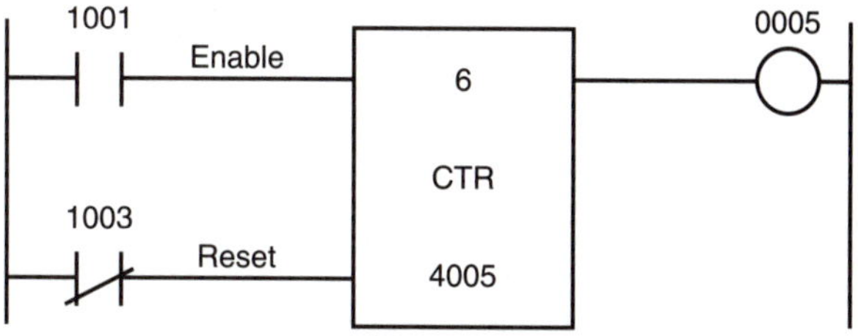

Figure 7–37 Gould Modicon counter.

Omron Counters

Two types of Omron counters are available: CNT and CNTR. CNT is a decrementing counter and CNTR is a reversible counter instruction. Timers/counters can be numbered from 0 to 511. Numbers cannot be duplicated. If you use number 1 for a counter, you may not use it for a timer. Both types of counters require a counter number, a set value (SV), inputs, and a reset input.

Figure 7–38 shows the use of an Omron counter (CNT). The set value is 20. For every low-to-high transition of the counter enable input (00001), the counter decrements the set value. When the count reaches zero, the counter output turns on. A low-to-high transition of the reset input will reset the count to the set value.

A reversible counter (CNTR) type is also available. This counter has three inputs. One input is used to make the counter count up. The second input is used to decrement the count. The third is used to reset the counter. When the reset input is high, the present value is set to 0000. Figure 7–39 shows the use of a CNTR counter.

Omron timers and counters can use external channels to receive their set values. Instead of giving a set value, the programmer assigns a channel from which the timer/counter will "look up" its set value. In this way a timer/counter set value can be made a variable.

AutomationDirect Counters

AutomationDirect counters are similar to the other brands of counters, with one exception (see Figures 7–40 and 7–41). The 405 series uses a normally open contact for the reset line. If the reset line goes high, the counter is reset. The programmer supplies a counter number and a preset value for each counter. The counter num-

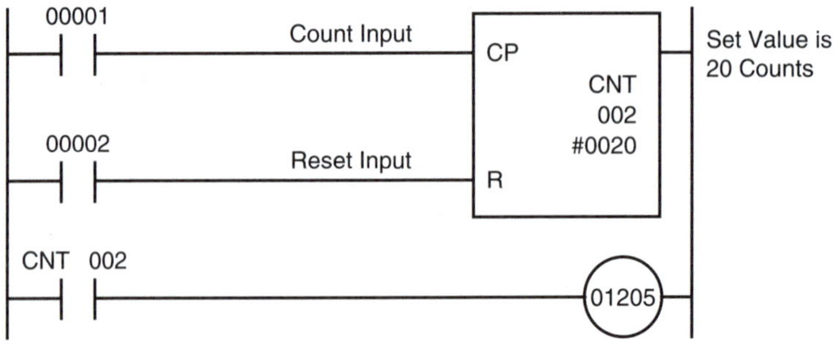

Figure 7–38 Use of an Omron CNT counter. Note that output 01205 turns on when the counter becomes true. The output will remain on until the reset input to the counter goes high and resets the set value to 20.

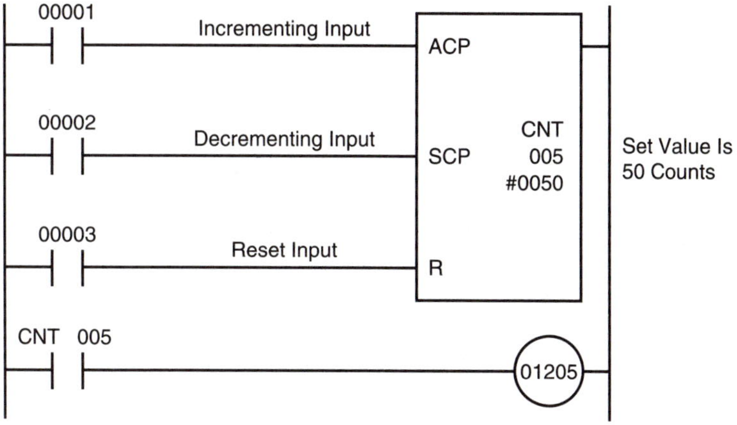

Figure 7–39 Use of a CNTR counter. The counter output turns on whenever the present count becomes 0000.

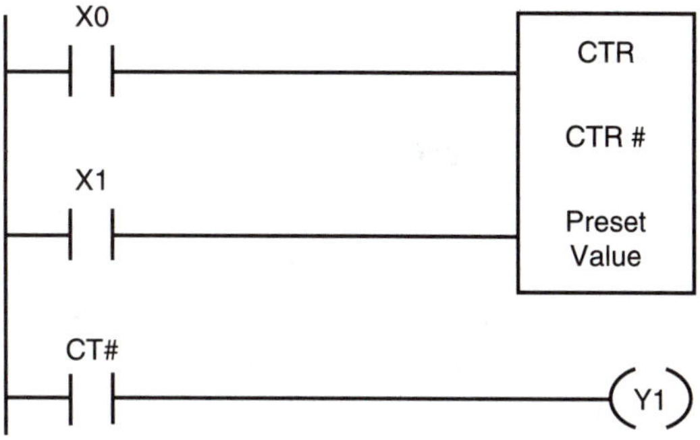

Figure 7–40 AutomationDirect counter. The programmer must provide a counter number and a preset value. Note that the reset input is a normally open contact. If the reset line becomes high, the counter accumulated value will be set to zero.

ber can then be used as a contact. When the counter accumulated count is equal to or greater than the preset value, the counter is true. The counter can be reset by making the reset input go high.

The preset value can be a constant (K) or a variable. If it is a variable, the CPU gets the preset value from that variable. If a constant is used, the letter K precedes it. K5 is a preset value (constant) of 5.

The memory locations that hold the accumulated values can be monitored or used in comparison instructions. The accumulated values are kept in variable memory.

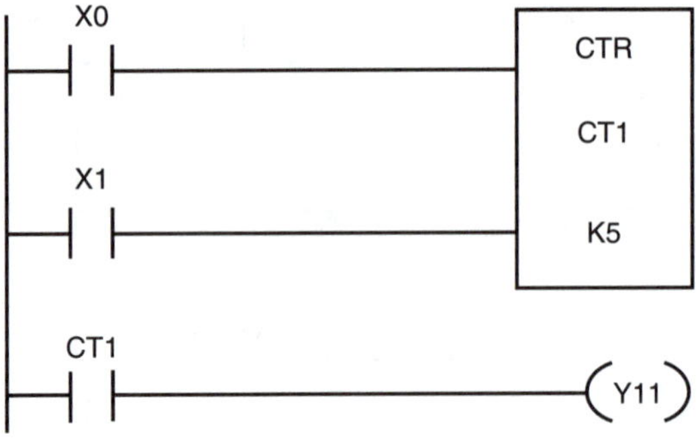

Figure 7–41 AutomationDirect counter. For every low-to-high transition of input X10, CT1 increments the count. When the accumulated count is equal to or greater than the preset value (5), counter CT1 is true. The counter is used as contact CT1. When counter CT1 becomes true, it energizes output Y11.

For counters, the variable numbers are V01000–V01177. Note that 128 counters are available, or 0 to 177 octal. Note also that 128 (V01000–V01177 octal) counter variables are available.

Rockwell Automation Counters

A Rockwell automation counter has a counter number, a preset, and an accumulated value. The counter is numbered like the timer except that it begins with a "C." The next number is a file number. The default file number for counters is 5. The third value is the counter number, in this case 4. (See Figure 7–42.) When programming a counter, enter C5: and the number of the counter for the timer name. For example, C5:4 would be counter 4 counter file 5.

Your ladder diagram can access counter status bits, presets, and accumulated values (see Figure 7–43). You may use the CU, CD, DN, OV, or UN bit for logic. Each will be covered later in the chapter. You can also use the preset (PRE) and the accumulated count (ACC).

Counter values are stored in three 16-bit words of memory (see Figure 7–44). The first eight bits of the first word are for internal use of the CPU only. The most

Figure 7–42 How counters are addressed.

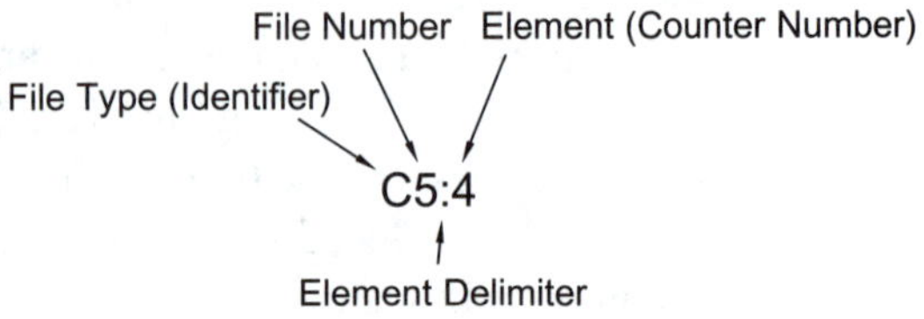

C5:4.DN	Use of the Done Bit
C5:4.PRE	Use of the Present Value
C5:4.ACC	Use of the Accumulated Value

Figure 7–43 Examples of the use of counter values.

15	14	13	12	11	10	9	8 7 6 5 4 3 2 1 0	Element
CU	CD	DN	OV	UN	UA		Internal Use	0 C5:0
Preset Value PRE								1 C5:0.PRE
Accumulated Value ACC								2 C5:0.ACC

Figure 7–44 How counter values and status bits are stored in memory.

significant bits of the first word are used to store the status of certain bits associated with the counter (see Figure 7-45). The count-up enable bit (CU bit 15) is used to indicate that the counter is enabled. The CU bit is reset when the rung becomes false or when it is reset by the RES instruction. When high, the count-up done bit (DN bit 13) indicates that the accumulated count has reached the preset value. It remains set even when the accumulated (ACC) value exceeds the preset (PRE) value. The DN bit is reset by a reset (RES) instruction.

The count-up overflow bit (OV bit 12) is set by the CPU to show that the count has exceeded the upper limit of +32,767. When this happens the counter accumulated value "wraps around" to −32,768 and begins to count up from there, back toward zero. (This has to do with the way computers store negative numbers. The preset value and accumulated value are stored as a two's-complement number.) The OV bit can be reset with a reset (RES) instruction.

Count-Up (CTU) Timer

Figure 7–46 shows the use of a count-up counter (CTU) in a ladder diagram. Each time input I:2/3 makes a low-to-high transition, the counter accumulated value is incremented by 1. The done bit of counter C5:0 (C5:0/DN) is used to turn output O:5/1 on when the accumulated value is equal to the preset value (ACC = PRE). The overflow bit of counter C5:0 (C5:0/OV) is used to turn on output O:5/2 if the count ever reaches +32,767. The last rung uses input I:2/1 to reset the accumulated value of counter C5:0 to zero.

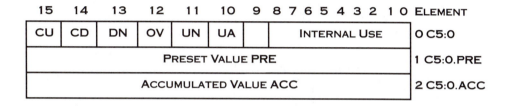

Bit	Set When	Remains Set Till
Count Up Overflow Bit (Bit 12 or OV)	Accumulated Value Wraps around to −32,768 (from +32,767) and Continues Counting Up from There	The RES instruction That Has the Same Address as the CTU Instruction Is Executed or the Count Is Decremented Less Than or Equal to +32,767 with a CTD Instruction
Done Bit (Bit 13 or DN)	The Accumulated Value Is Equal to or Greater Than the Preset Value	The Accumulated Value Becomes Less Than the Preset
Count Up Enable Bit (Bit 15 or CU)	Rung Conditions Are True	Rung Conditions Go False or the RES Instruction That Has the Same Address as the CTU Instruction Is Enabled

Figure 7–45 CTU counter bits.

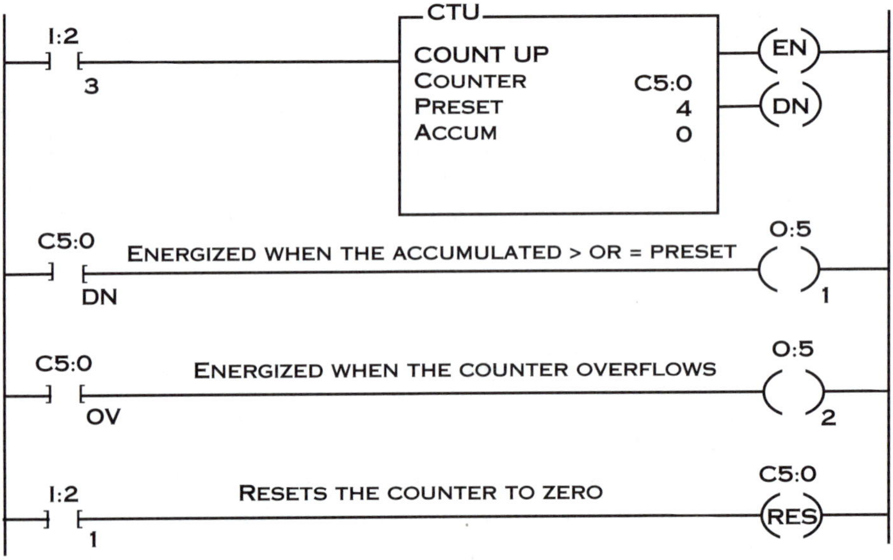

Figure 7–46 Use of a count-up counter (CTU) in a ladder diagram.

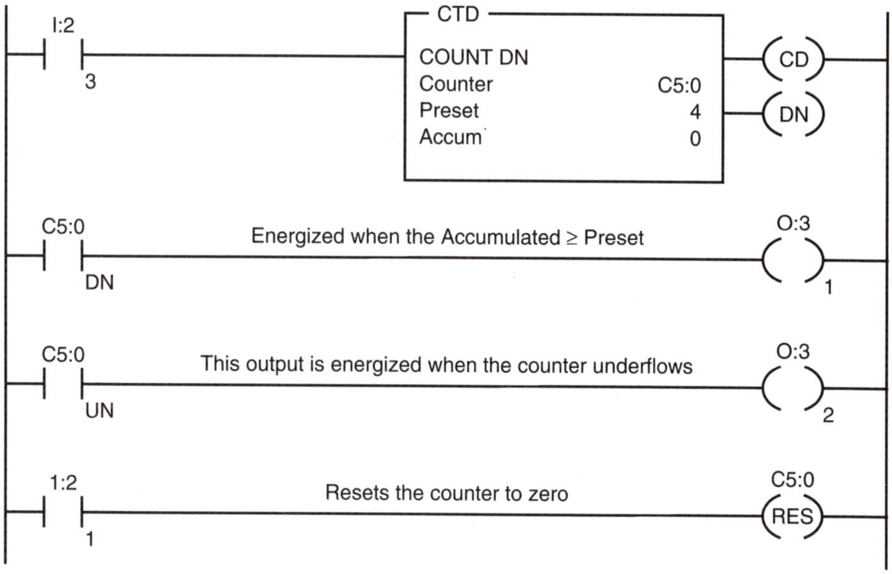

Figure 7–47 Use of a count-down counter (CTD).

Count-Down (CTD) Counter

Figure 7–47 shows the use of a count-down counter (CTD) in a ladder diagram. Each time input I:2/3 makes a false-to-true transition, the counter accumulated value is decremented by 1. The done bit of counter C5:0 (C5:0/DN) is used to turn output O:5/1 on when the accumulated value is equal to or exceeds the preset value (ACC ≥ PRE). The accumulated value of counters is retentive. They are retained until a reset instruction is used. The underflow bit of counter C5:0 (C5:0/OV) is used to set output O:5/2 on if the count ever underflows −32,768. Note the use of the reset instruction to reset the accumulated value of the counter to zero. The last rung uses input I:2/1 to reset the accumulated value of counter C5:0 to zero. Figure 7–48 shows the use of CTD counter bits.

Cascading Counters

A counter's DN bit can be used to increment another counter. For example, in Figure 7–49 counter C5:0 is used to count twenty-four cans for each case. When C5:0 reaches 24, it increments C5:1 and resets its own counts to zero with the reset instruction. C5:1 is used to count the number of cases of twenty-four cans. In this example 257 cases have been produced. Sometimes applications require higher counts than one counter can accomplish. Cascading can be used for this purpose.

BIT	SET WHEN	REMAINS SET TILL
COUNT DOWN UNDERFLOW BIT (BIT 11 OR UN)	ACCUMULATED VALUE WRAPS AROUND TO +32,768 (FROM −32,767) AND CONTINUES COUNTING DOWN FROM THERE	THE RES INSTRUCTION THAT HAS THE SAME ADDRESS AS THE CTD INSTRUCTION IS EXECUTED OR THE COUNT IS INCREMENTED GREATER THAN OR EQUAL TO +32,767 WITH A CTU INSTRUCTION
DONE BIT (BIT 13 OR DN)	THE ACCUMULATED VALUE IS EQUAL TO OR GREATER THAN THE PRESET VALUE	THE ACCUMULATED VALUE BECOMES LESS THAN THE PRESET
COUNT DOWN ENABLE BIT (BIT 14 OR CD)	RUNG CONDITIONS ARE TRUE	RUNG CONDITIONS GO FALSE OR THE RES INSTRUCTION THAT HAS THE SAME ADDRESS AS THE CTD INSTRUCTION IS ENABLED

Figure 7–48 CTD counter bits.

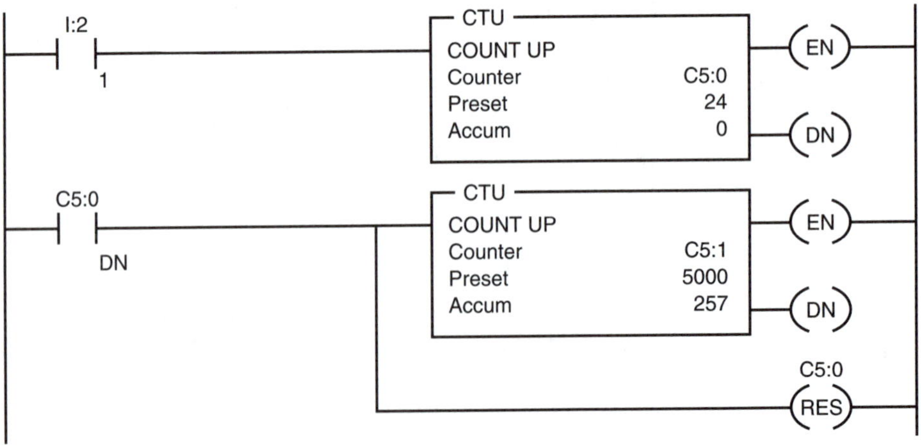

Figure 7–49 Cascading counters.

Programming Hints

Timers and counters are indispensable to application development. PLC manufacturers share much in common when it comes to implementation of timers and counters. Once the technician becomes familiar with one, the rest are easy to learn.

When you are asked to program a new brand or type of PLC, some logical steps can make the task less frustrating. First, make sure that the programming device is really communicating with the PLC. There are several ways to be sure. Most programming software will check and warn you if communications are not established. Some PLC communications modules also have LEDs that flash when communications are taking place. If there is a problem, the most likely place to look is at the cable. Is it the right cable, and is it properly attached to the correct port? The other possible problem is that the PLC may have been assigned a station address different from the one that the software is trying to access. Another potential problem is that the communication parameters were set up incorrectly. Check the baud rates, number of data bits, number of stop bits, and parity of the software and PLC.

If the PLC is now communicating, the next step is to see if a simple ladder diagram can be entered and executed. Keep it simple. One contact and one coil are adequate. Make the contact a normally closed contact. Execute the ladder. If the output LED turns on, you have accomplished several important tasks. You have entered a ladder and executed it successfully. You have also figured out the correct I/O numbering.

The next step is probably to get a timer to work, followed by a counter, and so on. Then you can develop the actual application quite quickly.

Questions

1. What are timers typically used for?
2. Explain the two types of timers and how each might be used.
3. What does the term *retentive* mean?
4. Draw a typical retentive timer and describe the purpose of the inputs.
5. What are counters typically used for?
6. How are counters and timers very similar?
7. Explain the two contacts usually required for a counter.
8. What is cascading?
9. You have been asked to program a system that requires that completed parts are counted. The largest counter available in the PLC's instruction set can count only to 999. We must be able to count up to 5000. Draw a ladder diagram that shows the method you would use to complete the task. *Hint:* Use two counters, one as the input to the other. The total count will involve looking at the total of the two counters.

10. List at least three reasons for using a flow diagram or pseudocode in program development.
11. The following figure is a partial drawing of a heat-treat system. You have been asked by your supervisor to troubleshoot the system. The engineer who originally developed the system no longer works for your company. He never fully documented the system. A short description of the system follows. You must study the drawings and data and complete this assignment.

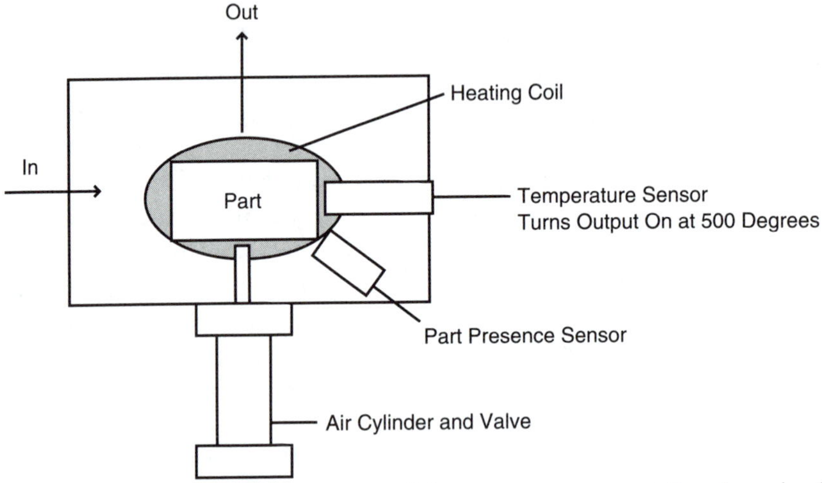

A part enters this portion of the process. The temperature must then be raised from room temperature to 500 degrees Fahrenheit. There are also a part presence sensor and a sensor that turns on when the temperature reaches 500 degrees. The part must then be pushed out of the machine. The cycle should take about 25 seconds. If it takes an excessive amount of time and the temperature has still not been reached, an operator must be informed and must reset the system. Study the system drawing, I/O chart, and ladder diagram, then complete the I/O table and answer the questions.

Complete the following I/O chart by writing short comments that describe the purpose of each input, output, timer, and counter. Make your comments clear and descriptive so that the next person to troubleshoot the system will have an easier task. Then refer to the accompanying figure and answer the following questions.

- What is the purpose of counter 1?
- What is the purpose of input X3?
- What is input X20 used for?
- What is the purpose of counter 2?
- What is the purpose of contact Y15?
- Part of the logic is redundant. Identify that part and suggest a change.

System I/O	
X3	Part presence sensor
X20	Operator reset
Y5	
Y8	
Timer 1	
Timer 9	
Counter 1	
Counter 2	

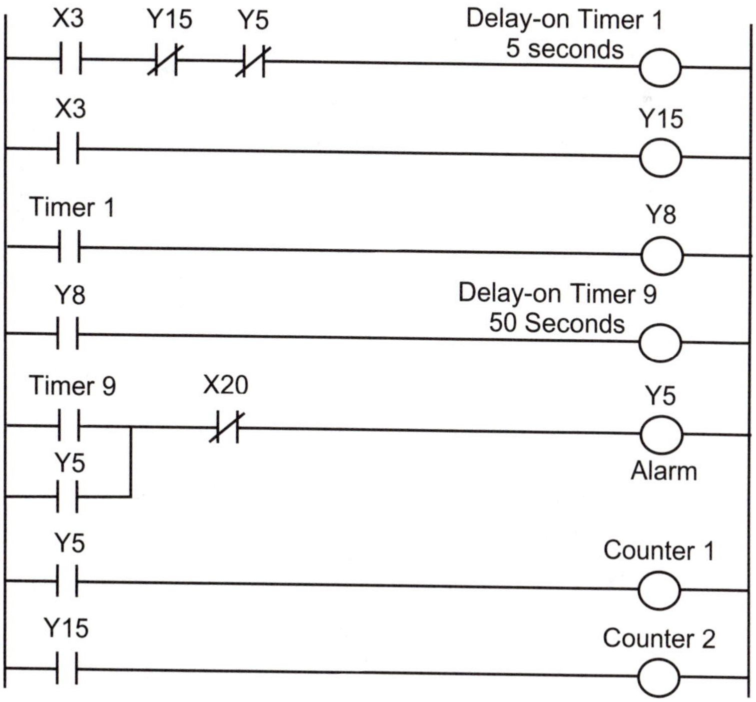

12. Examine the ladder diagram in the following figure. Assume that input 00001 is always true. What will this ladder logic do?

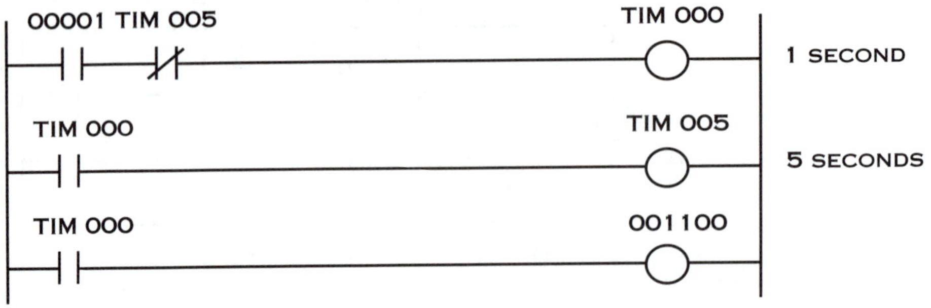

13. You have been assigned the task of developing a stoplight application. Your company thinks there is a large market in intelligent street corner control. Your company wants to develop a PLC-based system in which lights adapt their timing to compensate for the traffic volume. Your task is to program a normal stoplight sequence to be used as a comparison to the new system.

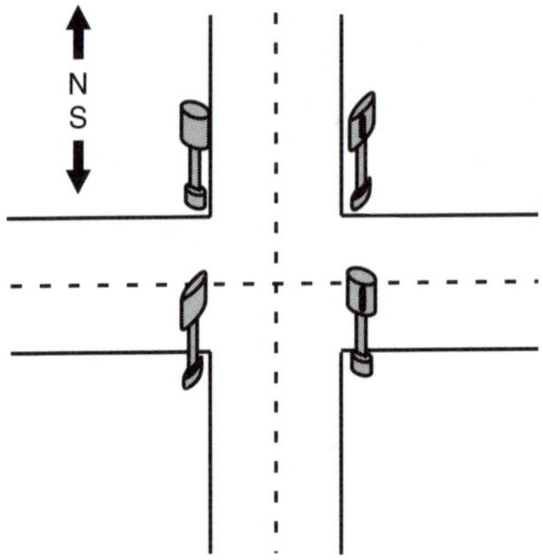

Note in the above figure that there are really two sets of lights. The north and south lights must react exactly alike. The east-west set must be the complement of the north-south set. Write a program that will keep the green light on for 25 seconds and the yellow one on for 5 seconds. The red will then be on for 30 seconds. You must also add a counter because the bulbs are replaced at a certain count for preventive maintenance. The counter should count complete cycles. (*Hint:* To simplify your task, do one small task at a time. Do not try to write the entire application at once. Write ladder logic to get one light working, then the next, then the next, before you even worry about the other stoplight. When you get one set done, the other is a snap. Remember: A well-planned job is half done.)

Write the ladder diagram. Thoroughly document the ladder with labels and rung comments.

chapter

8 Math Instructions

Arithmetic instructions are vital in the programming of systems. They can simplify the programmer's task. In this chapter we cover compare, add, subtract, multiply, and divide instructions. Several brands of PLC instructions will be covered.

OBJECTIVES

Upon completion of this chapter, you will be able to:

1. Describe typical uses for arithmetic instructions.
2. Explain the use of compare instructions.
3. Explain the use of typical arithmetic instructions.
4. Write ladder logic programs involving arithmetic instructions.

OVERVIEW

Many times contacts, coils, timers, and counters fall short of what the programmer needs. Many applications require mathematical computation. For example, imagine a furnace application that requires the furnace to be between 250 and 255 degrees (see Figure 8–1). If the temperature variable is between 250 and 255 degrees, we might turn off the heater coil. If the temperature is below 250, we turn on the heater coil. If the temperature is between 250 and 255 degrees, we turn on a green

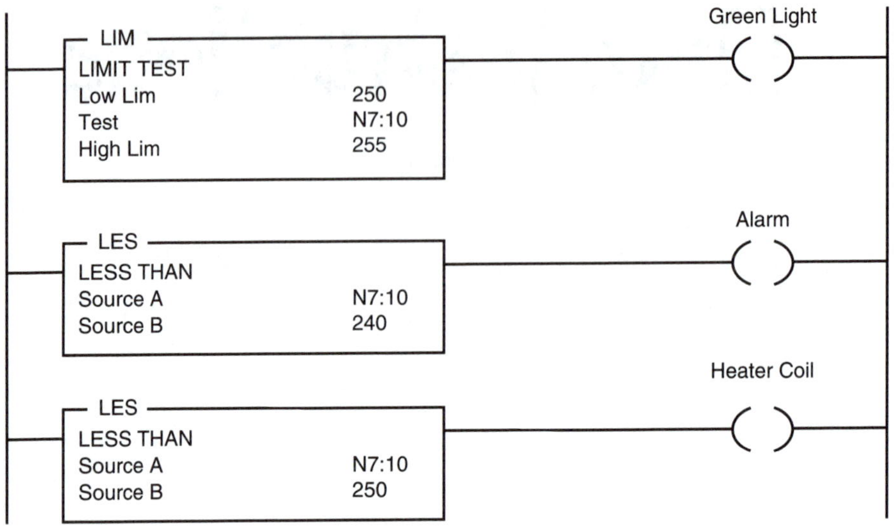

Figure 8–1 How comparison instructions can be used to program a simple application. In this application an integer (N7:10) contains the current temperature. The instructions are used to compare the current temperature (N7:10) to process limits. In the first rung, the green indicator lamp is on if the temperature is between 250 and 255 degrees. In the second rung, an alarm sounds if the temperature drops below 240 degrees. In the third rung, the heater coil is on if the temperature is below 250 degrees.

indicator lamp. If the temperature falls below 240 degrees, we might sound an alarm. (*Note:* Industrial temperature control is normally more complex than this example. Complex process control is covered in Chapters 12 and 13.)

This simple application requires the use of relational operators (arithmetic comparisons). The application involved tests of limits (250–255) and less than. The use of arithmetic statements makes this application easy to write. Many of the small PLCs do not have arithmetic instructions available. All of the larger PLCs offer a wide variety of arithmetic instructions.

Numbers often need to be manipulated. They need to be added, subtracted, multiplied, or divided. PLC instructions handle all these computations. For example, we may have a system to control the temperature of a furnace. In addition to controlling the temperature, we would like to convert the Celsius temperature to degrees Fahrenheit for display to the operator. If the temperature were to rise too high, an alarm would be triggered. Arithmetic instructions could do this very easily.

Next we will examine a few math instructions for various controllers. Many more are available. Check the programming manual for your brand of PLC for additional instructions.

GE Fanuc Functions

GE Fanuc Arithmetic Functions

GE Fanuc uses a function block for arithmetic instructions. Many arithmetic functions are available. Only a few are covered here. The allowed data types are shown with each description. INT stands for a signed integer, DINT stands for a double precision signed integer, BIT stands for bit, BYTE stands for byte, WORD stands for 16 bits of consecutive memory, BCD-4 is a 4-digit binary-coded decimal, and REAL stands for a floating-point number.

Add Function (INT, DINT, REAL) The GE Fanuc add instruction is shown in Figure 8–2. If input 1 (%I0001) becomes energized, it enables the add function. The value in I1 (the number in register %R0004) is then added to the number in I2 (the number in register %R0007). The result is put in register 3 (%R0003). If the add does not result in an overflow, the add function passes power and output 5 (%Q0005) is turned on.

Subtract Function (INT, DINT, REAL) The GE Fanuc subtract function is shown in Figure 8–3. If input 1 (%I0001) becomes energized, it enables the subtract function. The value in I2 (constant value 6 in this case) is then subtracted from the number in I1 (the number in register %R0005). The result is put in register 2 (%R0002). If the subtract does not result in an overflow, the subtract function passes power and output 5 (%Q0005) is turned on.

Multiply Function (INT, DINT, REAL) The GE Fanuc multiply function is shown in Figure 8–4. If input 1 (%I0001) becomes energized, it enables the multiply function. The value in I1 (the number in register %R0002) is then multiplied by the number in I2 (constant value of 3). The result is put in register 6 (%R0006). If

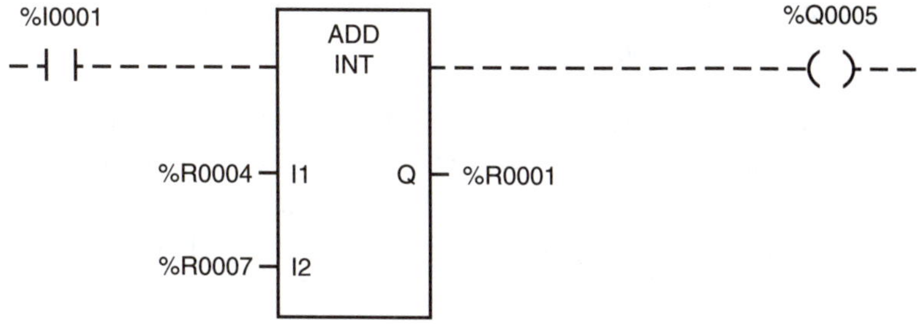

Figure 8–2 Example of the use of an add (ADD) function.

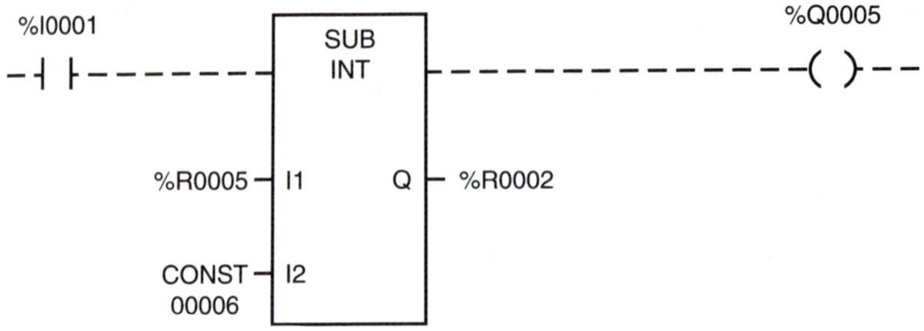

Figure 8–3 Example of the use of a subtract (SUB) function.

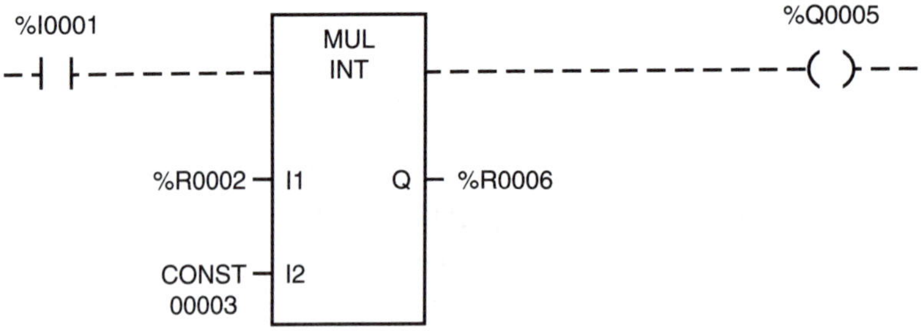

Figure 8–4 Example of the use of a multiply (MUL) function.

the multiply does not result in an overflow, the function passes power and output 5 (%Q0005) is turned on.

Divide Function (INT, DINT, REAL) This function divides one number by another, which yields a quotient. An example of the use of a DIV function is shown in Figure 8–5. The quotient is stored in Q:, in this case %R0006. The divide function passes power if the operation does not result in an overflow and if no attempt is made to divide by zero.

Square Root Function (INT, DINT, REAL) The square root function is used to find the square root of an input value. An example of the use of a square root function is shown in Figure 8–6. When the function receives power flow (%I0003), the function takes the square root of the value found in input IN (%R0002) and puts the result in Q (%R0004).

Modulo Function (INT, DINT) The modulo function is used to find the remainder from a division operation. An example of the use of a modulo function is

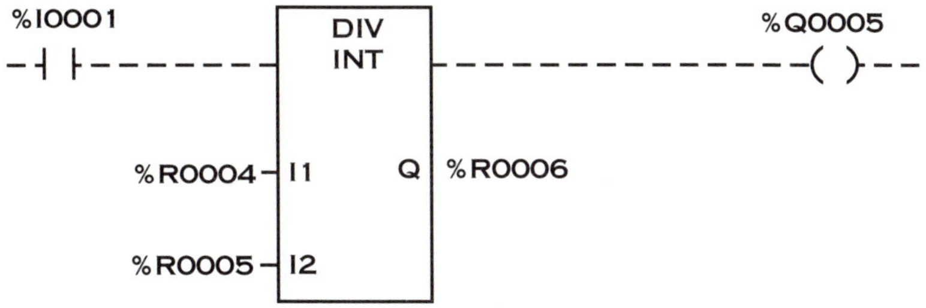

Figure 8–5 Example of the use of a divide (DIV) function.

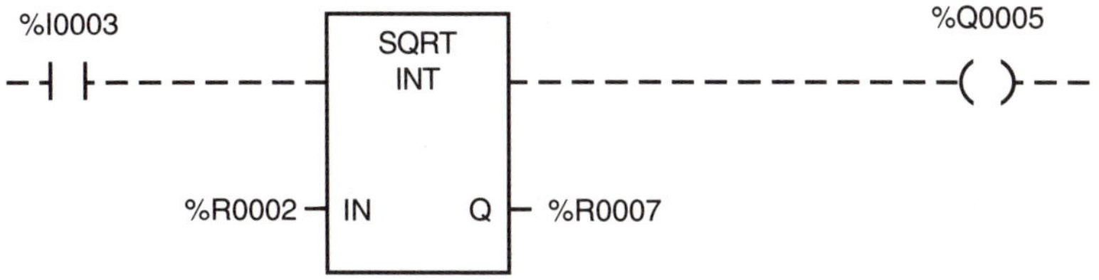

Figure 8–6 Example of the use of a square root (SQRT) function.

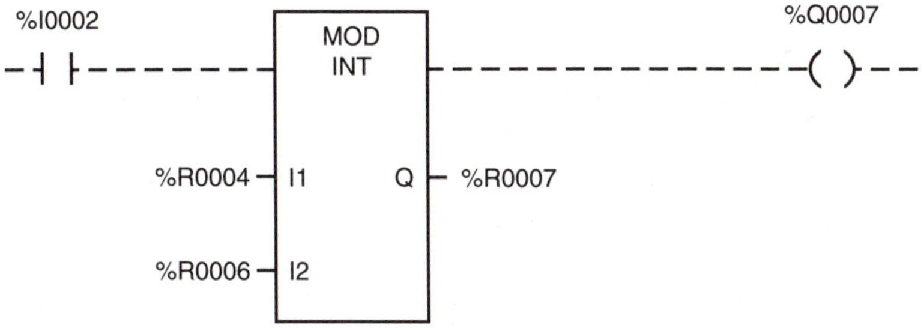

Figure 8–7 Example of the use of a modulo (MOD) function.

shown in Figure 8–7. When the function receives power flow (%I0002), the value found in input IN (%R0002) is converted to degrees and the result is placed in Q (%R0008).

Sine Function The sine function is used to find the trigonometric sine of the input value at IN. An example of the use of a SIN function is shown in Figure 8–8. If

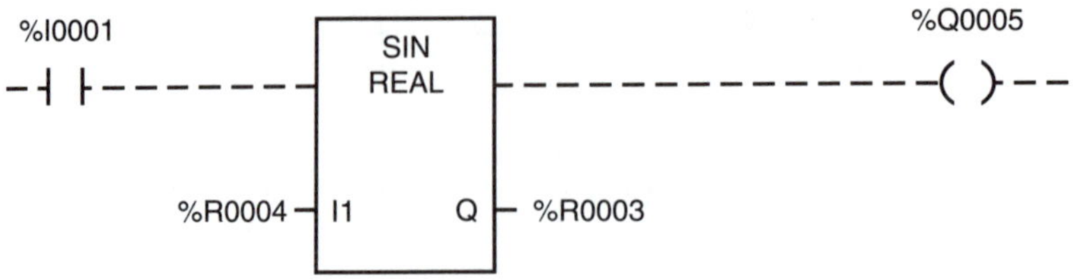

Figure 8–8 Example of the use of a sine (SIN) function.

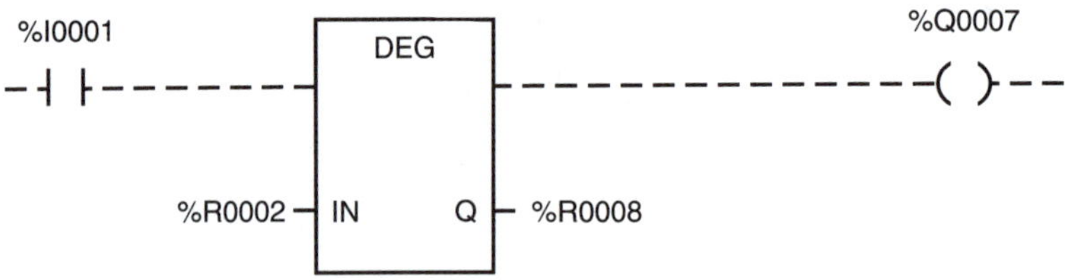

Figure 8–9 Example of the use of a degree (DEG) function.

input 1 (%I0001) is true, the sine function finds the trigonometric sine of the value held in input IN. In this case a register was used for the input value (%R0004).

Degree Function (REAL) The degree function is used to convert radians to degrees. An example of the use of a DEG function is shown in Figure 8–9. When the function receives power flow (%I0004), the radian value found in input IN (%R0002) is converted to degrees and the result is placed in Q (%R0008).

Figure 8–10 shows a list of arithmetic instructions for the GE Fanuc PLC. Figure 8–11 shows a list of trigonometric instructions for the GE Fanuc PLC.

Exponential Function (REAL) The GE Fanuc exponential function is shown in Figure 8–12. If input 1 (%I0001) becomes energized, it enables the exponential function. The value in I1 (the number in %AI001) is then raised to the power of the number in I2 (constant of 2.5 in this case). The result is put in register 1 (%R0001). If the exponential function does not result in an overflow, the function passes power and output 5 (%Q0005) is turned on, unless an invalid operation occurs and/or IN is not a number or is negative.

Instruction	Function	Description of Operation and Use
ADD	Addition	Adds two numbers. ADD functions pass power if the operation does not result in an overflow.
SUB	Subtraction	Subtracts one number from another. SUB functions pass power if the operation does not result in an overflow.
MUL	Multiplication	Multiplies two numbers. The MUL function passes power if the operation does not result in an overflow.
DIV	Division	Divides one number by another, yielding a quotient. The DIV function passes power if the operation does not result in an overflow and if no attempt is made to divide by zero.
MOD	Modulo division	Divides one number by another, yielding a remainder. The DIV function passes power if the operation does not result in an overflow and if no attempt is made to divide by zero.
SQRT	Square root	Finds the square root of an integer or real value. When the function receives power flow, the value of the output Q is set to the square root of the input IN.
LOG 10	Base 10 logarithm	When the function receives power flow, it finds the base 10 logarithm of the real value in input IN and places the result in output Q.
LN	Natural logarithm	When the function receives power flow, it finds the natural logarithm base (E) of the real value in input IN and places the result in output Q.
EXP	Power of E	When the function receives power flow, it finds the natural logarithm base (E) raised to the power specified by IN and places the result in Q.
EXPT	Power of X	When the function receives power flow, X is raised to the power specified by IN and places the result in Q.

Figure 8–10 Arithmetic instructions for the GE Fanuc PLC.

Instruction	Function	Description of Operation and Use
SIN	Sine	Finds the sine of the input. When the function receives power flow, it computes the sine of the value of IN, whose units are radians, and stores the result in output Q.
COS	Cosine	Finds the cosine of the input. When the function receives power flow, it computes the cosine of the value of IN, whose units are radians, and stores the result in output Q.
TAN	Tangent	Finds the tangent of the input. When the function receives power flow, it computes the tangent of the value of IN, whose units are radians, and stores the result in output Q.
ASIN	Inverse sine	Finds the inverse sine of the input. When the function receives power flow, it computes the inverse sine of the value of IN and stores the result in output Q.
ACOS	Inverse cosine	Finds the inverse cosine of the input. When the function receives power flow, it computes the inverse cosine of the value of IN and stores the result in output Q.
ATAN	Inverse tangent	Finds the inverse tangent of the input. When the function receives power flow, it computes the inverse tangent of the value of IN and stores the result in output Q.
DEG	Convert to degrees	When the function receives power flow, a RAD_TO_DEG conversion is performed on the real radian value of IN and the result is placed in output degree real value Q.
RAD	Convert to radians	When the function receives power flow, a DEG_TO_RAD conversion is performed on the real degree value in input IN and the result is placed in output real radian value Q.

Figure 8–11 Trigonometric instructions for the GE Fanuc PLC.

GE Fanuc Relational Functions

Equal to (EQ) Function (INT, DINT, REAL) GE Fanuc has many relational functions available (see Figure 8–13). The GE Fanuc EQUAL TO function is shown in Figure 8–14. If input 1 (%I0001) becomes energized, it enables the equal to function. The value in I1 (the number in register %R0004) is then added to the number in I2 (the number in register %R0007). The result is put in register 3

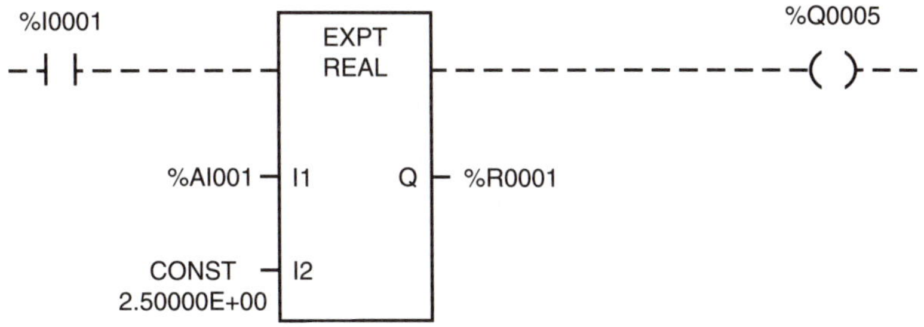

Figure 8–12 Example of the use of an exponential (EXPT) function.

INSTRUCTION	FUNCTION	DESCRIPTION OF OPERATION AND USE
EQ	EQUAL	THIS FUNCTION TESTS FOR EQUALITY BETWEEN TWO NUMBERS. THE EQ FUNCTION PASSES POWER IF THE TWO INPUTS ARE EQUAL.
NE	NOT EQUAL	THIS FUNCTION TESTS FOR INEQUALITY BETWEEN TWO NUMBERS. THE NE FUNCTION PASSES POWER IF THE TWO INPUTS ARE NOT EQUAL.
GT	GREATER THAN	THIS FUNCTION TESTS TO SEE IF ONE NUMBER IS GREATER THAN ANOTHER NUMBER. THE GT FUNCTION PASSES POWER IF THE FIRST NUMBER IS GREATER THAN THE SECOND.
GE	GREATER THAN OR EQUAL TO	THIS FUNCTION TESTS TO SEE IF ONE NUMBER IS GREATER THAN OR EQUAL TO ANOTHER NUMBER. THE GE FUNCTION PASSES POWER IF THE FIRST NUMBER IS GREATER THAN OR EQUAL TO THE SECOND.
LT	LESS THAN	THIS FUNCTION TESTS TO SEE IF ONE NUMBER IS LESS THAN ANOTHER NUMBER. THE LT FUNCTION PASSES POWER IF THE FIRST NUMBER IS LESS THAN THE SECOND.
LE	LESS THAN OR EQUAL TO	THIS FUNCTION TESTS TO SEE IF ONE NUMBER IS LESS THAN OR EQUAL TO ANOTHER NUMBER. THE LE FUNCTION PASSES POWER IF THE FIRST NUMBER IS LESS THAN OR EQUAL TO THE SECOND.
RANGE	RANGE	THIS FUNCTION TESTS THE INPUT VALUE AGAINST A RANGE OF TWO NUMBERS.

Figure 8–13 GE Fanuc comparison instructions.

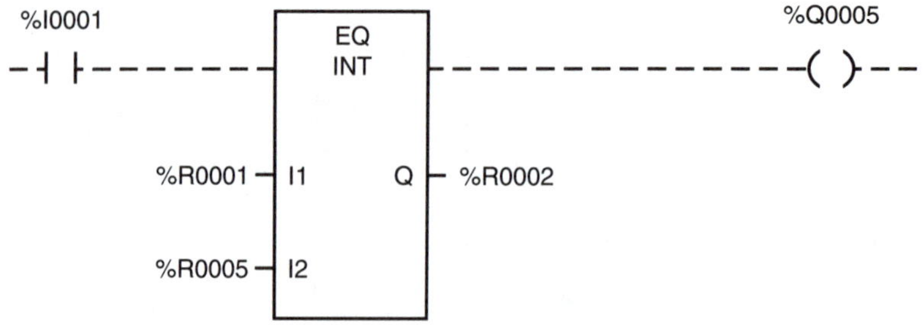

Figure 8–14 Example of the use of an equal to (EQ) function.

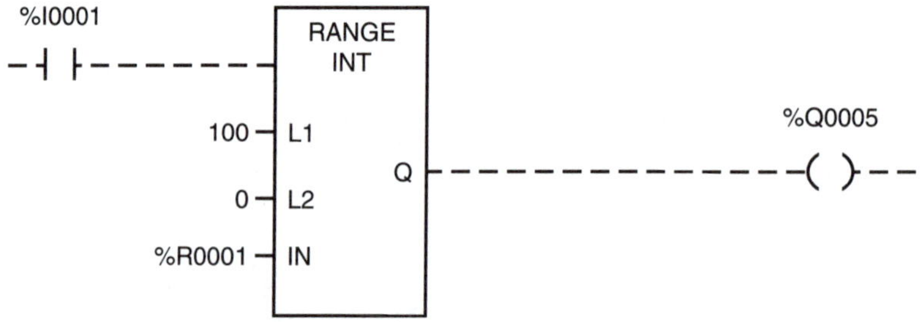

Figure 8–15 Example of the use of a range function.

(%R0003). If the add does not result in an overflow, the add function passes power and output 5 (%Q0005) is turned on.

Range Function (INT, DINT, WORD) The GE Fanuc range function is shown in Figure 8–15. If input 1 (%I0001) becomes energized, it enables the function. The range function compares the value in input parameter IN against the range specified by limits L1 and L2, inclusive. When IN is within the range specified by L1 and L2, output Q is turned on.

GE Fanuc Bit Functions

AND Function (WORD) GE Fanuc has many bit functions available (see Figure 8–16). The GE Fanuc AND function is shown in Figure 8–17. (The figure shows several logical bit instructions.) If input 1 (%I0001) becomes energized, it enables the function. If a bit string in I1 and the corresponding bit in bit string I2 are both 1, a 1 is placed in the corresponding location in output string Q (%R0001 in this

INSTRUCTION	FUNCTION	DESCRIPTION OF OPERATION AND USE
AND	LOGICAL AND	THIS INSTRUCTION IS A LOGICAL AND OF TWO BIT STRINGS.
OR	LOGICAL OR	THIS INSTRUCTION IS A LOGICAL OR OF TWO BIT STRINGS.
XOR	LOGICAL EXCLUSIVE OR	THIS INSTRUCTION IS A LOGICAL EXCLUSIVE OR OF TWO BIT STRINGS.
NOT	LOGICAL INVERT	THIS INSTRUCTION IS A LOGICAL INVERSION OF A BIT STRING.
SHL	SHIFT LEFT	THIS INSTRUCTION SHIFTS A BIT STRING LEFT.
SHR	SHIFT RIGHT	THIS INSTRUCTION SHIFTS A BIT STRING RIGHT.
ROL	ROTATE LEFT	THIS INSTRUCTION ROTATES A BIT STRING LEFT.
ROR	ROTATE RIGHT	THIS INSTRUCTION ROTATES A BIT STRING RIGHT.
BITTST	BIT TEST	THIS INSTRUCTION TESTS A BIT WITHIN A STRING.
BITSET	BIT SET	THIS INSTRUCTION SETS A BIT WITHIN A STRING TO TRUE.
BITCLR	BIT CLEAR	THIS INSTRUCTION TESTS A BIT WITHIN A STRING TO FALSE.
BITPOS	BIT POSITION	THIS INSTRUCTION LOCATES A BIT SET TO TRUE WITHIN A BIT STRING.
MASKCMP	MASKED COMPARE	THIS INSTRUCTION PERFORMS A MASKED COMPARE OF TWO ARRAYS.

Figure 8–16 GE Fanuc instructions.

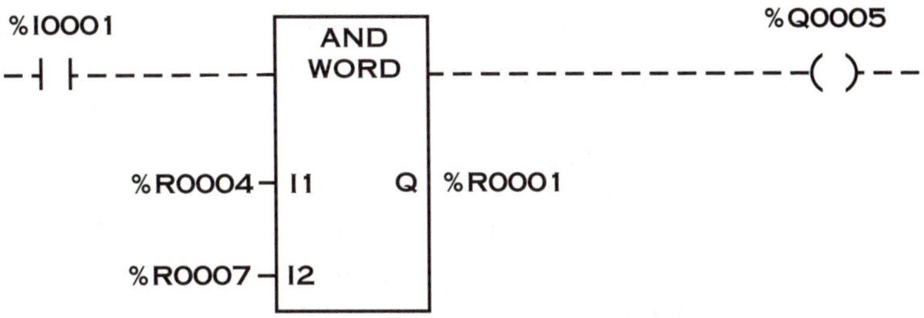

Figure 8–17 Example of the use of an AND function.

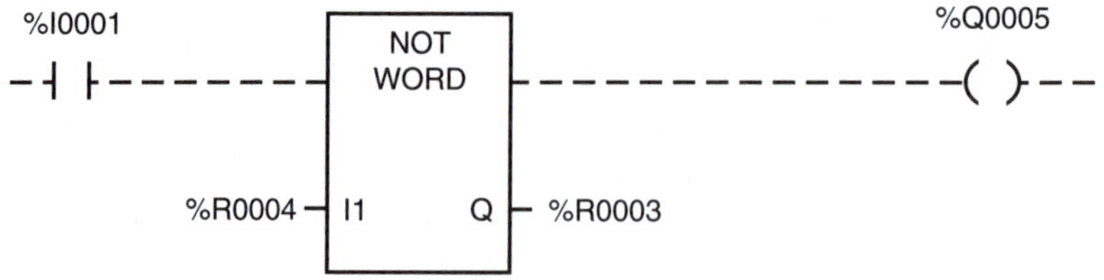

Figure 8–18 Example of the use of a not (NOT) function.

case). The function passes power and output 5 (%Q0005) is turned on whenever power is received by the function.

NOT Function (WORD) The GE Fanuc NOT function is shown in Figure 8–18. If input 1 (%I0001) becomes energized, it enables the function. The NOT function is used to set the state of each bit in the output bit string Q to the opposite of the state of the corresponding bit in bit string I. All bits are altered on each scan that power is received. Output string Q is the logical complement of I1. The function passes power whenever it receives power.

Shift Left Function (WORD) The GE Fanuc shift left function is shown in Figure 8–19. If input 1 (%I0001) becomes energized, it enables the function. The function is used to shift all the bits in a word or group of words to the left by a specified number of places. The specified number of bits is shifted out of the output string to the left. As the bits are shifted out of the high end of the string, the same number of bits are shifted into the low end. The function passes power whenever it receives power. A string length of 1 to 256 words can be used.

The number of bits to be shifted is placed in N. The bit to be shifted is located at B1. The bit to be shifted out is located at B2. The word to be shifted is located at IN. LEN is the length of the array. Output Q contains a shifted copy of the input string. If you want to shift the input string, you would use the same address for IN and Q.

Study Figure 8–19. If input %I0001 becomes true, the output string in %R0001 becomes a copy of IN (%R0005) left shifted by the number of bits specified in N (8 in this example). The status of the input bits is taken from input B1. The bits shifted out appear at B2.

Rotate Left Function (WORD) The GE Fanuc rotate left function is shown in Figure 8–20. If input 1 (%I0001) becomes energized, it enables the function. The function is used to rotate all the bits in a string to the left by a specified number of places. The specified number of bits is shifted out of the output string to the left and back into the string on the right. The function passes power unless the number of

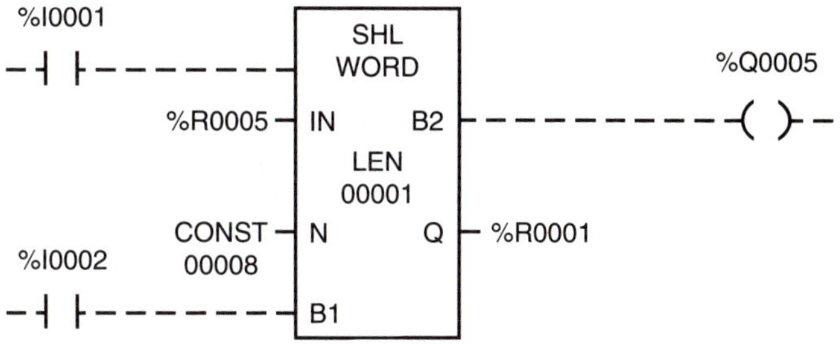

Figure 8–19 Example of the use of a shift left (SHL) function.

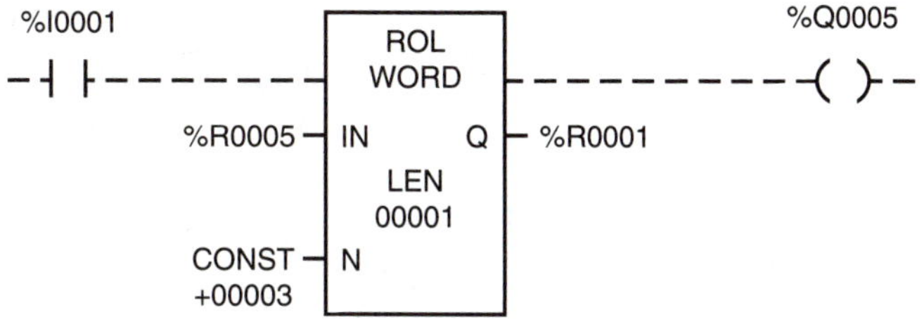

Figure 8–20 Example of the use of a rotate left (ROL) function.

bits specified to be rotated is greater than the total length of the string or less than zero. A string length of 1 to 256 words can be used.

GE Fanuc Data Move Functions

A wide variety of data move functions are available. They are shown in Figure 8–21.

Move Function (BIT, INT, WORD, REAL) The GE Fanuc move function is shown in Figure 8–22. If input 1 (%I0001) becomes energized, it enables the function. The function then moves 48 bits from memory location %M0001 to memory location %M0034. LEN specifies the number of words to be moved, in this case because move-word was used. Three words * 16 bits is equal to 48 bits. The data is copied in bit format so the new location does not have to be the same data type. The function passes power whenever it receives power.

INSTRUCTION	FUNCTION	DESCRIPTION OF OPERATION AND USE
MOVE	MOVE	COPY DATA AS INDIVIDUAL BITS AND MOVE TO ANOTHER LOCATION. THE MAXIMUM LENGTH IS 256 WORDS UNLESS THE MOVE_BIT FUNCTION IS USED. THE MOVE BIT LIMIT IS 256 BITS. DATA CAN BE MOVED TO A DIFFERENT DATA TYPE WITHOUT PRIOR CONVERSION.
BLKMOVE	BLOCK MOVE	COPIES A BLOCK OF SEVEN CONSTANTS TO A SPECIFIED MEMORY LOCATION. THE CONSTANTS ARE INPUT AS A PART OF THE FUNCTION.
BLKCLR	BLOCK CLEAR	REPLACES THE CONTENT OF A BLOCK OF DATA WITH ALL ZEROES. THIS CAN BE USED ON BIT AND WORD MEMORY. MAXIMUM LENGTH IS 256 WORDS.
SHFR	SHIFT REGISTER	SHIFTS ONE OR MORE DATA WORDS INTO A TABLE. THE MAXIMUM LENGTH IS 256 WORDS.
BITSEQ	BIT SEQUENCER	PERFORMS A BIT SEQUENCE SHIFT THROUGH AN ARRAY OF BITS. THE MAXIMUM LENGTH ALLOWED IS 256 WORDS.
COMMREQ	COMMUNICATIONS REQUEST	ALLOWS THE PROGRAM TO COMMUNICATE WITH AN INTELLIGENT MODULE.

Figure 8–21 GE Fanuc data move instructions.

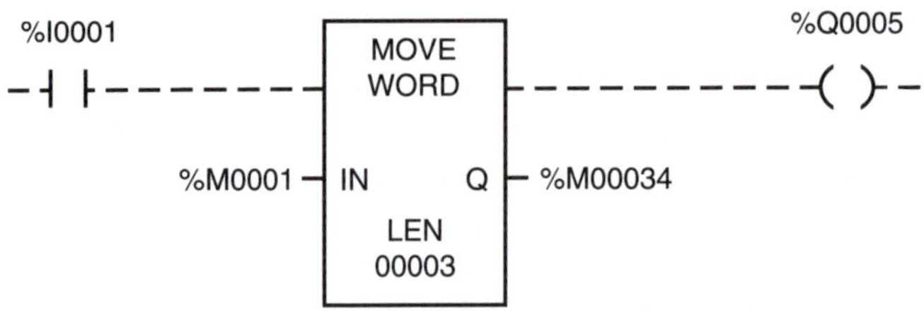

Figure 8–22 Example of the use of a move function.

GOULD MODICON ARITHMETIC INSTRUCTIONS

Gould Modicon uses a function block for arithmetic instructions. Many arithmetic instructions are available. Only a few are covered here.

Addition Instruction

The addition instruction is used to add two numbers and store the result to an address (see Figure 8–23). The addition instruction requires three entries from the

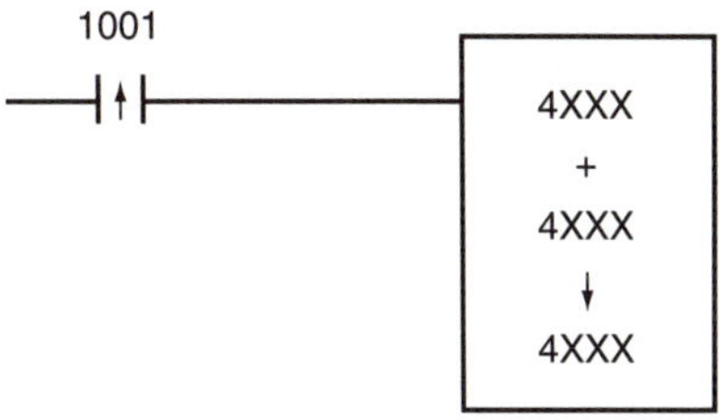

Figure 8–23 Gould Modicon addition instruction. The Xs represent digits.

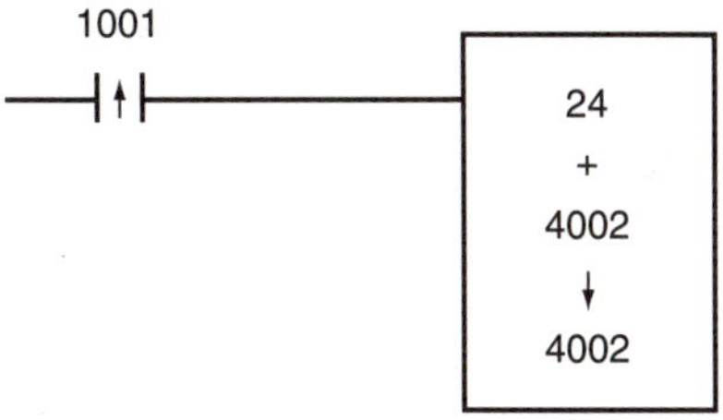

Figure 8–24 Gould Modicon addition instruction.

programmer. The top entry is added to the middle entry and the result is stored in the register shown by the bottom entry. The top two entries can be an actual number or a 4XXX or a 3XXX register. If actual numbers are used, they can be between 0 and 999. If registers are used for the top two entries, the instruction gets the numbers from those two addresses and adds them. The result is stored in the register shown by the bottom entry. The programmer must also supply an input condition. The instruction is executed every time the rung is scanned. The use of a transitional input can ensure that the instruction is executed only for transitions, instead of continuously if the input condition is true.

See Figure 8–24 for an example of the use of an addition instruction. Imagine that a PLC is controlling a bottling line. The manufacturer would like to count the number of bottles produced by the line. An addition instruction is used to keep track of the total number produced. Every time a case is full and it leaves the packing station, a sensor is triggered. The output from the sensor (input 1001) is used as an input coil to trigger the addition instruction. The addition instruction then adds 24 to the total count. The total count is held in register 4002. Note that one actual value and two addresses (both the same address) were used. The same address was used for both so that a running total of bottles can be kept. (If input 1 becomes true, 24 is added to the current total from address 4002 and then the new total is put back in 4002.)

Subtraction Instruction

The subtraction instruction is similar to the addition instruction. The subtraction instruction requires three values. The middle value is subtracted from the top value and

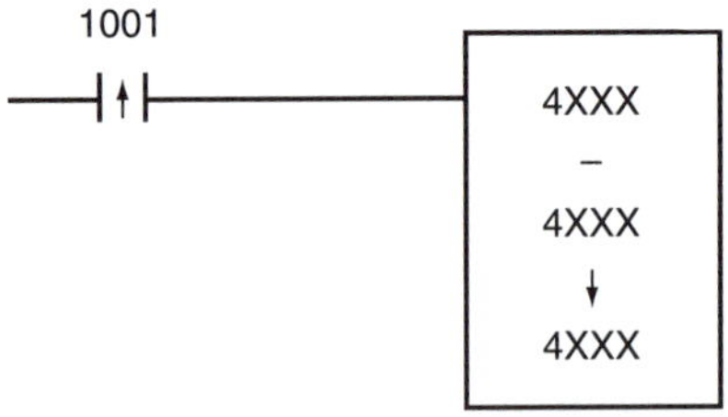

Figure 8–25 Gould subtraction instruction.

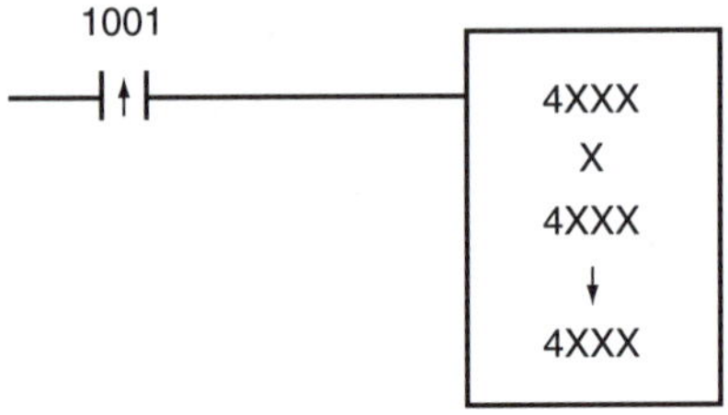

Figure 8–26 Gould multiplication instruction.

the result is stored in the register shown in the bottom entry (see Figure 8–25). The subtraction instruction is much more versatile than the addition instruction. It has three outputs available, and it can be used to compare the size of numbers. The top output is energized when the value of the top element is greater than the middle element. The middle value is energized when the top and middle elements are equal. The bottom output is energized when the top value is less than the middle value.

Multiplication Instruction

The multiplication instruction also uses a block format. Three entries are required. The top and middle entries are multiplied by each other and the result is placed in the register shown by the bottom entry. The top and middle entries can be an actual number or a register (see Figure 8–26). The third entry that the programmer makes is the register where the answer is to be stored. The Gould Modicon uses two addresses to store the result in case the number gets large. For example, if the programmer designated register 4002 for the bottom entry, the PLC would store the result in registers 4002 and 4003. The programmer should keep this in mind when using registers. Do not use the same register for two different purposes unless there is a valid reason.

Division Instruction

The division instruction also uses a block format (see Figure 8–27). There are three entries. The top and middle elements can be actual numbers or registers. The third

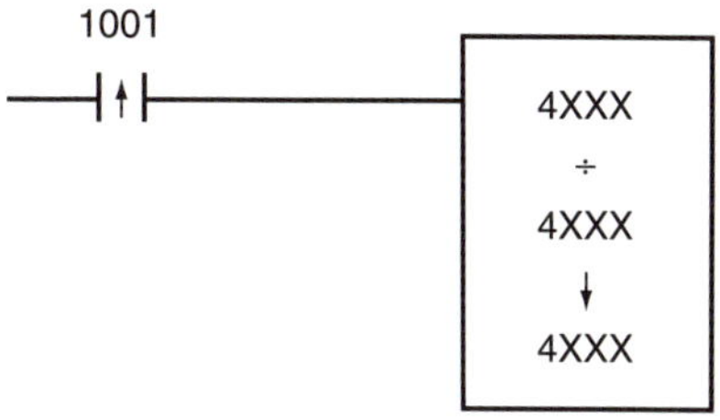

Figure 8–27 Gould divide instruction.

entry is where the answer is stored. The location must be a register. The division instruction divides the top number by the middle number. Note that a transitional input was used for the input to this instruction. This instruction uses two registers for the top entry. This means that if register 4002 were programmed for the top value, the number to be divided would be retrieved from registers 4002 and 4003. The output of the instruction is energized any time the result is too large for the storage register. The output can be used for error checking by the programmer.

Omron Arithmetic Instructions

Omron offers a wide variety of arithmetic instructions for binary and BCD numbers. For all these instructions, the programmer provides numbers, or sources of numbers, and the address where the result will be stored. Instructions are available for binary and BCD numbers. A few are described in the next few sections.

Omron C200H Memory Areas

Remember that Omron divides memory into areas (see Figure 8–28). There are several types of areas, each with a specific use.

Holding Relay Area The holding relay (HR) area is used to store and manipulate numbers. It retains the values even when modes are changed or if power fails. The address range for the HR area is 0000 to 9915. This memory area can be used by the programmer to store numbers that need to be retained.

Temporary Relay Area The temporary relay (TR) area is used for storing data at program branching points.

Auxiliary Relay Area The auxiliary relay (AR) area is used for internal data storage and manipulating data. A portion of this data area is reserved for system functions. The AR bits that can be written to by the user range from 0700 to 2215.

Figure 8–28 The memory areas for the Omron C200H series.

Omron Data Areas	
HR	Holding relay area
TR	Temporary relay area
AR	Auxiliary relay area
LR	Link relay area
TC	Timer/counter area
DM	Data memory area
IR	I/O and internal relay area
SR	Special relay area
UM	Program memory

Link Relay Area The link relay (LR) area is used for communications to other processors. If it is not needed for communications, it can be used for internal data storage and data manipulation. The address range for the LR area is 0000 to 6315.

Timer/Counter Area The timer/counter (TC) area is used to store timer and counter data. The timer counter area ranges from 000 to 511. Note that a number can be used only once. The same number cannot be used for a timer and a counter. For example, if the programmer uses the numbers 0 to 5 for timers, they cannot be used for counters.

Data Memory Area The data memory (DM) area is used for internal storage and manipulation of data. It must be accessed in 16-bit channel units. If a multiplication sign is used before the DM (*DM), indirect addressing is being used. Data memory ranges from addresses 0000 to 1999. The user can write only to addresses 0000 to 0999.

I/O and Internal Relay Area The internal relay (IR) area is used to store the status of inputs and outputs. Any of the bits that are not assigned to actual I/O can be used as work bits by the programmer. Channels 000 to 029 are allocated for I/O. The remaining addresses up to 24,615 are for the work area. The IR area is addressed in bit or channel units. The addresses are accessed in channels or channel/bit combinations. Channel/bit addresses are 5 bits long. The two least significant digits are the bit within the channel. For example, address 01007 would be channel 10, bit (or terminal) 7.

Special Relay Area The special relay (SR) area can be used to monitor the PLC's operation. It can also be used to generate clock pulses and to signal errors. For example, bit 25,400 is a 1-minute clock pulse. Bit 25,500 is a 0.1-second clock pulse. These special bits can be used by the programmer in ladder logic. Many bits are provided to signal errors that could occur. The programmer can use these bits in ladder logic to indicate problems.

Program Memory The program memory (UM) area is where the user's program instructions are stored. Memory is available in various sizes (RAM and ROM). These data areas are used simply by using the prefix (such as HR) followed by the appropriate address within that data area.

Omron Arithmetic Instructions

Binary Addition To program a binary add instruction (see Figure 8–29), the user chooses the ADB instruction (function 50). The programmer enters the function number to specify which instruction to use. The programmer then must supply the augend (Au) value, the addend (Ad) value, and the result (R) channel. If @ is used in front of the ADB (@ADB), the instruction is transitional. It will be active only when there is a transition on the input to it. The @ can be used with most of the instructions.

Figure 8–30 shows the data types that can be used in an add instruction. The # means that actual data are entered. A hex number would be entered.

The use of the ADB instruction in a ladder diagram is shown in Figure 8–31. This figure shows that an ADB (function 41) is used. If input 00000 comes true,

Figure 8–29 Binary add instruction.

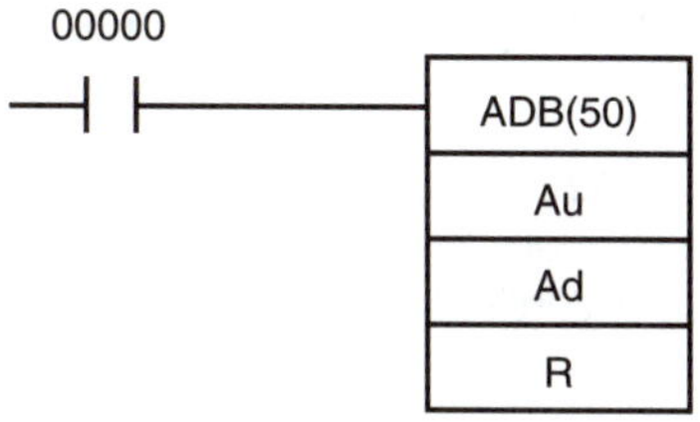

AU AND AD	R
IR,SR,HR,AR,LR,TC,DM,*DM,#	IR,HR,AR,LR,DM,*DM

Figure 8–30 Data types that can be used with a binary add instruction.

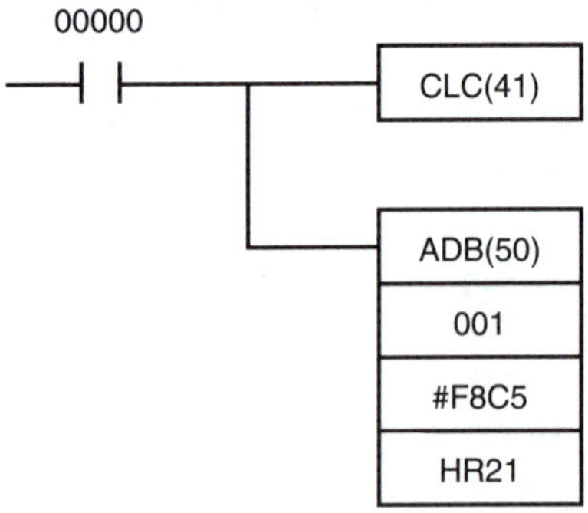

Figure 8–31 Use of the ADB (binary addition) instruction. Note the use of the clear carry instruction.

then the CLC (clear carry) instruction clears the carry bit. Input 00000 also causes the ADB instruction to execute. In this case the binary data from channel 1 is added to the hex number #F8C5. The result is stored in holding relay 21. The clear carry should be used to ensure that the addition is correct. The clear carry instruction should be used with any addition, subtraction, or shift instruction to ensure the correct result. Note that it can be programmed using the same input conditions as the actual add or subtract instruction. Figure 8–32 shows how the add instruction works. The Au and Ad are added, and the result is placed into the R address. In this case the binary data for Au comes from actual input channel 1. These data are added to the hex number F8C5. The result is FA4C. The result is stored in HR21 (holding relay 21).

Binary Subtraction A binary subtraction instruction is programmed just like an add instruction. The instruction's mnemonic is SBB (see Figure 8–33). If it is entered as @SBB, it will be active only when there is a transition on its input condition. The programmer enters a minuend (Mi), a subtrahend (Su), and an R address

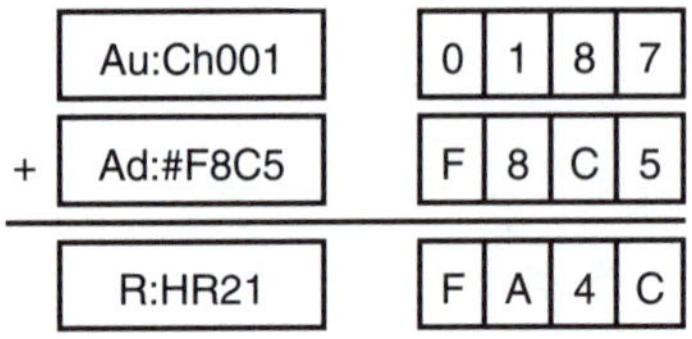

Figure 8–32 Results of a binary add instruction. The result is stored in address HR21.

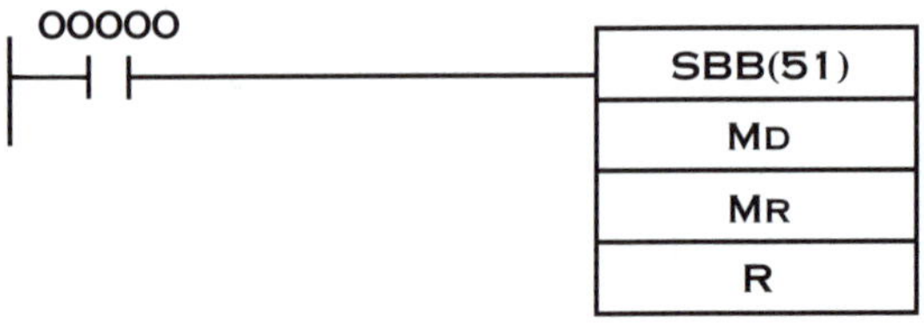

Figure 8–33 A binary subtract instruction.

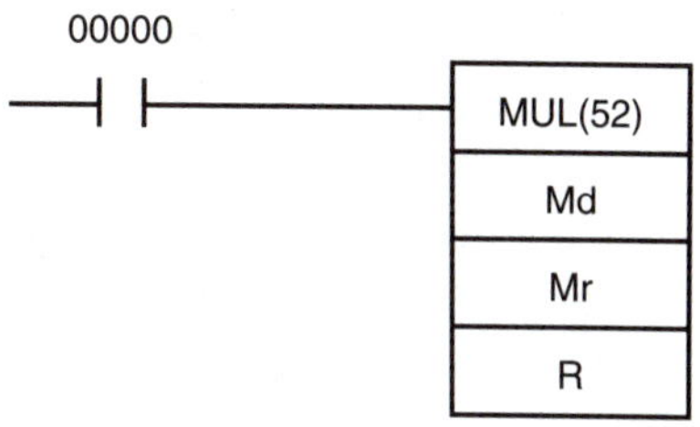

Figure 8–34 Binary multiply instruction.

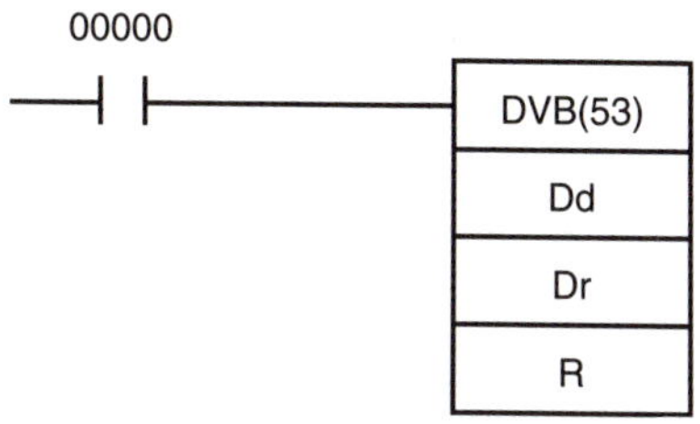

Figure 8–35 Binary divide instruction.

where the result will be stored. The same data types may be used for numbers to be subtracted as are used with the addition instruction. See Figure 8–30 for the possible types. The function code for this instruction is a 51. A BCD subtraction instruction is also available.

The subtrahend (Su) is subtracted from the minuend (Mi). The result is stored in the area specified by the R value. The same types of areas can be used for each of these, as was used for the addition instruction. Remember that a clear carry [CLC(41)] should be used to ensure that the correct answer is obtained.

Multiply Instructions The binary multiply (MLB) instruction can be used to multiply two numbers. The function number for this instruction is 52 (see Figure 8–34). The programmer supplies a multiplicand (Md) and a multiplier (Mr), and the result is stored in the area specified by R. The R specifies the first of two addresses that will be used to store the result of the multiplication. This instruction multiplies two 16-bit numbers together and stores a 32-bit answer at channel R and R+1. The same data types can be used for the numbers to be multiplied as for the addition instruction. A BCD multiply instruction is also available.

Binary Division The binary division [DVB (53)] instruction is programmed like the multiplication instruction. The programmer supplies a dividend (Dd), a divisor (Dr), and an R address. The R is the beginning address of the result. The quotient is stored in channel R, and the remainder of the division problem is stored in channel R+1. The same data types can be used as for the addition instruction. See Figure 8–35 for an example of the instruction.

AutomationDirect Arithmetic Instructions

Many arithmetic instructions are available for the programmer's use. There are binary and BCD instructions. Many of these instructions use the accumulator. The accumulator is a memory location used to store numbers temporarily. Think of it as a scratch pad that can be used to hold a number. Numbers can be "loaded" into it, or the number in the accumulator can be stored to a different memory location. For example, the states of inputs can be loaded into the accumulator. They can then be manipulated and sent to actual outputs. There are many uses for the accumulator (see Figures 8–36 and 8–37).

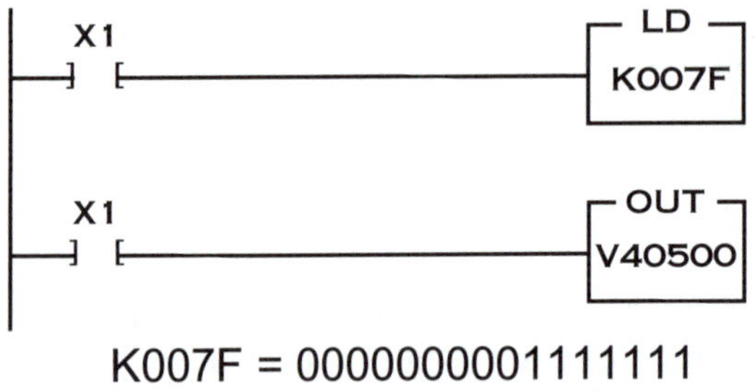

Figure 8–36 Use of a load instruction. If input X1 becomes true, the load instruction loads the accumulator with the hex number 007F. The K preceding the number stands for "constant." If X4 becomes true, the number in the accumulator (007F) is output to variable V40500. Variable V40500 is the word that controls the states of the first sixteen outputs. The number below the ladder logic shows that the first seven outputs are energized.

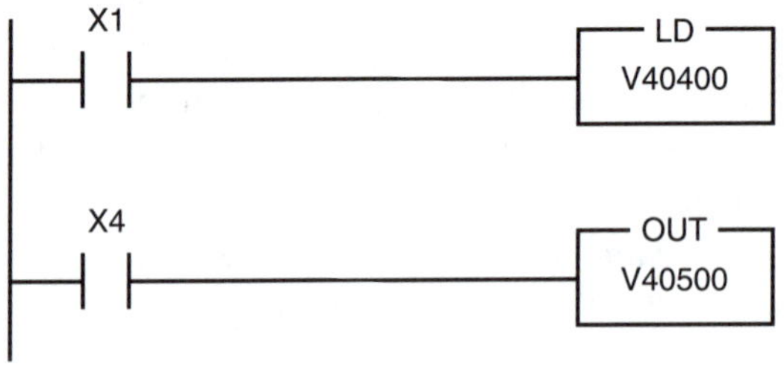

Figure 8–37 Use of a load instruction. If input X1 becomes true, the first sixteen input states are loaded. Variable V40400 is a word that contains the states of the first sixteen inputs. If input X4 becomes true, the number (word) held in the accumulator is output to variable V40500. Variable V40500 is the word that controls the states of the first sixteen outputs.

AutomationDirect arithmetic instructions utilize the accumulator. The use of an add instruction is shown in Figure 8–38. When input X5 becomes true, the hex number 1324 is added to whatever number is in the accumulator. The number that was in the accumulator is replaced by the result. If we want to add the 5 and 7, we would first use a load instruction to load 5 into the accumulator. We would then use an add instruction with a value of 7. The result of the instruction would be placed into the accumulator. We can use variables or constants with these instructions. In fact, the actual states of inputs, outputs, timers, and counters are all stored in variables. This means that we can utilize this data in instructions also. Study Figures 8–39 through 8–41 for examples of other arithmetic instructions.

Figure 8–38 Use of a add instruction. If input X5 becomes true, the hex number 1324 is added to the value in the accumulator. The result is then stored back in the accumulator. The value that was in the accumulator is lost when the new value is stored there.

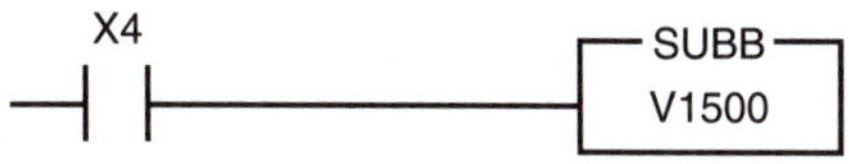

Figure 8–39 Use of a subtract instruction. If input X4 becomes true, the number in variable V1500 is subtracted from the number in the accumulator. The result is stored in the accumulator.

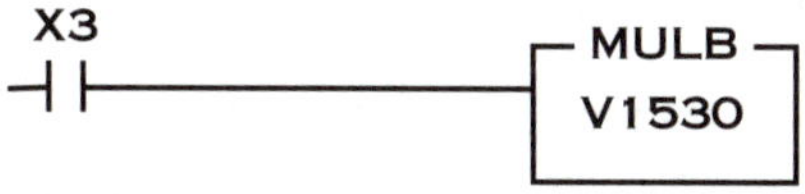

Figure 8–40 The use of a multiply instruction. If input X3 becomes true, the number in variable V1530 is multiplied by the number in the accumulator. The result is stored in the accumulator.

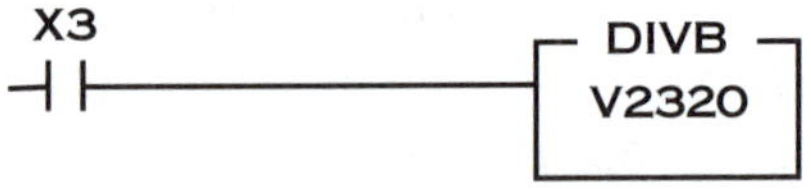

Figure 8–41 Use of a division instruction. If input X3 becomes true, the number in the accumulator is divided by the number in variable V2320. The result is stored in the accumulator.

Rockwell Automation PLC-5, SLC 500, and Micrologix 1000 Arithmetic Instructions

A multitude of arithmetic instructions are available in the PLC-5 and SLC-500 controllers. A few of them will be covered in this section. Figure 8–42 shows the arithmetic functions that are available. A few of the more common instructions are covered next.

Add The add (ADD) instruction is used to add two values together (source A + source B). The result is put in the destination address. Figure 8–43 shows an example of an add instruction. If contact I:5/2 is true, the add instruction adds the number from source A (N7:3) and the value from source B (N7:4). The result is stored in the destination address (N7:20).

Subtract The subtract (SUB) instruction is used to subtract two values. The subtract instruction subtracts source B from source A. The result is stored in the destination address. Figure 8–44 shows the use of a subtract instruction (SUB). If contact I:5/2 is true, the SUB instruction is executed. Source B is subtracted from source A; the result is stored in destination address N7:20.

Multiply The multiply (MUL) instruction is used to multiply two values. The first value, source A, is multiplied by the second value, source B. The result is stored in the destination address. Source A and source B can be either values or addresses of values. Figure 8–45 shows the use of a multiply instruction. If contact I:5/2 is true, source A (N7:3) is multiplied by source B (N7:4) and the result is stored in destination address N7:20.

Divide The divide (DIV) instruction is used to divide two values. Source A is divided by source B and the result is placed in the destination address. The sources can be values or addresses of values. Figure 8–46 shows the use of a divide instruction. If contact I:5/2 is true, the divide instruction divides the value from source A (N7:3) by the value from source B (N7:4). The result is stored in destination address N7:20.

INSTRUCTION	FUNCTION	DESCRIPTION OF OPERATION AND USE
ADD	ADD	ADDS SOURCE A TO SOURCE B AND STORES THE RESULT IN THE DESTINATION.
SUB	SUBTRACT	SUBTRACTS SOURCE B FROM SOURCE A AND STORES THE RESULT IN THE DESTINATION.
MUL	MULTIPLY	MULTIPLIES SOURCE A BY SOURCE B AND STORES THE RESULT IN THE DESTINATION.
DIV	DIVIDE	DIVIDES SOURCE A BY SOURCE B AND STORES THE RESULT IN THE DESTINATION AND THE MATH REGISTER.
DDV	DOUBLE DIVIDE	DIVIDES THE CONTENTS OF THE MATH REGISTER BY THE SOURCE AND STORES THE RESULT IN THE DESTINATION AND THE MATH REGISTER.
CLR	CLEAR	SETS ALL BITS OF A WORD TO ZERO.
SQR	SQUARE ROOT	CALCULATES THE SQUARE ROOT OF THE SOURCE AND PLACES THE INTEGER RESULT IN THE DESTINATION.
SCP	SCALE WITH PARAMETERS	PRODUCES A SCALED OUTPUT VALUE THAT HAS A LINEAR RELATIONSHIP BETWEEN THE INPUT AND SCALED VALUES.
SCL	SCALE DATA	MULTIPLIES THE SOURCE BY A SPECIFIED RATE, ADDS TO AN OFFSET VALUE AND STORES THE RESULT IN THE DESTINATION.
ABS	ABSOLUTE	CALCULATES THE ABSOLUTE VALUE OF THE SOURCE AND PLACES THE RESULT IN THE DESTINATION.
CPT	COMPUTE	EVALUATES AN EXPRESSION AND STORES THE RESULT IN THE DESTINATION.
SWP	SWAP	SWAPS THE LOW AND HIGH BYTES OF A SPECIFIED NUMBER OF WORDS IN A BIT, INTEGER, ASCII, OR STRING FILE.
ASN	ARC SINE	TAKES THE ARC SINE OF A NUMBER (SOURCE IN RADIANS) AND STORES THE RESULT (IN RADIANS) IN THE DESTINATION.
ACS	ARC COSINE	TAKES THE ARC COSINE OF A NUMBER (SOURCE IN RADIANS) AND STORES THE RESULT (IN RADIANS) IN THE DESTINATION.
ATN	ARC TANGENT	TAKES THE ARC TANGENT OF A NUMBER (SOURCE) AND STORES THE RESULT (IN RADIANS) IN THE DESTINATION.
COS	COSINE	TAKES THE COSINE OF A NUMBER (SOURCE IN RADIANS) AND STORES THE RESULT IN THE DESTINATION.
LN	NATURAL LOG	TAKES THE NATURAL LOG OF THE VALUE IN THE SOURCE AND STORES THE RESULT IN THE DESTINATION.
LOG	LOG TO THE BASE 10	TAKES THE LOG BASE 10 OF THE VALUE IN THE SOURCE AND STORES THE RESULT IN THE DESTINATION.
SIN	SINE	TAKES SINE OF A NUMBER (SOURCE IN RADIANS) AND STORES THE RESULT IN THE DESTINATION.
TAN	TANGENT	TAKES THE TANGENT OF A NUMBER (SOURCE IN RADIANS) AND STORES THE RESULT IN THE DESTINATION.
XPY	X TO THE POWER OF Y	RAISES A VALUE TO A POWER AND STORES THE RESULT IN THE DESTINATION.

Figure 8–42 Some of the Rockwell Automation math instructions.

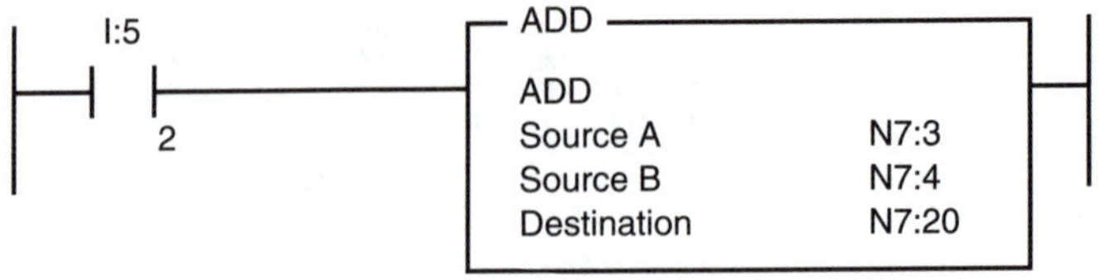

Figure 8–43 Use of an add instruction.

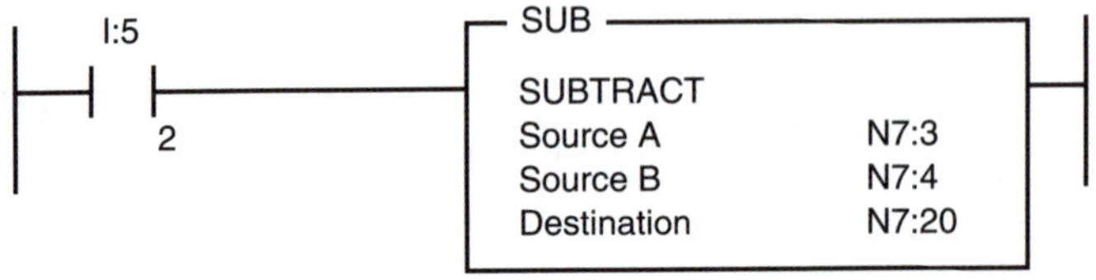

Figure 8–44 Use of a subtract (SUB) instruction.

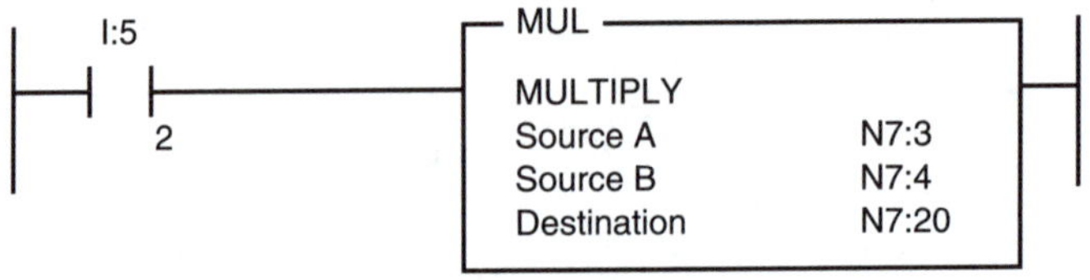

Figure 8–45 Use of a multiply (MUL) instruction.

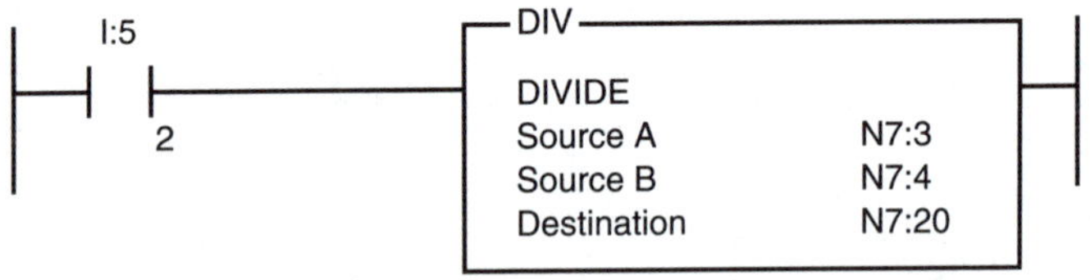

Figure 8–46 Use of a divide (DIV) instruction.

Negate The negate (NEG) instruction is used to change the sign of a value. If it is used on a positive number, it makes it a negative number. If it is used on a negative number, it will change it to a positive number. Remember that this instruction executes every time the rung is true. Use transitional contacts if needed. The use of

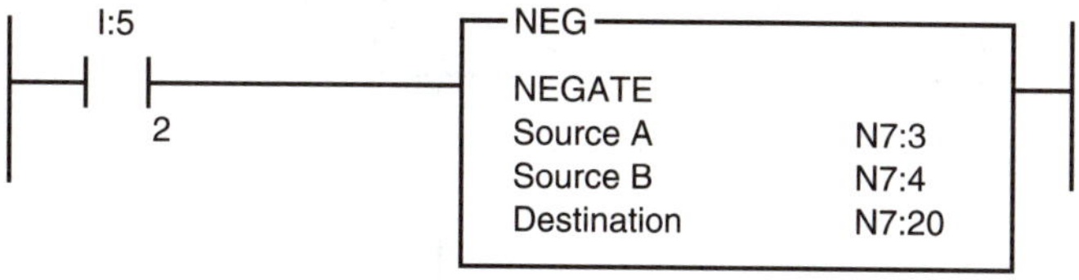

Figure 8–47 Use of a negate (NEG) instruction.

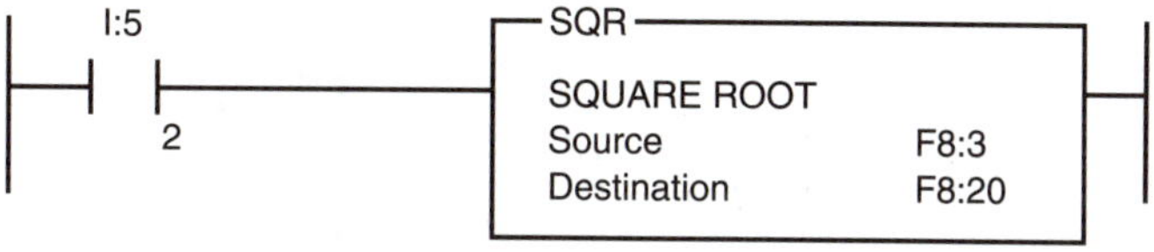

Figure 8–48 Use of a square root (SQR) instruction.

a negate instruction is shown in Figure 8–47. If contact I:5/2 is true, the value in source A (N7:3) is given the opposite sign and stored in destination address N7:20.

Square Root The square root instruction is used to find the square root of a value. The result is stored in a destination address. The source can be a value or the address of a value. Figure 8–48 shows the use of a square root instruction. If contact I:5/2 is true, the SQR instruction finds the square root of the value of the number found at the source address F8:3. The result is stored at destination address F8:20.

Compute The compute (CPT) instruction can be used to copy arithmetic, logical, and number conversion operations. The operations to be performed are defined by the user in the expression and the result is written in the destination. Operations are performed in a prescribed order. Operations of equal order are performed left to right. Figure 8–49 shows the order in which operations are performed. The programmer can override precedence order by using parentheses.

Figure 8–50 shows the use of a CPT instruction. The mathematical operations are performed when contact I:5/2 is true. In this case two floating point numbers (F8:1 and F8:2) are added to each other. Then the square root is calculated and stored in the destination address (F8:10).

Relational Operators

Figure 8–51 shows Rockwell Automation relational instructions.

ORDER	OPERATION	DESCRIPTION
1	**	EXPONENTIAL (X TO THE POWER OF Y)
2	-	NEGATE
	NOT	BITWISE COMPLEMENT
3	*	MULTIPLY
	/	DIVIDE
4	+	ADD
	-	SUBTRACT
5	AND	BITWISE AND
6	XOR	BITWISE EXCLUSIVE OR
7	OR	BITWISE OR

Figure 8–49 Precedence (order) in which math operations are performed. When precedence is equal, the operations are performed left to right. Parentheses can be used to override the order.

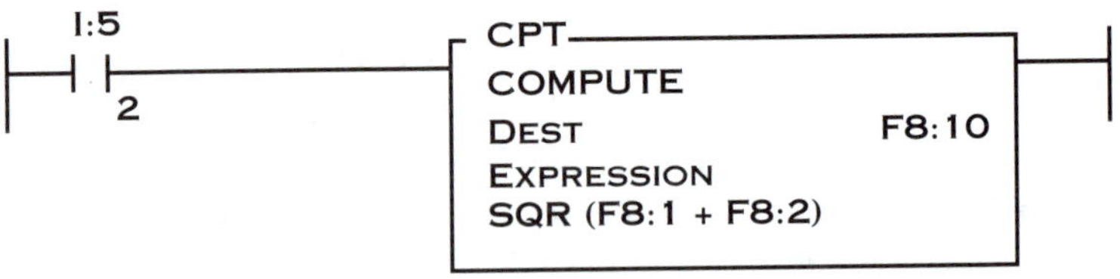

Figure 8–50 Use of a compute (CPT) instruction.

Equal To The equal to (EQU) instruction is used to test if two values are equal. The values tested can be actual values or addresses that contain values. An example is shown in Figure 8–52. Source A is compared to source B to see if they are equal. If the value in N:7:5 is equal to the value in N7:10, the rung is true and output O:5/01 is turned on.

Greater Than or Equal To The greater than or equal to (GEQ) instruction is used to test two sources to determine whether or not source A is greater than or equal to the second source. The use of a GEQ instruction is shown in Figure 8–53. If the value of source A (N7:5) is greater than or equal to source B (N7:10), output O:5/01 is turned on.

INSTRUCTION	FUNCTION	DESCRIPTION OF OPERATION AND USE
EQU	EQUAL	THE EQUAL INSTRUCTION IS USED TO TEST WHETHER TWO VALUES ARE EQUAL.
NEQ	NOT EQUAL	THE EQUAL INSTRUCTION IS USED TO TEST WHETHER TWO VALUES ARE NOT EQUAL.
LES	LESS THAN	THE LES INSTRUCTION IS USED TO TEST WHETHER ONE VALUE (SOURCE A) IS LESS THAN ANOTHER (SOURCE B).
LEQ	LESS THAN OR EQUAL TO	THE LEQ INSTRUCTION IS USED TO TEST WHETHER ONE VALUE (SOURCE A) IS LESS THAN OR EQUAL TO ANOTHER (SOURCE B).
GRT	GREATER THAN	THE GRT INSTRUCTION IS USED TO TEST WHETHER ONE VALUE (SOURCE A) IS GREATER THAN ANOTHER (SOURCE B).
GEQ	GREATER THAN OR EQUAL TO	THE GEQ INSTRUCTION IS USED TO TEST WHETHER ONE VALUE (SOURCE A) IS LESS THAN OR EQUAL TO ANOTHER (SOURCE B).
MEQ	MASKED COMPARISON FOR EQUAL	THE MEQ INSTRUCTION IS USED TO COMPARE DATA AT A SOURCE ADDRESS WITH DATA AT A COMPARE ADDRESS.
LIM	LIMIT TEST	THE LIMIT INSTRUCTION IS USED TO TEST FOR VALUES WITHIN OR OUTSIDE A SPECIFIED RANGE.

Figure 8–51 Rockwell Automation relational operators.

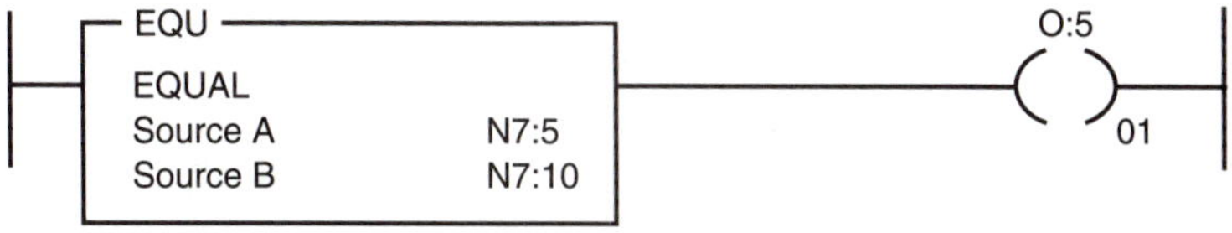

Figure 8–52 An equal to instruction (EQU).

Greater Than The greater than (GRT) instruction is used to see if a value from one source is greater than the value from a second source. An example of the instruction is shown in Figure 8–54. If the value of source A (N7:5) is greater than the value of source B (N7:10), output O:5/01 is set (turned on).

Less Than The less than (LES) instruction is used to see if a value from one source is less than the value from a second source. An example of the instruction is

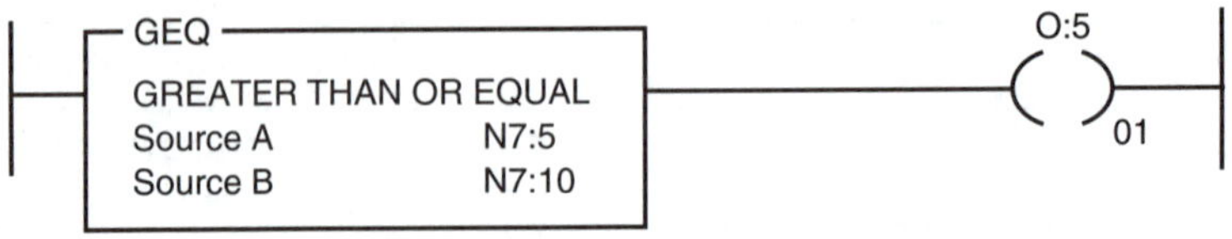

Figure 8–53 Use of a greater than or equal to (GEQ) instruction.

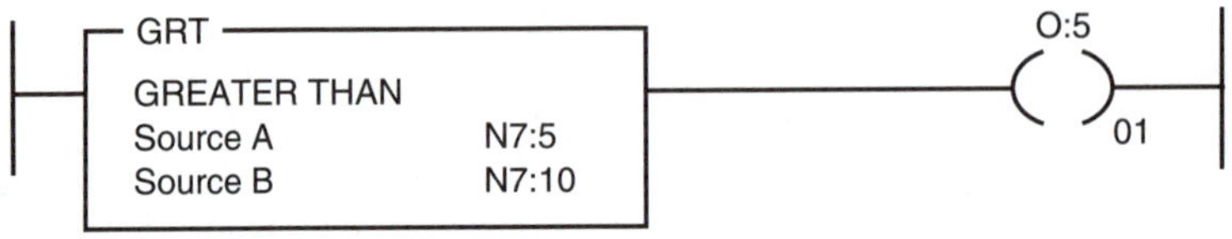

Figure 8–54 Use of a greater than (GRT) instruction.

shown in Figure 8–55. If the value of source A (N7:5) is less than the value of source B (N7:10), output O:5/01 is set (turned on).

Limit The limit (LIM) instruction is used to test a value to see if it falls within a specified range of values. The instruction is true when the tested value is within the limits. This instruction can be used, for example, to see if the temperature of an oven was within the desired temperature range. In this case the instruction would be testing to see if an analog value (a number in memory representing the actual analog temperature) is within certain desired limits.

The programmer must provide three pieces of data to the LIM instruction when programming. The programmer must provide a low limit. The low limit can be a constant or an address that contains the desired value. The address contains an integer or floating-point value (16 bits). The programmer must also provide a test value. This is a constant or the address of a value that is to be tested. If the test value is within the range specified, the rung is true. The third value the programmer must provide is the high limit. The high limit can be a constant or the address of a value.

Figure 8–56 shows the use of a limit (LIM) instruction. If the value in N7:15 is greater than or equal to the lower limit value (N7:10) and less than or equal to the high limit (N7:20), the rung is true and output O:5/01 is turned on.

Not Equal To The not equal to (NEQ) instruction is used to test two values for inequality. The values tested can be constants or addresses that contain values. An example is shown in Figure 8–57. If source A (N7:3) is not equal to source B (N7:4), the instruction is true and output O:5/01 is turned on.

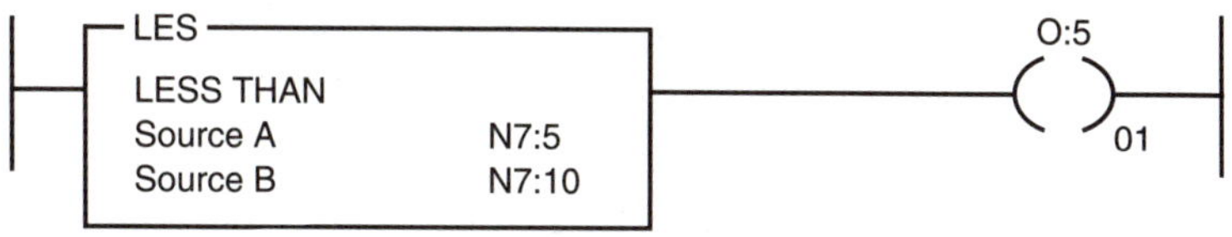

Figure 8–55 Use of a less than (LES) instruction.

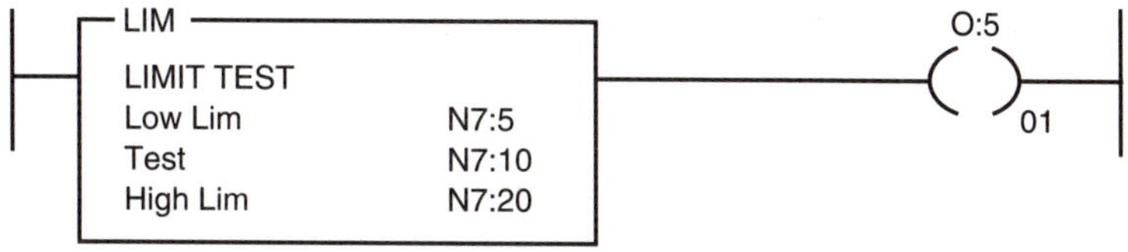

Figure 8–56 Use of a limit test (LIM) instruction.

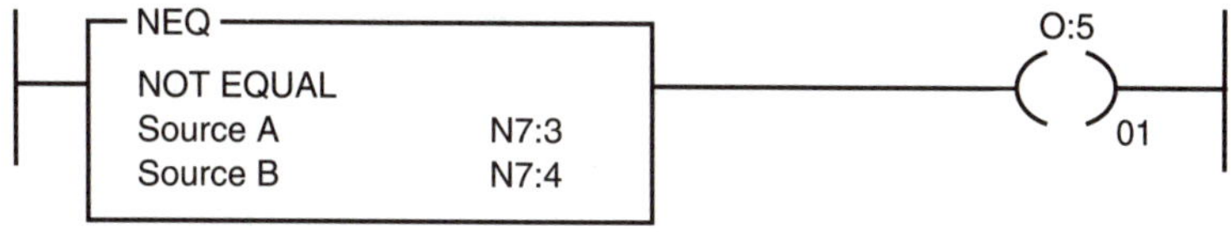

Figure 8–57 Use of a not equal to (NEQ) instruction.

Logical Operators

Several logical instructions are available. They can be useful to the innovative programmer. They can be used, for example, to check the status of certain inputs while ignoring others.

AND Several logical operator instructions are available. The AND instruction is used to perform an AND operation using the bits from two source addresses. The bits are ANDed and a result occurs. See Figure 8–58 for a chart that shows the results of the four possible combinations. An AND instruction needs two sources (numbers) to work with. These two sources are ANDed and the result is stored in a third address (see Figure 8–59). Figure 8–60 shows an example of an AND instruction. Address N7:3 and address N7:4 are ANDed. The result is placed in destination address N7:5. Examine the bits in the source addresses so that you can understand how the AND produced the result in the destination.

SOURCE A	SOURCE B	RESULT
0	0	0
1	0	0
0	1	0
1	1	1

Figure 8–58 Results of an AND operation on the four possible bit combinations.

Source A	N7:3	0	0	0	0	0	0	0	0	1	0	1	0	1	0	1	0
Source B	N7:4	0	0	0	0	0	0	0	0	1	1	1	0	1	0	1	1
Destination	N7:5	0	0	0	0	0	0	0	0	1	0	1	0	1	0	1	0

Figure 8–59 Result of an AND on two source addresses. The ANDed result is stored in address N7:5.

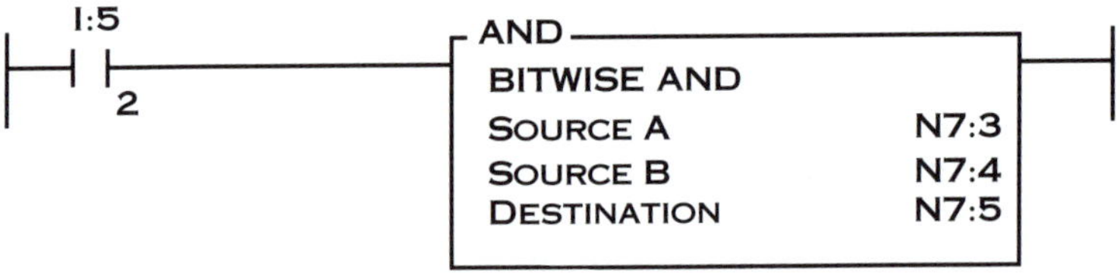

Figure 8–60 Use of an AND instruction. If input I:5/2 is true, the AND instruction executes. The number in address N7:3 is ANDed with the value in address N7:4. The result of the AND is stored in address N7:5.

NOT NOT instructions are used to invert the status of bits. A 1 is made a 0 and a 0 is made a 1. See Figures 8–61, 8–62, and 8–63.

OR Bitwise OR instructions are used to compare the bits of two numbers. See Figures 8–64, 8–65, and 8–66 for examples of how the instruction functions.

Figure 8–61 Results of a NOT instruction on bit states.

Source	Result
0	1
1	0

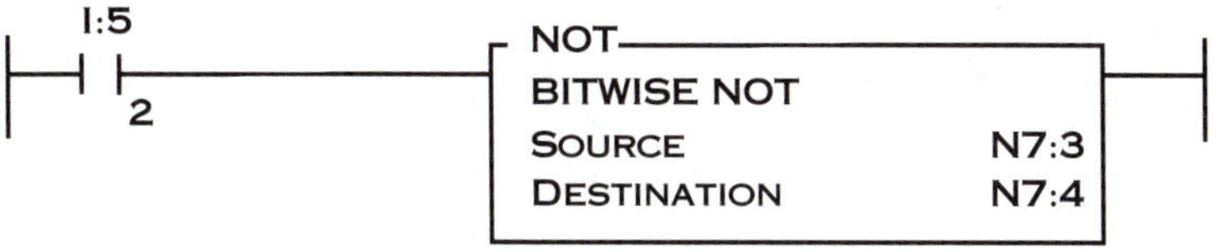

Figure 8–62 Use of a NOT instruction. If input I:5/02 is true, the NOT instruction executes. The number in address N7:3 is NOTed (one's complemented). The result is stored in destination address N7:4.

Source A	N7:3	0	0	0	0	0	0	0	0	1	0	1	0	1	0	1	0
Destination	N7:4	0	0	0	0	0	0	0	0	0	1	0	1	0	1	0	1

Figure 8–63 What would happen to the number 000000001010101010 if a NOT instruction were executed? The result is shown in destination address N7:4.

Figure 8–64 Result of an OR instruction on bit states.

SOURCE A	SOURCE B	RESULT
0	0	0
1	0	1
0	1	1
1	1	1

Scale with Parameters (SCP) Instruction

Another type of scale instruction is the SCP instruction. In the example, an SCP will be used to scale an input value so that an actual temperature can be displayed. In industrial control, many applications measure a physical property such as temperature. A temperature such as 72 degrees is measured by a thermocouple, which converts the temperature to a voltage. The voltage is received by an analog input module and

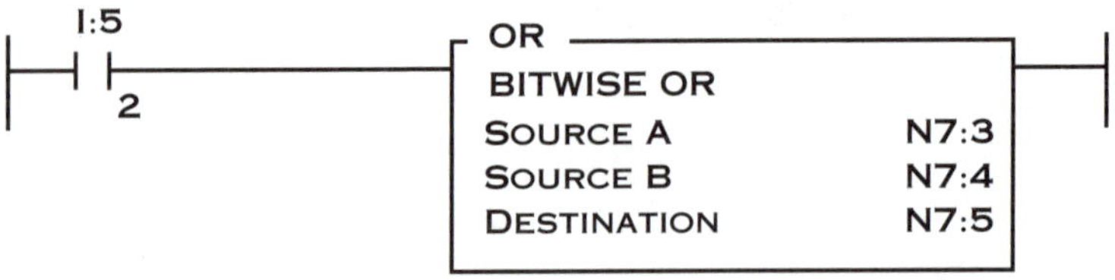

Figure 8–65 How an OR instruction can be used in a ladder diagram. If input I:5/2 is true, the OR instruction executes. Source A (N7:3) is ORed with source B (N7:4). The result is stored in the destination address.

Source A	N7:3	0	0	0	0	0	0	0	0	1	0	1	0	1	0	1	0
Source B	N7:4	0	0	0	0	0	0	0	0	1	1	1	0	1	0	1	1
Destination	N7:5	0	0	0	0	0	0	0	0	1	0	1	0	1	0	1	1

Figure 8–66 Result of an OR instruction on the numbers in addresses N7:3 and N7:4. The result of the OR is shown in the destination address N7:5.

converted to a bit count. For example, a 14-bit module converts an analog input to a number between 0 and 16,383. The user would rather see the actual temperature. A scale instruction can be used to convert the input bit count to a temperature.

The SCP instruction is shown in Figure 8–67. Imagine a system that measures a temperature between 32 and 21 degrees Fahrenheit. Next imagine that the analog module shows 8587 for 32 degrees and 13,032 for 212 degrees. The user would like to display the actual temperature on a touch screen. The user chose to use an SCP instruction to accomplish the task. The user can choose from among six values to enter. The first value to enter can be an integer or the actual address of the input. In this example the input is received from an analog input module. The address of the input is I:4.0 (input 0 on the analog module located in slot 4). The next two values to be entered are the input minimum and input maximum values. These values can be constants or addresses. These are the minimum and maximum values from the analog input. The value 8587 was entered for the minimum input value and 13,032 was entered for the maximum input value. The next two values are the scaled minimum and maximum values (in our example, 32 and 212). These can be constants or addresses. The last value to be entered is the scaled output. The scaled output can be a word address or an address of floating point data elements. In this example the user entered N7:20. Address N7:20 contains a scaled value between 32 and 212 based on the input value. The value of N7:20 can be displayed on the touch panel for an operator.

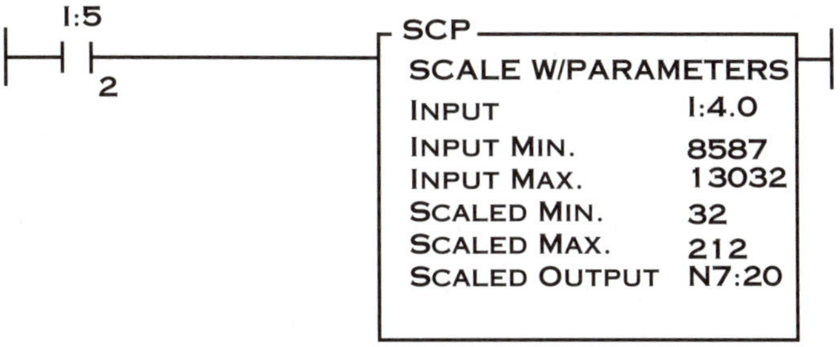

Figure 8–67 An SCP instruction.

QUESTIONS

1. Explain some of the reasons why arithmetic instructions are used in ladder logic.
2. What are LIM instructions used for?
3. What are comparison instructions used for?
4. Why would a programmer use an instruction that changes a number to a different number system?
5. Write a rung of ladder logic that compares two values to see if the first is greater than the second. Turn on an output if the statement is true.
6. Write a rung of ladder logic that checks to see if one value is equal to a second value. Turn on an output if the rung is true.
7. Write a rung of ladder logic that checks to see if a value is less than 20 or greater than 40. Turn on the output if the statement is true.
8. Write a rung of ladder logic to check if a value is less than or equal to 99. Turn on an output if the statement is true.
9. Write a rung of ladder logic to check if a value is less than 75 or greater than 100 or equal to 85. Turn on an output if the statement is true.
10. Write a ladder logic program that accomplishes the following. A production line produces items that are packaged twelve to a pack. Your boss asks you to modify the ladder diagram so that the number of items is counted and the number of packs is counted. A sensor senses each item as it is produced. Use the sensor as an input to the instructions you write to complete the task. (*Hint:* One way is to use a counter and at least one arithmetic statement.)

11. Write a ladder diagram program to accomplish the following. A tank level must be maintained between two levels (see the following figure).

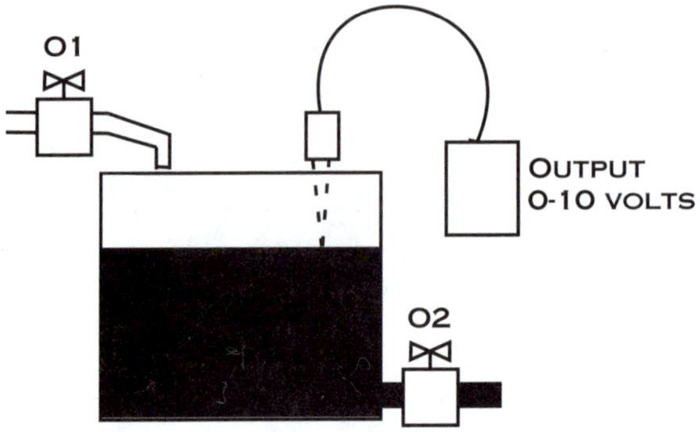

An ultrasonic sensor is used to measure the height of the fluid in the tank. The output from the ultrasonic sensor is 0 to 10 V (see the following figure).

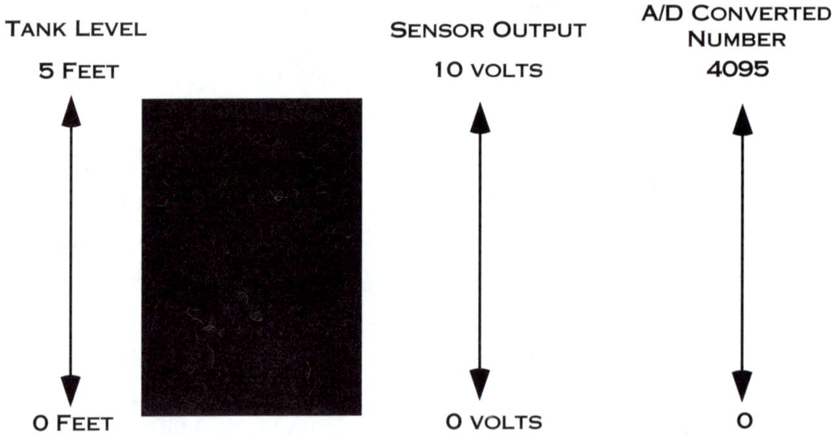

This directly relates to a tank level of 0 to 5 feet. The level must be maintained between 4.0 and 4.2 feet. Output 1 is the input valve. Output 2 is the output valve. The sensor output is an analog input to an analog input module.

chapter 9

Advanced Instructions

In this chapter we examine various instructions. We also learn about copying and moving memory, communicating between controllers, and other instructions that make the process of programming complex systems easier.

OBJECTIVES

Upon completion of this chapter, you will be able to:

1. Move and copy data utilizing SLC instructions.
2. Utilize SLC communication instructions.
3. Utilize special-purpose instructions such as copy, move, PID, and ramp.
4. Write ladder logic using sequencers.

Copy (COP) Instruction

A COP instruction is used to copy one range of memory to a different range of memory (Figure 9–1). The COP instruction has two values that must be entered: source and destination (Figure 9–2). The source is the beginning address of the file that you want to copy. The destination is the beginning address of the place where you would like the data copied. Length is the number of elements that you would like to copy. For example, if the address destination were a counter-type file (counters

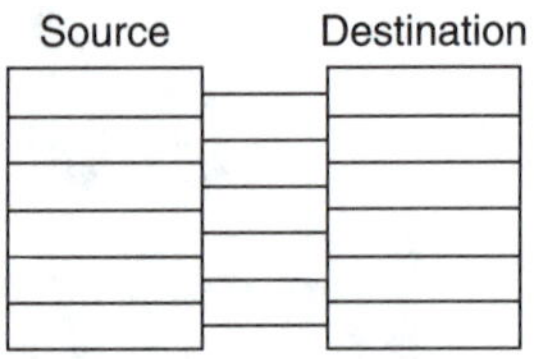

Figure 9–1 A copy instruction copies data from one place in memory to another.

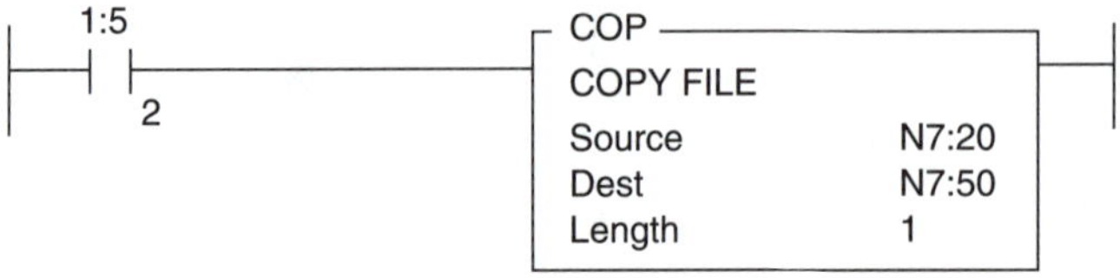

Figure 9–2 A COP instruction. In this example, the source is one element (word) from N7:20. This element will be copied to N7:50.

require three words per counter), then the maximum number you could copy would be 42. If you are copying words, the maximum would be 128. The entire source file is copied each program scan if the rung is true.

Move (MOV) Instruction

The MOV instruction is used to move the contents of one location to another location. This is a very useful instruction for many purposes. One example would be to change the preset value for a timer or counter. Two values must be entered by the programmer: source and destination (Figure 9–3). Source is a constant or the address of the data that you want to move. Destination is the address to which the data are moved. *Note:* If you want to move one word of data without affecting the arithmetic bits, use a COP instruction with a length of one word instead of a MOV.

Masked Move (MVM) Instruction

The MVM instruction is similar to the MOV instruction, except that the source data are moved through a mask before being stored into the destination. Three values must be entered by the programmer: source, mask, and destination (Figure 9–4). For example, if we are moving an integer (N7:5) and the first 8 bits in the mask are 1s and the second half are 0s in the mask, only the first 8 bits of N7:5 will be moved to the destination. The second 8 bits will not be changed in the destination. They will remain in the same state as they were before the move.

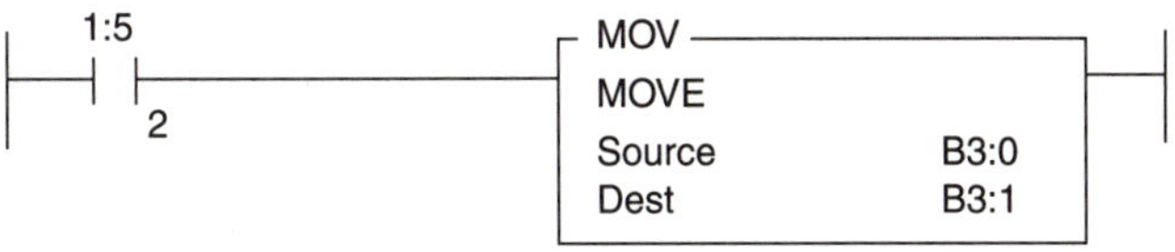

Figure 9–3 A MOV instruction. Note that in this example, word B3:0 was moved to destination B3:1. Note that a MOV instruction affects the math flags, a COP does not.

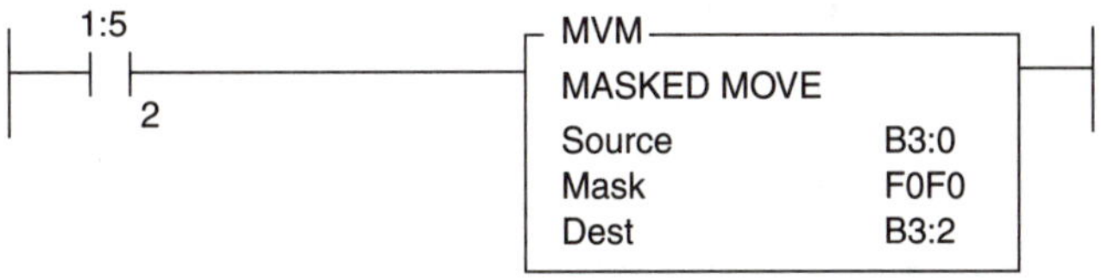

Figure 9–4 An MVM instruction.

File Fill (FLL) Instruction

The FLL instruction is used to fill a range of memory locations with a constant or a value from a memory address (Figure 9–5). The FLL instruction has two values that must be entered: source and destination (Figure 9–6). The source is the constant or the element address. The destination is the beginning address of the file you would like to fill. Length is the number of elements that you would like to fill. For example, if the address destination were a counter-type file (counters require three words per counter), the maximum number you could copy would be 42. If copying words, the maximum would be 128. The entire file fills each program scan if the rung is true.

Figure 9–5 An FLL instruction copies data from one location in memory to multiple locations.

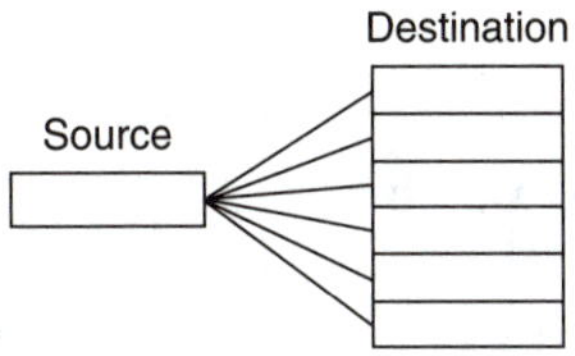

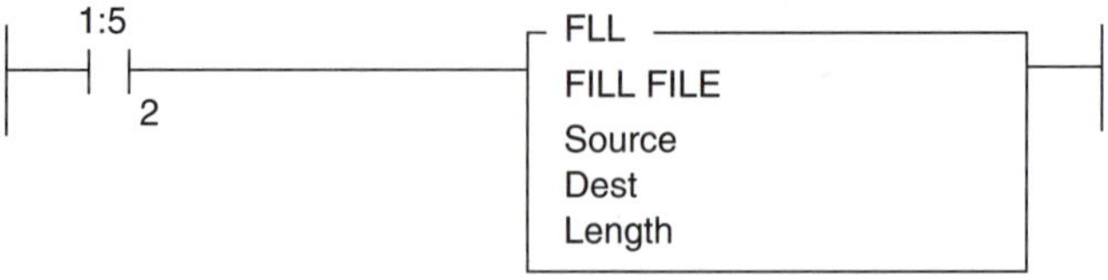

Figure 9–6 An FLL instruction.

Message (MSG) Instruction

The MSG instruction is used to communicate between PLCs. The length limit is 103 words.

The SLC 5/03 (OS301 and higher) and the SLC 5/04 processors service up to four message instructions per channel, for a maximum of eight message instructions. If you consistently enable more MSG instructions than the buffers and queues can accommodate, the order in which MSG instructions enter the queue is determined by the order in which they are scanned. This means MSG instructions closest to the beginning of the program enter the queue regularly, and MSG instructions later in the program may not enter the queue.

The MSG instruction initiates reads and writes through RS232 channel 0 when configured for the following protocols:

- DF1 Full-Duplex (Point-to-Point)
- DF1 Half-Duplex Master/Slave (Point-to-Multipoint)
- DH485 initiates reads and writes through
- DH485 channel 1 (SLC 5/03 processors only)
- DH+ channel 1 (SLC 5/04 processors only)

To invoke the MSG instruction, toggle its rung from false to true. Do not toggle the rung again until the MSG instruction has successfully or unsuccessfully completed the previous message indicated by the processor setting; that is, either the DN or ER bit. Below are some of the status bits available for the programmer:

- **Error (ER) bit (bit 12)** is set when message transmission has failed. The ER bit is reset the next time the associated rung goes from false to true. Do not set or reset this bit. It is informational only.
- **Done (DN) bit (bit 13)** is set when the message is transmitted successfully. The DN bit is reset the next time the associated rung goes from false to true. Do not set or reset this bit. It is informational only.

- **Start (ST) bit (bit 14)** is set when the processor receives acknowledgment (ACK) from the target device. The ST bit is reset when the DN, ER, or TO bit is set. Do not set or reset this bit. It is informational only. For SLC 5/05 Ethernet (channel 1) communications, the ST bit indicates internally that the Ethernet daughterboard has received a command and it is acceptable for a transmission attempt. The command has not yet been transmitted.
- **Enable (EN) bit (bit 15)** is set when rung conditions go true and the instruction is being executed. It remains set until message transmission is completed and the rung goes false.

Entering Parameters

After you place the MSG instruction on a rung, specify whether the message is to be a read or write. Then specify the target device and the control block for the MSG instruction. Figure 9–7 shows an MSG instruction.

- **Read/write:** Read indicates that the local processor (processor in which the instruction is located) is receiving data; write indicates that the processor is sending data.
- **Target device** identifies the type of device that will receive data.
- **Control block** is an integer file address that you select. It is a seven-element file containing the status bits, target file address, and other data associated with the message instruction.
- **Control block length** is fixed at seven elements. The MSG control block length increases from seven to fourteen words when changing from an SLC 5/02 to an SLC 5/03, SLC 5/04 (channel 0, DH485), or SLC 5/05 (channel 0, DH485) processor program. You must make sure that there are at least seven unused words following each MSG control block in your program.

Figure 9–7 MSG instruction.

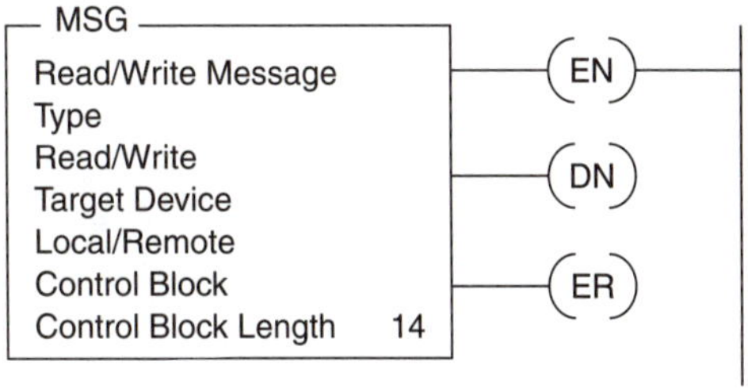

Block Transfer Instructions

The two block transfer instructions are block transfer read (BTR) and block transfer write (BTW). A block transfer instruction can be used with later models of 5/03, 5/04, and 5/05. Block transfer instructions can be used to transfer up to sixty-four words of memory to and from a remote device over a remote I/O link. A false-to-true transition initiates a BTW or BTR instruction. Figures 9–8 and 9–9 show BTR and BTW instructions.

The BTW instruction tells the processor to write data stored in the BTW data file to a device at the specified remote I/O (RIO) rack/group/slot address. The BTR instruction tells the processor to read data from a device at the specified RIO rack/group/slot address and store it in the BT data file. The data file may be any valid integer, floating-point, or binary data file. A total of thirty-two block transfer buffers are available. Each buffer is 100 consecutive words.

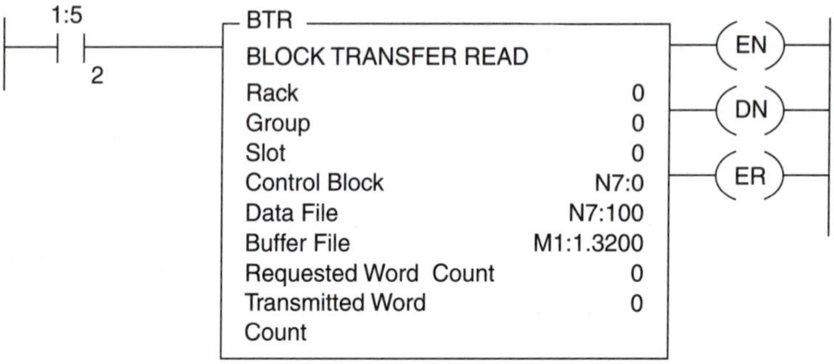

Figure 9–8 A BTR instruction.

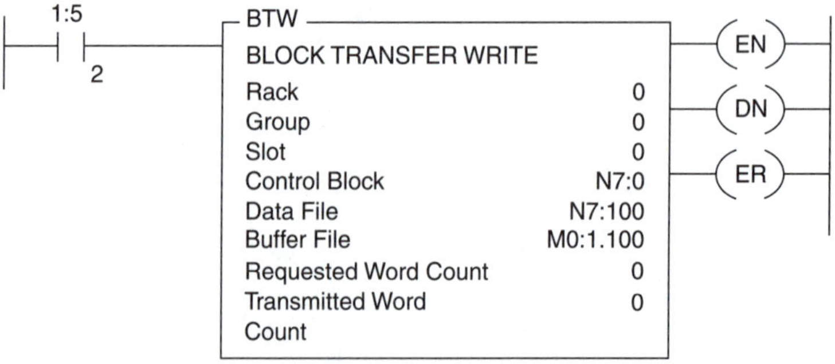

Figure 9–9 A BTW instruction.

Entering the Parameters for BTW and BTR Instructions

Data file is the address in the SLC's data file containing the BTW or BTR data. Bit integer or floating-point types can be used.

BTR/BTW buffer file is the file address used for the transfer buffer. For example, in M0:e.x00, *e* is the slot number of the scanner and *x* is the buffer number. The range of buffer numbers is 1 to 32. Each BTR and BTW instruction uses both the M1 and M0 files for a specific file number. M0 is used for BTR control and for BTW data. M1 is used for BTW status and BTR data.

Control block is an integer file address that stores block transfer control and status information. The control block is three words in length. You should provide the following information for the control structure:

- Rack—the I/O rack number 0–3 of the I/O chassis in which the target I/O module exists.
- Group—the I/O group number (0–7) of the position of the target module in the I/O chassis.
- Slot—the slot number (0 or 1) within the group. If you are using two-slot addressing, the left slot is 1 and the right slot is 0. If you are using one-slot addressing, the slot is always 0.
- Requested word count—the number of words to transfer. If the length is set to 0, the processor reserves sixty-four words for transfer. The block transfer module transfers the maximum words that the adapter can handle. If you set the length from 1 to 64, the processor transfers the number of words that are specified.

Time Stamp Instructions

Two instructions can be used together to time events, and they can be used in newer 5/03, 5/04, and 5/05 processors. The read high-speed clock (RHC) instruction is used to move the value of an internal 10-microsecond clock counter to either an integer (low 16 bits) or floating-point data location. The second instruction is the time difference (TDF) instruction. The TDF instruction is used to input two previously captured 10-microsecond clock values and return the elapsed time between them. This allows the user to time an event to within 10 microseconds. The user can utilize one event to trigger one RHC instruction to capture the counter time to N7:20, for example. A second RHC can be used to capture the time from a different event to address N7:21 (Figure 9–10). A TDF can then be used to compute the difference between the two events and store the difference in memory (N7:22 in this case). Figure 9–11 shows the use of a TDF instruction.

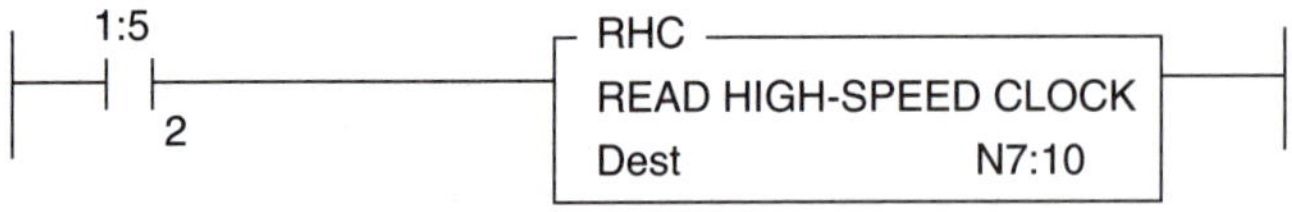

Figure 9–10 An RHC instruction.

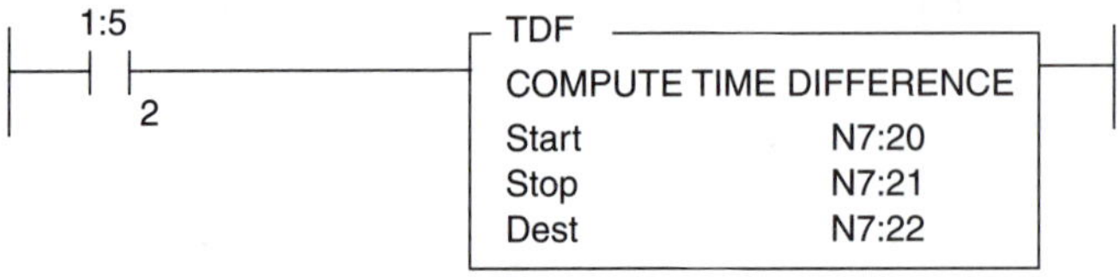

Figure 9–11 An TDF instruction.

Ramp Instruction

The ramp (RMP) instruction can be used to control motor speed or valve position. The RMP instruction allows the user to input time duration, beginning and ending output values, and a curve type (e.g., linear, acceleration, de-acceleration, and S-curve).

File Bit Comparison and Diagnostic Detect Instructions

The file bit comparison and diagnostic detect instructions can be used to compare large blocks of data. For example, they can be used for diagnostic data. The file bit comparison (FBC) instruction compares values in a bit file with values in a reference data file. The diagnostic detect (DDT) instruction is similar to the FBC except that when it finds a difference between the input file and the reference file, it changes the reference file. The user can develop a list of conditions for the process under operation. These conditions can be compared with actual conditions for diagnostics and troubleshooting. The FBC can be used to ensure that the actual conditions are the same as the desired conditions. The DDT can be used to record the actual conditions in the reference file for diagnostics.

Proportional, Integral, Derivative Instruction

The proportional, integral, derivative (PID) instruction is used to control processes. It can be used to control physical properties such as pressure, temperature, level, concentration, density, and flow rate. The PID instruction is straightforward: it takes one input and controls one output. The input normally comes from an analog input module. The output is normally an analog output, but it can be a time proportioning on/off output to drive a heating or cooling unit.

The PID instruction is used to keep a process variable at a desired setpoint. Figure 9–12 shows a tank-level example in which a level sensor outputs an analog signal between 4 and 20 mA. It outputs 4 mA if the tank is 0 percent full and 20 mA if the tank is 100 percent full. The output from this sensor becomes an input to an analog input module in the PLC. A variable valve on the tank controls the inflow to the tank. The valve is 0 percent open if it receives a 4-mA signal and 100 percent 4-mA open if it receives 20 mA. The PLC outputs an analog signal to the valve, and takes the input from the level sensor and uses the PID equation to calculate the proper output to control the valve.

Figure 9–12 shows how the PID system functions. The setpoint is set by the operator and is an input to a summing junction. The output from the level sensor becomes the feedback to the summing junction. The summing junction sums the setpoint and the feedback and generates an error. The PLC then uses the error as an input to the PID equation. There are three gains in the PID equation. The P gain is the proportional function and is the largest gain. It generates an output proportional to the error signal. If there is a large error, the proportional gain generates a large output. If the error is small, the proportional output is small. The proportional gain is based on the magnitude of the error, but it cannot completely correct an error.

The I gain is the integral gain, which is used to correct for small errors that persist over time. The P gain cannot correct for tiny errors, so the I gain is left to do the job.

The D gain is the derivative gain, and is used to correct for rapidly changing errors. The derivative looks at the rate of change in the error. When an error occurs, the proportional gain attempts to correct for it. If the error is changing rapidly (e.g., maybe someone opened a furnace door), the P gain is insufficient to correct the error and the error continues to increase. The derivative would “see” the increase in

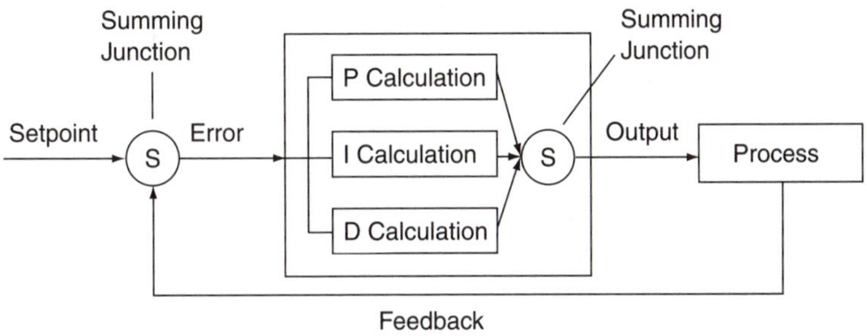

Figure 9–12 A block diagram showing a PID instruction.

the rate of the error and add a gain factor. If the error is decreasing rapidly, the D gain will damp the output. The derivative's damping effect enables the P gain to be set higher for quicker response and correction.

As you can see from Figure 9–12, the output from each of the P, I, and D equations is summed and an error (output) is generated. The output is used to bring the process back to setpoint.

Figures 9–13 and 9–14 show block diagrams of the whole system. As shown, the feedback from the process (tank level) is an input to an analog input module in the PLC. This input is used by the CPU in the PID equation, and an output is generated

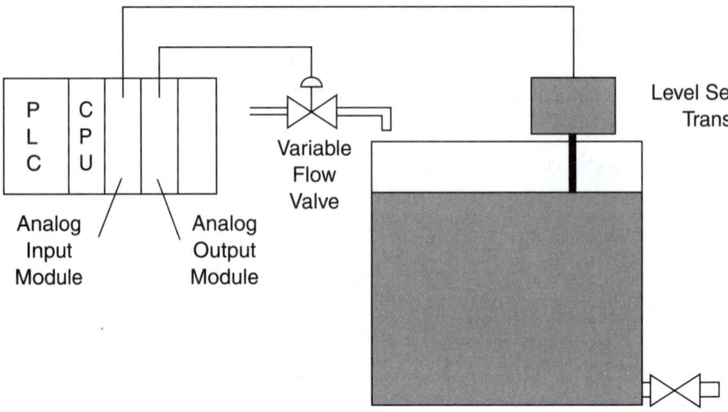

Figure 9–13 A level control system.

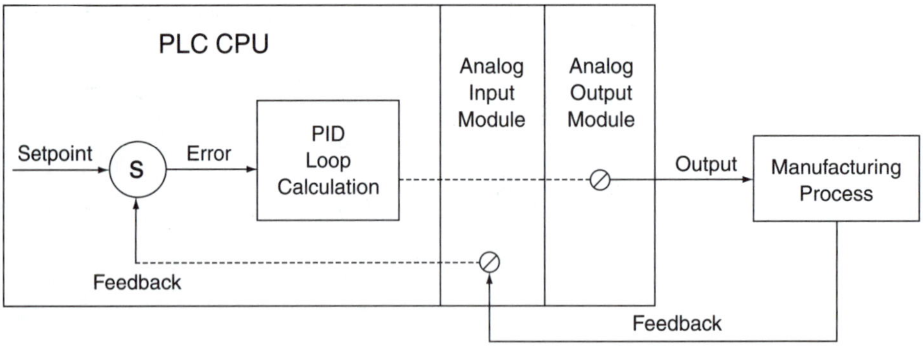

Figure 9–14 Block diagram of the use of one PLC and one PID instruction to control a process.

from the analog output module. This output is used to control a variable valve that controls the flow of liquid into the tank. Note that disturbances always affect the tank level. The temperature of the level affects the inflow and outflow. The outflow varies also due to density, atmospheric pressure, and many other factors. The inflow varies because of pressure of the fluid at the valve, density of the fluid, and many other factors. The PID instruction can account for disturbances and setpoint changes and control processes accurately.

Figure 9–15 shows the PID instruction. The user must provide some parameters. The first parameter is the control block. The control block is the file that will store the data used to operate the instruction. This file will occupy twenty-three words of memory. The user must enter the first address to be used. For example, if the user enters N7:10, the instruction will occupy addresses N7:0 through N7:22. Take care to use memory that is not used by other parts of the logic.

The second parameter that must be entered is the process variable (PV). Think of this variable as the feedback from the process. It is usually the address of an analog input, although it can be an integer address if you want to scale the real-world input from the analog module and store it in an integer. In this case, the user entered I:5.0 for the address of the analog input.

The third parameter is the control variable (CV). Think of this variable as the output to the process. The output ranges from 0 to 16,383. Output 16,383 is the 100 percent on value, which is normally an integer value, so that the user can scale the PID output range to the range needed for your application. In this case, the user entered N7:30 for the address.

Next the user must enter the rest of the PID parameters. Clicking on the setup screen in the instruction brings up another parameter screen (Figure 9–16).

The first parameter is used to choose the automatic or manual mode. If the bit is set, the instruction functions in manual mode. The user enters it here and it is located in word 0, bit 1 of the control register file that you specified. In manual mode, the user is setting the output value. When you are tuning, it is suggested that you make changes in manual mode and then return to the auto mode. Note that if you set an output limit, it is applied in the manual and auto mode.

Figure 9–15 A PID instruction.

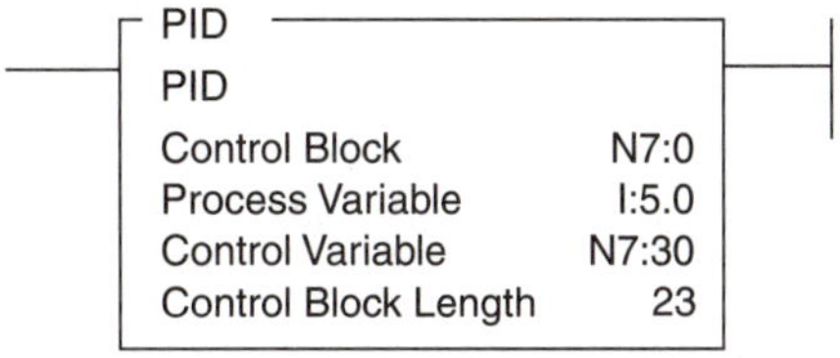

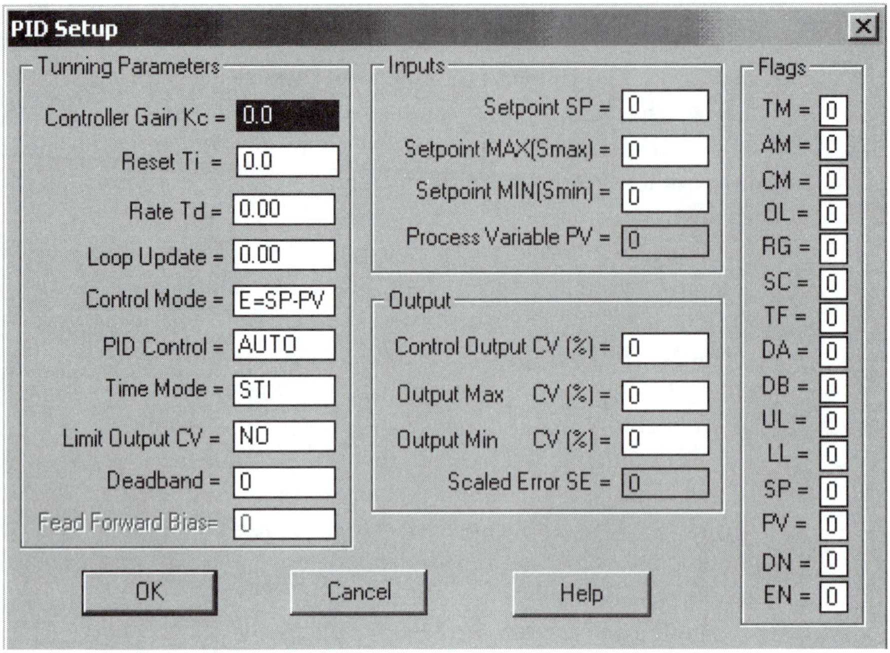

Figure 9–16 PID Setup screen.
(Courtesy Rockwell Automation Inc.)

The second parameter is the time mode (TM). The user can choose between timed and STI. The user enters it here and it is located in word 0, bit 0 of the control register file that you specified. Timed means that the PID updates its output at the rate the user specifies in the loop update parameter. Note that the scan time of your ladder logic should be at least ten times faster than your loop update time for the instruction to be accurate.

In the STI mode, the PID updates its output every time it is scanned. If you choose the STI mode, the PID instruction should be programmed in an STI subroutine, and the STI subroutine should have a time interval equal to the loop update parameter. The STI period is set in word S:30.

The next parameter is the control mode (CM) parameter. The control can be used to toggle between two modes: E=SP-PV (error = setpoint − present value) and E=PV-SP (error = present value − set point). E=PV-SP (direct acting) causes the output to increase when the input present value is larger than the setpoint. A cooling application might use direct acting. The other mode (E=SP-PV) is reverse acting. The output will increase when the input (PV) is less than the setpoint (SP). The user enters the choice here and it is located in word 0, bit 2 of the control register file that you specified.

The next parameter is the setpoint (SP). The setpoint is located in word 2 of the control register that you specified. You normally change this value with instructions in your ladder diagram. Your logic would move the desired value to word 3 of your control block. For example, if you had chosen N7:10 for the start address of your control register, you would move the desired setpoint value to N7:12. If you do not scale the range, this value is 0 to 16,383.

Gain (Kc) is located in word 3. It is the proportional gain, ranging from 0 to 3276.7 in most SLCs. A rule of thumb is to set the proportional gain to one-half the value that causes the output to oscillate when the reset (integral) and rate (derivative) are set to zero.

Reset (Ti) is located in word 4. It is the integral gain, with a valid range of values between 0 and 3276.7 minutes/repeat for most SLCs. A rule of thumb is to set the value equal to the natural period that is measured in the proportional gain calibration. Note that a value of 1 is the minimum integral term possible in the PID equation.

Rate (Td) is located in word 5. The rate is the derivative term, with a range of values from 0 to 3276.7 minutes for most SLCs. A rule of thumb is to set the value to one-eighth of the integral gain.

Maximum scaled (Smax) is located in word 7. If the setpoint is to be read in engineering units, this value corresponds to the value of the setpoint in engineering units when the control input is 16,383. In most SLCs the valid range is between −32,767 and +32,767.

Minimum scaled (Smin) is located in word 8. If the setpoint is read in engineering units, this parameter corresponds to the engineering value of the setpoint when the control output is 0. The valid range of values is between −32,767 and +32,767 for most SLCs.

Deadband (DB) is located in word 9. It must be a nonnegative value. The deadband extends above and below the value you enter. The deadband is entered at the zero crossing point of the PV and the SP. The deadband is in effect only after the PV enters the deadband and passes through the SP. The valid range is 0-scaled maximum or 0 to 16,383 if no scaling is done.

The loop update (word 13) is the time interval between PID calculations. The value is entered in 0.01-second intervals. The rule of thumb is to enter an update time 5 to 10 times faster than the natural period of the load. The natural period can be determined by setting the rate and reset parameters to 0 and then increasing the gain until the output begins to oscillate. If you are using the STI mode, this value must be equal to the STI time value. The valid range is between 0.01 and 10.24 seconds for most SLCs.

The next parameter is the scaled process (PV). It is located in word 14. This is the scaled value of the analog input PV. If it is not scaled, the value can be between 0 and 16,383.

Scaled error is used only for display. It is located in word 15. It is the scaled error as selected by the control mode parameter (E=SP-PV or E=PV=SP). The range is between −32,768 and +32,767. Errors smaller or larger than this cannot be represented.

Output CV (%) is located in word 16. This will display the actual output in terms of a percentage between 0 and 100. If you are in auto mode, this is only for display. If you are in manual mode, you can change the output percentage.

Output (CV) limit (OL) is located in word 0, bit 3. This parameter is used to toggle between yes and no. You would select yes if you want the output limited to the minimum and maximum values that you specify. The minimum output percentage is entered in the output (CV) minimum. The maximum output percentage is entered in the output (CV) maximum value.

The right column of parameters displays various flags associated with the PID instruction.

The TM bit (word 0, bit 0) specifies the PID mode. It is a 1 when the timed mode is selected and a 0 when the STI mode is in effect. You can set or clear this bit with your ladder logic.

The auto/manual (AM) bit is located in word 0, bit 1. This bit can be set or cleared by your ladder logic. If it is 0, the mode will be auto; if it is 1, the mode will be manual.

The CM bit is located in word 0, bit 2. If it is 1, the mode will be E=PV-SP; if it is 0, the mode will be E=SP-PV.

The output limiting enabled (OL) bit is in word 0, bit 3. It will be set if you have selected to limit the control variable. This bit can also be set or cleared in your ladder logic.

The reset and gain range enhancement (RG) bit is located in word 0, bit 4. If this bit is set, it causes the rest minute/repeat value and the gain multiplier to be enhanced by a factor of 10.

The scale setpoint flag (SC) bit is located in word 0, bit 5. It is cleared when setpoint scaling factors are specified.

The loop update time too fast (TF) bit is located in word 0, bit 6. It is set by the PID algorithm if the specified loop update time cannot be achieved because of scan time limitations.

The derivative (rate) action (DA) bit is located in word 0, bit 7. When this bit is set, it causes the derivative calculation to be evaluated based on the error instead of the PV. If the bit is clear, the derivative calculation is performed using the PV.

The DB is set when error in DB bit is located in word 0, bit 8. This bit is set when the process variable is within the deadband range.

The output alarm, upper limit (UL) bit is located in word 0, bit 9. This bit is set when the calculated control output CV exceeds the CV limit.

The output alarm, lower limit (LL) bit is located in word 0, bit 10. This bit is set when the calculated control output CV is less than the lower CV limit.

The setpoint out of range (SP) bit is located in word 0, bit 11. This bit is set when the setpoint exceeds the maximum scaled value or is less than the minimum scaled value.

The process variable out of range (PV) bit is located in word 0, bit 12. It is set when the unscaled process variable exceeds 16,383 or is less than 0.

The PID done (DN) bit is located in word 0, bit 13. It is set on scan where the PID algorithm is computed. It is computed at the loop update rate.

The PID enabled (EN) bit is set when the rung of the PID instruction is enabled.

Applications involving transport lags may require that a bias be added to the CV output in anticipation of a disturbance. This bias can be accomplished using the processor by writing a value to the feedforward bias element, the seventh element (word 6) in the control block file. Figure 9–17 shows a block diagram of how the feedforward gain is added. The value you write is added to the output, allowing a feedforward action to take place. You may add a bias by writing a value between

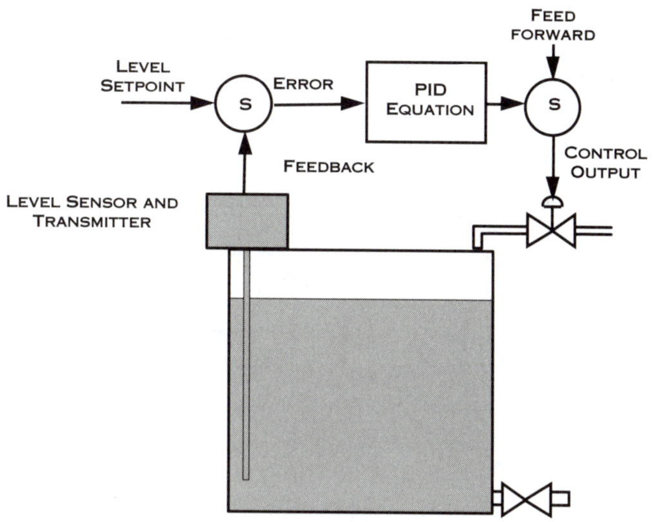

Figure 9–17 Addition of a feedforward gain to the level application.

−6,383 and +16,383 to word 6 with your ladder logic or directly from the computer or handheld programmer.

The instruction set manual on the CD has a PID tuning procedure in the section on the PID instruction.

Sequencers

Before PLCs there were many innovative ways to control machines. One of the earliest control methods for machines was punched cards, which date back to the earliest automated weaving machines. Punched cards controlled the weave. About twenty years ago, the main method of input to computers was punched cards.

Most manufacturing processes are highly sequential, meaning that they process a series of steps, from one to the next. Imagine a sequential bottling line: bottles enter the line, are cleaned, filled, capped, inspected, and packed. Many of our home appliances work sequentially. The home washer, dryer, dishwasher, breadmaker, and so on are examples of sequential control. Plastic injection molding, metal molding, packaging, and filling are a few examples of industrial processes that are sequential.

Many of these machines were (and some still are) controlled by a device called a drum controller. A drum controller functions just like a player piano. The player piano is controlled by a paper roll with holes punched in it. The holes represent the notes to be played. Their position across the roll indicates which notes should be played and when they are to be played. A drum controller is the industrial equivalent. It is a cylinder with holes around the perimeter and pegs placed in the holes (Figure 9–18). The pegs hit switches as the drum turns. Each peg then turns, clos-

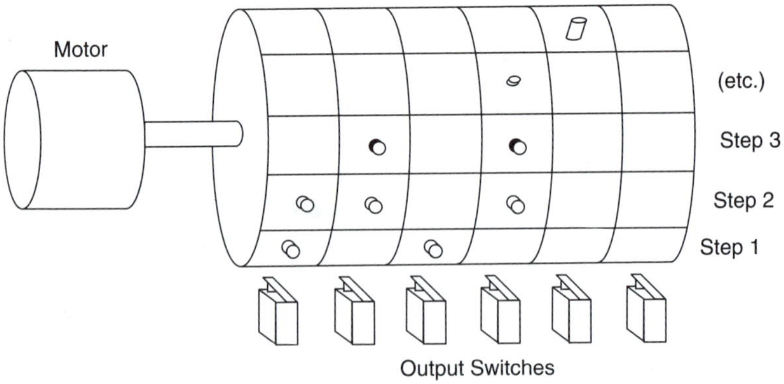

Figure 9–18 Drum controller. Note the pegs that activate switches as the drum is turned at slow speed by the motor. Figure 9–19 shows the output conditions for the steps.

ing the switch that it contacts and turning on the output to which it is connected. The speed of the drum is controlled by a motor. The motor speed can be controlled, but each step must take the same amount of time. If an output must be on longer than one step, then consecutive pegs must be installed.

The drum controller has several advantages. It is simple to understand, which makes it easy for a plant electrician to operate. It is also easy to maintain and program. The user makes a simple chart that shows which outputs are on in which steps (Figure 9–19). The user then installs the pegs to match the chart.

Many of our home appliances are also controlled with drum technology. Instead of a cylinder, some utilize a disk with traces (Figure 9–20). Think of a washing machine. The user chooses the wash cycle by turning the setting dial to the proper position. The disk is then moved slowly as a synchronous motor turns. Brushes make contact with traces at the proper times and turn on output devices such as pumps and motors. Many home appliances utilize a control that works like the one shown in Figure 9–18, except that a series of camlike disks are used that activate switches to control the sequence.

Although there are many advantages to the drum type of control, it also has some major limitations. The time for individual steps cannot be controlled individually. The sequence is set and it does not matter if something goes wrong. The drum will continue to turn, and turn devices on and off. In other words, it would be convenient if the step would not occur until certain conditions are met. PLC instructions have been designed to incorporate the good traits of drum controllers and also to overcome the weaknesses.

Step Number	Input Pump	Heater	Add Cleaner	Sprayer	Output Pump	Blower
1	On		On			
2	On	On		On		
3		On		On		
4				On		
5					On	

Figure 9–19 Output conditions for the drum controller shown in Figure 9–18.

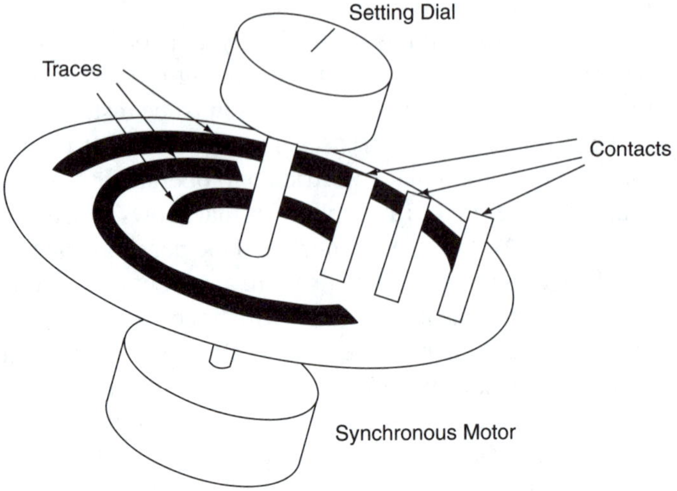

Figure 9–20 Disk-type drum controller. Some home appliances use these controllers. Note that when you turn the dial, you are actually turning the disk to its starting position.

Sequencer Instructions

Sequencer instructions can be used for processes that are cyclical in nature. They can be used to monitor inputs to control the sequencing of the outputs, and they can make programming many applications a much easier task. The sequencer is similar to the drum controller. Processes with defined steps can be easily programmed with sequencer instructions. Three main instructions are available in Rockwell Automation PLCs: sequencer output (SQO), sequencer compare (SQC), and sequencer load (SQL).

Imagine a sequential process of six steps. We can write the states of the outputs for every step (Figure 9–21). Note that in step 1, output 0 is on. In step 2, output 0 stays on and output 1 turns on.

This would be a good application for a sequencer instruction—an SQO instruction would be used. Figure 9–22 shows a simple ladder diagram with a timer and SQO instruction.

The first entry you need to make in the SQO is the file number to be used. File is the starting location in memory for our output conditions for each step. In this example, the starting file address is N7:0. The output states for each step are entered next, starting at N7:0 (Figure 9–23). This process has six steps, so they would be located in words N7:0, N7:1, N7:2, N7:3, N7:4, N7:5, and N7:6. Note that this is actually seven words in memory. The first word used is for step 0. Step 0 is not

Step Number	Output 5	Output 4	Output 3	Output 2	Output 1	Output 0
1						On
2					On	On
3				On		On
4			On			
5		On		On		
6	On			On		

Figure 9–21 Table showing which outputs are on in which steps.

actually part of our six operational steps. SQOs always start at step 0 the first time. When the SQO reaches the last step, however, it will reset to step 1 (position 1).

Next we enter a mask. The mask can be a file in memory or one word (16 bits) in memory. If an address is entered for the mask value, the mask will be a list of values starting at the address specified in the mask value. For example, if the user enters N7:20, then N7:20 through N7:25 would be the location of the mask values for each of the six steps. The user would have to fill these locations with the desired mask values. In this example there are six steps, so each of the six mask values would correspond to one step.

A constant can also be used. If it is, the SQO will use the constant as a mask for all of the steps. In our example, a hex number was entered (Figure 9–24). The four hex digits represent 16 bits. Each bit corresponds to one output in our SQO. If the bit is a 1 in the mask position, then the output is enabled and will turn on if the step condition tells it to be on. In other words, the bit conditions for the active step are ANDed with the mask. If the bit in the step is a 1 and the corresponding mask bit is a 1, the output will be set.

The destination is the location of the outputs. In this example the outputs are in O:5.0 (the output module in slot 5), which means that output 0 is O:5/0, output 1 is O:5/1, and so on.

Remember that the output states for each step are located in N7:1 through N7:6. When the ladder diagram is in the first step, output 0 will be on. In the next step, outputs 0 and 1 will be on. In the next step, outputs 0 and 2 will be on, and so it continues.

A file location must also be entered for control (Figure 9–22). Control is where the SQO stores status information for the instruction. An SQO or SQC uses three

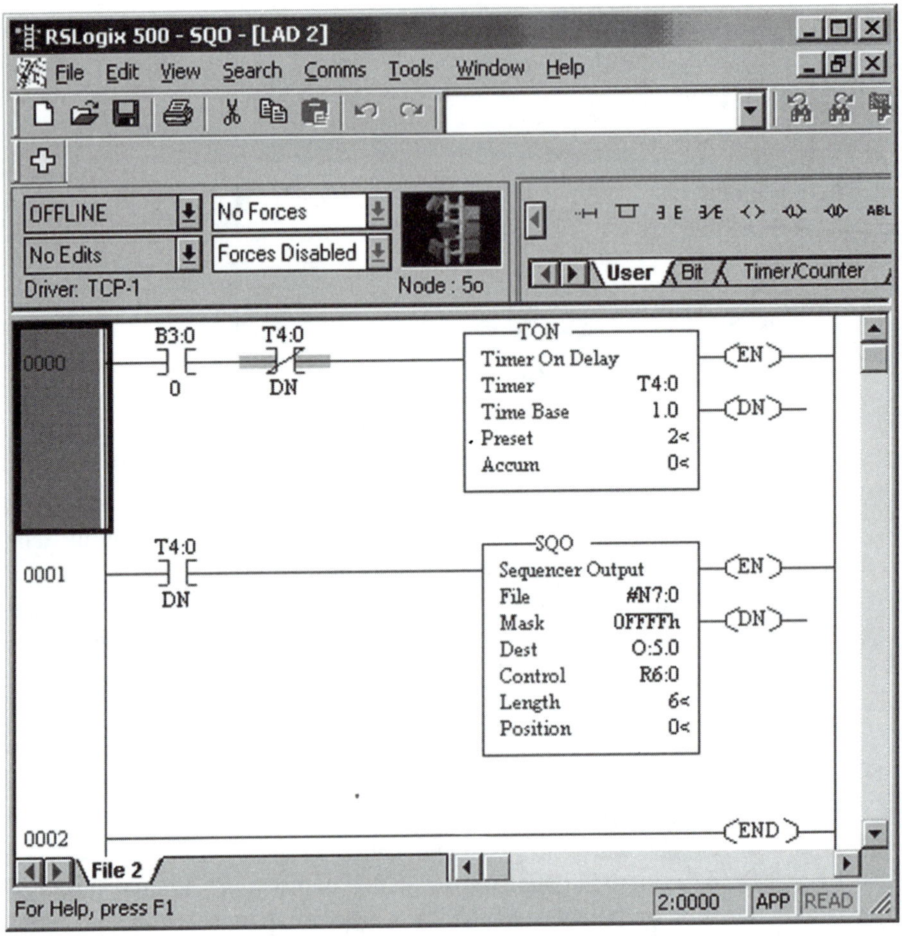

Figure 9–22 Simple example of ladder logic for an SQO.

words in memory (Figure 9–24). The three bits are EN, DN, and ER bits. Word 1 contains the length of the sequencer (number of steps). Word 2 contains the current position (step in the sequence).

Length is the number of steps in the process starting at position 1. The maximum length is 255 words. Position 0 is the start-up position. The first time the SQO is enabled, it moves from position 0 to position 1. The instruction resets to position 1 at the end of the last step.

Position is the word location (or step) in the sequencer file to and from which the instruction moves data. In other words, position shows the number of the step that is currently active.

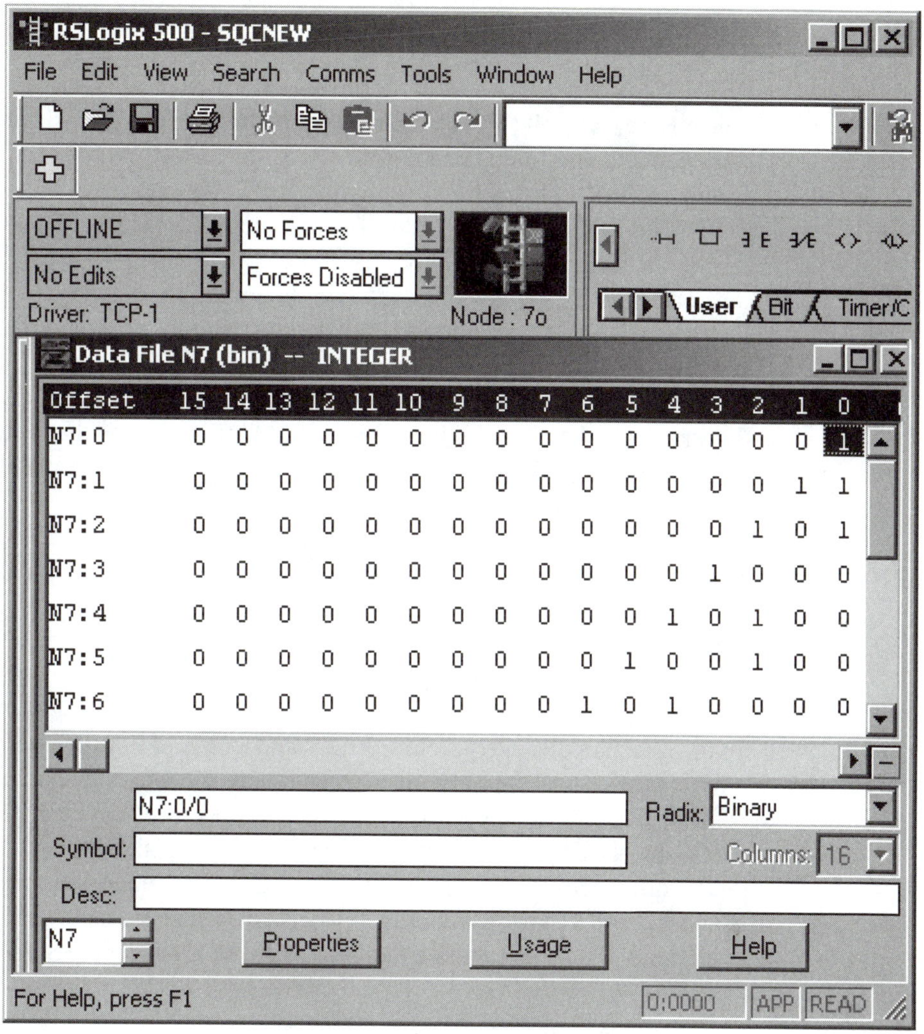

Figure 9–23 How desired output conditions are stored in memory.

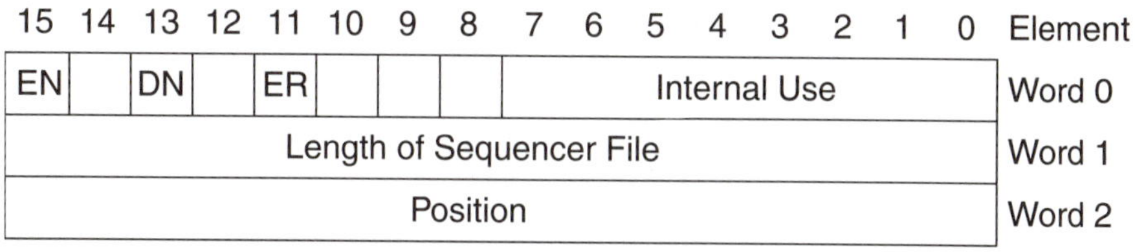
15 14 13 12 11 10 9 8 7 6 5 4 3 2 1 0
Element
EN
DN
ER
Internal Use
Word 0
Length of Sequencer File
Word 1
Position
Word 2

Figure 9–24 Memory organization for an SQO control memory.

The enable (EN) bit (bit 15) is set by a false-to-true rung transition and indicates that the SQO instruction is enabled.

The done (DN) bit (bit 13) is set by the SQO instruction after it has operated on the last word in the sequencer file. It is reset on the next false-to-true rung transition after the rung goes false.

The error (ER) bit (bit 11) is set when the processor detects a negative position value or a negative or zero length value. This results in a major error if it is not cleared before the END or TND instruction is executed.

Let us see how this process would work. First, assume that it is a simple process and every step takes 2 seconds. A 2-second timer is used as the input condition to the SQO instruction. Every time the timer accumulated value reaches 2 seconds, the timer done bit forces the SQO instruction to move to the next step. That step's outputs are turned on for 2 seconds until the timer done bit enables the SQO again. The SQO then moves to the next step and sets any required outputs. Note that the timer done bit also resets the accumulated time in the timer to zero and the timer starts timing to 2 seconds again. When the SQO is in the last step and receives the enable again, the SQO will return to step 1 (note that the first time the ladder is executed the SQO will start at step 0).

In the last example, every step was the same length of time, but we are not always that fortunate. In most applications we would have specific input conditions that need to be met before we can move to the next step of the process. We can use any logic we like to enable the SQO. It can get quite complex to set up all of the different conditions in logic. Fortunately, Rockwell Automation PLCs have a sequential compare (SQC) instruction that can make the task easy. If we make a list of the input conditions we need for every step, we can use those conditions as a way to trigger the SQO (Figure 9–25).

Step Number	Input 3	Input 2	Input 1	Input 0
1		On		On
2		On	On	
3	On			
4			On	
5		On		
6	On			On

Figure 9–25 Input condition table. Note that if we are presently in step 2, we would need inputs 1 and 2 to be true to move to step 3.

Note that in the SQC in Figure 9–22 we had to give the instruction for the starting file number. In this example, N7:10 was entered as the starting address. The desired input conditions to move from step 1 to step 2 are entered into N7:11 (Figure 9–26). Remember that the sequencer will initially begin with step 0, and the actual program steps begin in step 1. The rest of the desired input conditions are entered as shown in Figure 9–26.

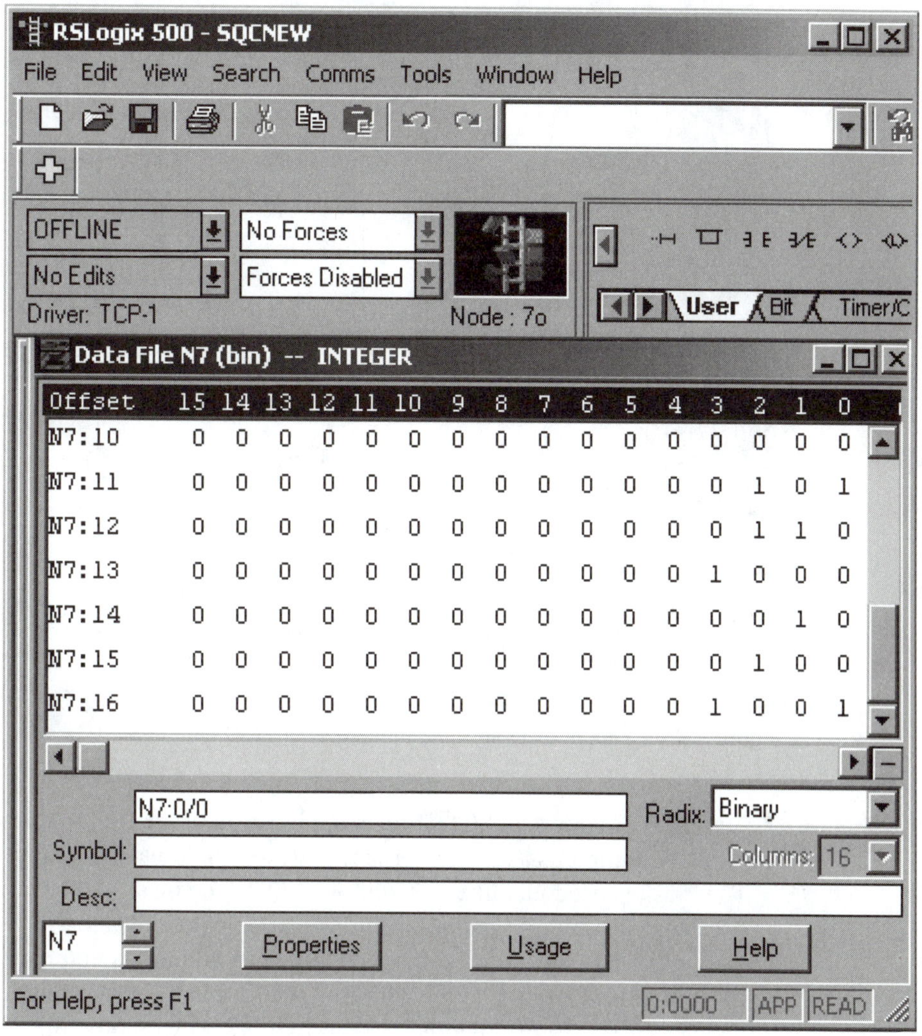

Figure 9–26 Desired input conditions in integer memory.

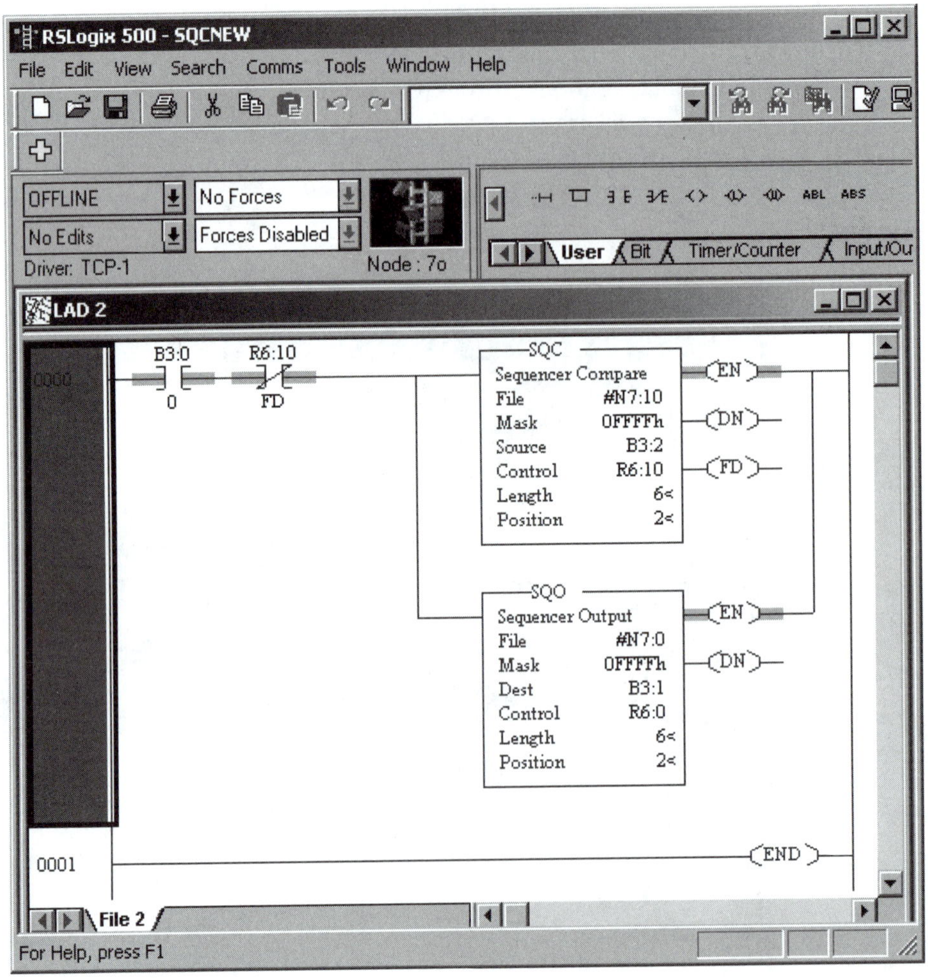

Figure 9–27 Ladder logic showing the use of an SQC and an SQO.

Figure 9–27 shows an SQC. File is the location of the first word of input conditions. Our example is a six-step process, so we would have six input words.

The mask is used just like it is with the SQO. The mask can be a file in memory or one word (16 bits) in memory. In our example, a constant will be used, but an address can be entered for the mask value. If an address is entered, the mask will be a list of values starting at the address specified in the mask value. For example, if the user enters N7:30, then N7:30 through N7:35 would be the locations of the mask values for each of the six steps. The user would have to fill these locations with the desired mask values. Each mask value would correspond to a step in the sequence. This example has six steps, so each of the six mask values correspond to one step.

Otherwise a constant can be used. In our case, a hex constant is entered. The four hex digits correspond to 16 bits. If the bit is a 1 in the mask, that input is enabled. The source word (usually real-world input conditions) is ANDed with the mask value and compared with the current step (word) in the file. In this case, our input mask is 000FH. In this application, we are interested only in the states of the first four inputs. The 000FH (H = HEX) means the first four inputs are used and that the last sixteen are ignored (0000 0000 0000 1111).

Next the source must be entered. The source is the address of the real-world inputs. In this example, they are at I:4.0. The address I:4.0 would be sixteen inputs in the module in slot 4. The states of these real-world inputs are compared with the desired inputs in the present step. If they are the same, the SQC position value is incremented to the next step and the FD bit is set in the control register. The FD bit (R6:10/FD in this case) was used to trigger the SQO to transition to the next step. This means that our desired input conditions can be used to control the transition from step to step in our process.

A file location must also be entered for control. Control is the location where the SQC stores status information for the instruction. An SQC uses three words in memory (Figure 9–28). The four bits that can be used are EN, DN, ER, and FD bits.

The enable (EN) bit (bit 15) is set by a false-to-true rung transition and indicates that the SQC instruction is enabled.

The done (DN) bit (bit 13) is set by the SQC instruction after it has operated on the last word in the sequencer file. It is reset on the next false-to-true rung transition after the rung goes false.

The error (ER) bit (bit 11) is set when the processor detects a negative position value or a negative or zero length value. This results in a major error if it is not cleared before the END or TND instruction is executed.

15	14	13	12	11	10	9	8	7	6	5	4	3	2	1	0	Element
EN		DN		ER			FD	Internal Use								Word 0
Length of Sequencer File																Word 1
Position																Word 2

Figure 9–28 Control register memory addressing for an SQC instruction.

The found (FD) bit (bit 8) is set when the status of all nonmasked bits in the source address match those of the corresponding reference word. This bit assessed each time the SQC instruction is evaluated while the rung is true.

Word 1 contains the length of the sequencer (number of steps). Word 2 contains the current position.

Length must also be entered. Length is the number of steps in the process starting at position 1. The maximum length is 255 words. Position 0 is the start-up position. The first time the SQO is enabled, it moves from position 0 to position 1. The instruction resets to position 1 at the end of the last step.

Position is the word location (or step) in the sequencer file to and from which the instruction moves data. In other words, position is the step that is currently active.

Sequencer Load Instruction

The sequencer load (SQL) instruction can be used to store source information (words) into memory. For example, input conditions can be stored into integer memory every time the SQL instruction sees a false-to-true transition. When it gets to the last position, it will transition back to step 1 when it sees the last false-to-true transition.

Shift Resister Programming

A shift register is a storage location in memory. These storage locations can typically hold 16 bits of data, that is, 1s or 0s. Each 1 or 0 can be used to represent good or bad parts, presence or absence of parts, or the status of outputs (Figure 9–29). Many manufacturing processes are linear in nature. Recall the bottling line exam-

Station Number	1	2	3	4	5	6	7	8
Part Present	1	0	0	0	1	0	0	1

Figure 9–29 What a shift register might look like when monitoring whether or not parts are present at processing stations. A 1 in the station location would be used to turn an output on and cause the station to process material. In this case, the PLC would turn on outputs at stations 1, 5, and 8. After processing has occurred, all bits would be shifted to the right. A new 1 or 0 would be loaded into the first bit, depending on whether a part was or was not present. The PLC would then turn on any stations that had a 1 in their bit. In this way, processing occurs only when parts are present.

ple used earlier. The bottles are cleaned, filled, capped, and so on, which is a linear process. There are sensors along the way to sense for the presence of a bottle, to check fill, and other steps. All of these conditions can easily be represented by 1s and 0s.

Shift registers essentially shift bits through a register to control I/O. In the bottling line are many processing stations, each represented by a bit in the shift register. We want to run the processing station only if there are parts present. As the bottles enter the line, a 1 is entered into the first bit. Processing takes place. The stations then release their product and each moves to the next station. The shift register also increments. Each bit is shifted one position. Processing takes place again. Each time a product enters the system, a 1 is placed in the first bit. The 1 follows the part all the way through production to make sure that each station processes it as it moves through the line. Shift register programming is applicable to linear processes.

Questions

1. What instruction can be used to fill a range of memory with the same number?
2. What instruction can be used to move an integer in memory to an output module?
3. What instruction can be used to make a copy of a range of memory and then copy it to a new place in memory?
4. What instruction can be used to move data if desired to mask some of the bits?
5. What does PID stand for?
6. What does the proportional gain do?
7. What does the integral gain do?
8. What does the derivative gain do?
9. Draw a PID instruction and set it up to use input 1 on an analog module in slot 3 and analog output 0 on an analog module in slot 4. Use appropriate addressing.
10. Describe the operation of an SQO.
11. Describe the operation of an SQC.
12. What are at least two advantages of SQO/SQC instructions over the traditional mechanical drum controller?

chapter

10

Advanced Programming

Many instructions are available to make the programming systems easier. In this chapter we examine a few of them. Many of these instructions were designed to "look" like the mechanical control devices they replaced. The purpose was to make them easy to understand.

OBJECTIVES

Upon completion of this chapter, you will be able to:

1. Define terms such as interlocking, stage programming, and step programming.
2. Explain the benefits and use of stage programming.
3. Explain the benefits and use of step programming.
4. Explain the benefits and use of fuzzy logic programming.
5. Explain the benefits and use of state logic programming.

STAGE PROGRAMMING

Stage programming is a different concept in PLC programming. Stage programming can be used with AutomationDirect PLCs, and it can make programming complex systems easier. This concept involves breaking the program into logical steps or stages. The stages can then be programmed individually without concern for how they will affect the rest of the program. Stage programming can be used to program AutomationDirect PLCs.

This method of programming can reduce programming time by up to about 70 percent. It can also drastically reduce the troubleshooting time by up to 85 percent. Much of the time and effort in writing a ladder logic program is spent programming interlocks to be sure that one part of the ladder does not adversely affect another. Stages help eliminate this problem.

A process or a manufacturing procedure is simply a sequence of tasks or stages. The PLC ladder diagram that we write must make sure that the process executes in the correct order. We have to design the ladder very carefully so that rungs execute only when they should. To do this we program interlocks. The process of designing the interlocks can take the majority of programming and debugging time. A ladder logic program that deals with process control may require that as much as 35 percent of the ladder be dedicated to interlocking. A ladder written to control a sequential process may require that up to 60 percent be devoted to interlocking.

Let's consider a simple example. The process is shown in Figure 10–1. It involves a conveyor and press operation. Figure 10–2 shows one ladder logic solution for the process. Remember that there are as many possible ladder logic programs as there are programmers. No two programmers would write the ladder

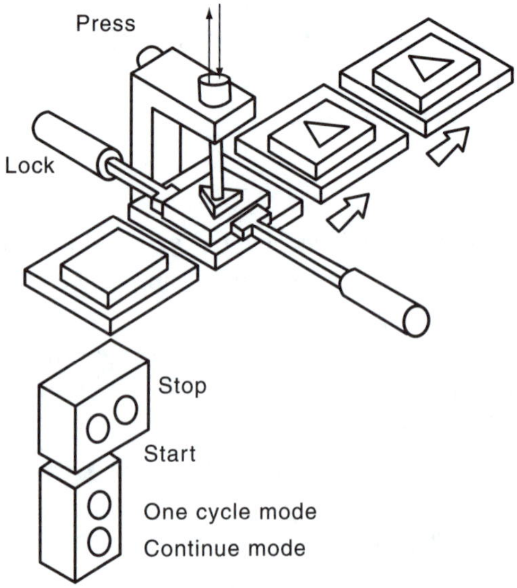

Figure 10–1 Simple press process. *(Courtesy AutomationDirect, Inc.)*

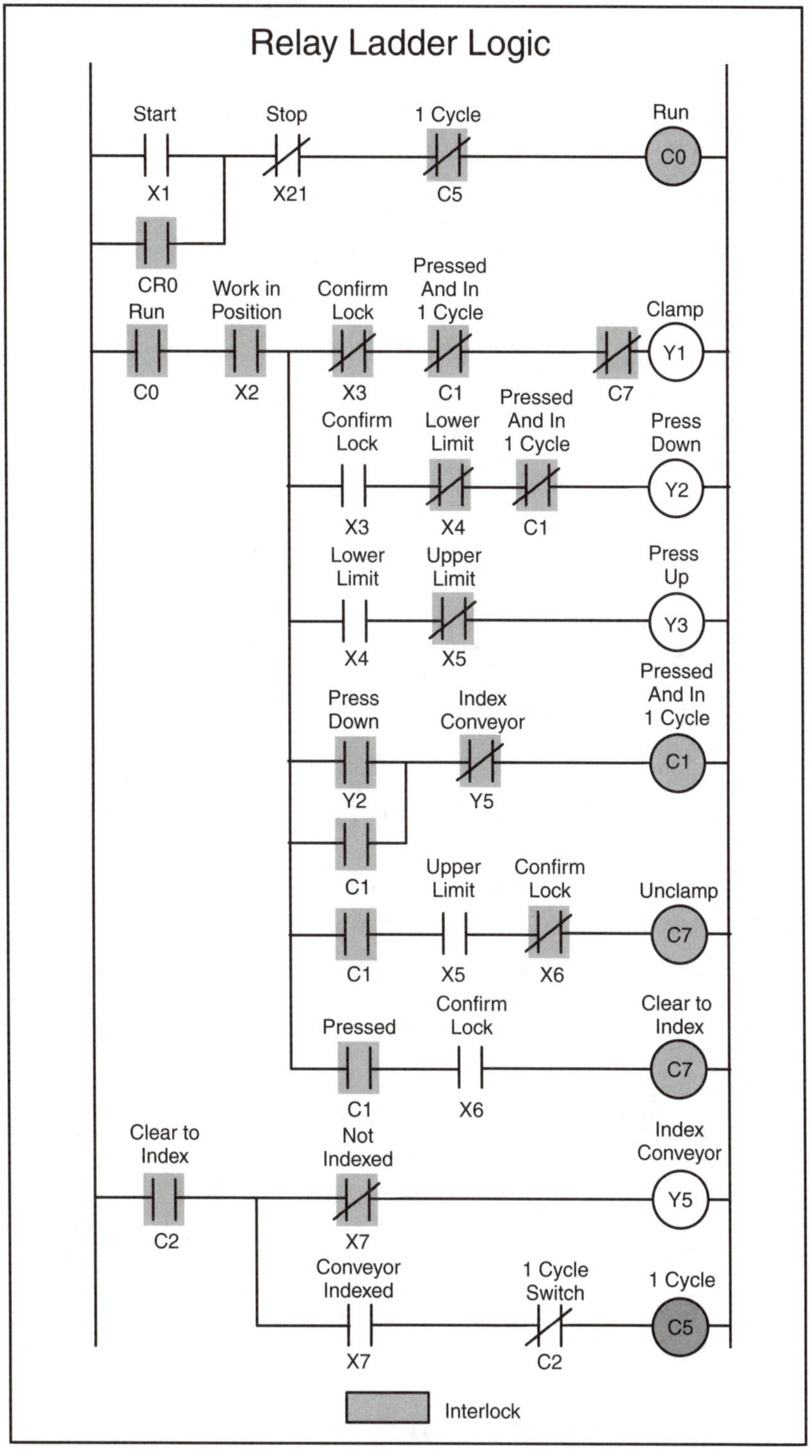

Figure 10–2 Ladder logic required to run the simple system shown in Figure 10–1. *(Courtesy AutomationDirect Inc.)*

in the same way. This makes it a little difficult when we are asked to troubleshoot or modify someone else's ladder. Even this simple process requires a fairly complex program and would require considerable time to understand. Also remember that a substantial portion of this ladder is devoted to interlocking.

What if we could program in logical blocks that sound like the English language? For example, we would have a block for starting the process, a block for checking for part presence, a block for locking the part, and so on. Then all we would have to do is break down any process into logical steps, and the steps (or stages) would be our program. Figure 10–3 shows a solution for the press process. It is much easier to understand than a ladder program.

When we turn on the PLC it will start in stage 0 (see Figure 10–3). When the start button is pushed, the PLC changes to stage 1.

- Stage 1 checks for part presence. When contact X2 closes, the PLC moves into stage 2.
- Stage 2 locks the part by turning on coil Y1. When contact X3 closes, the PLC moves into stage 3.
- Stage 3 turns on coil Y2, which activates the press. When the lower limit is reached, it closes contact X4 and the PLC enters stage 4.
- Stage 4 raises the press by turning on coil Y3. When the press reaches the upper limit, it activates the upper limit contact X5. The PLC then moves into stage 5.
- Stage 5 unlocks the part by turning coil Y4 on. When the confirm unlock contact X6 closes, the PLC moves into stage 6.
- Stage 6 moves the conveyor by turning on coil Y5. When X7 closes, the PLC will move into stage 7.

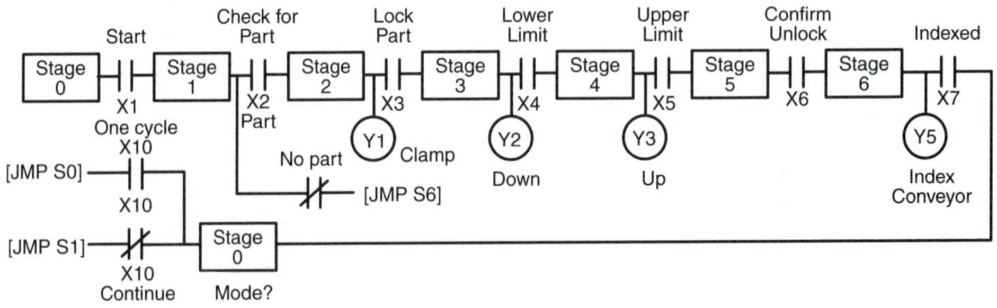

Figure 10–3 Same process as in Figure 10–1 but programmed using stage programming. *(Courtesy AutomationDirect Inc.)*

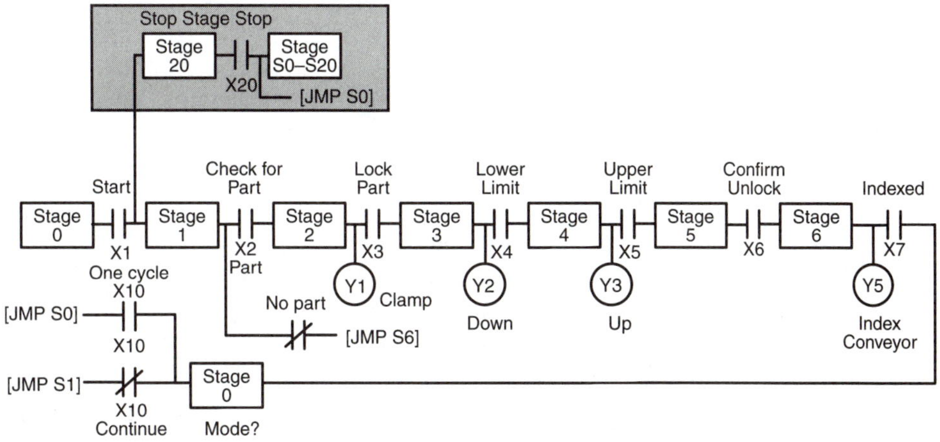

Figure 10–4 How a stop stage can be added to the stage program. *(Courtesy AutomationDirect Inc.)*

Let's add a stop stage to our program (see Figure 10–4). Note that when contact X1 is closed, the PLC enters stage 1 and stage 10. Both stages are active. In fact, stage 10 will be active during all the stages. This allows us to stop the program at any point during operation. If X20 is closed, stages 0 through 7 are reset and the PLC is told to jump to stage 0. Stage 10 is de-activated when the PLC is in step 0.

Rules of Stage Programming

Only instructions in active stages are executed. This eliminates the need for all the complex and sometimes devious interlocking. Stages are activated by one of the following:

1. Power flow makes contact with a stage label.
2. A jump to stage is executed.
3. The stage status bit is turned on by a set instruction.
4. The initial stage is executed when the PLC enters run mode.

Stages are de-activated by:

1. Power flow transitions from the stage.
2. Jumping from the stage.
3. A reset instruction.

Step Programming

Omron PLCs have a similar type of programming available. Omron calls it step programming. It is very similar to stage programming. The application to be written is broken into logical steps. This is very much the way that manufacturing processes function. The production of a product can be broken down into logical processing steps: step A, step B, step C "or" step D, step E, and so on. The beauty of step programming is that only the desired steps are active at that time, which means that almost no interlocking is required. The individual blocks (steps) can be written and will not affect other steps.

There are two types of instruction blocks: step and step next (SNXT) (see Figure 10–5). The step next is used to move from one block of instructions to the next. The step instruction is used to show the beginning of a block of ladder logic. The block of instructions is then marked by a step next (SNXT) instruction.

Input conditions are used with step next instructions to switch between blocks (steps) (see Figure 10–6). If input 0001 becomes true, the PLC will move to step HR0001. It will evaluate the logic only between step HR0001 and the following step next instruction. It will stay in this block of instructions until input 0002 becomes true. At that point, the PLC will move to step HR0002 and evaluate the logic for that step. Note that only the logic between the step and the step next is evaluated. This eliminates all of the interlocking that normally occupies about two-thirds of the ladder and two-thirds of the programmer's time.

Figure 10–7 shows the diagram of a two-process manufacturing line. Product enters and is weighed on the input conveyor. It is then routed to one of two possible processes, depending on the weight of the product. After the individual processing, all product is printed at the last processing station. Note that sensors are used to sense when product enters and leaves a process.

Figure 10–8 shows a block diagram of the process. Note that sensor input conditions have been used to route the product through the various possible production steps. Also note that parallel processing is possible; steps do not have to be serial.

Figure 10–9 shows an example of what the ladder logic would look like for this process. Switches A1 and B1 have been used as input conditions to step next in-

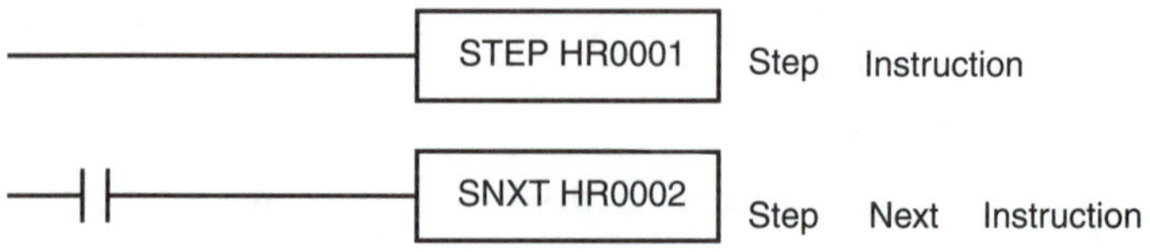

Figure 10–5 Two types of step instruction.

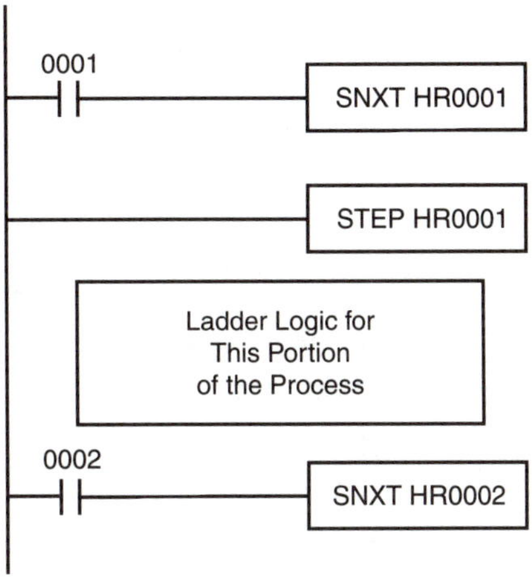

Figure 10–6 How a step instruction is used in a ladder diagram. *(Courtesy of Omron Electronics.)*

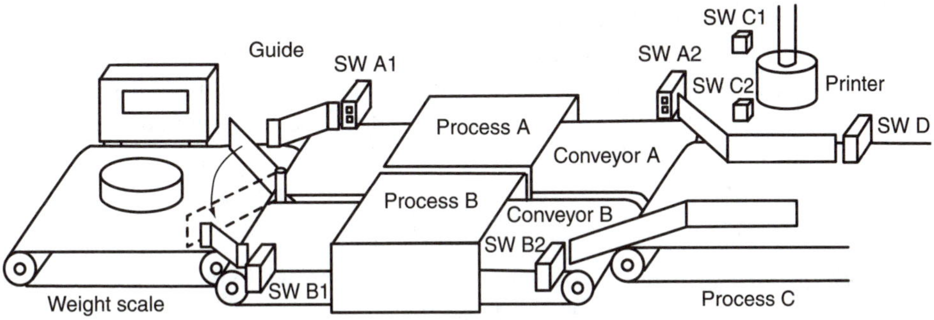

Figure 10–7 Conveyor system. The processing of the product depends on its weight. *(Courtesy of Omron Electronics.)*

structions. If switch A1 is true, the PLC will jump to step HR0000. This is process A. The logic of step 0000 is then active until switch A2 senses the product leaving process A. Switch A2 is then true, which makes the conditions true for step next to jump to HR0002. The PLC then evaluates step HR0002 (process C). When the product leaves process C, it makes switch D true. Switch D makes the step next 24614 true. This sets a bit in memory, which indicates that process C is complete and has completed a part. This bit can be used by the programmer to allow further processing.

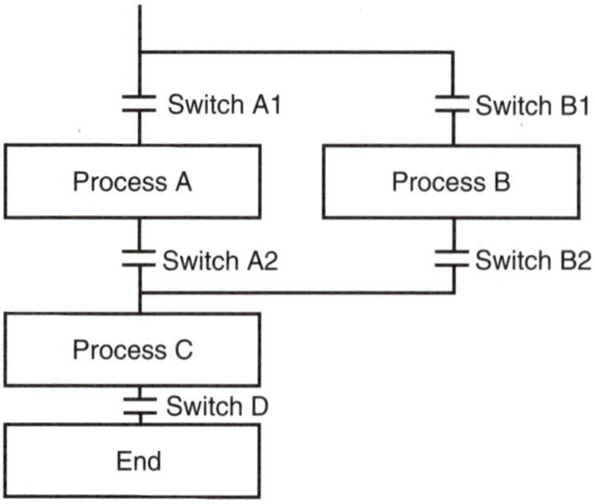

Figure 10–8 A block diagram of the conveyor processing system, including the input conditions. *(Courtesy of Omron Electronics.)*

If the product weight had indicated that the product was a process B type, the product would have been routed through process B and then C. While this is a relatively simple example, it would require quite a lengthy ladder diagram if it were written with normal ladder logic. Which would you rather troubleshoot, a regular ladder or a ladder written in logical steps? In addition to being logically organized into processing steps, the step ladder logic is probably one-fifth to two-thirds shorter.

Fuzzy Logic

Fuzzy logic is a control method that attempts to make decisions as a human being would. When we make decisions, we consider all the data we have available to us based on the rules we have formulated and the present conditions. We do not use hard and fast rules; we can weight each rule as to its importance. This means that we do not use one fixed mathematical formula to make our decision. Fuzzy logic is an attempt to mimic human decision making. Video recorders with fuzzy logic can differentiate wanted movement of the camera from unwanted movement and thus stabilize the picture. Many industrial applications are appropriate for fuzzy logic.

The basis of fuzzy logic is the "fuzzy" set. If we thought of people's height or weight, we could easily see an image of normal height and weight. To us it seems straightforward. If you think about it, though, what is normal? A bell curve can be used to show the relationship of people's heights (see Figure 10–10). Let's assume for the sake of discussion that 5 feet 9 inches is a normal height. We are not trying to say that someone who is 5 feet 1 inch or 6 feet 5 inches is not of normal height, however. Let's agree that anyone shorter than 5 feet 0 inches is short and anyone

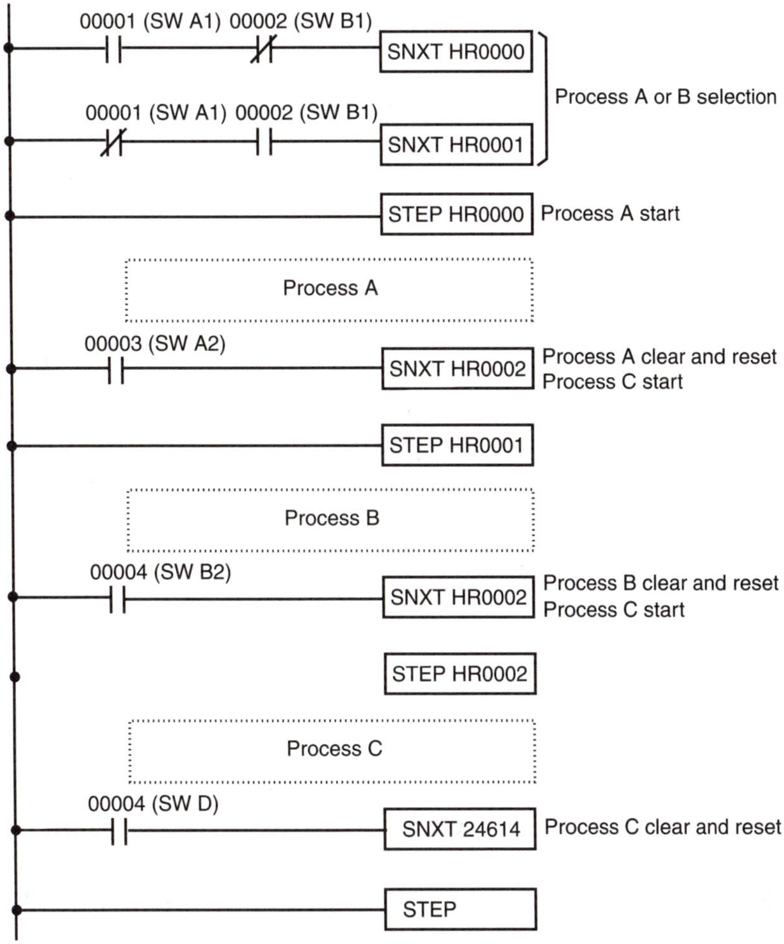

Figure 10–9 Ladder diagram for the conveyor processing system. Note that the actual ladder logic for each process has not been shown. *(Courtesy of Omron Electronics.)*

taller than 6 feet 6 inches is tall. As you can see, the definition of normal height is actually quite complex.

Consider Figure 10–10 again. A grade between 1.0 and 0.0 has been assigned to show how strongly we feel that the height is "normal." For example, 5 feet 9 inches has been assigned the value 1.0. We do not feel quite as strongly about 5 feet 3 inches being normal height. We assign a value of only 0.2 to it. The graph (bell curve) can also be called a *membership function.*

Fuzzy logic decision making can be divided into two steps: the inference step and the "defuzzifier" step (see Figure 10–11). The inference process is made up of

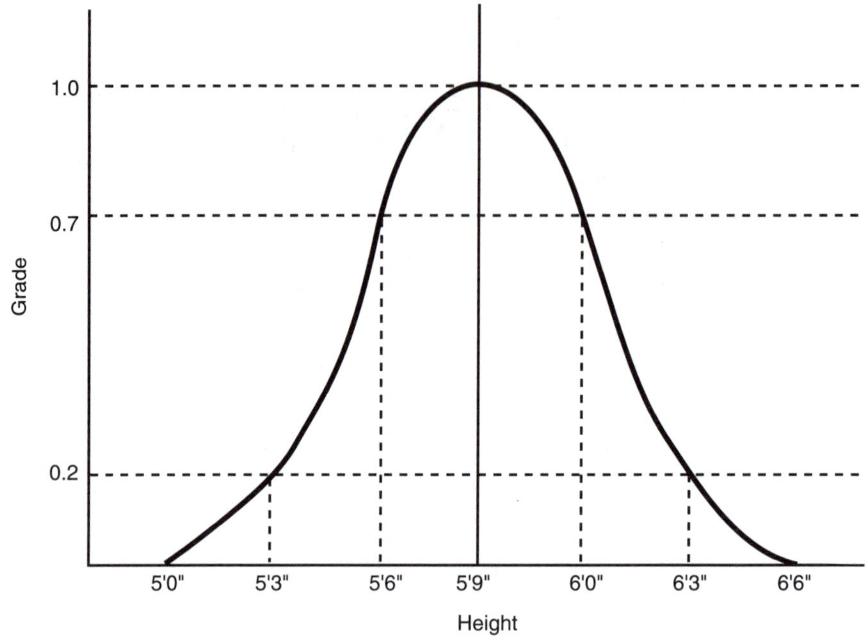

Figure 10–10 Chart of people's heights. *(Courtesy of Omron Electronics.)*

several rule-based decisions that are summed to a single logical sum. Each rule is composed of input conditions (antecedent block) and a conclusion (consequent block). The logical products of each of these are summed.

Each rule is independent. They are analyzed independently. They are separate until they are combined to produce the logical sum. This combination of simple rules allows very complex decisions to be made. The rules are analyzed in parallel and a logical sum is generated. This produces a much better result than a simple formula to control all situations.

Imagine a cart with a stick (see Figure 10–12). The cart must be moved to balance the stick. Earlier, when we thought about height, we used terms such as *short, normal,* and *tall.* We can call these words *codes* or *labels.* We use these labels to express degree, such as moderate, almost, or a little. This example utilizes seven labels to express degree. More or fewer labels can be used.

Figure 10–13 seems complex at first glance, but it is not. The figure shows the antecedent membership functions and their labels. Note that the figure actually shows a series of triangles. A triangular membership function is used instead of the bell curve we used in the height example. Originally the bell-shaped membership function was used but due to the complexity of calculation, the triangular function

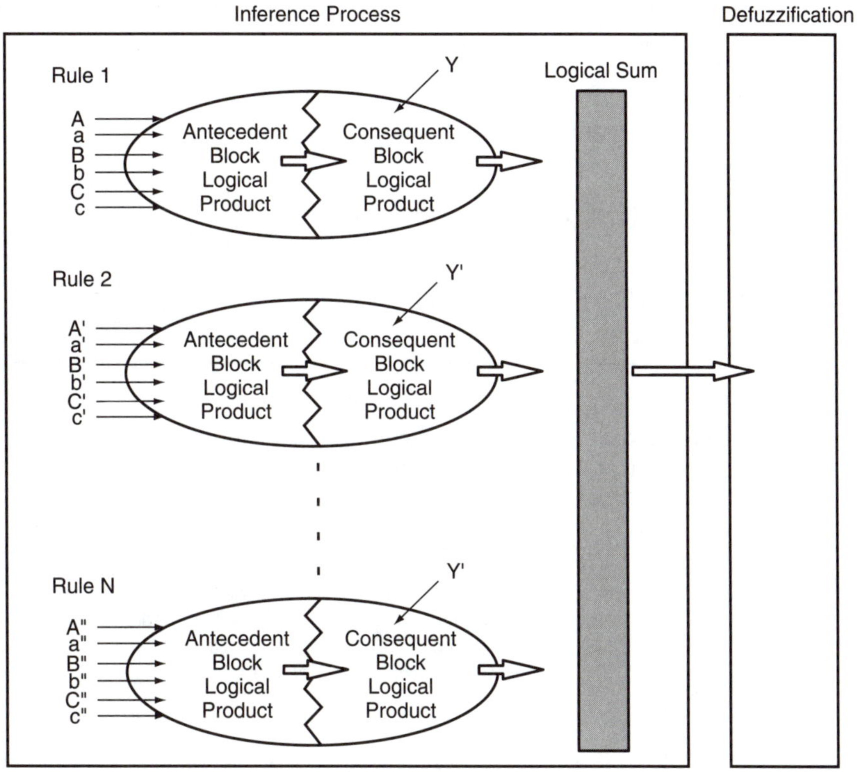

Figure 10–11 Basic fuzzy logic decision-making process. The rules are shown on the left. Decisions are made based on the individual rules. These decisions are summed to provide a logical sum. The sum is "defuzzed" and used to control the system. *(Courtesy of Omron Electronics.)*

is now used most often. The results of both are comparable. These triangular membership functions (antecedent membership) are composed so that the labels overlap (see Figure 10–13). This overlap permits reliable readings even when the level is not distinct or when the input from sensors is continually changing.

Developing the Rules in Code

The inclination of the stick from vertical is θ, and the speed with which the inclination is changing (the angular velocity) is $d\theta$. Both θ and $d\theta$ are inputs from sensors. The sensors might be an encoder and a tachometer. The tachometer would give input on the angular velocity, and the encoder would give input on the inclination. These two variables can be used to write the production rules. The change in speed of the platform on which the stick is mounted is ΔV.

$$m*\frac{d^2}{dt^2}[s(t) + L*\sin\phi(t)] = H(t)$$

$$m*\frac{d^2}{dt^2}[L*\cos\phi(t)] = V(t) - m*g$$

$$J*\frac{d^2}{dt^2}\phi(t) = L*V(t)*\sin\phi(t) - L*H*\cos\phi(t)$$

$$M*\frac{d^2}{dt^2}s(t) = \mu(t) - H(t) - F*\frac{d}{dt}s(t)$$

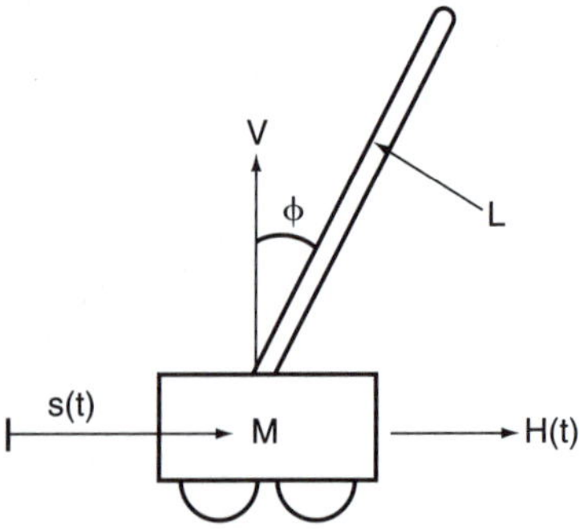

Figure 10–12 System controlled in this example. The cart must be moved just the right amount and at the right velocity to balance the stick. *(Diagram and example courtesy of Omron Electronics.)*

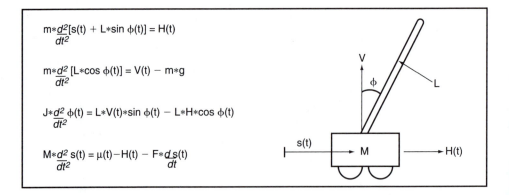

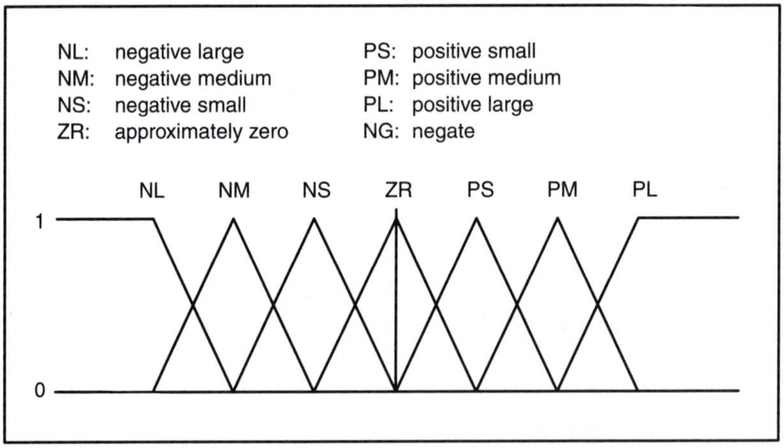

Figure 10–13 The triangular membership functions used to represent this system. Note that seven triangular functions are shown.

The antecedent block is usually composed of more than one variable linked by ANDs. The rules are linked by ORs. In this example we use seven rules. Refer to Figure 10–14. There are seven possible states for each input (NL for negative large, NM for negative medium, etc.).

Our example has two inputs: inclination and angular velocity. This means that if there are seven rules, we can have 49 potential combinations. (The number of possible combinations was calculated by raising 7 to the second power.) Each combination can be a rule. For this example seven rules are used. Only important rules need to be defined.

Examine Figure 10–14. Only five of the seven states describe the angle of inclination. Only three describe the angular velocity. This reduces the possible combinations to 15 (3×5) antecedent blocks. Of the fifteen remaining combinations, only seven are useful in describing how the system must operate. This is very similar to the way we think. We discard the rules that are not applicable to a particular decision and use only the relevant ones.

The following explanation of fuzzy logic evaluation might seem confusing at first. You must study the graphs carefully as you read to gain an understanding of the process. Then it will seem simple and straightforward. Study Figure 10–15. This figure shows the two variables (inclination and angular velocity) at one point in time. The input value at that precise point in time is shown by the vertical line.

Rule	Antecedent Block	Consequent Block
Rule 1	It the stick is inclined moderately to the left and is almost still	then move the hand to the left quickly
Rule 2	If the stick is inclined a little to the left and is falling slowly	then move the hand to the left slowly
Rule 3	If the stick is inclined a little to the left and is rising slowly	then keep the hand as it is
Rule 4	If the stick is inclined moderately to the right and is almost still	then move the hand to the right quickly
Rule 5	If the stick is inclined a little to the right and is falling slowly	then move the hand moderately to the right slowly
Rule 6	If the stick is inclined a little to the right and is rising slowly	then keep the hand as it is
Rule 7	If the stick is almost vertical and is almost still	then keep the hand as it is

Figure 10–14 The seven rules that will be used for the decision-making process. Note the consequent blocks. The consequent block shows what should be done if the rule is true.

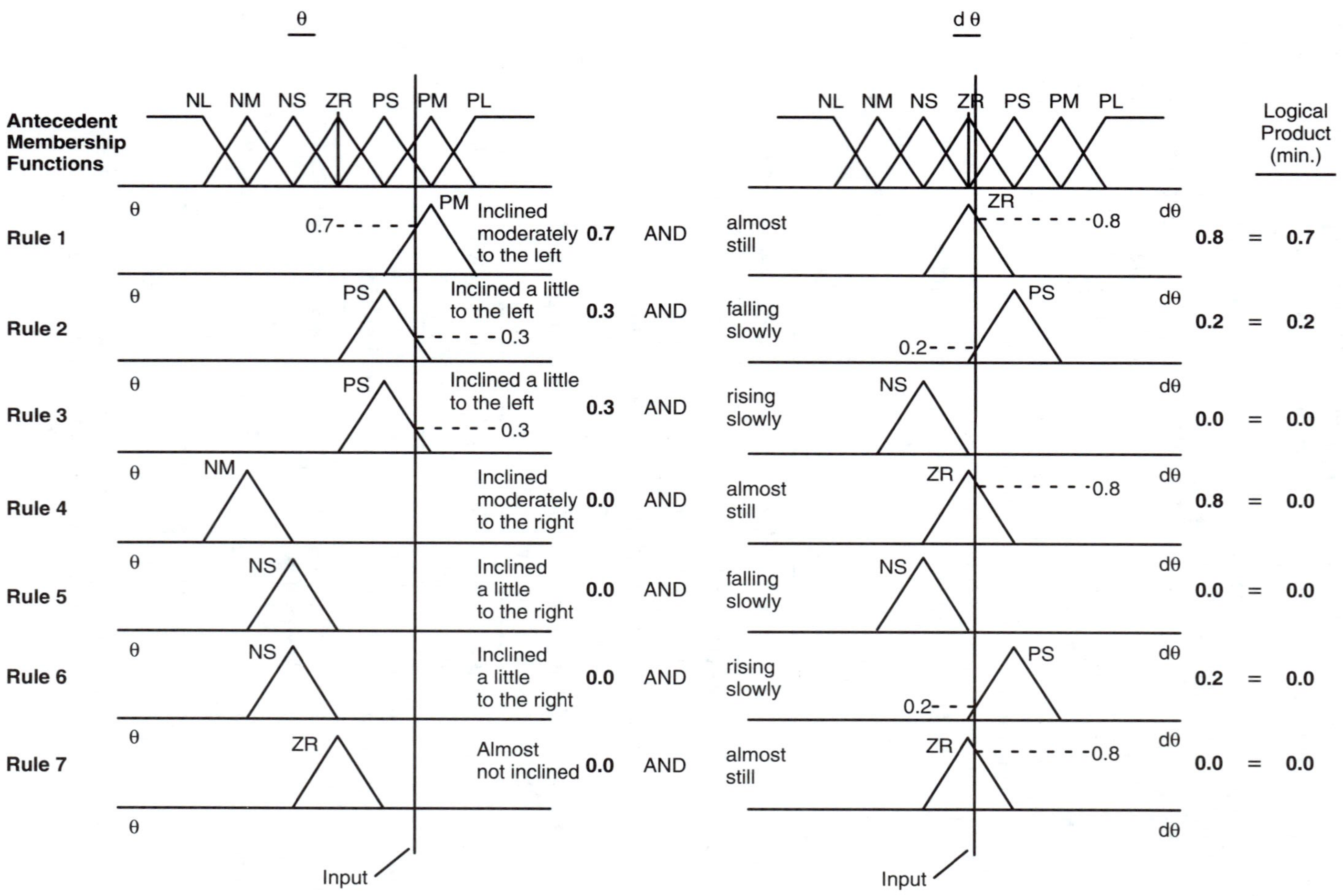

Figure 10–15 How the two variables were evaluated by each rule at one point in time. *(Courtesy of Omron Electronics.)*

Each rule evaluates the input based on its membership function and assigns a value. For our example, rule 1 evaluates the input based on the PM label (positively medium inclination to the right). The rule assigns a value of 0.7 based on where the input intersects the triangular membership function (see Figure 10–15). Rule 1 is also evaluated for the angular velocity and a value of 0.8 is assigned.

The values of rule 1 inclination and rule 1 angular velocity are then evaluated to find the logical product (minimum). The minimum value is used. In this case rule 1 inclination was 0.7 and rule 1 angular velocity was 0.8. The logical product is the smallest value, or 0.7 for rule 1. Note that each of the seven rules is evaluated based on the two inputs (inclination and angular velocity). Logical minimums are found for each rule. Note that each rule is evaluated based on where the input intersects (or does not intersect) its membership function.

The next step is to find the logical sum. The logical sum is the combination of the results of the rule evaluations. Study Figure 10–16. This figure shows how the logical sum is derived. Again it may look complex at first. Study rule 1. The logical product of rule 1 was 0.7. The area comparable to 0.7 is filled in the triangular membership function for the rule 1 consequent block. The result of this first evaluation says that the cart should be moved moderately to the left quickly. Rule 2 shows that the cart should be moved to the left a little quickly (value of 0.2). The products of rules 3 to 7 were zero, so they do not affect the outcome.

The logical sum of the consequent blocks is shown at the bottom of the figure. The membership functions for rules 1 and 2 are simply combined to give the result shown at the bottom of the figure. A decision must now be made about how to move the cart (how far and how fast). This is done by calculating the center of gravity for the logical sum of the rules. The result becomes the output value.

As you have seen, seven rules and two inputs were used for this simple example. The evaluation resulted in an output value that was based on an evaluation of the rules and the status of the inputs at that particular time. The advantage of fuzzy logic is that it is more flexible than mathematical models of systems. PID is a mathematical model. With the wide variety of industrial systems and operating conditions, it is difficult to develop an accurate mathematical model for all industrial systems.

Fuzzy logic helps solve the problem. Fuzzy logic breaks any system down into more humanlike rules. These simple rules are evaluated and compensate for the actual conditions at that time. Fuzzy logic is also easier to understand because the operator's knowledge and thought process is reflected in the control process. PID, on the other hand, is often difficult for people to understand.

In 1965, Professor L. A. Zadeh of the University of California at Berkeley presented a paper outlining fuzzy theory. At first the theory was met with indifference and even hostility. Around 1970 fuzzy logic began to be used in Japan, Europe, and China. It began to show positive results. There will be room for both fuzzy logic and

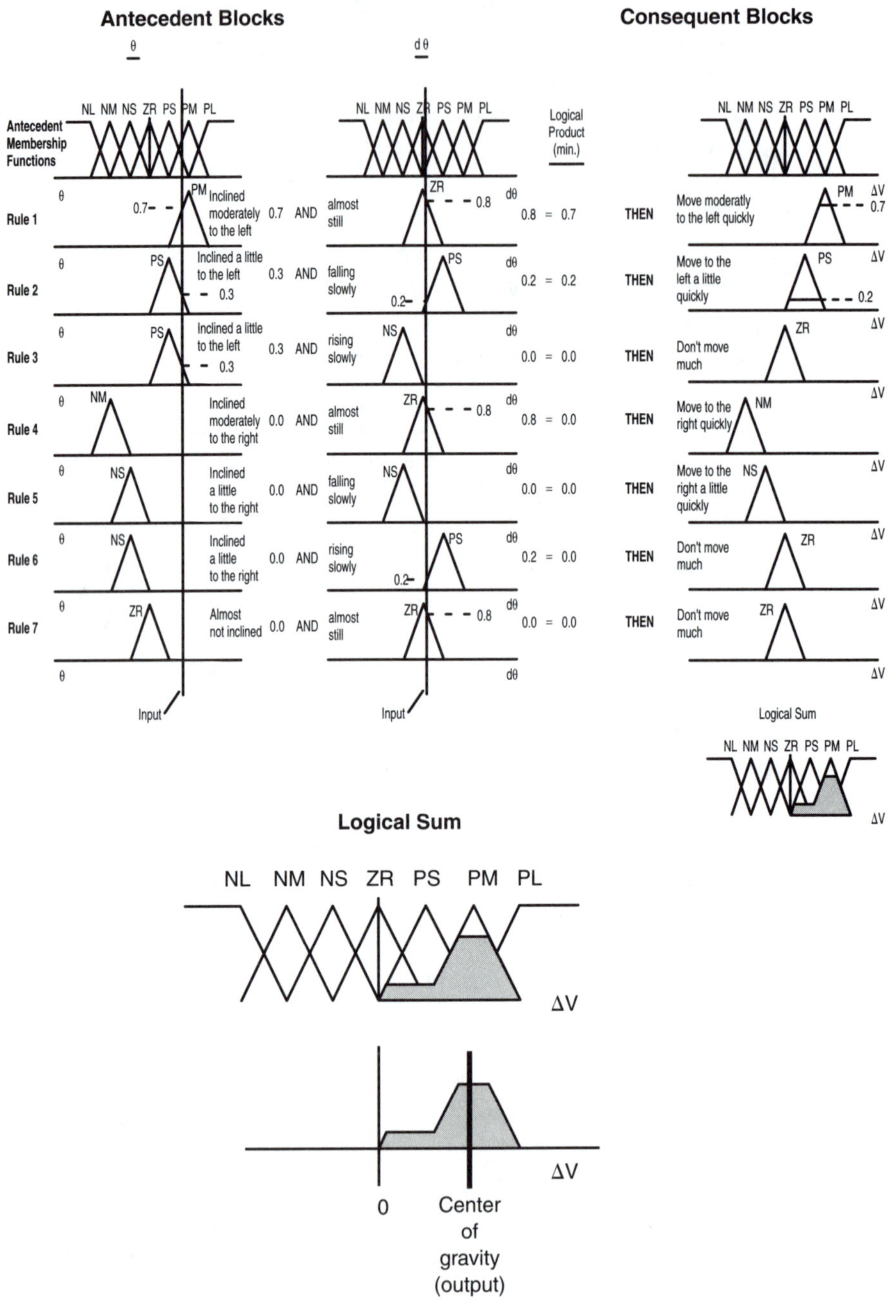

Figure 10–16 How the logical sum is derived.

other control systems, such as PID. Each will have its own niche. The future of fuzzy logic does appear bright, however. Although fuzzy technology is still young, it has already contributed significantly to commercial and industrial control applications. Fuzzy technology can be applied through software, dedicated controllers, or fuzzy microprocessors in products. This flexibility and simplicity of fuzzy logic may make it a basic control and information-processing technology in the twenty-first century.

Here are a few examples of applications that could benefit from fuzzy logic control:

- Nonlinear systems such as process control, tension control, and position control.
- Systems with gross input deviations or insufficient input resolution.
- Difficult-to-control systems that require human intuition and judgment.
- Systems that require adaptive signal processing to overcome changing environmental or process conditions.
- Processes that must balance multiple inputs or that have conflicting constraints.

State Logic

State logic is one of many new approaches to programming manufacturing systems. Many new languages are being developed to program control systems. They are an attempt to make programming an easier task. They do not use ladder logic. The commands typically used in these new languages are straightforward English statements. State logic is presented here as an example of one of these emerging languages. It represents a different approach to control logic. State logic does not use ladder logic. It breaks processes into states and tasks. The actual logic is written in plain English. The resulting program is easy to understand.

Study Figures 10-17 and 10-18. The system is shown in Figure 10–17 and the logic is shown in Figure 10–18. Note that the logic consists of small sections called tasks. Each task is one logical portion of the entire process. Each task is then divided into states that control a portion of the task. Each state has a statement associated with it to make decisions based on real inputs and/or values and control outputs. Note that the statements are close to English language constructions.

Example System

Cans are filled with two chemicals and mixed as they move along a conveyor belt. When a can is placed on the conveyor line, it trips the can_in_place limit switch, and if neither the fill nor the mix tasks are currently active, the conveyor will start.

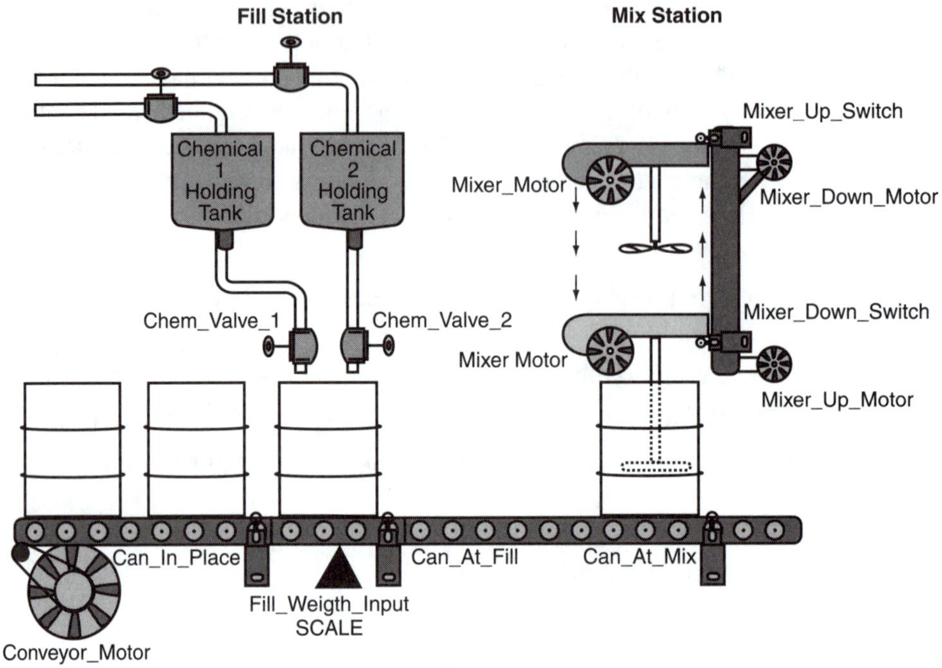

Figure 10–17 Example of a fill-and-mix system. *(Figure and description courtesy of Adatek.)*

The conveyor runs until a can arrives at either the fill or mix stations, at which time the conveyor stops and waits for another can and any filling or mixing tasks to be completed.

When a can arrives at the fill station, it trips the can_at_fill limit switch. Chemical valve 1 opens until the fill weight is above 20 lbs. Then chemical valve 2 opens until the fill weight is above 30 lbs. Then the fill station waits for another can to arrive to begin another cycle.

When a can arrives at the mix station, it trips the can_at_mix limit switch. The mixer down motor runs until the mixer down limit switch is tripped. The mixer motor then starts. After the mixer motor has run for 30 seconds, the mixer up motor runs until the mixer up switch is tripped. A counter is incremented to keep track of the number of completed cans. (The counter is called Can_Inventory.) The message "Batch Cycle Complete" is written to the operator panel every time a can is mixed. The actual logic (program) is shown in Figure 10–18.

State logic is a very high-level programming language for programming systems. It is based on finite state machine theory. State logic is essentially a framework for modeling real-world processes. It is a language designed to control systems.

Project: Batching System

```
Task: Fill_Station
        State: PowerUp
          When Can_At_Fill is on, go to the Batch_Chem_1 State.
        State: Batch_Chem_1
          Open Chem_Valve_1
          When Fill_Weight is above 20 pounds, go to Batch_Chem_2.
        State: Batch_Chem_2
          Open Chem_Valve_2 until Fill_Weight is more than 30 lbs,
          then go to the Batch_Complete State.
        State: Batch_Complete
          When Can_At_Fill is off, go to the PowerUp State.
Task: Mix_Station
        State: PowerUp
          If Can_At_Mix is on, go to the Lower_Mixer State
        State: Lower_Mixer
          Run the Mixer_Down_Motor until the Mixer_Down_Switch is tripped,
          then go to the Mix_Chemicals State.
        State: Mix_Chemicals
          Start the Mixer_Motor.
          When 30 seconds have passed, go to the Raise_Mixer State
        State: Raise_Mixer
          Run the Mixer_Up_Motor until the Mixer_Up Switch is tripped,
          then go to the Batch_Complete.
        State: Batch_Complete
          When Can-At-Mix is off, go to the Update_Inventory State.
        State: Update_Inventory
          Add 1 to Can_Inventory.
          Write "Batch Cycle Complete" to the Operator_Panel and go to PowerUp.
Task: Conveyor
        State: PowerUp
          When Can_In_Place is on and
          (Fill_Station Task is in the PowerUp or Batch_Complete) and
          (Mix_Station Task is in the PowerUp or Batch_Complete),
          go to Start_Cycle
        State: Start_Cycle
          Start the Conveyor_Motor.
          When Can_In_Place is off, Go to Index_Conveyor State.
        State: Index_Conveyor
          Run Conveyor_Motor.
          When Can_At_Fill or Can_At_Mix is on, go to the PowerUp State.
```

Figure 10–18 The actual logic for the system shown in Figure 10–17. The state logic control language can presently be used on personal computers and some PLCs. *(Logic and explanation courtesy of Adatek.)*

The State Logic Model

All real-world processes move through sequences of states as they operate. Every machine or process is a collection of real physical devices. The activity of any device can be described as a sequence of steps in relation to time. For example, a cylinder can exist in only one of three states: extending, retracting, or at rest. Any desired action for that cylinder can be expressed as a sequence of these three states. Even a continuous process goes through start-up, manual, run, and shut down phases. All physical activity can be described in this manner. It is not difficult to express an event or condition that can be used to cause the cylinder (or other device) to change states. For example, if the temperature is over 100°F, turn the warning light on and go to the shutdown procedure. Time and sequence are natural dimensions of the state model just as they are natural dimensions of the design and operation of every control system, process, machine, and system. State logic control uses these attributes (time and sequence) as the components of program development. As a result, the control program is a clear snapshot of the system being controlled. State logic is a hierarchical programming system that consists of tasks, states, and statements.

Tasks Tasks are the primary structural elements of a state logic program. A task is a description of a process activity expressed sequentially and in relation to time. If we were describing an automobile engine, the tasks would include the starting system task, the fuel system task, the charging system task, the electrical system task, and so on. Almost all processes contain multiple tasks operating in parallel. Tasks operating in parallel are necessary because most machines and processes must do more than one task at a time. State logic provides for the programming of many tasks that are mutually exclusive in activity yet interactive and joined in time.

States States are the building blocks of tasks. The activity of a task is described as a series of steps called states. A state describes the status or value of an output or group of outputs. These are the outputs of the control system and thus are inputs to the process. Every state contains the rules that allow the task to transition to another state. A state is a subset of a task that describes the output status and the conditions under which the task or process will change to another state. The states, when taken in aggregate, provide a description of the sequence of activity of the process or machine under control. They also provide an unambiguous specification of how that portion of the process responds in all conditions.

Statements Statements are the user's commands set to create state descriptions. The desired output-related activity of each state can be described by using statements. Statements can initiate actions or can base an output status change on a conditional statement or a combination of conditions. Any input value or variable can be used in conditional statements. Variables can include state status from other tasks as well as typical integer, time, string, analog, and digital status variables. State logic can be used to program computers and some PLCs. An example of a state logic module for a PLC is shown in Figure 10–19.

State logic allows the programmer to write the control program in natural English statements. State logic and other types of system languages are sure to increase in acceptance and popularity because of their simplicity and ease of use.

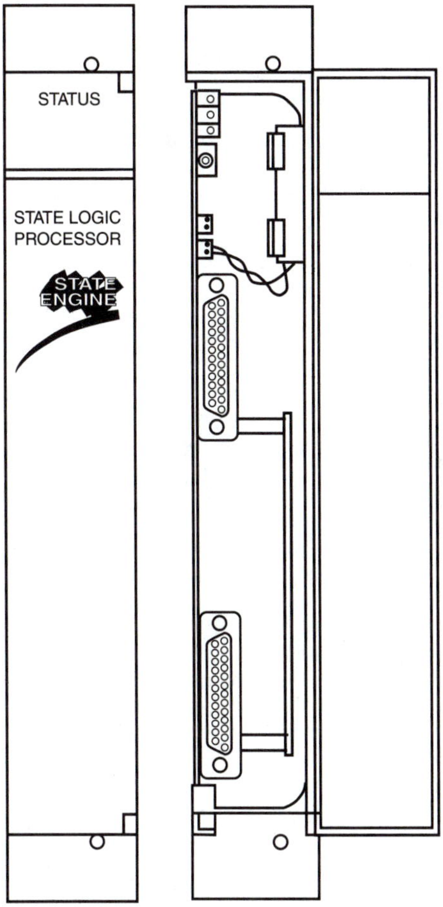

Figure 10–19 A GE Fanuc state logic processor module for a series 90-30 PLC. *(Courtesy GE Fanuc Inc.)*

Questions

1. What is stage/step programming?
2. List at least three advantages of stage/step programming.
3. Why does stage programming reduce the need for interlocking logic?
4. What is fuzzy logic?
5. Fuzzy logic decision making can be divided into two steps: the _____step and the _____ step.
6. What are rules?
7. What types of applications are appropriate for fuzzy logic?
8. What is state logic programming?
9. List at least three advantages of state logic programing.
10. What is a task in state logic programming?

Chapter 11

Industrial Sensors

This chapter examines the types and uses of industrial sensors. It discusses digital and analog sensors and wiring of sensors.

OBJECTIVES

Upon completion of this chapter, you will be able to:

1. Describe at least two ways in which sensors can be classified.
2. Choose an appropriate sensor for a given application.
3. Describe the typical uses of digital sensors.
4. Describe the typical uses of analog sensors.
5. Explain common sensor terminology.
6. Explain the wiring of load and line-powered sensors.
7. Explain how field sensors function.
8. Explain the principle of operation of thermocouples.
9. Explain common thermocouple terms.

The Need for Sensors

Sensors have become vital in industry, and manufacturers are moving to integrating pieces of computer-controlled equipment. In the past, operators were the brains of the equipment because they were the source of all information about the operation of a process. The operator knew whether parts were available, which parts were ready, whether they were good or bad, whether the tooling was acceptable, whether the fixture was open or closed, and so on. The operator could sense problems in the operation by seeing, hearing, feeling (vibration, etc.), and smelling problems.

Industry is now using computers (in many cases PLCs) to control the motions and sequences of machines. PLCs are much faster and more accurate than an operator at these tasks. PLCs cannot see, hear, feel, smell, or taste processes by themselves but use industrial sensors to give industrial controllers these capabilities.

The PLC can use simple sensors to check whether parts are present or absent, to size the parts, and even to check if the product is empty or full. The use of sensors to track processes is vital for the success of the manufacturing process and to ensure the safety of the equipment and operator. In fact, sensors perform simple tasks more efficiently and accurately than people do. Sensors are much faster and make far fewer mistakes.

Studies have been performed to evaluate how effective human beings are in doing tedious, repetitive inspection tasks. One study examined people inspecting table tennis balls. A conveyor line brought table tennis balls by a person. White balls were considered good, and black balls were considered scrap. The study found that people were only about 70 percent effective at finding the defective Ping-Pong balls. Certainly, people can find all of the black balls, but they do not perform mundane, tedious, repetitive tasks well; they become bored and make mistakes, whereas a simple sensor can perform simple tasks almost flawlessly.

Sensor Types

Contact versus Noncontact

Sensors are classified in several ways; one common classification is contact or noncontact. The use of a noncontact sensor, also called a *proximity sensor,* is a simple way to identify a sensor. If the device must contact a part to sense it, the device is a *contact sensor.* A simple limit switch on a conveyor is an example. When the part moves the lever on the switch, the switch changes state. The contact of the part and the switch creates a change in state that the PLC can monitor.

Noncontact sensors can detect the part without touching it physically, which avoids slowing down or interfering with the process. *Noncontact sensors* (electronic) do not operate mechanically (i.e., they have no moving parts) and are more reliable and less likely to fail than mechanical ones. Electronic devices are also much faster than mechanical devices, so noncontact devices can perform at very high production rates.

The remainder of the chapter examines noncontact sensors.

Digital Sensors

Another way to classify sensors is as digital or analog. Industrial applications need both digital and analog sensors. A *digital sensor* has two states: on or off. Most applications involve presence versus absence and counting, which a digital sensor does perfectly and inexpensively. Digital sensors are simpler and easier to use than analog ones, which is a factor in their wide use. Computers are digital devices that actually work with only 1s or 0s (on or off).

Digital output sensors are either on or off. They generally have transistor outputs. If the sensor senses an object, the transistor turns on and allows current to flow. The output from the sensor is usually connected to a PLC input module.

Sensors are available with either normally closed or normally open output contacts. Normally open contact sensors are off until they sense an object and turn on. Normally closed contact sensors are on until they sense an object, when they turn off. When photo sensors are involved, the terms *light-on* and *dark-on* are often used. *Dark-on* means that the sensor output is on when no light returns to the sensor, which is similar to a normally closed condition. A *light-on* sensor's output is on when light returns to the receiver, similar to a normally open sensor.

The current limit for most sensors' output is quite low. Usually output current must be limited to less than 100 milliamps. Users should check the sensor before turning on the power. Output current must be limited or the sensor can be destroyed! This is usually not a problem if the sensor is being connected to a PLC input because the PLC input limits the current to a safe amount.

Analog sensors, also called *linear output sensors,* are more complex than digital ones but can provide much more information about a process.

Think about a sensor used to measure the temperature. A temperature is analog information. The temperature in the Midwest is usually between 0 and 90 degrees. An analog sensor could sense the temperature and send a current to the PLC. The higher the temperature, the higher the output from the sensor. The sensor may, for example, output between 4 and 20 milliamps, depending on the actual temperature, although there are an unlimited number of temperatures (and thus current outputs). Remember that the output from the digital sensor is on or off. The output from the analog sensor can be any value in the range from low to high. Thus, the PLC can monitor temperature very accurately and control a process closely. Pressure sensors are also available in analog style. They provide a range of output voltage (or current), depending on the pressure.

A 4–20 milliamp current loop system can be used for applications when the sensor needs to be mounted a long distance from the control device. A 4–20 milliamp loop is good for about 800 meters. A 4–20 milliamp sensor varies its output between 4 and 20 milliamps. The sensor must be adjusted for range and sensitivity so that it can measure the required value of the characteristic, such as temperature.

Digital Sensors

Digital sensors come in many types and styles, which are examined next.

Optical Sensors

All optical sensors use light to sense objects in approximately the same manner. A light source (emitter) and a photodetector sense the presence or absence of light. Light emitting diodes (LEDs), which are semiconductor diodes that emit light, are typically used for the light source because they are small, sturdy, and very efficient, and can be turned on and off at extremely high speeds. They operate in a narrow wavelength and are very reliable. LEDs are not sensitive to temperature, shock, or vibration and have an almost endless life. The type of material used for the semiconductor determines the wavelength of the emitted light.

The LEDs in sensors are used in a pulsed mode. The photo emitter is pulsed (turned off and on repeatedly). The "on time" is extremely small compared to the "off time." LEDs are pulsed for two reasons: to prevent the sensor from being affected by ambient light and to increase the life of the LED. This pulsed mode can also be called *modulation.*

The photodetector senses the pulsed light. The photo emitter and photo receiver are both "tuned" to the frequency of the modulation. The photodetector essentially sorts out all ambient light and looks only for the correct frequency. Light sources chosen are typically invisible to the human eye and wavelengths are chosen so that the sensors are not affected by other lighting in the plant. The use of different wavelengths allows some sensors, called *color mark sensors,* to differentiate between colors. Visible sensors are usually used for this purpose. The pulse method and the wavelength chosen make optical sensors very reliable.

Some sensor applications utilize ambient light emitted by red-hot materials such as glass or metal. These applications can use photo receivers sensitive to infrared light.

All of the various types of optical sensors function in the same basic manner. The differences are in the way in which the light source (emitter) and receiver are packaged.

Light/Dark Sensing Optical sensors are available in either light or dark sensing, also called *light-on* or *dark-on.* In fact, many sensors can be switched between light and dark modes. Light/dark sensing refers to the normal state of the sensor and whether its output is on or off in its normal state.

Light Sensing (Light-On) The output is energized (on) when the sensor receives the modulated beam of light. In other words, the sensor is on when the beam is unobstructed.

Dark Sensing (Dark-On) The output is energized (on) when the sensor does not receive the modulated beam. In other words, the sensor is on when the beam is blocked. Street lights are examples of dark-on sensors. When it gets dark outside, the street light is turned on.

Timing Functions Timing functions are available on some optical sensors. They are available with on-delay and off-delay. *On-delay* delays turning on the output by a user-selectable amount. *Off-delay* holds the output on for a user-specified time after the object has moved away from the sensor.

Types of Optical Sensors

Reflective Sensors One of the more common types of optical sensors is the *reflective* or *diffuse reflective type.* The light emitter and receiver are packaged in the same unit. The emitter sends out light, which bounces off the product to be sensed. The reflected light returns to the receiver where it is sensed (Figure 11–1). Reflective sensors have less sensing distance (range) than other types of optical sensors because they rely on light reflected off the product.

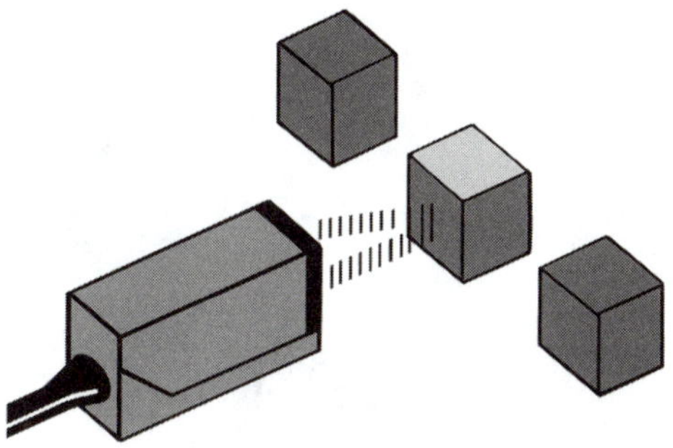

Figure 11–`1 Reflective-type sensor. The light emitter and receiver are in the same package. When the light from the emitter bounces off an object, the receiver senses it and the sensor's output changes state. The broken-line style of the arrows represents the pulsed mode of lighting, which is used to ensure that ambient lighting does not interfere with the application. The sensing distance (range) of this style is limited by how well the object can reflect the light back to the receiver. *(Courtesy ifm efector inc.)*

Polarizing Photo Sensor A special photo sensor senses shiny objects using a special reflector. The reflector actually consists of small prisms that polarize the light from the sensor. The reflector vertically polarizes the light and reflects it back to the sensor receiver. The photo sensor emits horizontally polarized light. Thus, if a very shiny object moves between the photo sensor and reflector and reflects light back to the sensor, it is ignored because it is not vertically polarized (see Figure 11–2).

Retroreflective Sensor The retroreflective sensor (see Figure 11–3) is similar to the reflective sensor. The emitter and receiver are both mounted in the same package. The difference is that the retroreflective sensor bounces the light off a reflector instead of the product. The reflector is similar to those used on bicycles. Retroreflective sensors have more sensing distance (range) than do reflective (diffuse) sensors but less sensing distance than that of a thru-beam sensor. They are a good choice when scanning can be done from only one side of the application, which usually occurs when space is a limitation.

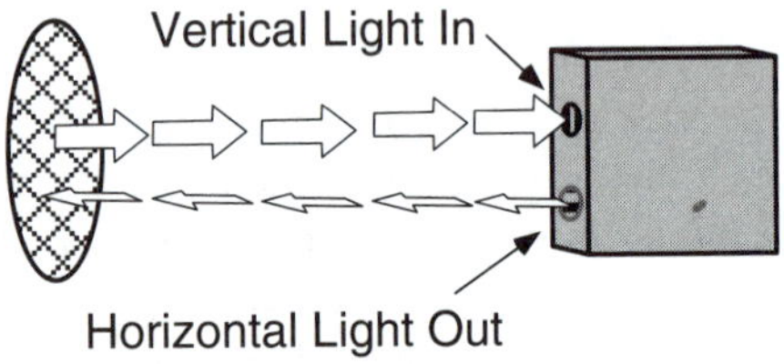

Figure 11–2 A polarizing photo sensor. Note the use of the special reflector.

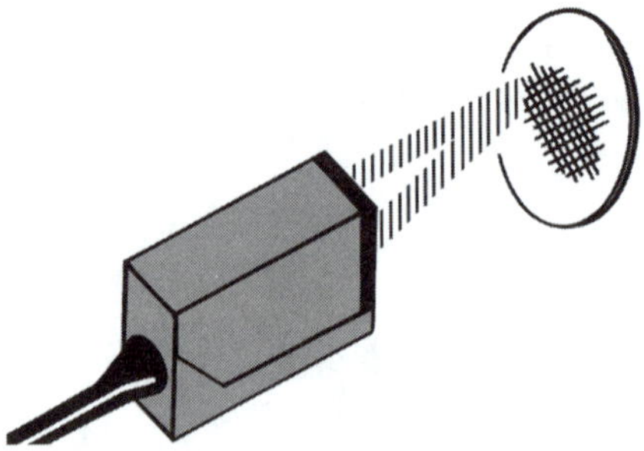

Figure 11–3 Retroreflective sensor. The light emitter and receiver are in the same package. The light bounces off a reflector and is sensed by the receiver. If an object obstructs the beam, the output of the sensor changes state. The excellent reflective characteristics of a reflector give this sensor more sensing range than a typical diffuse style does. The broken-line rules represent the pulsed method of lighting that is used. *(Courtesy ifm efector inc.)*

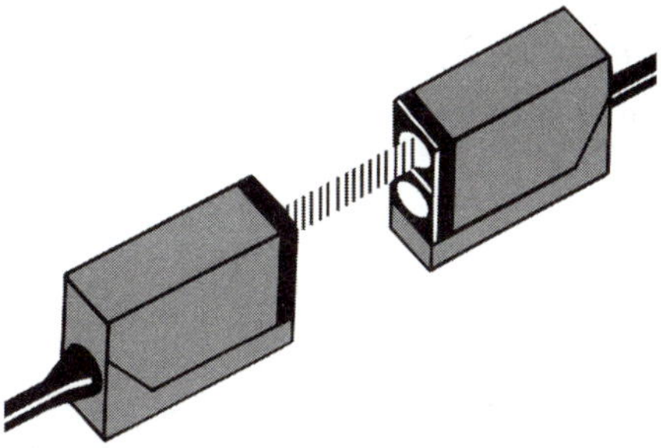

Figure 11–4 Thru-beam sensor. The emitter and receiver are in separate packages. The broken-line rule symbolizes the pulsed mode of the light that is used in optical sensors. *(Courtesy ifm efector inc.)*

Thru-Beam Sensor Another common sensor is the thru-beam (Figure 11–4), also called the *beam break mode.* In this configuration, the emitter and receiver are packaged separately. The emitter sends out light across a space and the receiver senses the light. If the product passes between the emitter and receiver, it stops the light from hitting the receiver, telling the sensor that a product is present. This is probably the most reliable sensing mode for opaque (nontransparent) objects.

Convergent Photo Sensor A convergent photo sensor, also called *focal length sensor,* is a special type of reflective sensor. It emits light to a specific focal point. The light must be reflected from the focal point to be sensed by the sensor's receiver (see Figure 11–5).

Fiber-Optic Sensor A fiber-optic sensor is simply a mix of the other types. The actual emitter and receiver are the same but with a fiber-optic cable attached to each. Fiber-optic cables are very small and flexible and are transparent strands of plastic or glass fibers used as a "light pipe" to carry light. The cables are available in both thru-beam and reflective configuration (see Figure 11–6). The light enters the end of the cable attached to the receiver, passes through the cable, and is sensed by the receiver. The light from the emitter passes through the cable and exits from the other end.

Color Mark Sensor A color mark sensor is a special type of diffuse reflective optical sensor that can differentiate between colors; some can even detect contrast between colors. It is typically used to check labels and to sort packages by color mark. A sensitivity adjust function can make fine adjustments. The object's background color is an important consideration. Sensor manufacturers provide charts for the proper selection of color mark sensors for various colors.

Laser Sensor A laser sensor is also used as a light source for optical sensors that perform precision-quality inspections requiring very accurate measurements.

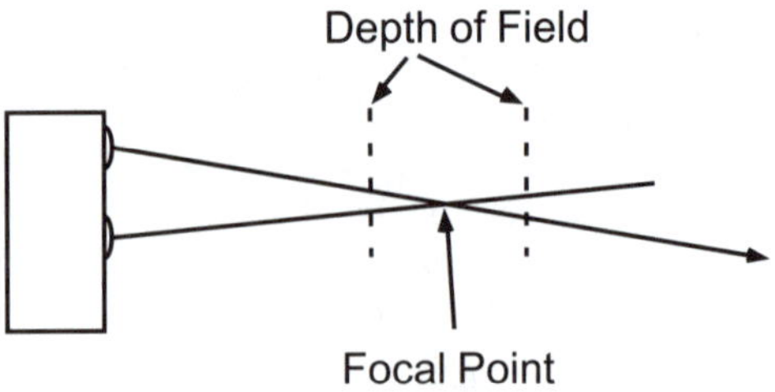

Figure 11–5 A convergent-style photo sensor.

Figure 11–6 Fiber-optic sensor. *(Courtesy ifm efector inc.)*

Resolution can be as small as a few microns. A laser LED is used as the source. Outputs can be analog or digital. The digital outputs can be used to signal pass/fail or other indications. The analog output can be used to monitor and record actual measurements.

Encoder Sensor An encoder sensor is used for position feedback and in some cases for velocity feedback. The two main types of encoders are incremental and absolute; incremental is the most common type. The resolution of an encoder is determined by the number of lines on the encoder disk. The more lines, the higher the resolution. Encoder increments with 500, 1000, or even more lines are common. The light from an LED shines through the lines on the encoder disk and a mask and is then sensed by light receivers (photo transistors). See Figure 11–7 for an enlarged view of an encoder.

Incremental Encoder

An incremental encoder (Figure 11–8) creates a series of square waves. Incremental encoders are available in various resolutions, which are determined by the number of slits for light to pass through. For example, a 500-count encoder produces 500 square waves in one revolution or 250 pulses with a 180-degree turn. The two main types of incremental encoders are tachometer (single-track) and quadrature (multitrack).

Tachometer Encoders Sometimes called a *single-track encoder,* a tachometer encoder has only one output and cannot detect the direction of travel. Its output is a square wave; its velocity can be determined by the frequency of pulses. The disk in Figure 11–9 shows a tachometer-style encoder disk.

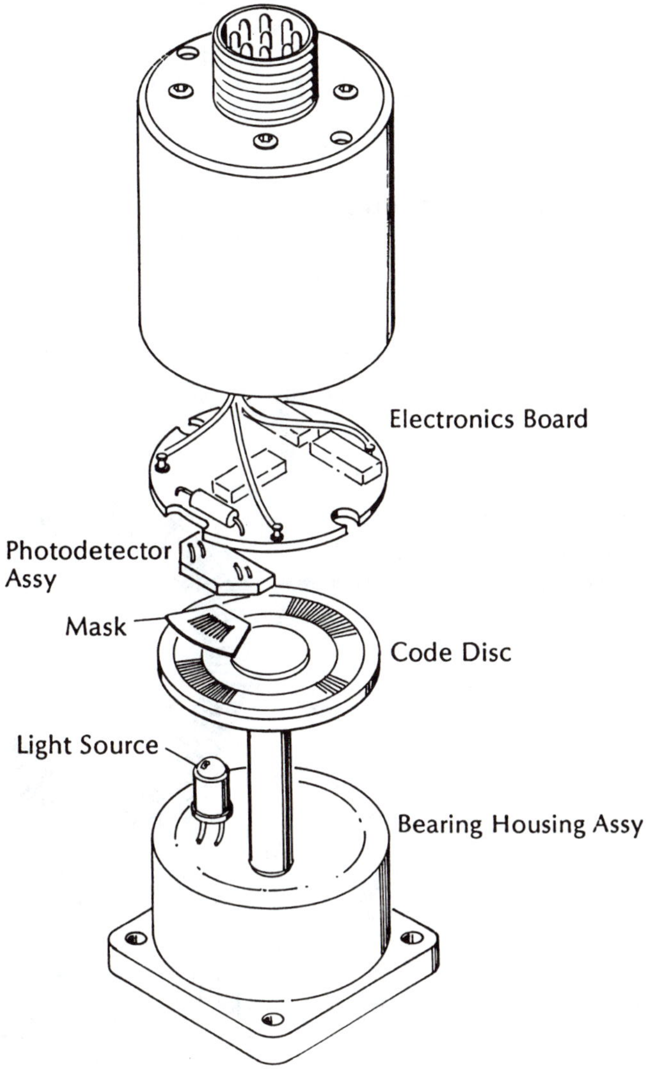

Figure 11–7 An enlarged view of an encoder. *(Courtesy BEI Sensors & Systems Company.)*

Quadrature Encoder Quadrature is a configuration in which the two output channels have an angular separation of 90 degrees. Most encoders use two quadrature output channels for position sensing. Figure 11–10 shows what the A and B pulses would look like on an oscilloscope. Note that the A and B pulses are square

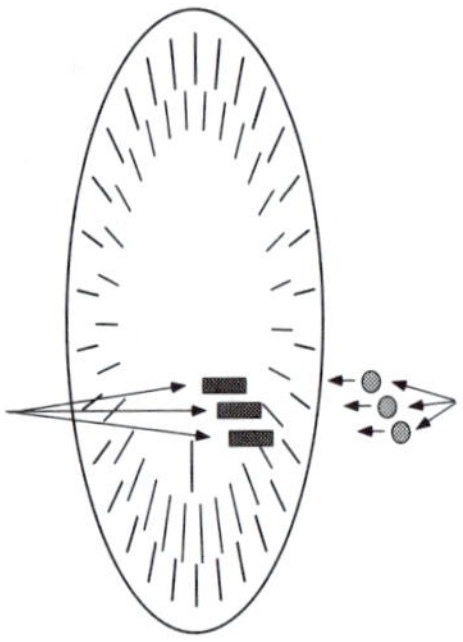

Figure 11–8 An incremental encoder.

Figure 11–9 A disk from a tachometer-style incremental encoder. *(Courtesy BEI Sensors & Systems Company.)*

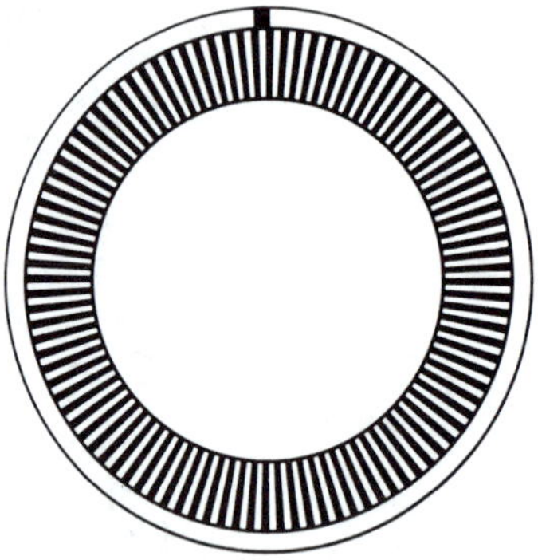

Figure 11–10 A and B ring output from the encoder on an oscilloscope.

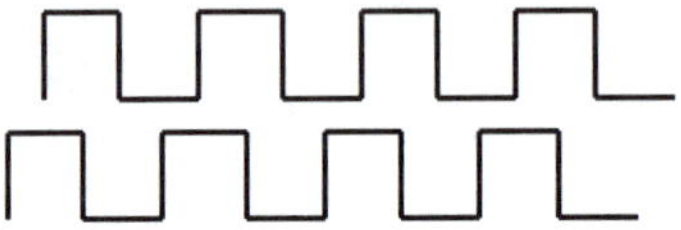

waves. Every slit in the encoder disk represents a high pulse. Note the relationship between the A and B pulses.

The transitions from high to low or low to high on these channels can be sensed by a computer, PLC, or special controller such as a CNC, or robot controller. Most incremental quadrature encoders have three tracks (see Figure 11–11). The A and B tracks have the same number of slits. The index track has only one pulse per rev-

Figure 11–11 An incremental encoder.

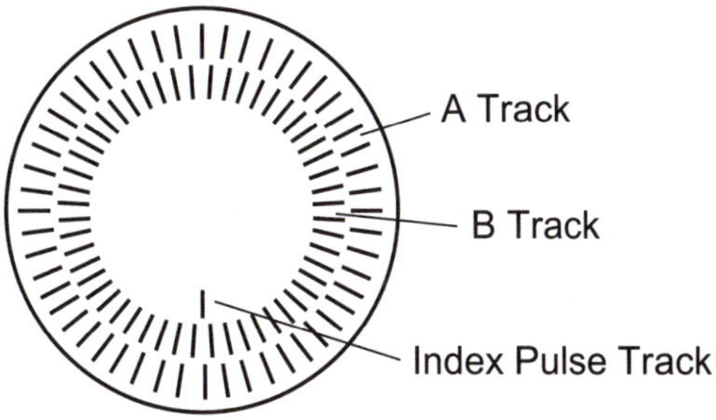

olution, usually called a *zero,* or *index, pulse,* which is used to establish the home position for position controllers.

The A and B channels can be compared to determine the direction of rotation. By using one channel as a reference, we can tell whether the transition on B leads or lags the A transition, indicating the direction of rotation. Study Figure 11–12 to see how direction can be determined by comparing the two channels.

A quadrature encoder might have a total of six lines: three for the A, B, and home pulse signals; two to supply power to the encoder; and one for case ground. A 13-bit absolute encoder has thirteen output wires plus two power lines. If the absolute encoder is provided with complementary outputs, the encoder requires twenty-eight wires.

The number of transitions identifies the amount of rotation. The quadrature encoder has a distinct advantage over the tachometer encoder for position sensing. Imagine what would happen if a tachometer encoder stopped right on a transition and the machine vibrated so that the encoder continued to output transitions due to the vibration, forcing the unit back and forth across a transition edge, without rotation. This cannot happen with a quadrature encoder. Even if the encoder stopped on a transition and vibration caused the output to show transitions on one channel, there would be no transition on the other channel. By using quadrature detection on a two-channel encoder and viewing the transition in its relationship to the condition of the other channel, we can generate accurate and reliable position and direction information.

Pulses from the A and B channels can be fed to an up-down counter or programmable controller input card. Many PLC high-speed input modules have the capability of accepting encoder input. With quadrature detection, we can derive one, two, or four times the encoder's basic resolution. For example, if a 1000-count encoder is used, its basic resolution is 1000 pulses per revolution, or 360 degrees/1000. If the detection circuit looks at the high-to-low transition and the high-to-low transition of the B channel, we would have 2000 pulses per revolution. If the detection

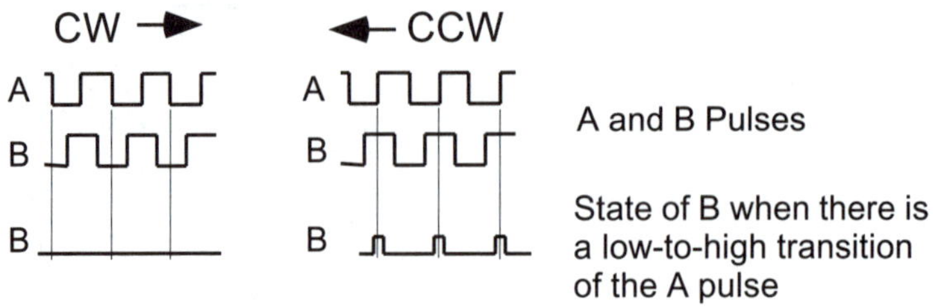

Figure 11–12 Determination of rotation direction. Encoder pulses for clockwise and counterclockwise rotation.

circuit looks at the high-to-low transition and the low-to-high transition for the A and the B channels, we would have 4000 pulses per revolution. With a quality encoder, four times the signal will be accurate to better than 1/5 pulse.

Absolute Encoders The absolute encoder provides a whole word of output with a unique pattern that represents each position (see Figure 11–13). The LEDs and receivers are aligned to read the disk pattern (see Figure 11–14). Many types of coding schemes can be used for the disk pattern; the most common patterns are gray, natural, binary, and binary-coded decimal (BCD). Gray and natural binary are available up to about 256 counts (8 bits). Gray code is popular because it is a nonambiguous code. Only one track changes at a time. This feature allows any indecision that may occur during any edge transition to be limited to a ±1 count. Natural binary code can be converted from gray code through digital logic. If the output changes while it is being read, a latch option locks the code to prevent ambiguities.

Encoder Wiring Encoders are available in two wiring types: single ended and differential. Single-ended encoders have one wire for each ring (A, B, and X) and a common ground wire. A differential encoder has a separate ground for each ring and is more immune to noise because the rings do not share a common ground. The differential type is the most common in industrial applications.

Ultrasonic Sensor An ultrasonic sensor uses high-frequency sound to measure distance by sending out sound waves and measuring how long it takes them to return. The distance to the object is proportional to the length of time it takes to return. See Figure 11–15. An ultrasonic sensor can make very accurate measurements; their accuracy for objects as small as 1 millimeter can be ±0.2 millimeter. Some cameras use ultrasonic sensing to determine the distance to the object to be photographed.

Figure 11–13 Disk from an absolute encoder. *(Courtesy BEI Sensors & Systems Company.)*

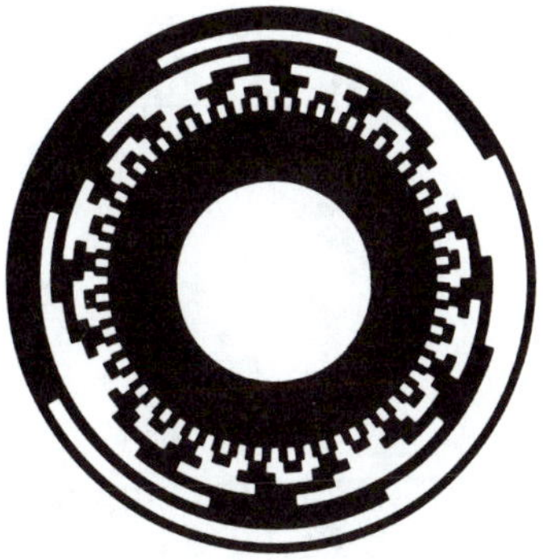

8 Bit Absolute Disc

Figure 11–14 Photodetectors read the position of an absolute encoder. *(Courtesy BEI Sensors & Systems Company.)*

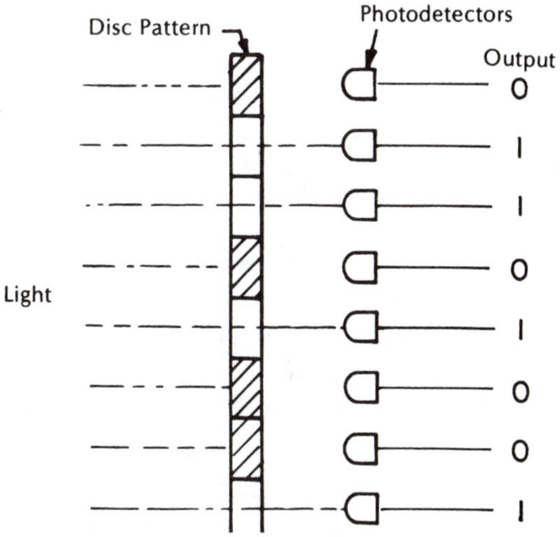

Figure 11–15 An ultrasonic sensor. Note the use of the "gate" sensor to notify the PLC when a part is present. *(Courtesy of Omron Electronics.)*

Measuring height of different objects on a conveyor using external gate input to coordinate inspection

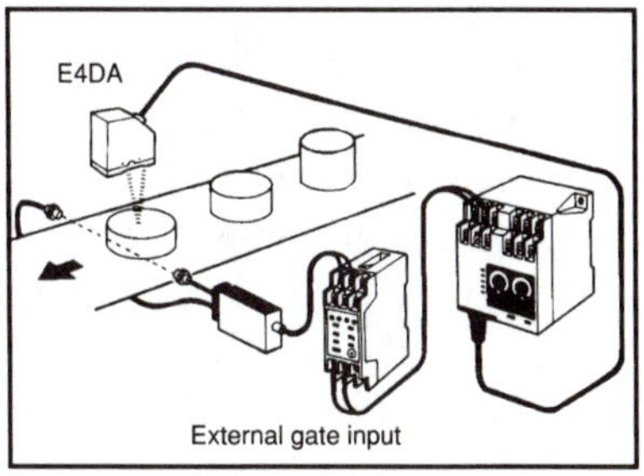

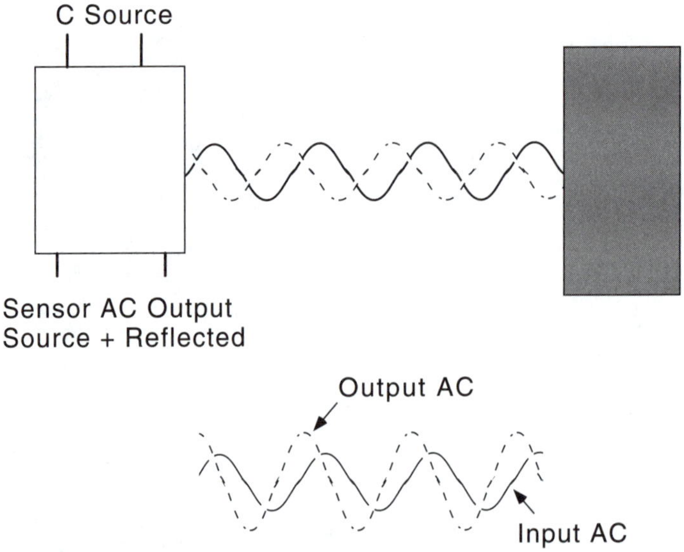

Figure 11–16 Ultrasonic sensor.

An *interferometer sensor* sends out a sound wave that is reflected back to the sensor. See Figure 11–16. The transmitted wave interferes with the reflected wave. If the peaks of the two waves coincide, the amplitude is twice that of the transmitted wave. If the transmitted wave is 180 degrees out of phase with the reflected wave, the amplitude of the result is zero. Between these two extremes, the amplitude is between zero and twice the amplitude, and the phase will be shifted between 0 and 180 degrees.

Interferometer sensors can detect distances to within a fraction of a wavelength. This is a very fine resolution indeed because some lights have wavelengths in the size range of .0005 millimeter.

Electronic Field Sensors

Electronic field sensors sense objects by producing an electromagnetic field. If the field is interrupted by an object, the sensor turns on. Field sensors are a better choice in dirty or wet environments than a photo sensor, which can be affected by dirt, liquids, or airborne contamination.

The two most common types of field sensors, capacitive and inductive, function in essentially the same way. Each has a field generator and a sensor to sense when the field is interfered with. The field generator puts out a field similar to the magnetic field from a magnet.

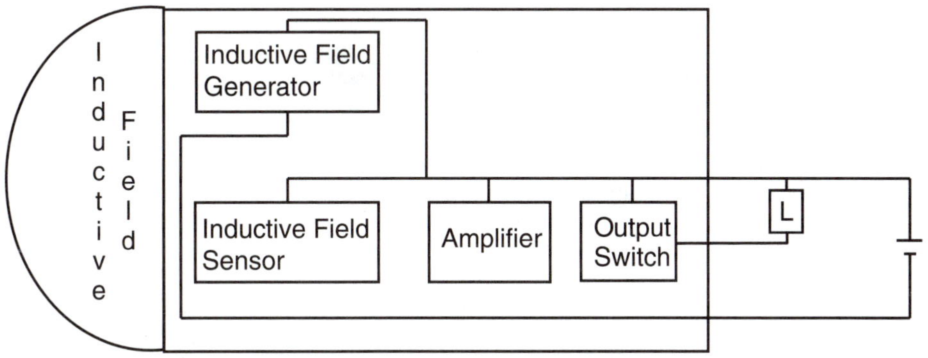

Figure 11–17 Block diagram of an inductive sensor. The inductive field generator creates an inductive field in front of the sensor; the field sensor monitors this field. When metal enters the field, the disruption in the field is sensed by the field sensor, and the sensor's output changes state. The sensing distance of these sensors is determined by the size of the field: the larger the required sensing range, the larger the diameter of the sensor.

Inductive Sensor

Used to sense metallic objects, the inductive sensor works by the principle of electromagnetic induction (see Figure 11–17). It functions in a manner similar to the primary and secondary windings of a transformer. The sensor has an oscillator and a coil; together they produce a weak magnetic field. As an object enters the sensing field, small eddy currents are induced on the surface of the object. Because of the interference with the magnetic field, energy is drawn from the sensor's oscillator circuit, decreasing the amplitude of the oscillation and causing a voltage drop. The sensor's detector circuit senses the voltage drop of the oscillator circuit and responds by changing the state of the sensor.

Sensing Distance Sensing distance (range) is related to the size of the inductor coil and whether the sensor coil is shielded or nonshielded (see Figure 11–18). This sensor coil is shielded in this case by a copper band. This prevents the field from extending beyond the sensor diameter and reduces the sensing distance. The shielded sensor has about half the sensing range of an unshielded sensor. Sensing distance generally varies by about 5 percent due to changes in ambient temperature.

Hysteresis Hysteresis means that an object must be closer to a sensor to turn it on than to turn it off (see Figure 11–19). Direction and distance are important. If the object is moving toward the sensor, it must move to the closer point to turn on. Once the sensor turns on (at the *operation point* or *on-point*), it remains on until the object moves to the *release point* (*off-point*). Hysteresis causes this differential gap,

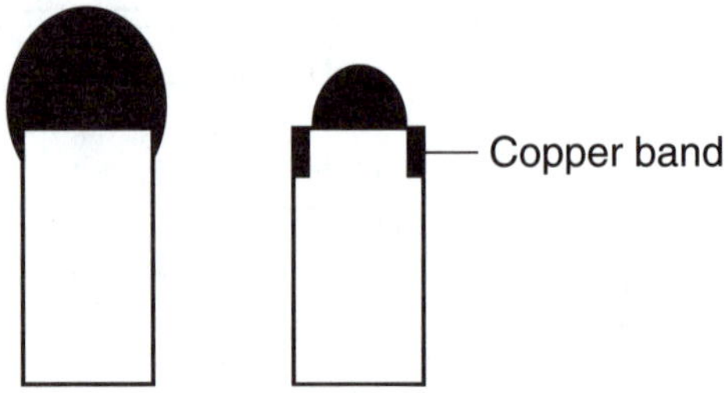

Figure 11–18 A copper band in a shielded field sensor. Note that the sensing distance is reduced. If the unshielded sensor were mounted flush, it would detect the object in which it was mounted.

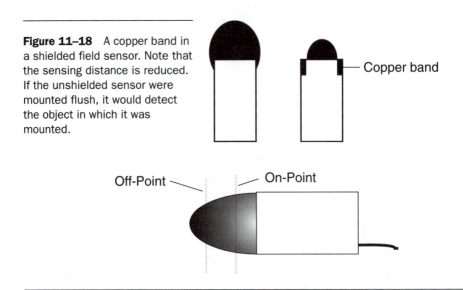

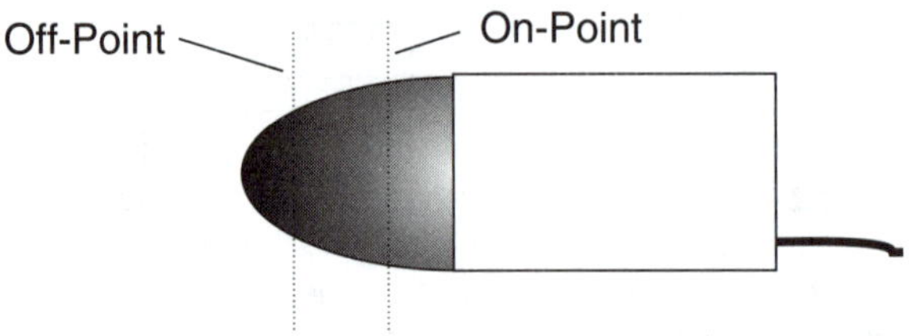

Figure 11–19 Example of hysteresis. Note that the on-point and off-point are different.

or differential travel. The principle is used to eliminate the possibility of "teasing" the sensor. The sensor is either on or off.

Hysteresis is a built-in feature in proximity sensors that helps stabilize part sensing. Imagine a bottle moving down a conveyor line. Vibration causes the bottle to wiggle as it moves along the conveyor. If the on-point were the same as the off-point and the bottle were wiggling as it went by the sensor, it could be sensed many times as it wiggles in and out past the on-point. When hysteresis is involved, however, the on-point and off-point are at different distances from the sensor.

To turn the sensor on, the object must be closer than the on-point. The sensor output remains on until the object moves farther away than the off-point, preventing multiple unwanted reads.

Capacitive Sensors

Capacitive sensors (Figures 11-20 and 11-21) can sense both metallic and nonmetallic objects as well as product inside containers (see Figure 11–22). These sensors are commonly used in the food industry and to check fluid and solid levels inside tanks.

Capacitive sensors operate on the principle of electrostatic capacitance in a manner similar to the plates of a capacitor. The oscillator and electrode produce an electrostatic field (remember that the inductive sensor produced an electromagnetic field). The target (object to be sensed) acts as a capacitor's second plate.

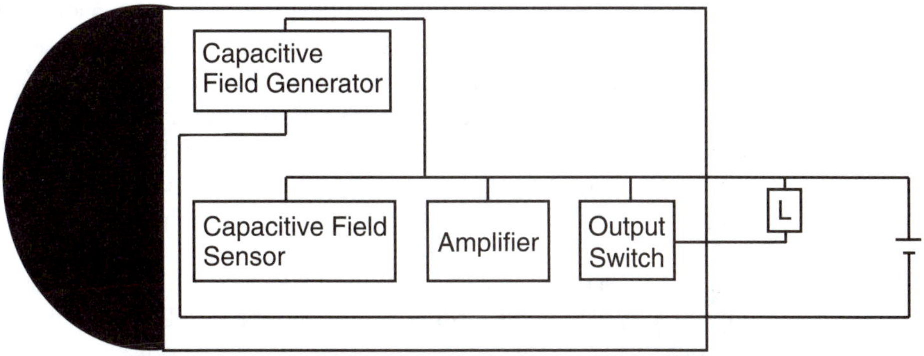

Figure 11–20 Block diagram of a capacitive sensor.

Figure 11–21 Capacitive sensors. *(Courtesy ifm efector inc.)*

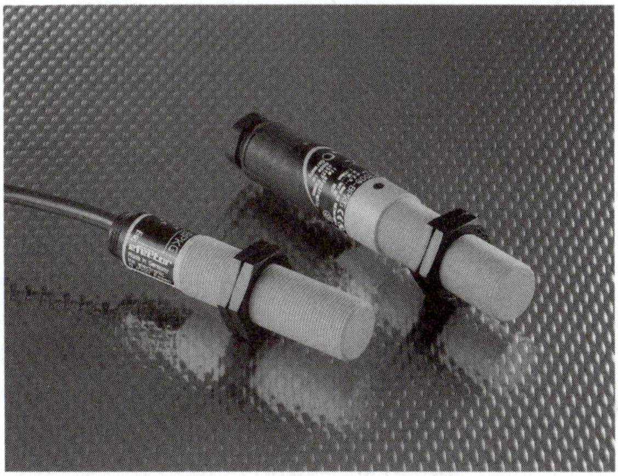

Figure 11–22 Capacitive sensor checking inside boxes. *(Courtesy ifm efector inc.)*

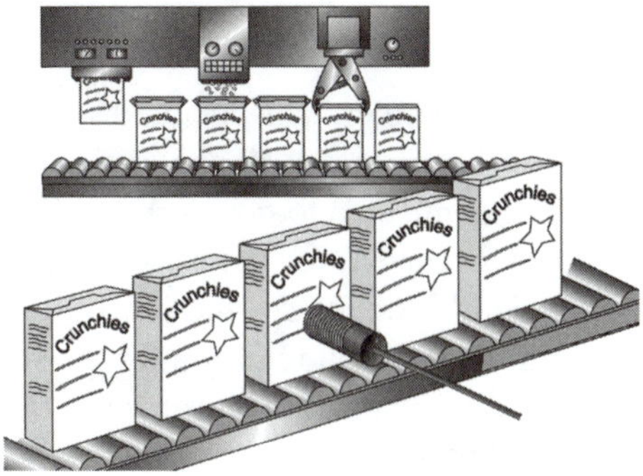

An electric field is produced between the target and the sensor. As the amplitude of the oscillation increases, the oscillator circuit voltage increases, and the detector circuit responds by changing the state of the sensor.

A capacitive sensor can sense almost any object. The object acts as a capacitor to ground. The entrance of the target (object) in the electrostatic field disturbs the DC balance of the sensor circuit, starting the electrode circuit oscillation and maintaining the oscillation as long as the target is within the field.

Sensing Distance Capacitive sensors are unshielded, nonembeddable devices (see Figure 11–21), which means that they cannot be installed flush in a mount because they would then sense it. Conducting materials can be sensed farther away than nonconductors because the electrons in conductors are freer to move. The target mass affects the sensing distance: the larger the mass, the larger the sensing distance.

Capacitive sensors are more sensitive to temperature and humidity fluctuation than are inductive sensors, but capacitive sensors are not as accurate as inductive ones. Repeat accuracy can vary by 10 to 15 percent in capacitive sensors. Sensing distance for capacitive sensors can fluctuate as much as ±15 to 20 percent.

Some capacitive sensors are available with an adjustment screw, which can be adjusted to sense a product inside a container (see Figure 11–22). The sensitivity can be reduced so that the container is not sensed but the product inside is.

Sensor Wiring

Both AC- and DC-powered sensors have basically two wiring schemes: *load-powered* (two-wire) and *line-powered* (three-wire). The most important power consideration is to limit the sensor's output current to an acceptable level. If the output current exceeds the output current limit, the sensor will fail. The sensor's output device is normally a transistor. A sensor with a transistor output can generally handle up to about 100 milliamps. Some sensors have a relay output and can handle 1 amp or more.

Load-Powered Sensors

In two-wire or load-powered sensors, one wire is connected to power, and the other wire is connected to one of the load's wires (see Figure 11–23). The load wire is then connected to power. Wiring diagrams can usually be found on the sensor.

The current for the sensor is typically under 2.0 milliamps (see Figure 11–24), which is enough for the sensor to operate but not enough to turn on the PLC's input. The current required for the sensor to operate must pass through a *load,* which is anything that will limit the current of the sensor output. Think of the load as an input to a PLC. The small current that must flow to allow the sensor to operate is

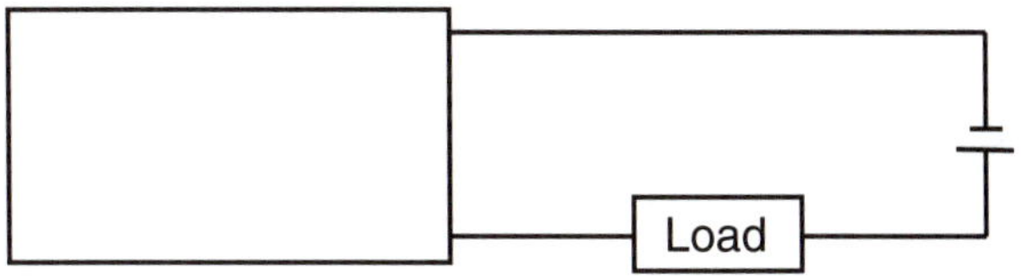

Figure 11–23 Two-wire (load-powered) sensor. The load represents whatever the technician monitors the sensor output for, normally a PLC input. The load must limit the sensor's current to an acceptable level, or the sensor output blows.

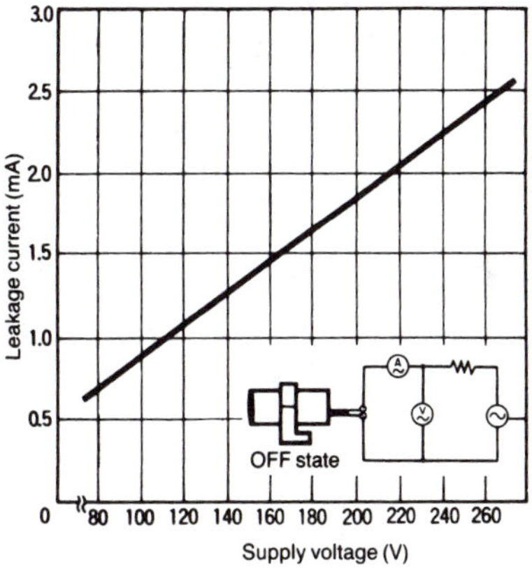

Figure 11–24 Leakage current versus supply voltage.

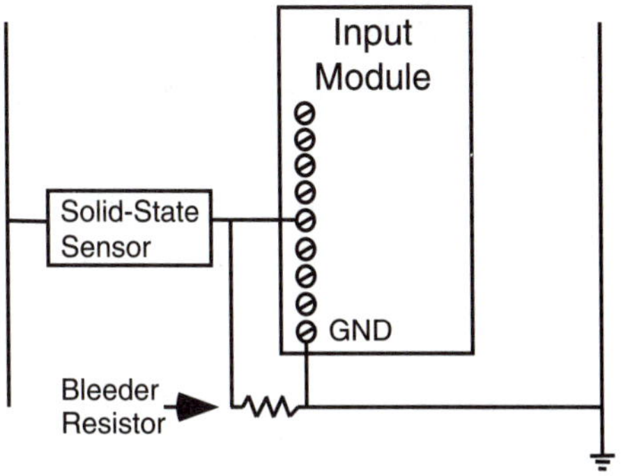

Figure 11–25 Use of a two-wire (load-powered) sensor. In this case, the leakage current was large enough to cause the input module to sense an input when there was none. A resistor was added to bleed the leakage current to ground so that the input could not sense it.

called *leakage current* or *operating current.* (The leakage current is usually not enough current to activate a PLC input. If it is enough, it must be connected to a bleeder resistor as shown in Figure 11–25.) When the sensor turns on, it allows enough current to flow to turn on the PLC input.

Figure 11–26 Three-wire (line-powered) sensor. The load must limit the current to an acceptable level.

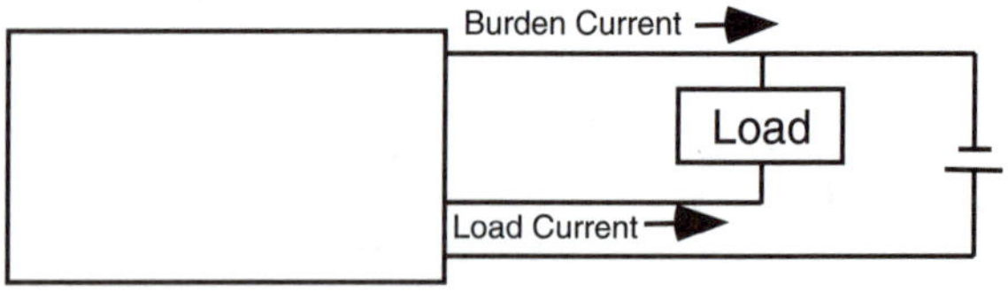

Response time, which can be crucial in high-demand production applications, is the lapsed time between sensing the target and changing the output state. Sensor specification sheets provide response time information.

Line-Powered Sensors

Although line-powered sensors are usually of the three-wire type (see Figure 11–26), they can have either three or four wires. The three-wire variety has two power leads and one output lead.

The line-powered sensor needs a small current, called *burden current* or *operating current,* to operate. This current flows whether the sensor output is on or off. The load current is the output from the sensor; if it is on, there is load current. This load current turns on the load (PLC input). The maximum load current is typically between 50 to 200 milliamps for most sensors. The output LED on a sensor can function even though the output is bad.

Sourcing and Sinking Sensors

Sourcing Sensor (PNP)

In the case of a sourcing sensor (PNP), when there is an output current from the sensor, the sensor sources current to the load (see Figure 11–27). Conventional current flow goes from plus to minus. When the sensor is off, the current does not flow through the load.

Sinking Sensor (NPN)

In the case of sinking (NPN) sensors, when the sensor is off (nonconducting), no current flows through the load. When the sensor is on (conducting), a load current flows from the load to the sensor (see Figure 11–28). The choice of whether to use an NPN or a PNP sensor depends on the type of load. In other words, choose a sensor that matches the PLC input module requirements for sinking or sourcing.

Sinking = path to supply ground (2)
Sourcing = path to supply source (1)

Figure 11–27 A sourcing switch connected to a PLC input. *(Courtesy PLC Direct by Koyo.)*

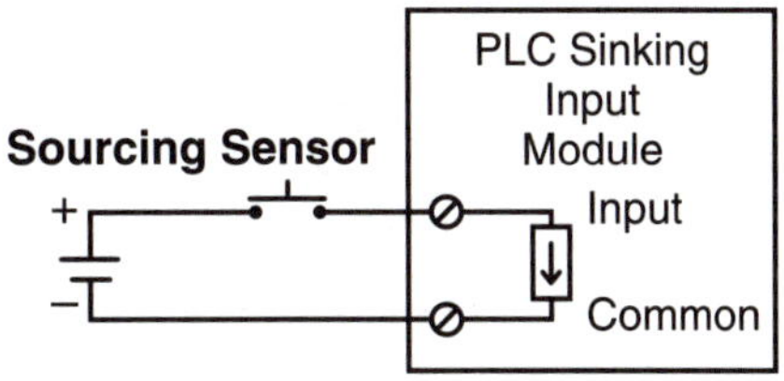

Figure 11–28 A sinking switch connected to a PLC input. *(Courtesy PLC Direct by Koyo.)*

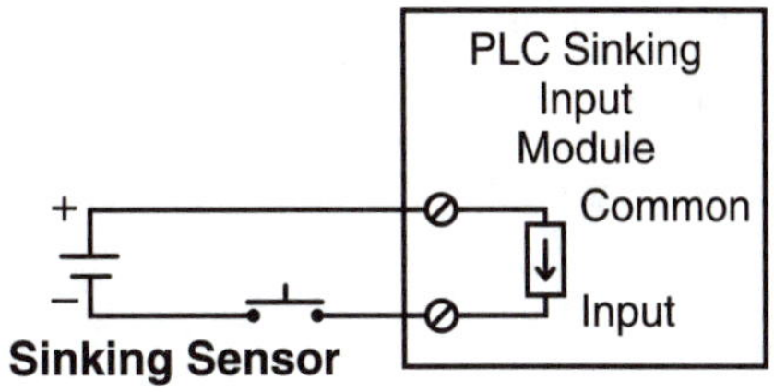

Analog Sensors

Many types of analog sensors are available. Many of the types already discussed are available with digital and analog output. Photo sensors and field sensors are available with analog output.

Analog Considerations

Analog sensors provide much more information about a process than digital sensors do. Their output varies depending on the conditions being measured.

Accuracy, Precision, and Repeatability

Accuracy can be defined as how closely a sensor indicates the true quantity being measured. In a temperature measurement, accuracy is defined as how closely the sensor indicates the actual temperature being measured. *Precision* refers to how closely each sensor in a group of sensors measures the same variable. For example, when measuring a given temperature, the output of each should be the same. Precision is very important in many applications. *Repeatability* is a sensor's ability to repeat its previous readings. For example, in measuring temperature, the sensor's output should be the same every time it senses a specific temperature.

Thermocouples

The principle of thermocouple was discovered by Thomas J. Seebeck in 1821. The *thermocouple* is one of the most common devices for temperature measurement in industrial applications. A *thermocouple* is a very simple device that has two pieces

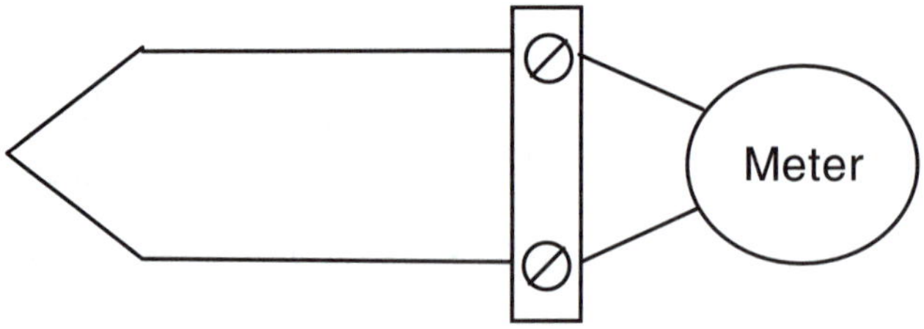

Figure 11–29 Simple thermocouple. A thermocouple can be made by twisting the desired type of wire together and silver soldering the end. To measure the temperature change, the wire is cut in the middle to insert a meter. The voltage is proportional to the difference in temperature between the hot and cold junctions.

of dissimilar metal wire joined at one or both ends. The typical industrial thermocouple is joined at one end (see Figure 11–29). The other ends of the wire are connected via compensating wire to the analog inputs of a control device such as a PLC (see Figure 11–30). The operation principle is that joining dissimilar metals produces a small voltage. The voltage output is proportional to the difference in temperature between the cold and hot junctions. Thermocouples are color-coded for polarity and for type. The negative terminal is red, and the positive terminal is a different color that can be used to identify the thermocouple type (see Figure 11–31).

A cold junction is assumed to be at ambient temperature (room temperature). In reality, temperatures vary considerably in an industrial environment. If the cold junction varies with the ambient temperature, the readings are inaccurate, which is unacceptable in most industrial applications. Industrial thermocouple tables use 75 degrees Fahrenheit for the reference temperature (see Figure 11–32), but it is too complicated to try to maintain the cold junction at 75 degrees, so industrial thermocouples must be compensated. Figure 11–32 also shows a comparison of the voltage output versus temperature for a J-type and K-type thermocouple.This is normally accomplished with the use of temperature-sensitive resistor networks. The resistor used in the network has a negative coefficient of resistance. Because resistance decreases as the temperature increases, the voltage automatically adjusts so that readings remain accurate. PLC thermocouple modules automatically compensate for temperature variation.

The resolution of a thermocouple, which provides accurate measurements, is determined by the device that takes the output from the thermocouple. The device is normally a PLC analog module. The typical resolution of an industrial analog module is 12 bits; 2 to the twelfth power is 4096. This means that if the range of temperature to be measured were 1200 degrees, the resolution would be 0.29296875 degrees/bit (1200/4096 = 0.29296875), which means that the PLC could tell the temperature to

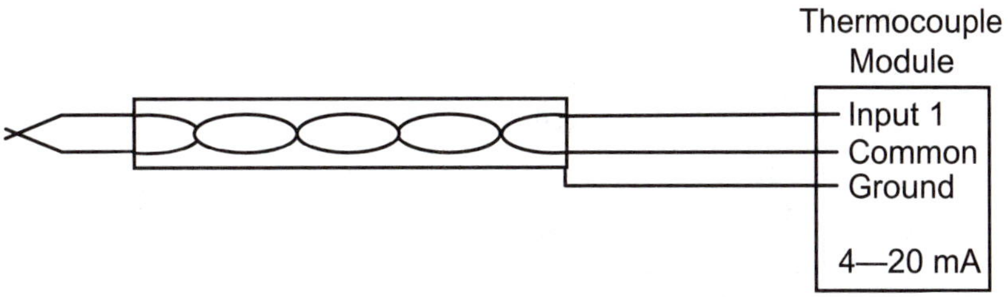

Figure 11–30 Thermocouple. The wire connecting the thermocouple to the PLC module is a twisted-pair (two wires twisted around each other) shielded cable. The shield around the twisted pair helps eliminate electrical noise as a problem. Twisted-pair wiring also helps reduce the effects of electrical noise. The shielding is grounded only at the control device. Typical PLC modules allow four or more analog inputs.

Type	+lMeta	– Meta	Average Microvolt Change per Degree Farenheit	Useful Range in Degrees Centigrade
J	Iron (white)	Constantan (red)	2 29	0 0–08
K	Chromel (yellow)	Alumel (red)	5 21	0 0–120
R	Platinum (black)	Rhodium (red)	7 6	500–150
T	Copper (blue)	Constantan (red)	5 22	0–100–40

Figure 11–31 Thermocouple color table.

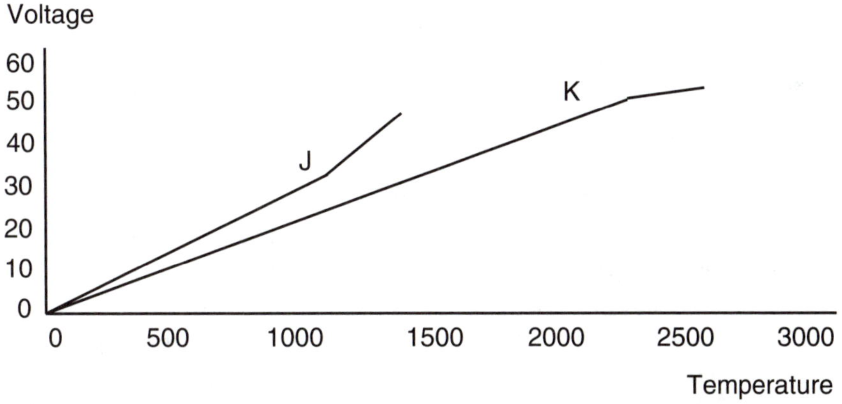

Figure 11–32 Voltage output versus temperature for J- and K-type thermocouples. The relationship is approximately linear. For example, if the output voltage were 20 millivolts with a J-type thermocouple, the temperature would be approximately 525 degrees (600 – 75). Note the voltage produced by the difference in temperature between the cold and hot junctions. The chart assumes a 75 degree ambient temperature.

Thermocouple Types
Nickel-chromium versus copper-nickel (chromel-constantan)
Iron versus copper-nickel (iron-constantan)
Nickel-chromium versus nickel-aluminum (chromel-alumel)
Copper versus copper-nickel (copper-constantan)
Tungsten 5 percent rhenium versus tungsten 26 percent rhenium
Tungsten versus tungsten 26 percent rhenium
Tungsten 3 percent rhenium versus tungsten 25 percent rhenium
Platinum versus platinum 13 percent rhodium
Platinum versus platinum 10 percent rhodium
Platinum 6 percent rhodium versus platinum 30 percent rhodium

Figure 11–33 Composition of various thermocouple types.

about one-fourth of one degree. This is reasonably good resolution, close enough for the vast majority of applications, and higher-resolution analog modules are available for higher requirements. See Figure 11–33 for the composition of some of the various thermocouples available. Figure 11–34 also shows several types of thermocouples and their temperature ranges. The user chooses the correct thermocouple for the range of temperature in the application.

Resistance Temperature Device (RTDs)

A resistance temperature device (RTD) is a precision resistor whose resistance changes with temperature. RTDs are more accurate than thermocouples. The most common RTD resistance is 100 ohms at zero degrees Celsius; others are available in the 50 to 100 ohm range. One of the basic properties of metals is that their electrical resistivity changes with temperature. Some metals have a very predictable change in resistance for a given change in temperature. The chosen metal is made into a resistor with a nominal resistance at a given temperature. The change in temperature can then be determined by comparing the resistance at the unknown temperature to the known nominal resistance at the known temperature. Tables of the temperature-resistance relationships for various metals used in RTDs are available.

Type	Temperature (°F)	Temperature (°C)
J	–50 to +1400	–40 to +760
K	–100 to +2250	–80 to +1240
T	–200 to +660	–130 to +350
E	–200 to +680	–130 to +1250
R	0 to +3200	–20 to +1760
S	0 to +3200	–20 to +1760
B	+500 to +3200	+260 to +1760
C	0 to +4200	–20 to +2320

Figure 11–34 Temperature ranges for various types of thermocouples. Note the output for the difference in temperature between the hot and cold junctions. The table is based on a cold junction temperature of 75 degrees Fahrenheit.

RTDs are made from a pure wire-wound metal, usually one that occurs naturally, that has a positive temperature coefficient. As the temperature increases, the very small change in resistance to current flow increases.

RTDs come in two configurations, a wire-wound construction and a thin film metal layer construction deposited on a nonthermally conductive substrate. The wire-wound coiled construction allows for a greater change in resistance for a given change in temperature, increasing the sensor's sensitivity and improving the resolution. The coil is wrapped around a nonconductive material such as ceramic, which enables the wire to respond more quickly to temperature change. The coil is then covered by a sheath made from very temperature-conductive materials to protect it from abuse and the environment. The sheath is filled with a dry thermal-conducting gas to increase the temperature conductivity.

Platinum is the most popular material for RTDs. It has a linear change in resistance versus temperature and has a wide operating range. Platinum is a stable element, which ensures long-term stability. Platinum sensors are now being made with very thin film resistance elements that use only a small amount of platinum, which makes the platinum RTDs more competitive in price. Other materials include copper, nickel, balco, tungsten, and iridium.

The second RTD configuration is in the form of a thin film metal layer deposited on a nonthermally conductive substrate, which is often ceramic. These RTDs can

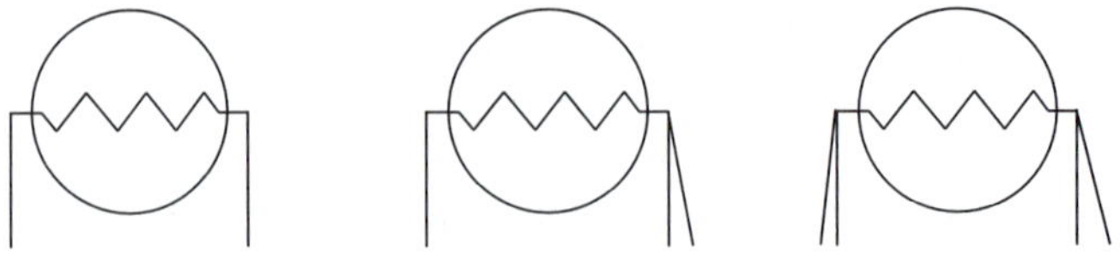

Figure 11–35 Three RTD wiring configurations.

be very small and do not require an external sheath because the ceramic acts as a protective sheath. Simple circuitry can be used to linearize RTDs.

Wiring RTDs can have three different wiring configurations (see Figure 11–35). The lead wires from the RTD can affect its accuracy because they create additional uncompensated resistance. A third wire can be added to compensate for lead wire resistance. Four-wire RTDs are also available but are generally used only in laboratory applications. The four-wire RTD can be used with a three-wire device but this does not improve a three-wire RTD's accuracy. A three-wire RTD cannot be used with a four-wire device. It is best to use the appropriate RTD for the measurement device.

Thermistors

A *thermistor* is a temperature-measuring sensor constructed of human-made materials. It is more sensitive to temperature than an RTD. A thermistor has a negative temperature coefficient. The resistance decreases as the temperature increases. Because a thermistor is a semiconductor, it cannot operate above about 300 degrees Celsius. Major advantages of a thermistor are precision, stability, and production of a large change in resistance for a small change in temperature. If the range of temperature to be measured is relatively small, the thermistor is a good device. One of the thermistor's main problems is that its output is linear only within a small temperature range (see Figure 11–36). Its resistance does not vary proportionately with a change in temperature, although thermistor networks have very linear voltage change with temperature change.

Different package styles of thermistors are available for various applications, such as monitoring the temperature of electric motors. The thermistor is fastened to the housing of the motor and connected to a bridge circuit whose output is compared to a reference voltage. The reference voltage is chosen to be a safe value for the motor's maximum operating temperature. It can then be used to shut off the motor circuit.

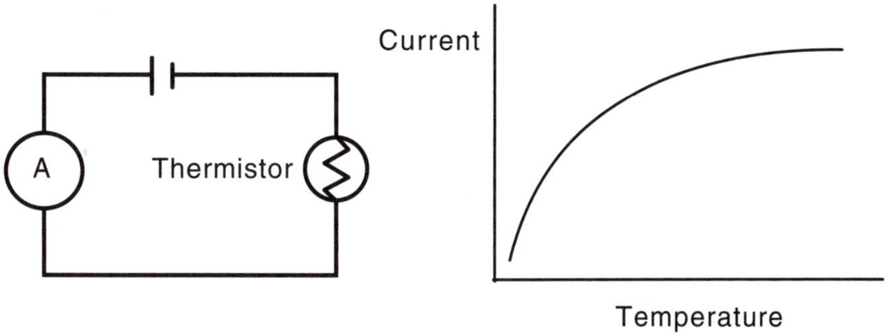

Figure 11–36 Graph of current versus temperature change for a thermistor.

Integrated Circuit (Semiconductor) Temperature Sensors

Some integrated circuit semiconductors can be used to sense temperature change. They respond to temperature increases by increasing their reverse bias current across PN junctions and generating a small, detectable current or voltage that is proportional to the change in temperature (see Figure 11–37).

Magnetic Reed Sensors

Magnetic reed sensors usually have two sets of contacts, a normally closed set and a normally open set. When a small magnet is brought close to the reed sensor, the common contact is drawn toward the normally open contact and moves away from the normally closed contact. These magnetic reed sensors are commonly used for sensing whether a piston in a cylinder is extended or retracted for feedback in a system.

Figure 11–37 Graph of voltage versus temperature change for an integrated circuit (semiconductor).

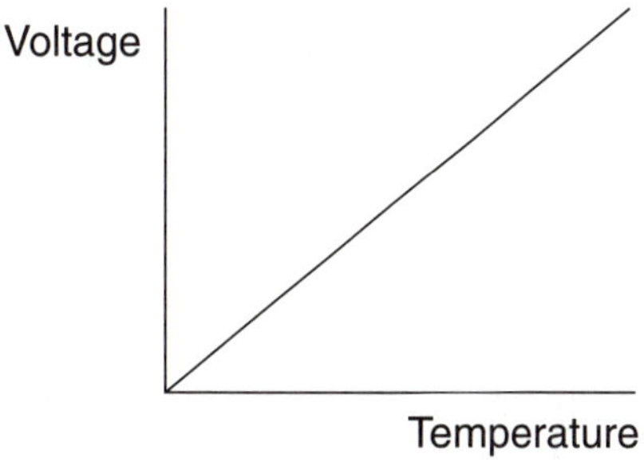

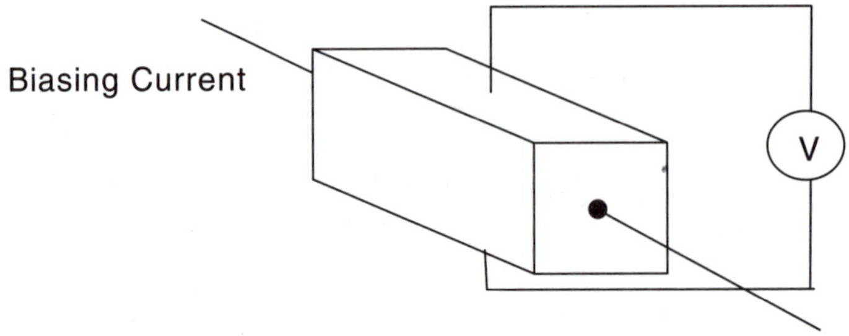

Figure 11–38 Basic principle of the hall effect sensor.

Hall Effect Sensor

The hall effect sensor is based on a principle discovered shortly after the transistor was developed in the 1950s. A hall effect sensor is biased with a current. If a magnetic field is introduced to the sensor material, it allows a small current to flow through the sensor and then can be sensed (see Figure 11–38). Hall effect sensors are available as both digital and analog.

Hall effect sensors have many applications. They are used in cruise (speed) control for cars, as blood flow sensors in the medical field, as linear displacement devices, and as sensors of rotary motion for speed. They are often used to detect the velocity of a rotating shaft and to determine displacement of a shaft.

Strain Gages

Strain gages measure force based on the principle that the thinner the wire is, the higher the resistance. A wire with a smaller diameter would have higher resistance than a wire with a larger diameter (see Figure 11–39). If we had an elastic wire, we could stretch it and measure its change in resistance because its diameter decreases in the middle, increasing the resistance. The change in resistance compared to the change in voltage is quite linear. If a constant current is supplied to a strain gage and the force applied to the strain gage varies, the resistance of the strain gage will vary and the voltage change will be proportional to the change in force.

Figure 11–39 Two pieces of wire. Which one will have lower resistance? The larger the wire diameter, the lower the resistance.

Strain gages have several uses. They are used for pressure measurement by bonding them to a membrane that is then exposed to the pressure and for weight measurement. They are also used in accelerometers. Most strain gages are zigzag wire-mounted on a paper or membrane backing (see Figure 11–40). They must be mounted correctly because they are sensitive to change in only one direction. Strain gages normally have an arrow that indicates the direction in which they should be mounted. They are normally mounted with adhesives.

They are often set up in a bridge configuration to increase the change in resistance and to compensate for temperature fluctuations (see Figure 11–41). To do this,

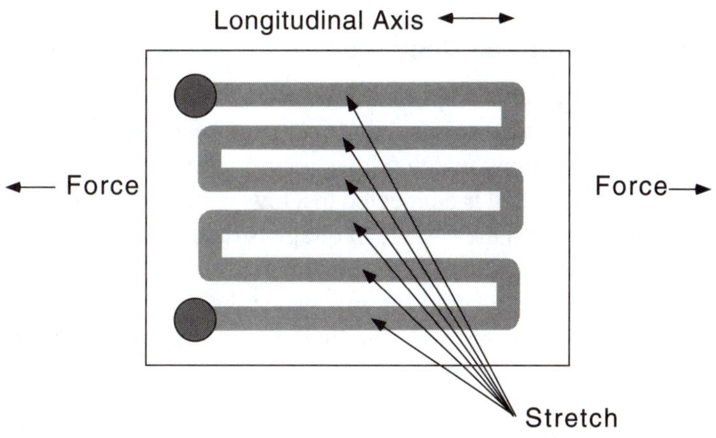

Figure 11–40 Strain gage zigzag mounted on paper or membrane backing.

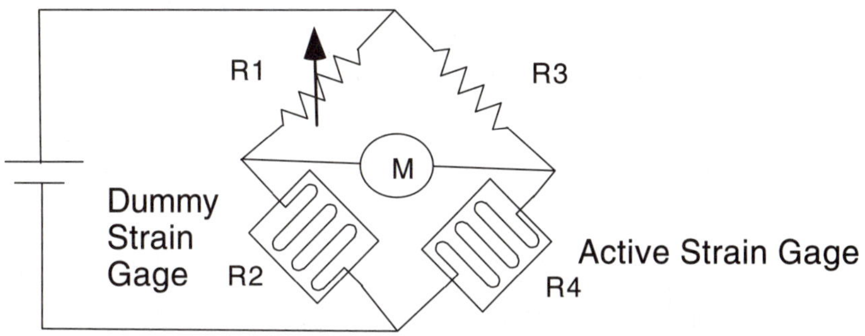

Figure 11–41 Bridge configuration of a strain gage to compensate for temperature fluctuations.

a dummy strain gage is inserted and is not subjected to the strain but to the same temperature change. This eliminates the effect of temperature fluctuation in the stress measurement. Because strain gages are made from long metal strands (or metallic paths) and metal responds to temperature change, strain gages are susceptible to variation in temperature, which can cause a variance in the output.

Strain gages can be set up in another way to overcome fluctuation due to variation in temperature (see Figure 11–42). In this case, four strain gages and two balancing resistors (R5 and R6) are used to compensate for thermal drift and to balance the bridge. All of the strain gages adapt to any temperature fluctuation and thus cancel any temperature-induced fluctuation.

Linear Variable Displacement Transformer (LVDT)

A linear variable displacement transformer (LVDT) uses the transformer principle to sense position. It is a cylindrical sensor with three fixed windings. The cylinder's center is hollow, which allows a core to slide in the cylinder (see Figure 11–43). The primary coil is connected to a low-voltage AC source (a sine wave). As the core moves through the LVDT, a voltage is induced in the other two windings. The voltage is proportional to the core position.

LVDTs can be wired in two ways: the series-adding mode or the series opposition mode. In the series-adding mode, the secondary coils are wired in series,

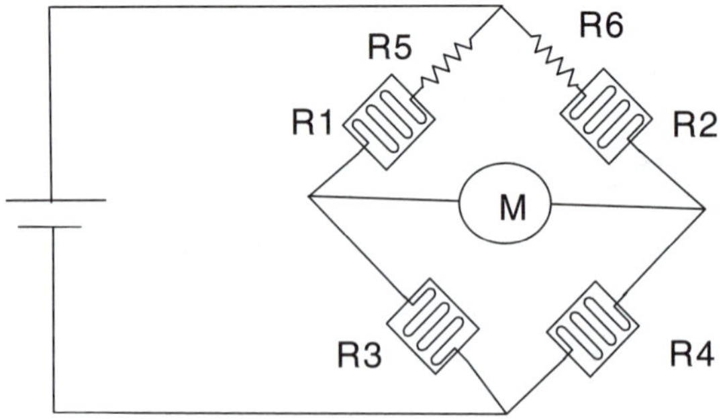

Figure 11–42 Strain gage set up to compensate for temperature fluctuation.

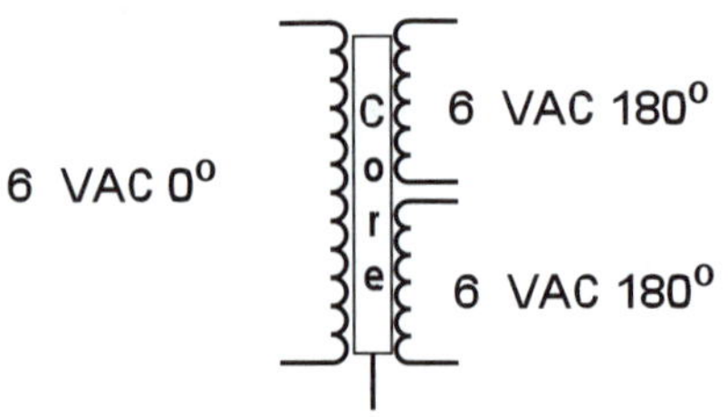

Figure 11–43 Typical LVDT. The core is aligned with both windings. Each of the output windings is 6 VAC with a phase of 180 degrees.

adding their output. In the series opposition the leads from one secondary are reversed and then the secondaries are wired in series. The output voltages are then in opposition. The series-adding mode increases the voltage smoothly as the core moves into and through the core (see Figures 11-44 and 11-45). The voltage in the series opposition mode is zero when the core is centered and increases as the core moves from the centered position (see Figure 11–46). The signal also changes phase on each side of the centered position. Figures 11-47, 11-48, and 11-49 show examples of the output for an LVDT that is wired in the series opposition mode.

Figure 11–44 Series-adding method of use with output of 12 VAC 180 degrees.

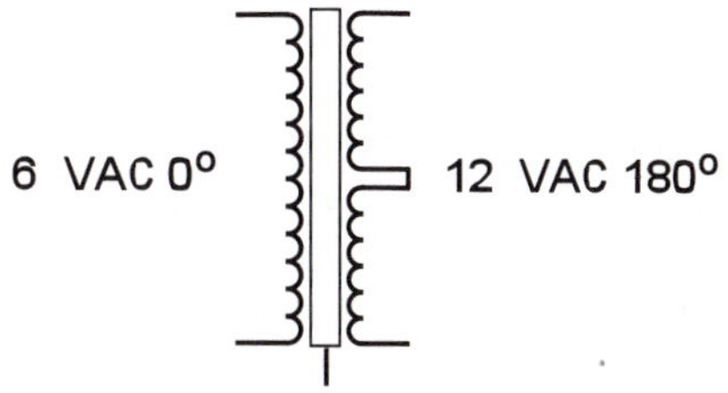

Figure 11–45 Total output in the series-adding mode whose core is located halfway into the windings is 6 VAC 180 degrees.

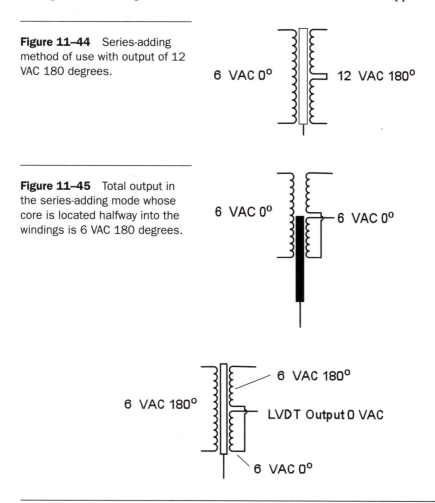

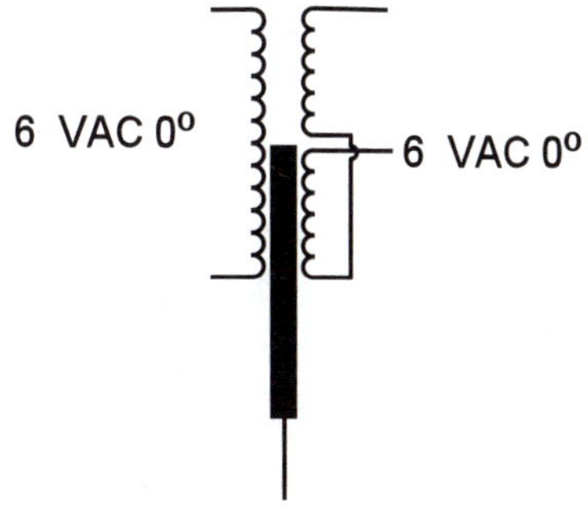

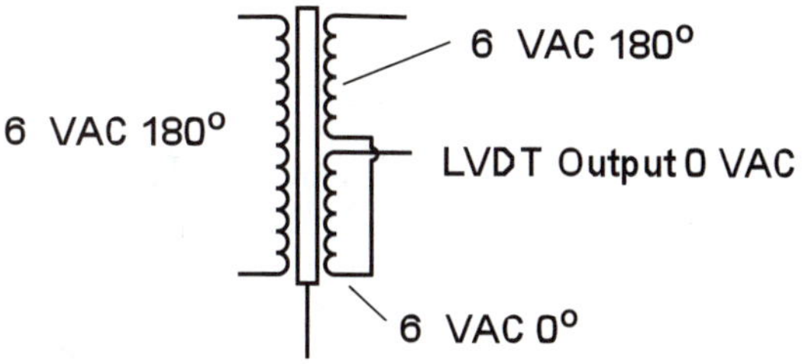

Figure 11–46 Series opposition method with the core centered. The output of the top winding is 6 VAC 180 degrees. The bottom winding is 6 VAC 0 degrees because, in effect, the leads have been reversed. The output when the core is centered is 0 volts.

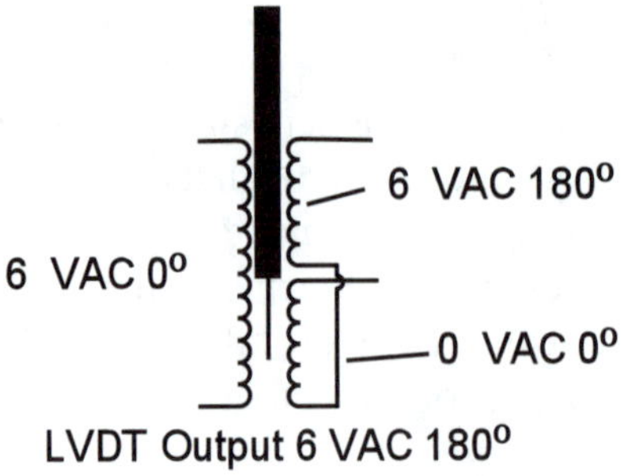

Figure 11–47 LVDT core is aligned with the top output winding. The output is 6 volts 180 degrees.

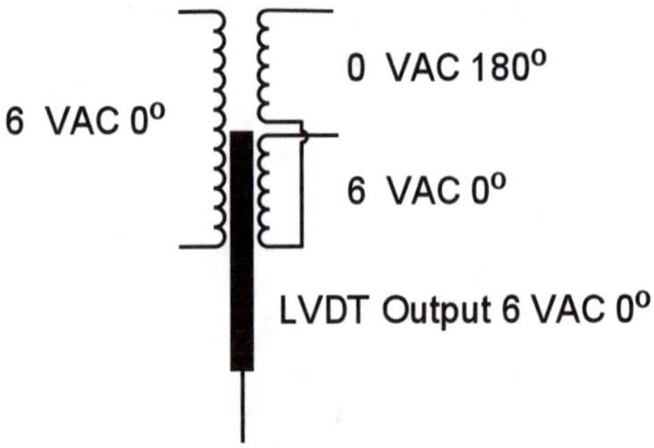

Figure 11–48 LVDT whose core is aligned with the lower winding. The output is 6 VAC 0 degrees.

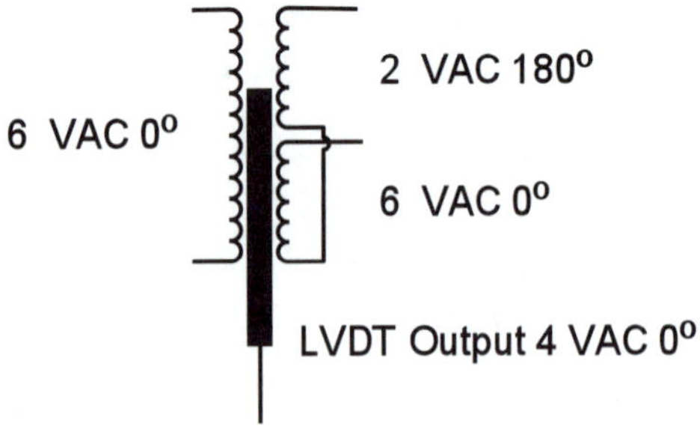

Figure 11–49 Core aligned with the lower winding and part of the upper winding. The output of the LVDT in the series opposition mode is 4 VAC 0 degrees.

The output from an LVDT can be rectified to provide a DC signal (see Figure 11–50) because it is easier for PLCs to work with a DC signal than an AC signal. This is called signal conditioning. LVDTs can be purchased with DC output.

Resolvers

A resolver is an analog position transducer that uses the principle of the transformer to operate. It is a very reliable device. The bearings are really the only element that can fail.

Figure 11–50 Output from an LVDT can be rectified to provide a DC signal that can be read by an analog input card.

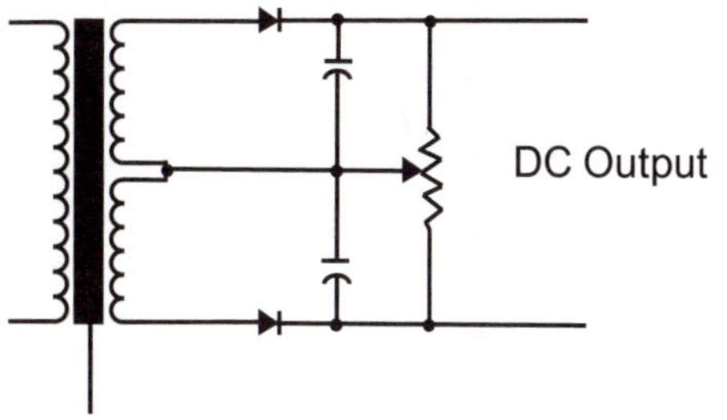

In a resolver, a 0-degree sine wave (AC voltage) is applied to one stationary coil (see Figure 11–51). A 90-degree sine (cosine) wave is applied to a second stationary winding. The frequency of these signals is typically about 2 kilohertz. The resolver's rotor has a third winding that rotates with the rotor. Because of the transformer action, this winding produces an output. Usually the rotor has half as many windings as there are on the stator windings, so the output voltage will be one-half the input voltage. If the rotor is aligned with the 0-degree sine winding, the output is 6 volts 180 degrees out of phase (remember the transformer). So if we put the resolver rotor winding output on a scope, it would be 6 volts 180 degrees out of phase. See Figure 11–52. If the rotor is rotated one half-turn (180 degrees), it aligns with the 0-degree sine winding again, but the output is 6 volts 0 degrees (see Figure 11–52).

When the rotor is aligned with the 90-degree stator, the output is 6 volts 270 degrees; the output is 180 degrees out of phase with the input 90-degree sine wave (see Figure 11–53). When the rotor is rotated one half-turn (180 degrees), the output is 6 volts 90 degrees (see Figure 11–54).

What if the rotor is not aligned with either stator but is located at some other angle? Examine Figure 11–55. In this case, the rotor is at a 45-degree angle to the 0- and 90-degree stators. In this case the rotor winding output is a combination of the field created by the 0- and 90-degree stator. The output is 6 volts 225 degrees. If the rotor is now turned one half-turn (180 degrees), the output is 6 volts

Figure 11–51 The rotor is aligned with stator winding 1. The rotor output is 6 VAC (one-half the number of windings on the rotor) 180 degrees.

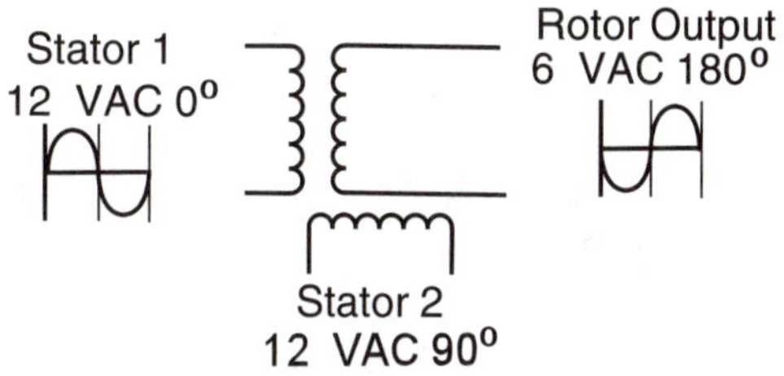

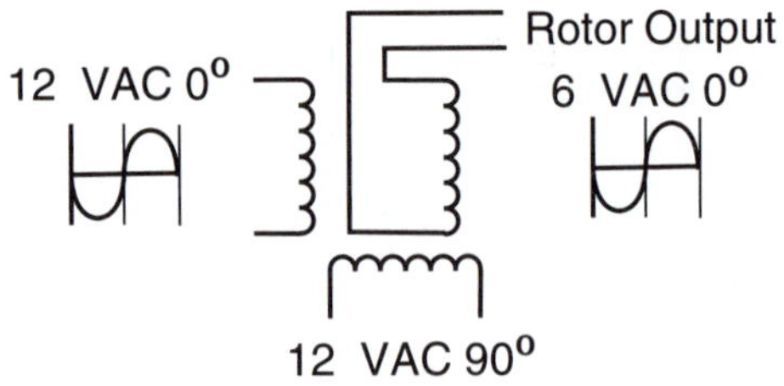

Figure 11–52 Rotor is aligned with the 0-degree stator but has been rotated 180 degrees.

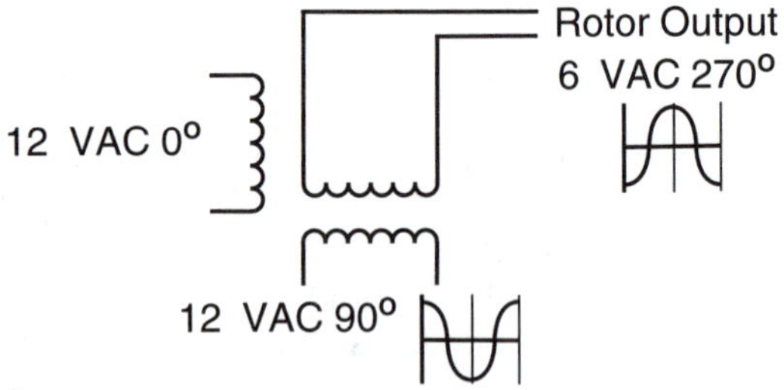

Figure 11–53 Rotor lined up with the 90-degree stator.

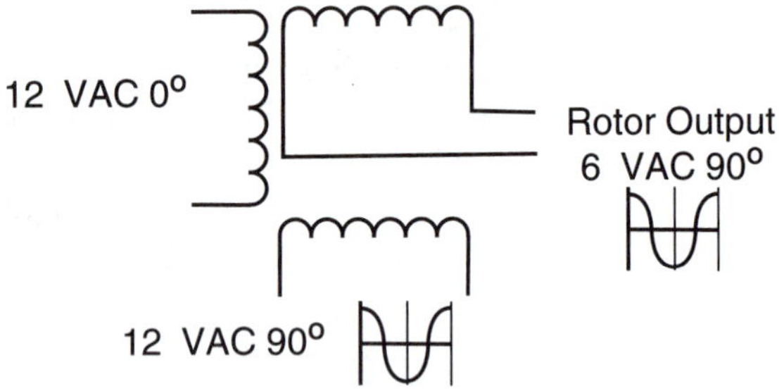

Figure 11–54 Output with the rotor aligned with the 90-degree stator but revolved 180 degrees.

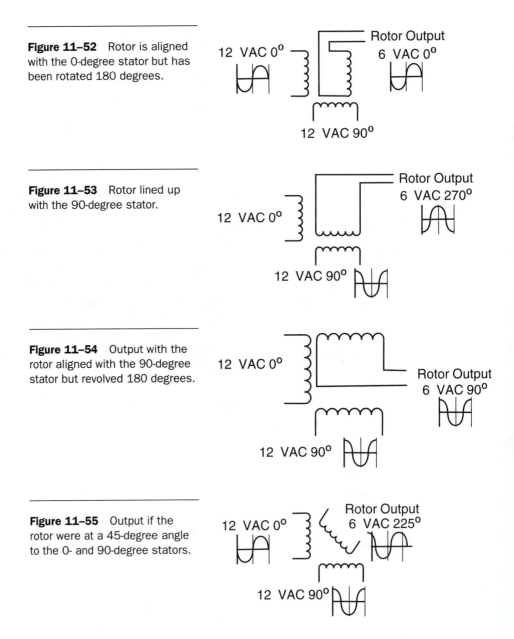

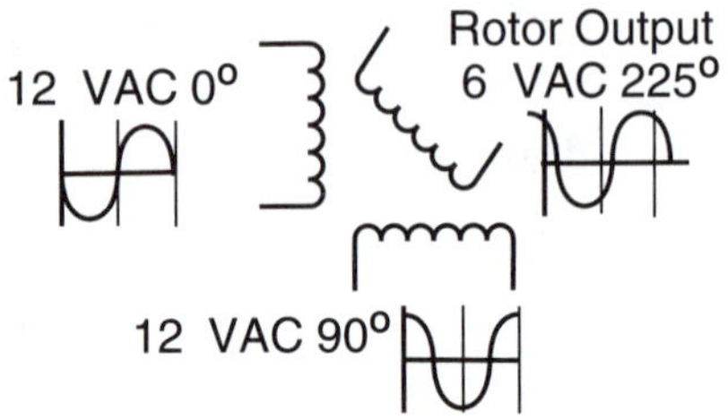

Figure 11–55 Output if the rotor were at a 45-degree angle to the 0- and 90-degree stators.

45 degrees. The rotor output is always 6 volts. The phase of the rotor output is determined by the orientation of the rotor in relation to the stator windings. A computer is capable of comparing the 0-degree sine wave to the output phase. By comparing them, the computer always knows the position of the rotor. One man-

ufacturer of position controllers uses a clock that pulses 4000 times per cycle to compare the phase of the input and output and to break the cycle into 4000 pieces. This gives a resolution of 4000 for 360 degrees of rotation, which is very fine.

Resolvers are also used in the ratiometric tracking method, which is the reverse of the type just studied. The ratiometric tracking method applies a sine-wave input to the rotor. This causes an output voltage and phase on the stators. As the rotor turns, the stator voltage and phase vary. The computer can compare the two stator outputs and determine position (see Figure 11–56).

Pressure Sensors

Pressure sensors typically measure and control fluids such as gases and liquids. Some pressure sensors operate through a change in resistance, some through a change in capacitance, and some through changes in inductance. The strain gage pressure sensor discussed earlier has attached a strain gage to a membrane that

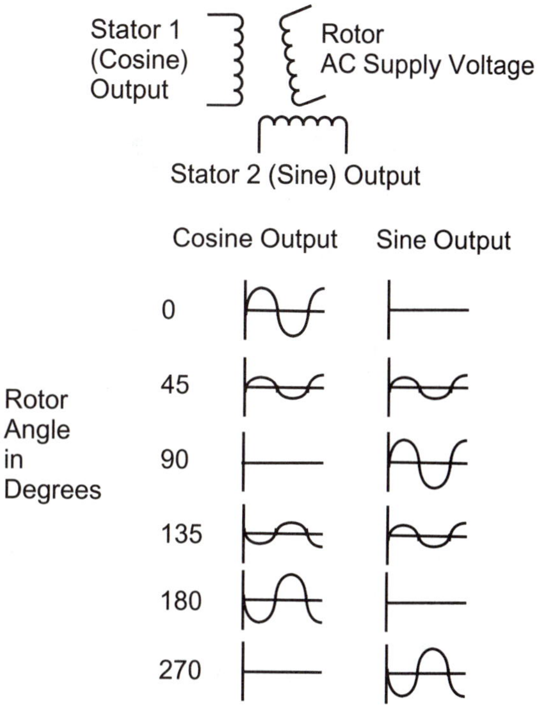

Figure 11–56 Output for a resolver using the ratiometric method.

stretches in proportion to the pressure applied to it (see Figure 11–57). If a constant current is applied to the strain gage, its output voltage changes according to the change in pressure.

Another pressure sensor utilizes an LVDT (see Figure 11–58). Pressure applied to the bellows causes it to expand and move the core in the LVDT. This change in core position can then be measured by the output of the LVDT.

Industrial networks (buses) can minimize installation and maintenance costs of sensors. Figure 11–59 shows an example of an industrial sensor network.

Figure 11–57 Strain gage pressure sensor. The circle represents a membrane to which the pressure is applied.

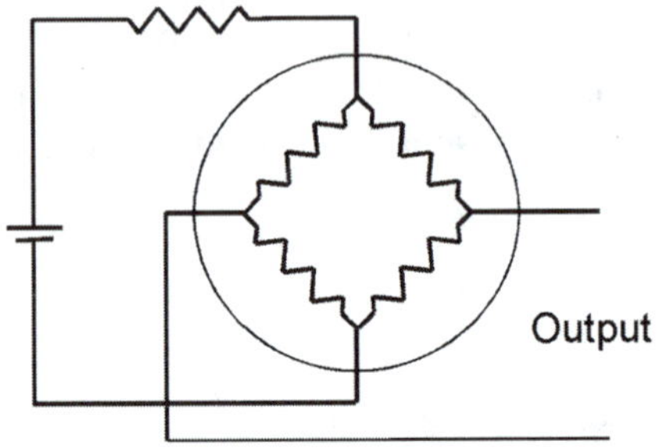

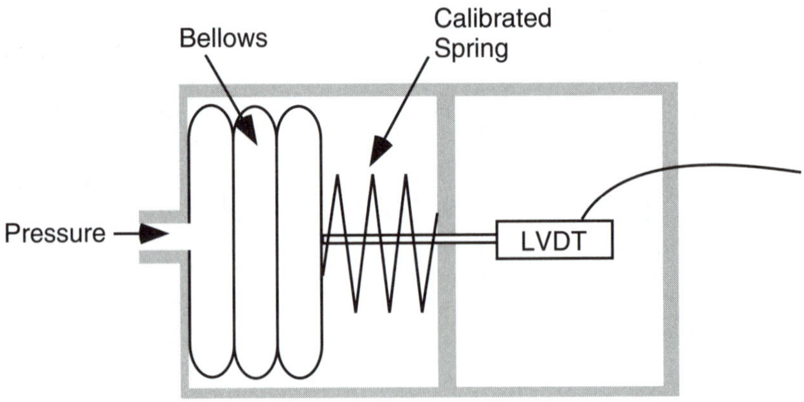

Figure 11–58 Bellows-style pressure sensor. Any change in pressure affects the bellows and moves the core in the LVDT.

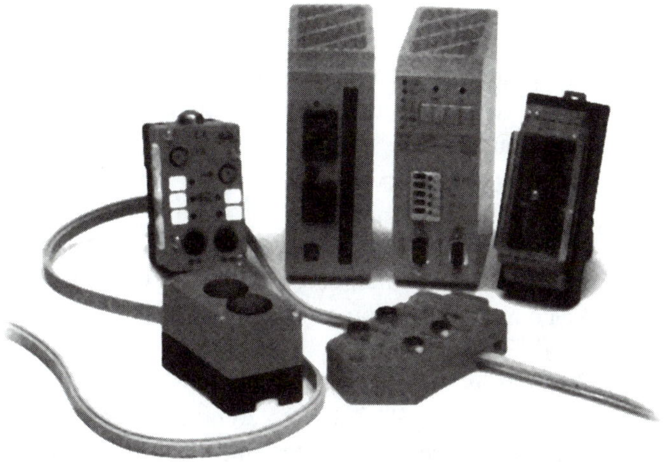

Figure 11–59 An industrial sensor network. *(Courtesy ifm efector inc.)*

Sensor Installation Considerations

Electrical

The main installation consideration for sensors is to limit the load current. The output (load) current must be limited on most sensors to a very small output current. The output limit is typically between 50 and 200 milliamps. If the load draws more than the sensor current limit, the sensor fails and must be replaced. More sensors probably fail because of improper wiring than from use. It is crucial that current be limited to a level that the sensor can handle. PLC input modules limit the current to acceptable levels, and sensors with relay outputs can handle higher load currents (typically 3 amps).

If high-voltage wiring is run in close proximity to sensor cable, the cable should be run through a metal conduit to prevent the sensor from false sensing, malfunction, or damage. The other main consideration is to buy the proper polarity sensors. If the PLC module requires sinking devices, they should be installed.

Mechanical

Mechanical sensors should be mounted horizontally whenever possible to prevent the buildup of chips and debris on the sensor, which could cause misreads. In a vertical position, chips, dirt, oil, and so on, can gather on the sensing surface and cause the sensor to malfunction. In a horizontal position, the chips fall away. If the sensor must be mounted vertically, provision must be made to remove chips and dirt periodically using air blasts or oil baths.

Care should be taken so that the sensor does not detect its own mount. For example, an inductive sensor mounted improperly in a steel fixture might sense the fixture. If two sensors are mounted too close together, they can interfere with each other and cause erratic sensing.

Typical Applications

One of the most common uses of a sensor is in product feeding when parts move along a conveyor or in some type of parts feeder. The sensor notifies the PLC when a part is in position and is ready to be used. This is typically called a *presence/absence check.* The same sensor can also provide the PLC with additional information that the PLC uses to count the parts as they are sensed. The PLC can also compare the completed parts and elapsed time to compute cycle times and can track scrap rates, production rates, and cycle time.

One simple sensor allowed the PLC to determine three conditions:

Parts present

Parts that have been used

Cycle time for each part

Simple sensors can be used to decide which product is present. Imagine a manufacturer that produces three different package sizes on the same line. The product sizes move randomly along a conveyor. When each package arrives at the end of the line, the PLC must know which size of product is present. This can be done very easily with three simple sensors. If only one sensor is on, the small product is present. If two sensors are on, the midsize product is present. If three sensors are on, the product is the large size. The same information could then be used to track production for all product sizes and cycle times for each size.

Sensors can be used to check whether containers have been filled. Imagine aspirin bottles moving along a conveyor with the protective foil and the cover on. Simple sensors can sense right through the cap and seal and make sure that the bottle is filled. One sensor, often called a *gate sensor,* senses when a bottle is present. A gate sensor shows when a product is in place. The PLC then knows that a product is present and can perform other checks.

A second sensor senses the aspirin in the bottle. If a bottle is present, but the sensor does not detect the aspirin inside, the PLC knows that the aspirin bottle is not filled.

As discussed earlier, sensors can monitor temperature. Imagine for example, a sensor that can monitor the temperature in an oven in a bakery. The PLC can then control the heater element in the oven to maintain the ideal temperature.

Pressure is vital in many processes. Plastic injection machines force heated plastic into a mold under a given pressure. Sensors can monitor the pressure, which must be maintained accurately or the parts will be defective. The PLC can monitor the sensor and control the pressure.

Flow rates are important in process industries such as papermaking. Sensors can monitor the flow rates of fluids and other raw materials. The PLC can use these data to adjust and control the system's flow rate. The water department monitors the flow rate of water to calculate bills.

Figure 11–60 is an illustration of a temperature and flow application. The flow sensor on the upper left of the machine monitors the proper flow of cooling water into the chiller. A built-in display displays the flow setpoint and the flow status. The sensor on the lower right of the machine is a temperature sensor that can be set to the proper temperature and alarm point.

The applications noted here are a few of the simple uses for sensors. Innovative engineers or technicians invent many other uses. The data from one sensor can be used to provide many different types of information (e.g., presence/absence, part count, cycle time).

When choosing the sensor to use for a particular application, several important considerations, such as the material in the object to be sensed, are crucial. Is the

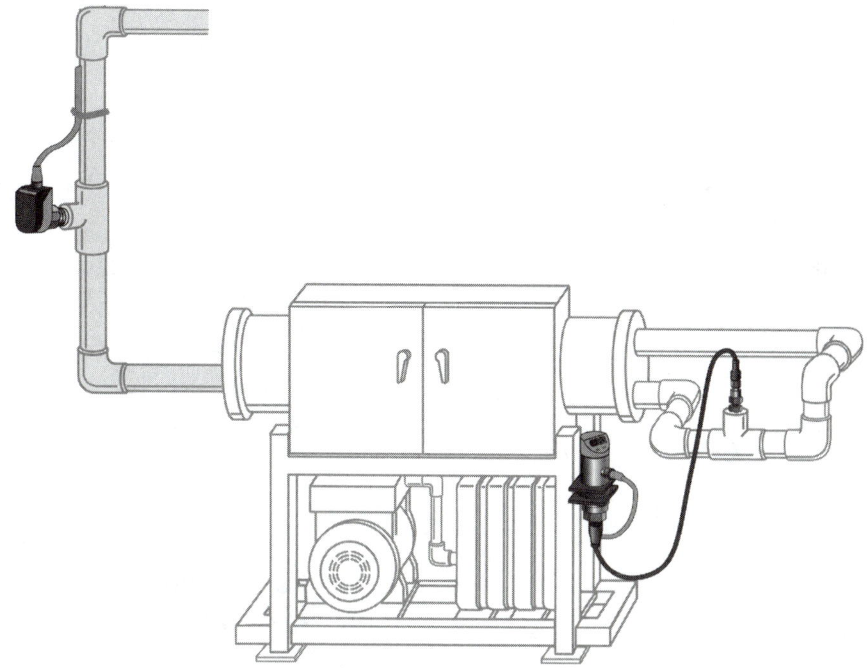

Figure 11–60 Temperature and flow application. *(Courtesy ifm efector inc.)*

material plastic? Is it metallic? Is it a ferrous metal? Is the object clear, transparent, or reflective? Is it large or very small?

The specifics of the physical application determine the sensor type to use. Is a large area available on which to mount the sensor? Are contaminants a problem? What speed of response is required? What sensing distance is required? Is excessive electrical noise present? What accuracy is required? Answering these questions helps narrow the choice, which must be made based on criteria such as the sensor's cost and reliability as well as the cost of failure. The cost of failure is usually the guide to how much sensing must be done. If the cost is high, sensors should be used to notify the PLC of problems.

Figure 11–61 shows a smart-level sensor. The sensor's microprocessor and push button are used to teach the PLC to recognize the container's empty and full conditions.

Two typical applications utilize ultrasonic sensors to measure distances. See Figure 11–62. The application on the left uses the sensor's analog output to control the web precisely. The application on the right measures the height of objects. The fiber optic sensor is used as a gate to indicate part presence. The ultrasonic sensor then takes a reading and the height of the part is known.

Numerous applications are shown in the following figures:

- Figure 11–63 shows a package-sorting application that uses color-mark sensors.
- Figure 11–64 is a photosensor used in a bottle-capping application. This sensor's output could be used to distinguish between bottles that are capped and those that are not. Note that a field sensor can work in hostile environments.

Figure 11–61 Sensor that senses empty and full levels in a container. *(Courtesy ifm efector inc.)*

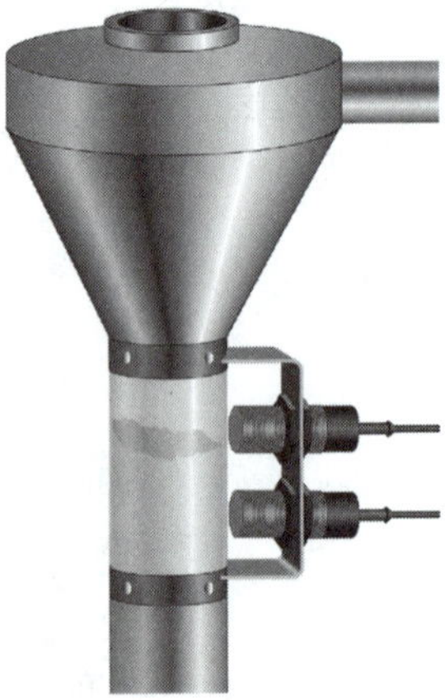

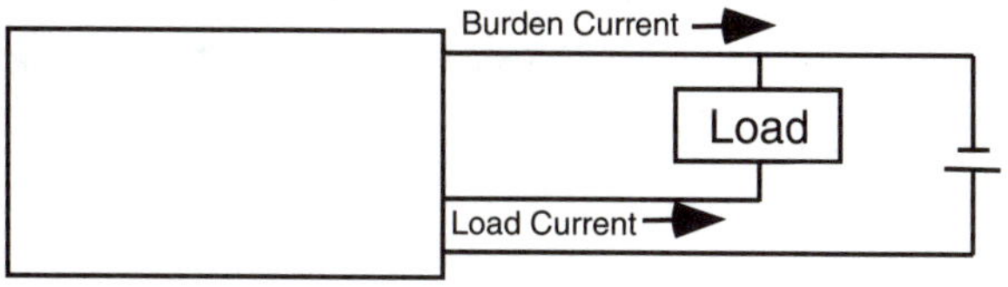

Figure 11–62 caption: **Figure 11–62** Two typical applications that use ultrasonic sensors to measure distances. *(Application concept courtesy of Omron Electronics.)*

Figure 11–63 Sorting packages using color-mark sensors. *(Application concept courtesy of Omron Electronics.)*

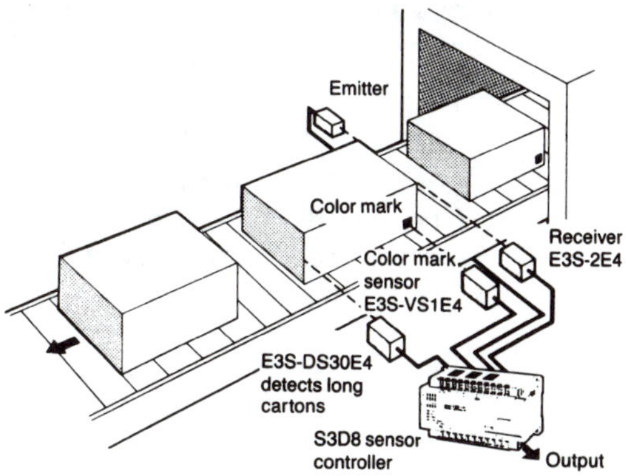

Figure 11–64 A photo sensor used to sense the proper capping of a container. *(Application concept courtesy of Omron Electronics.)*

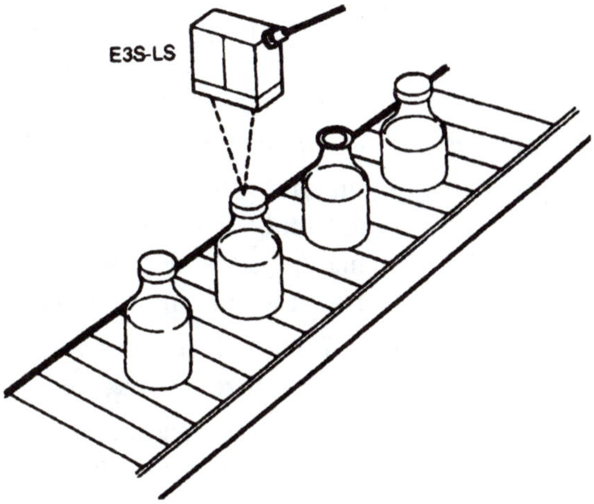

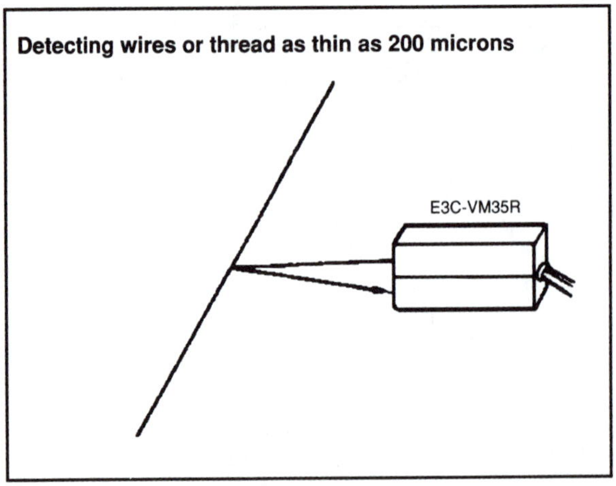

Figure 11–65 Wire can be sensed by a laser sensor. *(Application concept courtesy Omron Electronics.)*

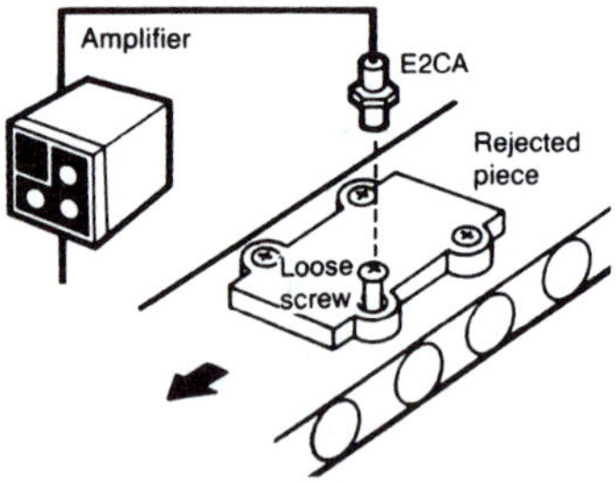

Figure 11–66 Small objects can be sensed by a laser sensor. *(Application concept courtesy of Omron Electronics.)*

- Figure 11–65 shows the use of a laser sensor with a 5-micron resolution. The sensor uses a laser LED to emit a beam 10 millimeters wide, with a maximum sensing distance of 300 millimeters (11.8 inches). In this case the analog output is used to detect wire breakage.
- Figure 11–66 shows an inductive sensor being used to check the correct assembly of screws. This inductive sensor provides digital outputs for high/pass/low detection and could be used to pass or fail the parts into pass bins, too tall bins, and too short bins.
- In Figure 11–67 the sensor is mounted perpendicular to the transparent film to maximize the reflected light to the receiver as it detects the film.
- Figure 11–68 shows three fiber-optic applications used to sense small objects. In the first, the sensor checks for part presence. The second uses a

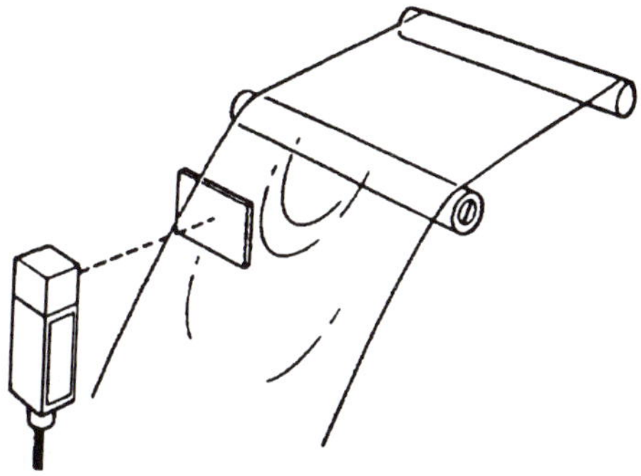

Figure 11–67 Detection of transparent film. *(Application concept courtesy of Omron Electronics.)*

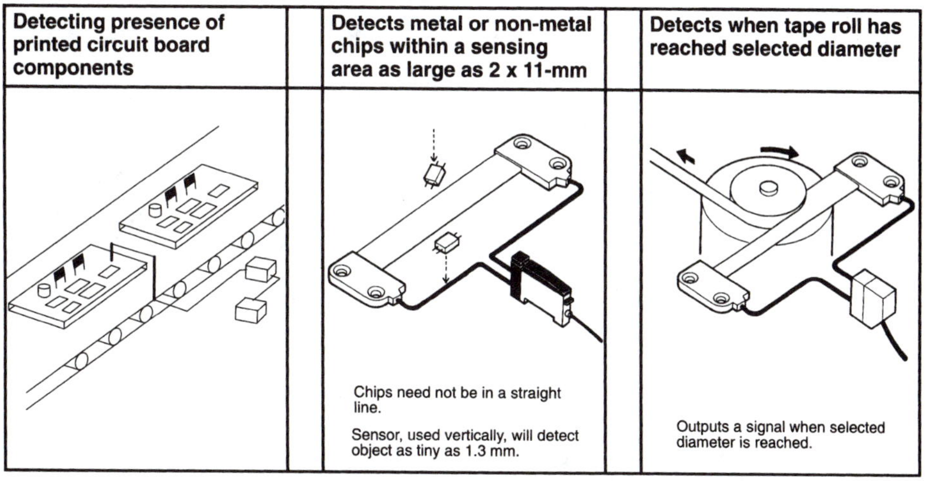

Detecting presence of printed circuit board components
Detects metal or non-metal chips within a sensing area as large as 2 x 11-mm
Chips need not be in a straight line.
Sensor, used vertically, will detect object as tiny as 1.3 mm.
Detects when tape roll has reached selected diameter
Outputs a signal when selected diameter is reached.

Figure 11–68 Detection of small metal objects. *(Application concept courtesy of Omron Electronics.)*

special fiber-optic head to spread the beam. The third checks the diameter in a tape-winding application.

- Figure 11–69 is an example of a hydraulic pressure and level application. The pressure sensor on the upper left of the hydraulic power unit monitors the hydraulic pressure. The sensor on the lower right detects the level of hydraulic fluid in the power unit. The sensor is mounted to a sight glass with a strap but it ignores the sight glass and senses only the liquid.
- Figure 11–70 is an application utilizing a fiber-optic sensor in a confined space.
- Figure 11–71 shows capacitive sensors checking a box to determine that all nine bottles have been loaded into the carton.

Figure 11–69 Hydraulic application. *(Courtesy ifm efector inc.)*

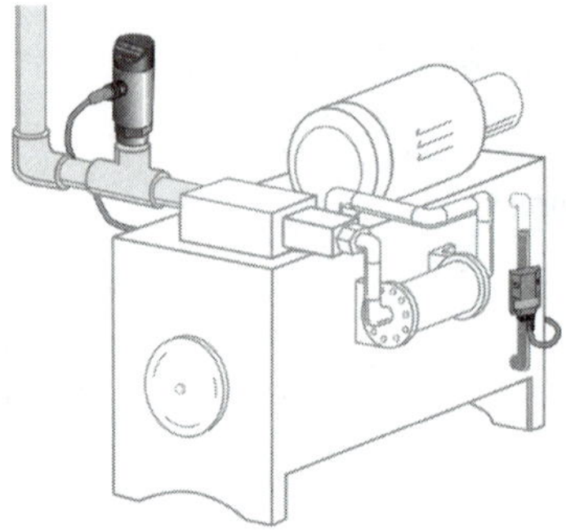

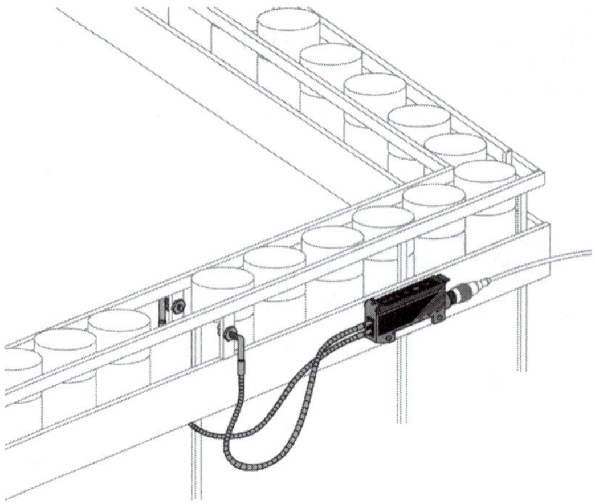

Figure 11–70 Fiber-optic application in confined space. *(Courtesy ifm efector inc.)*

Figure 11–71 Capacitive sensor application. *(Courtesy ifm efector inc.)*

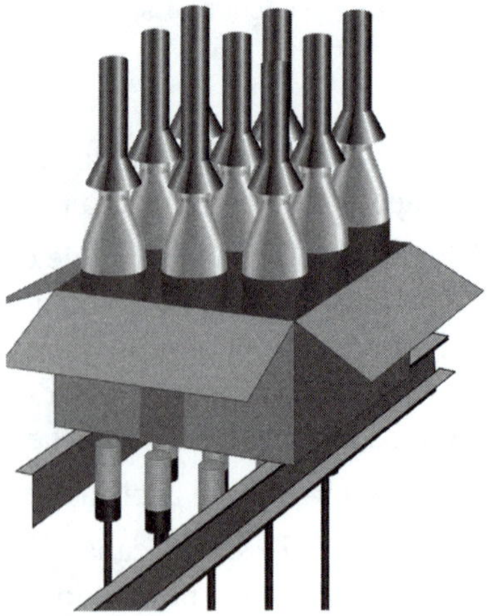

- Figure 11–72 is a laser sensor used to check very small glass parts. The laser provides a very small diameter coherent beam to sense even very small parts.
- Figure 11–73 shows a pressure-sensing application that provides feedback to a controller for safety and proper operation. Digital and analog pressure sensors are available for this type of application.

Figure 11–72 Laser sensor application of small objects. The part is spherical, not flat. Only a very small area can reflect light back to the sensor. *(Courtesy ifm efector inc.)*

Figure 11–73 Pressure-sensing application. *(Courtesy ifm efector inc.)*

Figure 11–74 Inductive sensor used to sense each tooth as it passes. *(Courtesy ifm efector inc.)*

- Figure 11–74 is an inductive sensor used to sense gear teeth as the gear rotates. The user can tell the position by the number of gear teeth that passes or the velocity by the number of teeth that passes the inductive sensor in a given amount of time.

Choosing a Sensor for a Special Application

Applications always present special problems for the technician or engineer. The sensor may fail to sense every part or fail randomly. In this case, a different type or different model of sensor may be needed. Sales representatives and applications engineers from sensor manufacturers can be very helpful in choosing a sensor to meet a specific need. They have usually seen the particular problem before and know how to solve it. Magazines devoted to the topic of sensors are good sources of information.

The number of sensor types and the complexity of their use in solving application problems increases daily. New sensors are constantly being introduced to solve needs. Almost as many sensor types are available now as there are applications.

The innovative use of sensors can help increase the safety, reliability, productivity, and quality of processes. The technician must be able to choose, install, and troubleshoot sensors properly.

Questions

1. Describe at least four uses of digital sensors.
2. Describe at least three analog sensors and their uses.
3. List and explain at least four types of optical sensors.
4. Explain how capacitive sensors work.

5. Explain how inductive sensors work.
6. Explain the term *hysteresis.*
7. Draw and explain the wiring of a load-powered sensor.
8. Draw and explain the wiring of a line-powered sensor.
9. What is burden current? Load current?
10. Why must load current be limited?
11. Explain the basic principle on which a thermocouple is based.
12. How are changes in ambient temperature compensated for?
13. What is the temperature range for a type-J thermocouple?
14. Describe at least three electrical precautions as they relate to sensor installation.
15. Describe at least three mechanical precautions as they relate to sensor installation.

chapter 12

IEC 61131-3 Programming

IEC 61131-3 is a standard for programming. It promises to revolutionize PLC and other controller programming.

OBJECTIVES

Upon completion of this chapter, you will be able to:

1. Describe the purpose of the IEC 61131-3 standard.
2. List and explain each of the languages specified by IEC 61131-3.
3. Explain how the languages can be integrated in programs.
4. Explain the terms *function blocks, functions,* and *statements.*
5. Explain how IEC 61131-3 will affect industrial automation.

OVERVIEW OF IEC 61131-3

The International Standard IEC 61131 is a complete collection of standards on programmable controllers and their associated peripherals. It consists of the following parts: Part 1, general information; Part 2, equipment requirements and tests; Part 3, programming languages; Part 4, user guidelines; Part 5, communications; Part 6, reserved for future use; Part 7, fuzzy control programming; and Part 8, guidelines for the application and implementation of programming languages.

What Is the 6 in IEC 61131?

You may be more familiar with the number IEC 1131. The International Electrotechnical Commission, IEC, is a worldwide standardization body. Nearly all countries have their own national standards organizations. These organizations have agreed to accept the IEC approved and published standards. At local publication, the standard was often published under a local number. This local number often had no match to the number of the IEC published standard. To correct this discrepancy, they searched for a worldwide numbering system. This is where the 6 came in.

The IEC 61131-3 standard was developed to meet a very natural need. Users have struggled for years with the problem of programming different brands of PLCs. Each brand has slightly different ladder logic. In some cases, different models from the same manufacturer often have different languages; thus users need multiple programming packages. It has also required users to learn the software and logic differences. Another change is that special-purpose controllers and languages emerged to fill the ladder logic gaps. Languages for more complex control and for special purposes, such as motion control, emerged, as did languages that attempted to simplify programming.

Ladder logic has some weaknesses. First, it is hard to reuse. Complex programs are little use in developing new programs. Each application is essentially created from scratch. Ladder logic can be quite inconvenient and cumbersome to perform mathematical computations and comparisons. It is also difficult to segment ladder logic. Every line affects every other line, which means that it is difficult to control the execution of ladder diagrams (they are normally executed top to bottom). It would be advantageous to control which sections execute and in which order.

IEC 61131-3 emerged to meet these needs and establish a standard for programming. IEC 61131-3 was built using programming techniques that were already well established. The IEC standard specifies the following programming languages: ladder diagram, instruction list, function block diagram, structured text, and sequential function chart. The standard also allows the user to select any of the languages and mix their use in the program. The user can choose the best language for each part of the application. Three of the languages are graphics-based and two are text-based. The standard should ensure that the vast majority of software written for one PLC will run on PLCs from other manufacturers.

The first revision of IEC 61131-3 was published in 1993. Programs can be written using any of the IEC languages. A program is typically a collection of function blocks that are connected. Programs can communicate with other programs and can control I/O.

The IEC 61131-3 standard defines a program as "a logical assembly of all programming elements and constructs necessary for the intended signal processing required for the control of a machine or process by a programmable controller system."

Function Blocks

Function blocks are one of the keys to IEC 61131-3 programming. They allow a program to be broken into smaller, more manageable blocks. Function blocks can even be used to create new, more complex function blocks. They take a set of input data, perform actions or an algorithm on the data, and produce a new set of output data. Functions can hold data values between execution. You can think of a function block as a special-purpose integrated chip. We find special-purpose chips in many consumer products today. They take a set of inputs, process the data, and produce outputs. Think about the processors in automobiles and antilock braking systems. For these examples, a function block can be plugged into programs and can perform a portion of the needed logic and control. Standard functions can be used off the shelf, or users can develop and use their own functions.

Function blocks can be used to solve control problems such as PID control. Temperature and servo controls are important in many industrial applications. The temperature and servo controls are typically only a portion of the control problem, however. Function blocks can be used to develop a PID or a fuzzy logic algorithm.

A function block defines the purpose of input and output data. These data can be shared with the rest of the control program. Only input and output data can be shared. A function block can also have an algorithm. These algorithms are run every time the block is executed. The algorithms process the current input data values and produce new output values.

Function blocks also have the ability to store data. Data can be used locally or may be used globally. Input or output data can be shared, but internal variables are not accessible to the rest of the program. This is an important feature because it allows function blocks either to be modular and independent or to share their data.

The standard specifies some standard function blocks, for example, counters, timers (Figure 12–1), and clocks. Function blocks can also be developed by the user. Users can build new blocks using existing function blocks and other software logic.

Function blocks are crucial because they allow the user to develop very logical programs that are easy to understand and maintain.

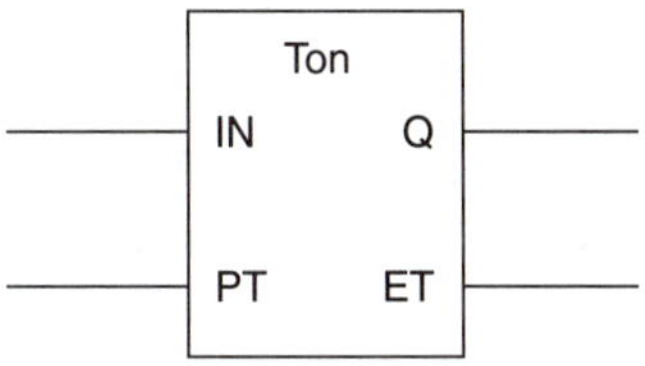

Figure 12–1 A Ton function block timer.

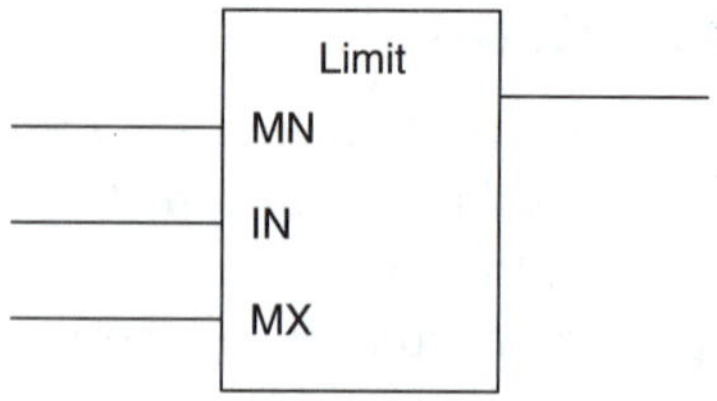

Figure 12–2 Example of a limit function.

Functions

IEC 61131-3 also specifies functions. Functions are different from function blocks. Functions are designed to perform common processing tasks such as trigonometric functions. They take a set of inputs and produce a result. Given the same input, the output will always be the same.

All standard trigonometric calculations are available as functions. All common arithmetic calculations such as add and subtract are also available as functions, as are a variety of bit functions such as shifts and rotates, Boolean operators, and a set of selection type functions. Figure 12–2 shows an example of a limit function, which limits the value of an input data value between the values at minimum (MN) and maximum (MX) and sends out the result.

There are also comparison functions such as greater than (GT), greater than or equal to (GE), or equality (EQ). Character string function is also specified, as well as a wide variety of functions for working with the time and date.

Structured Text Programming

Structured text (ST) is a high-level language that has some similarities to Pascal programming. The IEC standard defines structured text as a language in which statements can be used to assign values to variables, however, the use of structured text is broader than the IEC definition.

Expressions

Structured text uses expressions to evaluate values that are derived from other constants and/or variables. Expressions can also be used to calculate values based on the values of other variables or constants. When using expressions, the user must use a variable data type that will match the result of the expression.

Assignment Statements

Assignment statements are used to assign or change the value of a variable. An assignment statement in ST is similar to an assignment statement in any programming language, particularly Pascal.

An example of an assignment statement is:

TEMP := Z

This statement would assign the value of whatever is in Z to TEMP. If the value of Z was equal to 78, the statement would put the value 78 into the variable called TEMP. (We could use a constant instead of Z.)

Operators

Figure 12–3 shows the many arithmetic operators available. They are shown in their order of precedence, from the highest to the lowest.

Operator	Description
(....)	Parentheses used to group for precedence
Function (....)	Used for function parameter list and evaluation
**	Exponentiation
−	Negation
NOT	Boolean complement
*	Multiplication
/	Division
MOD	Modulus
+	Addition
−	Subtraction
<,>, <=,> =	Comparison
=	Equality
<>	Inequality
AND, &	Boolean AND
XOR	Boolean exclusive OR
OR	Boolean OR

Figure 12–3 Structured text arithmetic operators.

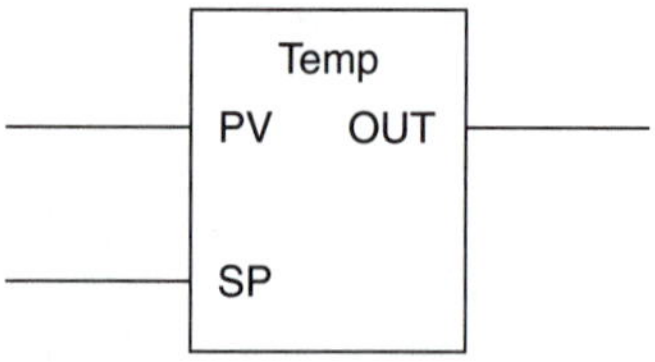

Figure 12–4 A Temp function.

Statements

Structured text language has various statements that can be used to call functions, perform iteration (loops), or perform conditional evaluation.

Function blocks can be run using statements, for example:

```
Temp (PV := 74, SP := 83);
```

In this example a statement is used to call the Temp function (Figure 12–4). The Temp function requires two input values, PV and SP. Values are sent to the function by specifying them therein. In this case, 74 will be sent to the PV input of the function and 83 will be sent to the SP of the function. The function would then calculate the output value. The output value is always available. It can be used in any assignment statement, for example:

```
Tval := Temp.Out;
```

This statement would assign the output of the Temp function to variable Tval.

Conditional Statements

Conditional statements are used to control which statements are to be executed. IF, THEN, ELSE, and CASE statements are allowed.

IF, THEN, ELSE Statements

The conditional statements IF, THEN, and ELSE evaluate a Boolean expression to determine whether to execute the logic they control. Consider the following example:

```
IF Tval = 87 THEN
  Var1 := 35;
  Var2 := 12;
ELSE
  Var1 := 54;
  Var2 := 17;
END_IF;
ELSIF can also be used.
IF Var1 < 83 THEN
  A:= 5;
```

```
ELSIF VAR1 = 83 THEN
  A:= 6;
ELSIF VAR1 > 83 THEN
  A:= 7;
END_IF;
```

CASE Statements

The CASE statement is also conditional. It is a useful statement when the user needs to execute a set of statements that are conditional on the value of an expression that returns an integer. In other words, a CASE statement evaluates an integer expression and executes the portion of the CASE code that matches the integer value. Here is an example of a CASE statement:

```
CASE Var1 OF
1 : Var2 := 5;
2 : Var2 := 5; Pump1 := ON;
3,4 : Var2 := 5; Pump2 := ON;
5&. . . 7 : Var2 := 5; Alarm := ON;
END_CASE;
```

Loops

Several types of loops are available.

REPEAT . . . UNTIL Loop The REPEAT UNTIL loop is used to execute one or more statements while a Boolean expression is true. The Boolean expression is tested after execution of the statements. If it is true, the statements are executed again, for example:

```
Count := 1;
REPEAT
  Count := Count + 1;
  Pack := 5;
  Pump1 := OFF;
UNTIL Count = 5
END_REPEAT
```

FOR . . . DO Loop The FOR DO loop allows execution of a set of statements to be repeated based on the value of a loop variable, for example:

```
FOR I := 1 TO I < = 50 BY 1 DO
  Temp (PV := Var1, SP := Var2);
END_FOR;
```

WHILE . . . DO Loop The WHILE DO loop permits repeated execution of one or more statements while a Boolean expression remains true. The expression is tested before executing the statements. If the expression is false, the statements are not executed, for example:

```
WHILE Var1 < 95 DO
  Temp (PV := Var1, SP := Var2);
END_WHILE
```

RETURN and EXIT Statements

RETURN and EXIT statements can be used to end loops prematurely. The return statement can be used within functions and function block bodies. RETURN is used to return from the code. Consider the following example:

```
FUNCTION_BLOCK TEMP_TEST
VAR_INPUT
  PV, SP : REAL;
END_VAR
VAR_OUTPUT
  OUT : REAL;
END_VAR
IF PV > 100 THEN
  ALARM := TRUE; RETURN;
END_IF;
IF SV > 100 THEN
  ALARM := TRUE; RETURN;
END_IF;
END_FUNCTION_BLOCK;
```

If any IF is true, ALARM is set to TRUE and the RETURN statement is executed to end the execution of this function block.

EXIT Statement

The EXIT statement can be used to end loops before they would ordinarily end. When an EXIT statement is reached, the execution of the loop is ended and execution continues from the end of the loop. The following example shows a loop that increments through a single-dimensional array named TEMP:

```
FOR I:= 1 TO 10 DO
  IF TEMP[I] > 90 THEN
    EXIT;
  END_IF;
END_FOR;
```

As the loop increments, the IF statement checks if the value of each element of the TEMP array is greater than 90. If it is, the loop is exited. Program execution would then proceed to the next statement following this loop.

Function Block Diagram Programming

Function block diagram (FBD) programming is one of the graphical languages specified by the standard. FBDs can be used to show how programs, functions, and function blocks operate. An FBD looks like an electrical circuit diagram, where lines are used to show current flow to and between devices. There are typically inputs to and outputs from each device. Figure 12–5 shows the overall system and the interrelationships of the components. This is similar to FBDs. Many people find function block diagrams easier to program than language-based languages such as structured text.

FBD programming is used when the application involves the flow of signals between control blocks. Remember that we examined functions and function blocks earlier in the chapter. Figure 12–6 shows an example of a function block with inputs on the left and outputs on the right. The function block type name (e.g., TempControl) is shown inside the block. Remember that the name shown in the block is the type name. The name shown above the block in an FBD is the name of the function

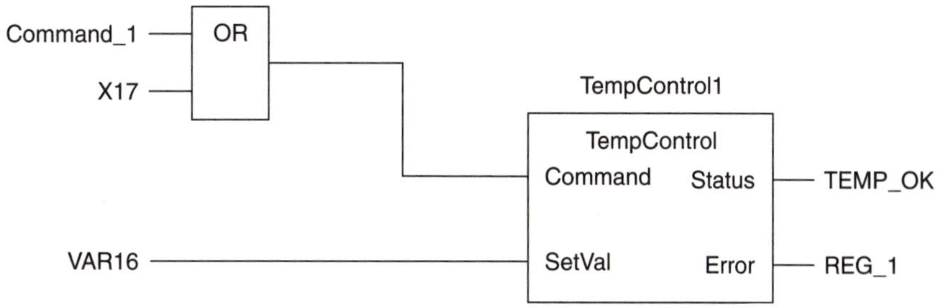

Figure 12–5 Use of Boolean operators in a simple FBD.

Figure 12–6 An example of a function block.

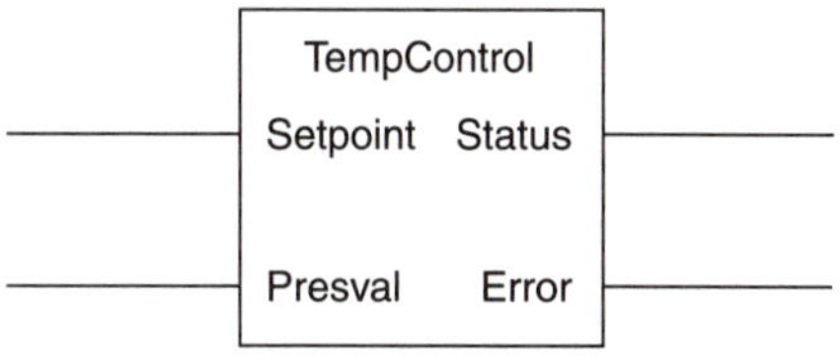

block instance. Again, the names of the inputs are shown on the left of the block and the output names are shown on the right.

Figure 12–5 utilizes Boolean functions. Remember that we can utilize standard functions such as Boolean, comparison, string, selection, time, mathematical, and numerical as well as user-created function blocks in FBDs.

Figure 12–7 shows an example of a function block diagram. This is a temperature control application, using two temperature control blocks. Note that the two blocks are of the same type. These blocks needed to be created only once. Note also that the name above each is different because each is a different use of the same function block type and the name identifies each particular instance. Note also that a block called TempMonitor has been used. It is providing the current temperature reading to the input called Presval in each of the TempControl blocks.

Variables or constants can be used as inputs to function blocks. Note also that the outputs from the function blocks are sent to variables. Outputs are used to supply values to variables and function block inputs (also variables).

The IEC 61131-3 standard also allows for values to be fed back to create feedback loops. Figure 12–8 shows an example of feedback. This feature allows values to be used as inputs to previous blocks. This means that the output value of Status from MachControll is used as an input value to the input (ComVal) of LoopCont1. What happens if LoopCont1 is evaluated before MachControl1?

Figure 12–9 shows a list of operators that are reserved for use with standard function blocks. The standard states that implementations may provide a facility so

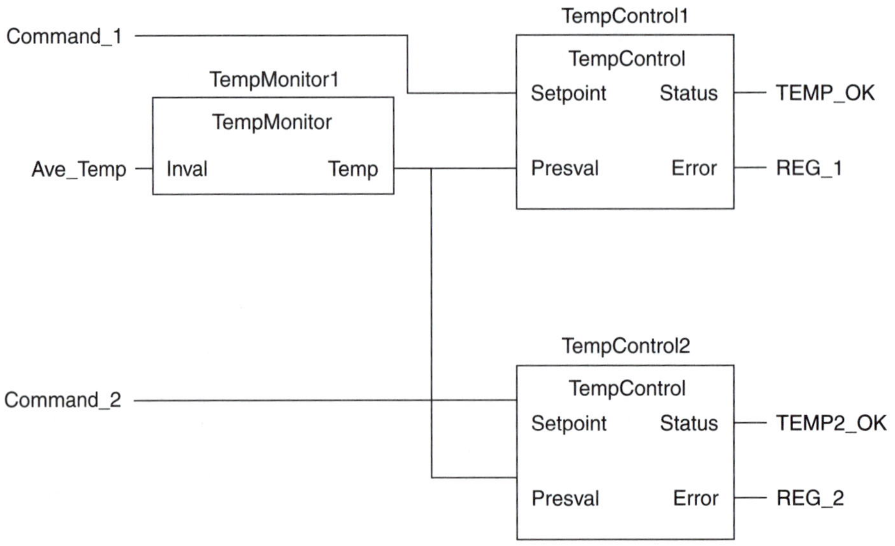

Figure 12–7 A small function block diagram.

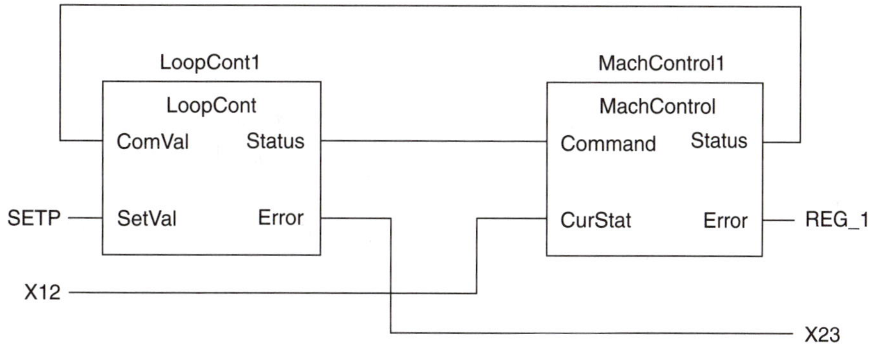

Figure 12–8 How a value can be used to create feedback.

Operator	Function Block Type	Use
CD, LD, PV	Down Counter CTD	Parameters for the CTD, to count down (CD), load (LD), and set the count preset (PV)
CU, R, PV	Up Counter CTU	Parameters for the CTU, to count up (CU), reset, (R), and load present value (PV)
CU, CD, R, LD, PV	Up/Down Counter	Same as for up and down counters
IN, PT	Pulse Timer TP	Parameters are IN to start timing and PT to set up the pulse time
IN, PT	On-Delay Timer TON	Parameters include IN to start timing and PT to set up the delay time
IN, PT	Off-Delay Timer TOF	Parameters include IN to start timing and PT to set up the delay time
CLK	R_Trig, Rising-Edge Detector	Clock input to the rising-edge detector function block
CLK	F-Trig, Falling-Edge Detector	Clock input to the falling-edge detector function block
S1, R	SR Bistable	Set and reset the SR bistable
S, R1	RS Bistable	Set and reset the RS bistable

Figure 12–9 A list of operators that are reserved for use with standard function blocks.

that the order in which functions are evaluated in a network can be defined. The standard does not define the method to be used. This means that different companies that sell FBD programming software can define their own method of determining how functions would be evaluated. One method is a list to specify evaluation order.

Ladder Diagramming

Ladder diagramming is also a graphical language that IEC 61131-3 defines. Ladder diagramming was a logical choice as a programming language for one of the standard languages because it is the most widely used. Ladder diagramming can also be used in combination with all of the other programming methods that IEC 61131-3 specifies. Figure 12–10 shows the contacts specified in IEC 61131-3. Figure 12–11 shows the coils specified in IEC 61131-3. Ladder logic can also be used with the other IEC 61131-3 languages. Figure 12–12 shows an example of ladder logic used with function blocks.

Figure 12–10 Contacts specified in IEC 61131-3.

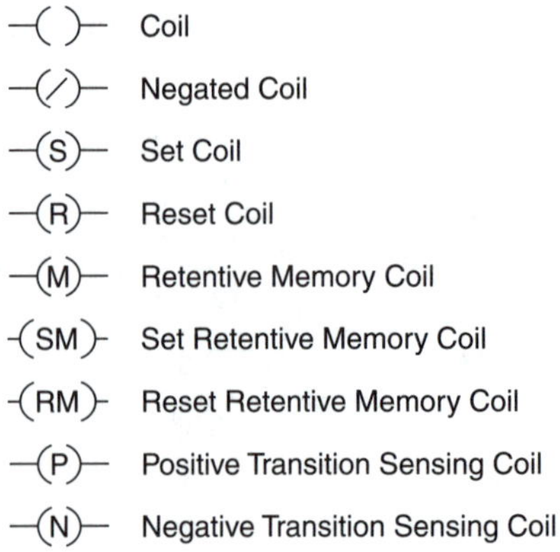

Figure 12–11 Coils specified in IEC 61131-3.

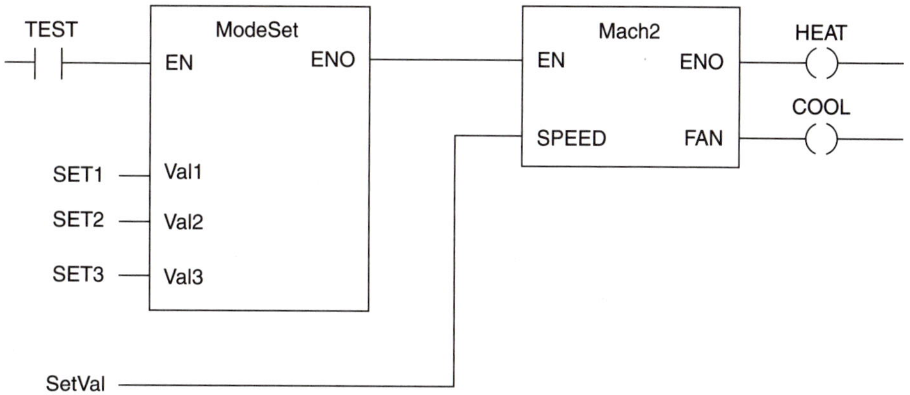

Figure 12–12 Use of ladder logic to combine function blocks.

Jumps

The standard for ladder logic programming specifies jumps. A jump can be used to jump from one section of a ladder diagram to another, based on the condition of a rung of logic. This is accomplished by having the rung of logic end in a label identifier. The label identifier is a name that is then used to identify the start of the logic to which the user wishes to jump. Figure 12–13 shows an example of the use of a jump.

Instruction List Programming

Instruction list (IL) programming is a low-level programming language, which means it is typically not very user-friendly. The higher the level of the language, the easier it is to use. Instruction list programming is much like assembly language programming. Figure 12–14 shows an example of IL programming.

IL programs can be used by experienced programmers to develop efficient, fast code. IL language is considered by many to be the base language for IEC-compliant PLCs. It is the language into which all of the other languages can be converted. IEC 61131-3 does not specify that any language is the base language, however. Figure 12–15 shows an example of arithmetic and Boolean instructions that can be used in IL programs. Figure 12–16 shows the comparison instructions that are available in IL programming.

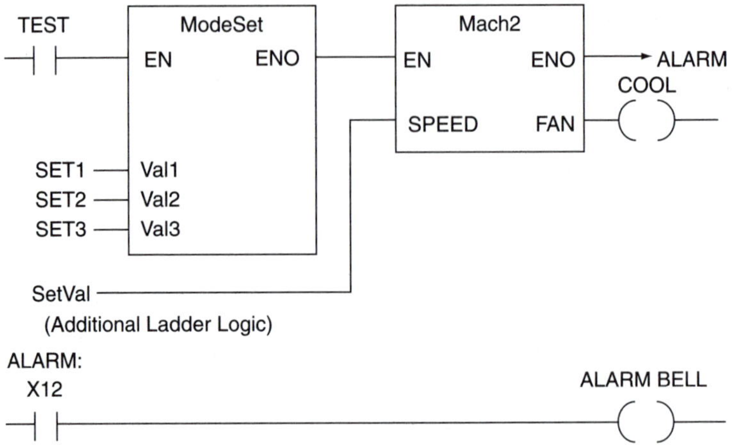

Figure 12–13 Jumps in ladder logic.

Label	*Operator*	*Operand*	*Comments*
	LD	Temp	(* Load Temperature *)
	LT	90	(* Test if Temperature < 90 *)
	JMPCN	Temp_Hi	(* Jump to Temp_Hi if Temp is not < 90 *)
	LD	Count	(* Load Count *)
	DIV	6	(* Divide by 6 *)
	ST	CASES	(* Store Result of Division to Cases Variable *)
	LD	0	(* Load 0 *)
	ST	Y12	(* Turns Output Y12 Off *)
	ST	Y23	(* Turns Output Y23 Off *)
	LD	1	(* Loads 1 *)
	ST	Y5	(* Turns on Output 5 *)
Temp_Hi:	LD	1	(* Loads 1 *)
	ST	ALARM	(* Stores a 1 in ALARM Variable *)

Figure 12–14 Instruction list program example.

Let us compare the IL program in Figure 12–14 to what it would look like in structured text:

```
IF Temp < 90 THEN
  CASES := Count / 6;
  Y12 := 0;
  Y23 := 0;
  Y5 := 1;
END_IF
ALARM : = 1;
```

Operator	Operand	Purpose
ADD	Any Type	Add
SUB	Any Type	Subtract
MUL	Any Type	Multiply
DIV	Any Type	Divide
LD	Any Type	Load
ST	Any Type	Store
S	Boolean	Set operand true
R	Boolean	Reset operand false
AND / &	Boolean	Boolean AND
OR	Boolean	Boolean OR
XOR	Boolean	Boolean exclusive OR

Figure 12–15 The arithmetic and Boolean instructions available in IL programming.

Operator	Operand	Purpose
GE	Any Type	Greater than or equal to
GT	Any Type	Greater than
EQ	Any Type	Equal
NE	Any Type	Not equal
LE	Any Type	Less than or equal to
LT	Any Type	Less than
CAL	Name	Call function block
JMP	Label	Jump to label
RET		Return from a function or function block
)		Execute last deferred operation

Figure 12–16 The comparison instructions available in IL programming.

Functions and function blocks can be called by using the CAL operator, for example:

```
CAL TEMP1( SETPT := 85, CYC := 5 )
```

This would call and execute a function block called TEMP1. It would send the values of 85 to the SETPT input parameter and 5 to the CYC input parameter. These are the inputs that the TEMP1 function block requires. As you can see, IL can be combined with the other types of programming languages that IEC 61131-3 specifies.

There is another way to load parameters and make a function block call. In the following example, the values are loaded and then stored to the input parameters for the function before the function is called:

```
LD 85.0
ST TEMP1.SETPT
LD 5
ST TEMP1.CYC
CAL TEMP1
```

The LD function is used to load the value, and then the ST is used to store the value into the parameter. The first ST stores 85 into the SETPT input parameter for the TEMP1 function block (TEMP1.SETPT). The same method is then used to load 5 into the CYC parameter. The function block is then called with the CAL operator.

Figure 12–17 shows an example of a function block diagram for a start sequence. The equivalent logic is as follows:

```
LD Command_1
OR X17
ST StartRS.S
Load X12 Str\art.R1
LD StartRS.Q1
ST Start
```

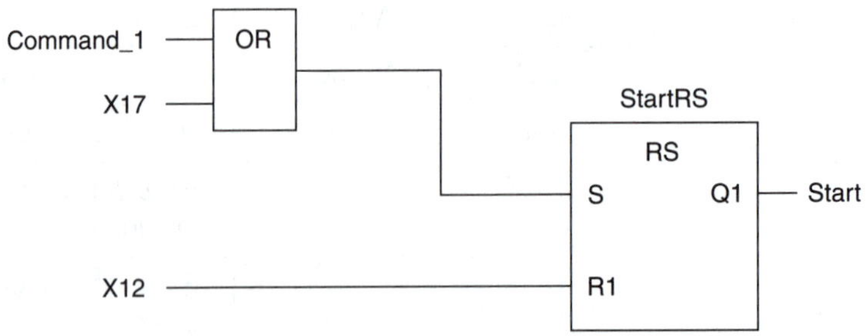

Figure 12–17 Example of the use of an OR.

SEQUENTIAL FUNCTION CHART PROGRAMMING

Sequential function chart (SFC) programming is a graphical programming method. SFCs are a useful method for describing sequential type processes. Figure 12–18 shows an example of a simple linear-type sequence. There are five steps in the sequence, and a condition must be fulfilled before moving from one step to the next. For example, to move from step 1 to step 2, S1 must be true. S1 in this case might be a start switch in the process. It could be anything that would evaluate to either a true (1) or false (0). There are conditions between each step in this example. Transitions can be defined by name.

Sequential function charts can also be used for processes that have portions of the applications whose steps are dependent on conditions. For example, imagine a bottling application (Figure 12–19). Two types of bottles come down a line at random.

A sensor mounted at the fill station identifies which bottle is present. The sensor has a tagname of Prod. If Prod is equal to 0, it should be filled with one product; if it is equal to 1, it should be filled with a different product. Most of the processes (steps) in Figure 12–19 are the same. The fill processes are the only difference in this process. After the delay, the bottles are either filled in the Fill1 process or the Fill2 process. The processes following the fill processes are the same in this application. Sequential function charts handle these types of applications easily. It could have been much more complicated than this; for example, there could have been many alternative processes, or each path could have had multiple steps.

Figure 12–18 Example of a simple linear process.

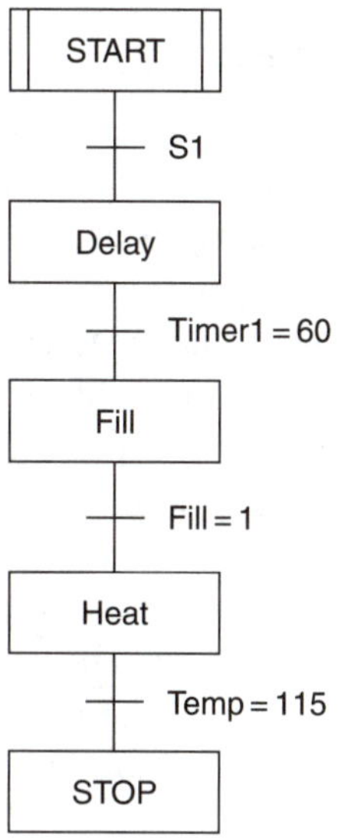

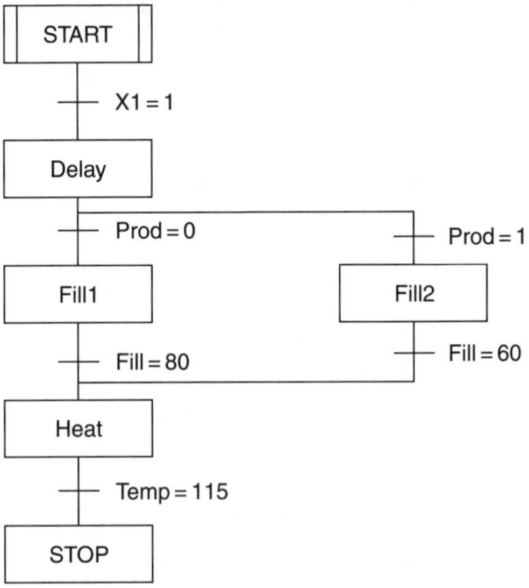

Figure 12–19 A simple process with alternate steps that are dependent on a condition.

Branching

Branches can also be used to alter the sequence processing during operation (Figure 12–20). In this case, X17 is evaluated after the second step. If X17 is equal to 1, the sequence continues to the next step. If X17 is equal to 0, sequencing branches back to a position immediately following the start block.

Concurrent Processing

In many applications the processing is not linear, and several processes may occur at the same time (Figure 12–21). In this case, the three processes are all active at the same time. The first process has a Fill and a Heat step. The second process is an assembly process. The third contains two steps. Note that double lines are used to show where concurrent processes begin and end, and that there can be multiple steps in each process and conditions in each to control movement between steps. Concurrent processing is a powerful tool. As processors gain more power and speed, they have tremendous capability to control complex or even multiple processes.

Certain rules apply to how steps and transitions can be used. Transitions cannot be linked directly. There must be a step between any two transitions. If a transition leads to two or more steps, then all of the steps are executed independently and simultaneously. Two steps cannot be linked directly either. Steps must be separated by a transition.

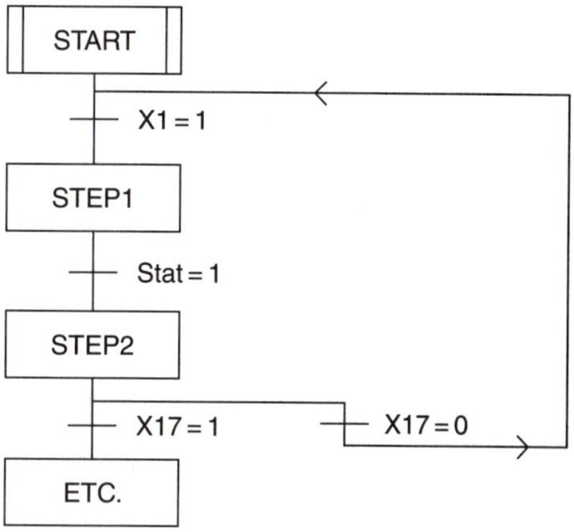

Figure 12–20 An example of branching in a sequence.

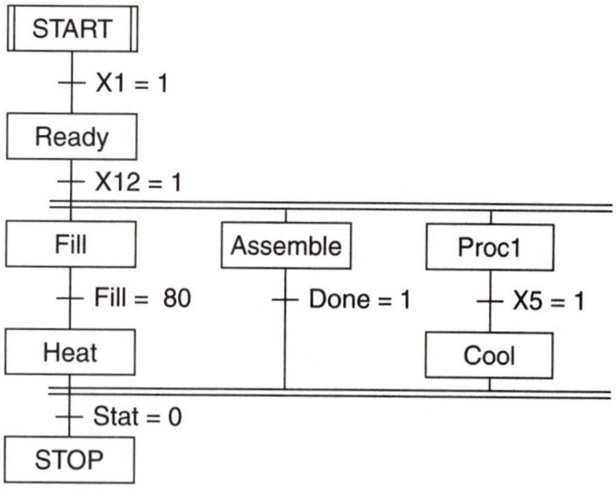

Figure 12–21 An example of concurrent processing.

Step Variables

Two variables are associated with every step that the programmer can utilize. The first variable is set to a 1 while the step is active. This can be useful for monitoring and for logic. It is called the step active flag, named .X. To use it, the programmer uses the name of the step and adds the .X. There is a step named Ready in Figure 12–22. If the programmer wants to use the active flag, he or she would simply use Ready.X. If the step is active, Ready.X would be equal to 1. If the step is inactive, Ready.X would be equal to 0. The active flag can be used in another way. Figure 12–23 shows how the active flag can be used directly. In this case, Y21 is set to a 1 when this step is active.

The second type of step variable is a time variable. Every step has a variable that contains the time that the variable has been active. The elapsed time variable can

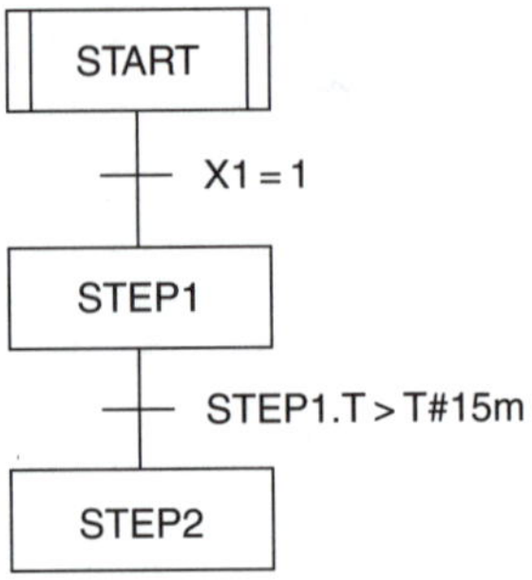

Figure 12–22 Use of the elapsed time flag for STEP1.

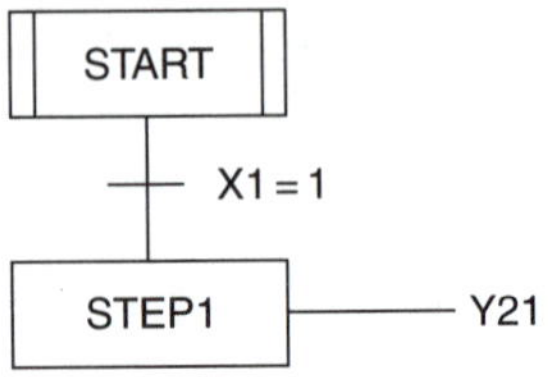

Figure 12–23 Use of the active flag to set Y21 when STEP1 is active.

be used by giving the step name and adding a .T. Figure 12–22 uses the elapsed time variable to control the transition between Step1 and Step2. The logic essentially says that if the elapsed time of Step1 is greater than 15 minutes, move to Step2 (Step1 > T#15m).

Ladder logic can be used as a transition condition between steps (Figure 12–24). In this example, when the rung is true, the program will move from the START step to STEP1.

Function blocks or FBDs can be used to control transitions between steps as long as their output is discrete (1 or 0). Figure 12–25 shows an example of a function block being used to control the transition from START to STEP1.

Ladder logic that ends in a transition connection can be used to control transitions between steps. Figure 12–26 shows a ladder diagram rung that ends in a transition connector named Alt1. Alt1 is also used as a transition condition in the sequential function chart.

Structured text expressions can be used as transition conditions. In Figure 12–27, an ST expression controls the transition between STEP1 and STEP2. The ST expression must result in a true or false. In this case, if Var1 is less than 45 or Var5 is greater than 35, the expression will result in a true (1) and the process will move from STEP1 to STEP2.

Transitions can also be defined by using IL programming. Figure 12–28 shows the use of an IL program to control a transition. In this case, the transition in the sequential function chart is called Cond1. Note that in the IL program, the keyword

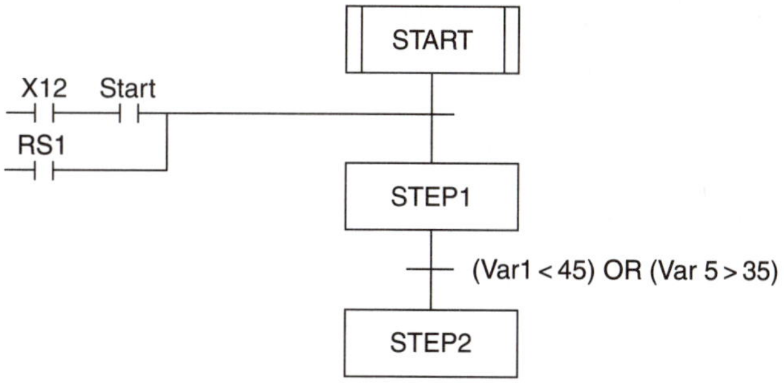

Figure 12–24 An example of ladder logic used as a transition condition.

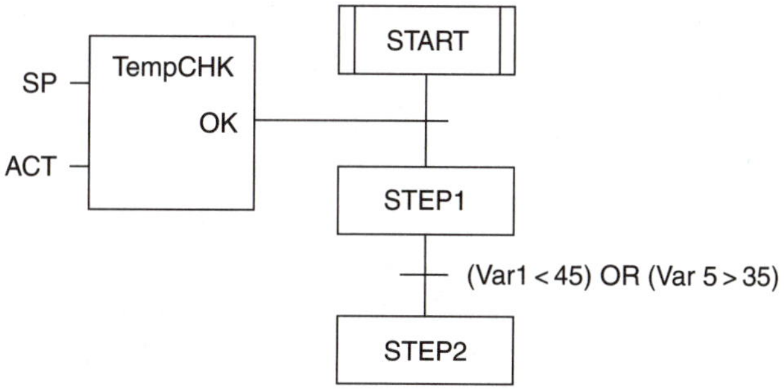

Figure 12–25 An example of a function block used as a transition condition between steps.

Figure 12–26 An example of a ladder diagram rung that ends in a transition connector named Alt1.

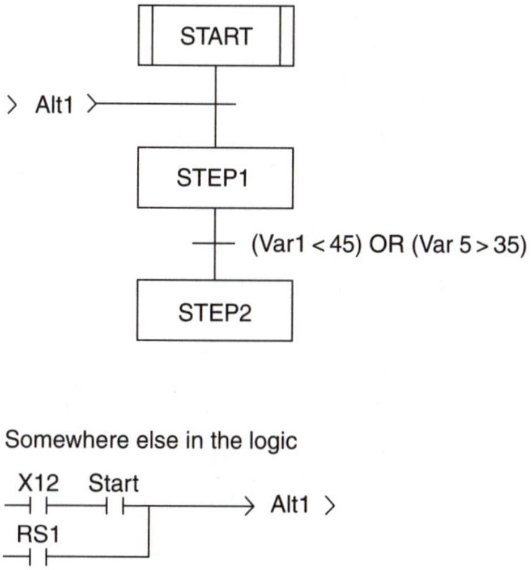

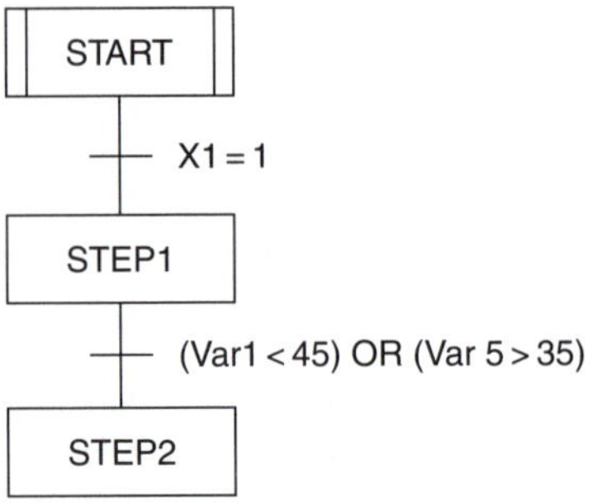

Figure 12–27 Use of a structured text expression to control a transition.

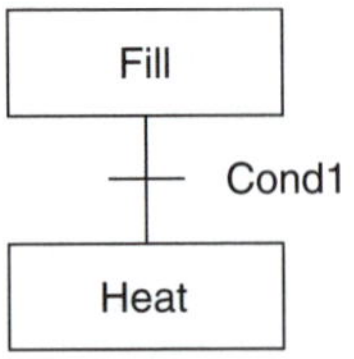


```
TRANSITION Cond1 :
   := X1 AND X15 OR X5;
END_TRANSITION
```

Figure 12–28 An example of the use of an IL program to control a transition.

```
TRANSITION Cond1 :
   LD X12
   OR X18
END_TRANSITION
```

Figure 12–29 An example of instruction list programming to control a transition condition.

TRANSITION is used to declare Cond1 as the name of the transition that is being defined. The conditions in the IL program are then evaluated when the program runs. If X12 or X15 is true, this IL program will result in a 1 and Cond1 will be true and cause the program to change from the Fill step to the Heat step.

Instruction list programming can also be used to program transition conditions. Figure 12–29 shows an example. The IL program starts with the keyword TRANSITION followed by the name that the programmer assigns to the transition condition. In this case, the programmer chose Cond1 for the name; the logic would be as follows: if X12 OR X18 is true, then Cond1 will be set to a 1, the transition condition between two steps will be met, and the program will move from one step to the next.

Ladder logic can also be used to control transitions. The logic shown in Figure 12–30 is an example. Note that the transition name appears at the right of the controlling logic.

Figure 12–30 Example of ladder logic to create a transition element.

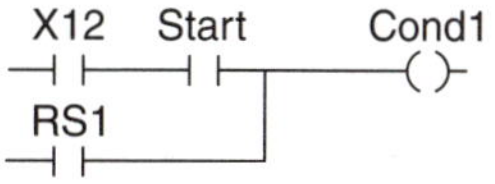

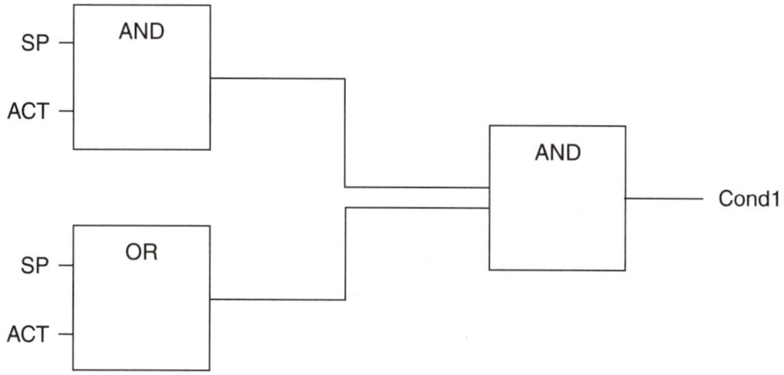

Figure 12–31 Example of an AND block ANDed with an OR block to define the transition logic.

FBD language can also be used to define transitions. Figure 12–31 shows an example of an AND block ANDed with an OR block to define the transition logic.

Step Actions

Steps are used to control actions that occur during that step of the sequence. Steps can control multiple actions. Figure 12–32 shows the general format for an action. The action block is shown to the right of the step. The first part of the action block can contain a qualifier. The qualifier controls how the action is performed. The qualifier in this case is an N. An N means that the action will be executed while the associated step is active. Additional qualifiers are shown in Figure 12–33. The second part of an action block is the action to be performed. In this case the action is StartSeq. The third part of the action block is optional. The user may use a variable in the third part of the action block. The variable is called an indicator variable. The variable will indicate when the action has completed its execution. In this example, the variable is called Stat. When StartSeq has finished its execution, Stat will be set to a 1.

Figure 12–34 shows an example of the use of an action. The action is associated with STEP1. The qualifier is N, so the action will be executed while STEP1 is active. There is no indicator variable. The action name is StartMotor. StartMotor is a

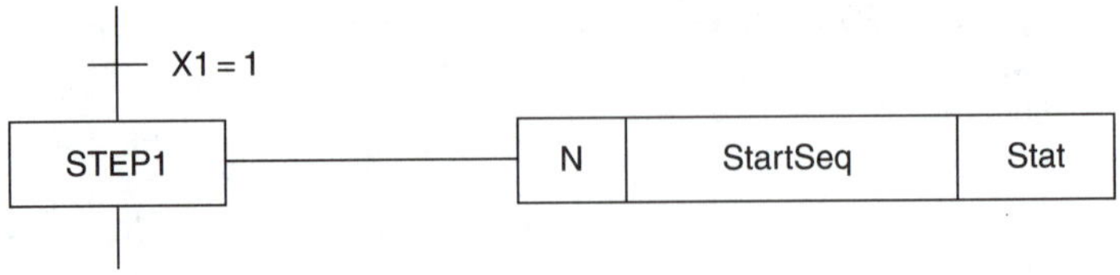

Figure 12–32 General format for an action.

Qualifier	Use
N	Not stored, executes while associated step is active.
None	Default, not stored, executes while associated step is active.
R	Resets a stored action.
S	Sets an action active.
L	Terminates after a given time period.
D	Starts after a given time period.
P	Pulse action that occurs only once when the step is activated and once when the step is deactivated.
SD	Time delay and stored. The action is set active after a given period, even if the associated step has been deactivated before the time period elapsed.
DS	Action is stored and time delayed. If the associated step is deactivated before the time period elapses, the action is not stored.
SL	Time limited and stored. Action is started and executes for a given time period.

Figure 12–33 Qualifiers that can be used with actions.

simple action that is used to turn Motor1 on and Fan off. This has been specified in the block under the StartMotor action name. Only two outputs are associated with this action, so the user programmed them with the action block. The action could have been defined on another page if it was more complex. Another valuable asset of action blocks is that, once they are defined, they can be used repeatedly with the

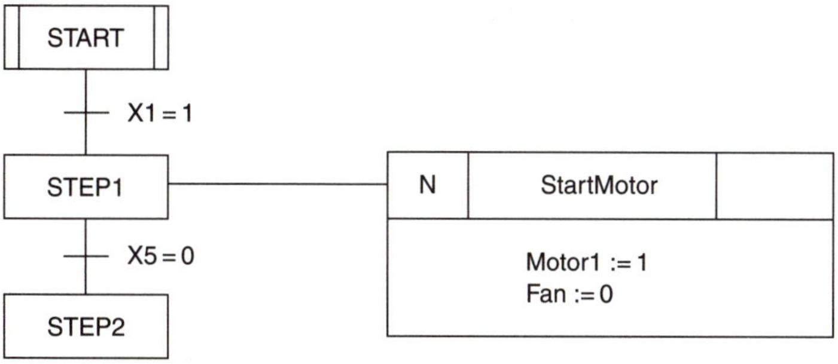

Figure 12–34 Example of the use of an action.

Figure 12–35 Example of the use of FBD language to define an action.

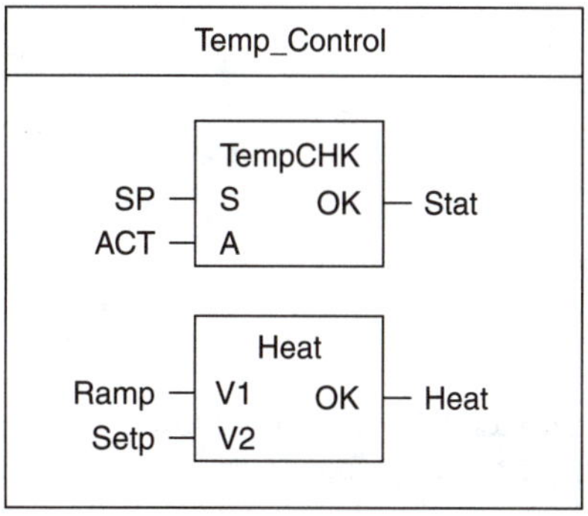

qualifier and action name. So StartMotor can be used anywhere in our program now without redefining the actual I/O to be performed.

Actions can also be defined by using the FBD language. Figure 12–35 shows an example used to define an action. The FBD logic that the user creates is enclosed in a box with the name of the action that it defines at the top.

Ladder logic can be used to define what the action does. Figure 12–36 shows an example of the use of ladder logic for this purpose. The ladder logic is enclosed in a rectangle with the name of the action at the top.

Sequential function charts can also be used to describe the behavior of an action. Figure 12–37 shows how complex action behavior can be defined simply with SFC. In this case there are three basic steps in the SFC—Fill, Mix, and Pour. Also note the action associated with each step.

Figure 12–38 shows an example of the use of structured text to define the behavior of an action. Note that the keyword ACTION starts the structured text program. Then the name of the action that is being defined is shown. The name of the action is Stir. This action decides if Type is equal to 1 or not. If the Type is 1, then

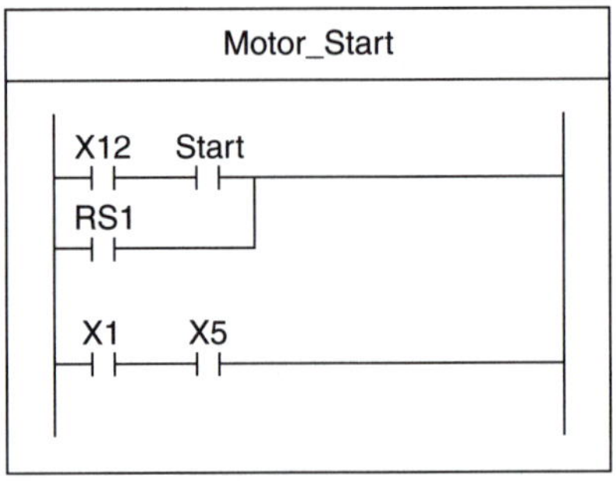

Figure 12–36 Example of the use of ladder logic to define an action.

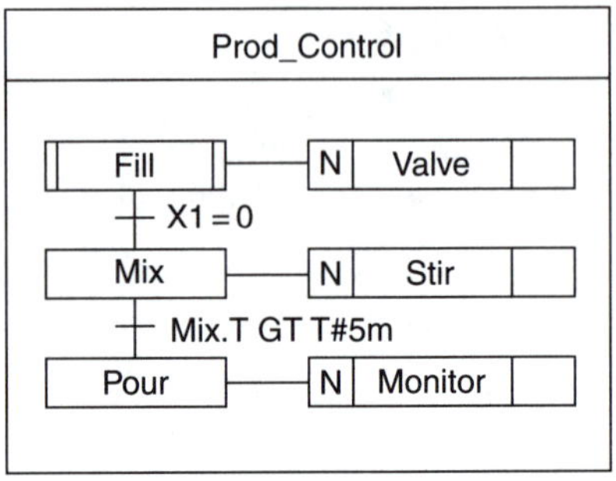

Figure 12–37 Example of how complex action behavior can be defined with SFC programming.

Figure 12–38 Structured text program used to define an action.

```
ACTION Stir :
IF Type = 1 THEN
   Speed1 = 75;
   Motor = ON;
ELSE
   SPEED1 = 25;
   Motor = ON;
   FAN = ON;
END_ACTION
```

Speed1 is set to 75 and Motor is turned on. If Type is not equal to 1, then Speed1 is set to 25, Motor is turned on, and Fan is turned on. The program ends with the keyword END_ACTION.

Figure 12–39 shows the use of instruction list programming to define an action's behavior. The keyword ACTION begins the IL program, followed by the name of the action. The IL program loads the state of X1 in this case and stores it to Y5. The keyword END_ACTION ends the IL program.

Use of Action Blocks in Graphical Languages Action blocks can be used in ladder logic (Figure 12–40). When there is power flow into the action block, it is

Figure 12–39 Instruction list program used to define an action.

```
ACTION Monitor :
  LD X1
  ST X5
END_ACTION
```

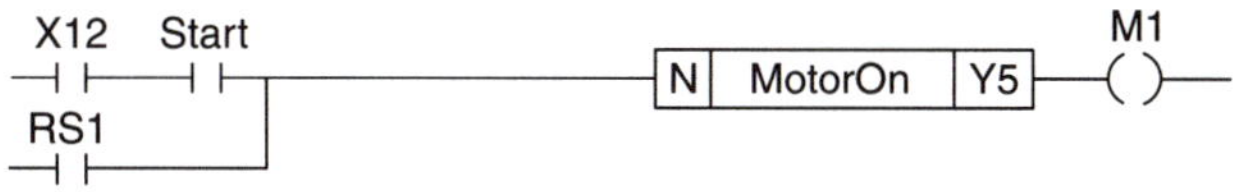

Figure 12–40 Example of the use of an action block in ladder logic.

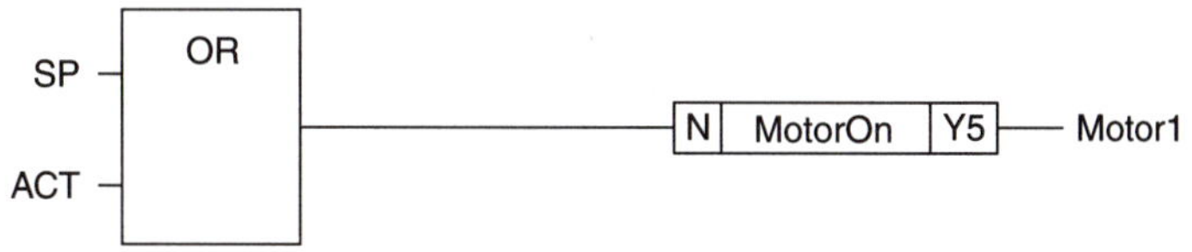

Figure 12–41 Use of an action in FBD language.

active. The indicator variable (Y5) can be used to indicate when the action is complete.

Figure 12–41 shows the use of an action in the FBD language. Note that the indicator variable (Y5) can be used elsewhere to indicate when the action is complete.

Summary

This chapter has been a quick introduction to IEC 61131-3 programing. It is by no means complete. A book could be written on each of the programming methods.

Questions

1. Explain why the IEC 61131-3 standard was developed.
2. Explain structured text programming.
3. Explain function block diagram programming.
4. Explain IEC 61131-3 ladder logic programming.
5. Explain instruction list programming.
6. Explain sequential function chart programming.
7. Explain how the languages mentioned in Questions 1 through 6 can be combined in control programs. Examples might be helpful.

chapter 13

Supervisory Control and Data Acquisition

Supervisory control and data acquisition software has become prevalent in industrial automation. This chapter will examine the use of one of the more common supervisory control and data acquisition systems.

OBJECTIVES

Upon completion of this chapter, you will be able to:

1. Describe a typical SCADA application.
2. Define terms such as *driver, topic, item, recipe,* and *historical log.*
3. Explain how typical SCADA software is programmed.
4. Explain how SCADA software can be used to improve the performance of an enterprise.
5. Explain how a SCADA system can be integrated with the business software in an enterprise.

OVERVIEW OF SUPERVISORY CONTROL AND DATA ACQUISITION (SCADA)

Many different technologies have come together in a relatively short time to revolutionize manufacturing. Almost all of these technologies are computer based. The rapid advancements in microcomputers in the last twenty years have given them the power that only mainframes had in the past. Networking technology has been developed to enable these microcomputers to become even more powerful through shared resources. The reduction in the price of memory and the increase in the speed of microprocessors have allowed incredibly powerful software to be developed and used.

In the past software development was proprietary. Software developers reaped large monetary rewards from developing, installing, and servicing proprietary software. When an industry decided to automate a process, they normally hired a system integrator who would develop the necessary software to operate the system. Software that did exist was too complex for most enterprises to develop their own automation applications.

Slowly software developers started to partner with other software developers to offer software packages that worked together. While this was occurring, a revolution was taking place in the offices of enterprises. The mainframe and "dumb" terminal had been the prominent technology. Microcomputers began to appear for many tasks. Spreadsheets, databases, word-processing packages, and computer-aided drafting (CAD) began to be used in offices. Productivity increased drastically. The old "dumb" ASCII terminals displayed text only. Microcomputers had the ability to display graphics and colors. They were also user-friendly and more intuitive.

Two main microcomputer systems battled for supremacy: Apple and IBM. Both were leaders in some fields. Apple was particularly strong in graphics; IBM gained prominence in business and industrial use. Apple developed an easy-to-use, intuitive operating system. Microsoft developed the Windows operating system, which addressed the need for a user-friendly, intuitive operating system. Meanwhile industrial controllers were gaining capability also. Faster processors, more networking capability, and more powerful communications modules were all being offered. These changes began to make it easier to communicate with industrial controllers. Until this time microcomputers were typically used to program the PLCs or upload and download programs to them. Communication was still proprietary, however. Each PLC had its own communications protocol.

Until the end of the 1980s, few software development companies had written Windows-based industrial software. Most still utilized the DOS environment. One of the first software packages developed for the industrial automation world was a package called Wonderware InTouch. It was designed to be a human-machine interface (HMI) package. It was one of the first to make effective use of the Windows graphical user interface (GUI), which made it user-friendly and intuitive. It also broke new ground in graphics capability in an industrial software package. HMI is also called supervisory control and data acquisition (SCADA) software. It was intended to be easy to use so that companies could develop and modify their own applications.

About the same time, the Windows dynamic data exchange (DDE) protocol began to be used for a new type of I/O driver. These DDE servers made it easy and convenient to acquire data from industrial control devices such as PLCs. DDE used reports by exception techniques to create databases that were then used to analyze processes. The data could be gathered and analyzed online. The data could be displayed graphically to show historical trends and to analyze the process. DDE was also designed to share data with other Windows applications. This allowed production data obtained from industrial controllers to be shared automatically with other Windows-based spreadsheets, word processors, databases, and many other types of software. For example, a graph could be created in a spreadsheet program that would show production for the day compared with the past week. The data used in the graph is retrieved by the HMI software and shared via DDE with the spreadsheet. Data is updated whenever it changes in the system so the graph is always current. All of these steps occur automatically. The same can be done with a production report created in word-processing software. The report can be updated automatically with current production data and is available instantly.

Meanwhile, Windows-based software flourished. All Windows-based software had a common user interface. The software was also highly graphical in nature, making it very intuitive. Once a person learned to use one Windows application, it was very easy to learn a new Windows software package. Windows-based software drastically reduced the learning curve.

Industrial software developers also started to make their products user-friendly by utilizing the Windows environment. Wonderware's InTouch is a good example of user-friendly industrial software. InTouch was designed to make it easy for an applications person to develop industrial applications that could communicate with industrial controllers. The basis of InTouch is a graphical interface. By drawing graphics on the screen and then answering a few configuration questions specific to the particular industrial controller, the application is developed and can communicate with the controller. If we decide to switch to a different brand of controller, the application is still usable. All we would have to do is reconfigure the driver and tagnames for the new controller. Notice that with InTouch, developing an industrial application did not require a programmer. An applications person can easily develop applications. InTouch made it possible for manufacturers to develop their own applications.

SCADA was traditionally used for data collection from PLCs and plant floor controllers. SCADA systems were also used for monitoring and supervisory control of processes. The role of SCADA systems has expanded. Now they are a vital part of many manufacturers' information systems. They provide manufacturing data to many other software systems in the typical manufacturing enterprise.

As companies automated and evolved to improve productivity, quality changes also occurred that affected the workforce. Workers are expected to have broader capabilities these days. Maintenance personnel are trained in many skills that were not traditionally part of their job duties. This is sometimes called cross-training. For example, electricians may be taught some mechanical skills, while mechanical maintenance personnel are taught some electrical skills. Cross-training makes workers more valuable in today's manufacturing environment. Production workers are also expected to be responsible for more than they were previously. In the past a worker was most likely responsible for running only one machine. Now that same worker is probably responsible for several computer-controlled machines or one or more automated systems. The result is that more is expected of both the skilled trades personnel and the production personnel. This has made SCADA systems more important also. When we broaden workers' responsibilities, they have less time and experience with each particular machine or technology. This makes operator information crucial. A SCADA system can be used to provide operator information, prompts for required action and/or input, alarms, detailed instructions, plans, and so on. This information can appear on the operator's screen, in the maintenance department, or with today's technology and the Internet anywhere in the world, if it is desired. The alarm can notify the maintenance department that a repair is needed. It can notify them which part probably needs repair and even give detailed instructions on video about how to repair or replace the part. Any of these capabilities can be designed into SCADA systems today.

SCADA systems have also been given recipe capability. Recipes for various products can be stored and downloaded to controllers as needed. SCADA systems have also gained the capability to share their information over the Internet. It is possible today for a manufacturer to watch graphics and have data from a process anywhere in the world.

SCADA systems are continually gaining capability and can handle many of the functions associated with an enterprise's business systems, such as inventory tracking, scheduling, etc. Today's SCADA systems integrate easily with a wide variety of other types of enterprise software.

The graphical user interface (GUI) is the key to manufacturing systems and in fact all software applications. As previously mentioned, Windows has given most software a common interface and "feel." To be successful software systems must have user-friendly interfaces (GUIs) and must be easy to use for developing applications. The key to all technology will increasingly be ease of use.

Sample Application

The best way to understand a SCADA system is to see how a simple industrial application is developed. We will develop a simple temperature control application. Note that this example is not intended to teach you every key to press. It is intended to give you a broad, overall understanding of how SCADA applications are developed.

Figure 13–1 shows a simple temperature control system. It consists of a conveyor to move product through two heat chambers. There are two temperature controllers in the system. The product moves along the conveyor line and through each furnace for a controlled time. The first furnace acts as a preheat chamber and is controlled at a lower setpoint than the second heat chamber.

Figure 13–2 shows a table that contains the temperature setpoints and PID values for various products produced by this process. Four values need to be changed

Figure 13–1 A simple temperature control system. A conveyor and two ovens are to be controlled.

Product	Setpoint Chamber 1	Setpoint Chamber 2	P	I	D
P134	180	225	90	10	30
P135	192	238	85	3	30
P136	163	207	97	12	32
P137	193	267	76	23	25
P138	215	237	85	12	24
P139	199	332	90	10	25
P140	137	183	87	8	22

Figure 13–2 The parameters for different products that are loaded into the temperature controllers.

in each temperature controller for each different product run. A Rockwell Automation PLC takes input from sensors and controls the conveyor.

At this point we need to use some specific hardware and software to develop our application. We will use a personal computer, Wonderware InTouch, Omron temperature controllers (see Figure 13–3), and a Rockwell Automation SLC 5/04. Figure 13–4 is a table that shows all the controllers and I/O used in the system. The Omron temperature controllers in this example have RS-485 communications modules installed.

We will use communications port one (serial port 1) from our microcomputer to talk to the Omron controllers. The Omron controllers have RS-485 communications modules, so we will need to convert our RS-232 computer output to RS-485 output. This is done with an RS-232 to RS-485 converter (see Figure 13–5). Such converters are inexpensive; they typically cost less than $50.

There are two temperature controllers, so we need to give each a unique name or address. This is called a unit address by Omron. We set the unit address of the first temperature controller to 1 and the second temperature controller to 2.

DDE is an abbreviation for dynamic data exchange. It is a communications protocol designed by Microsoft to enable Windows applications to send and receive

Figure 13–3 Omron temperature controllers.

CONTROL DEVICE	I/O ADDRESS OR NAME	TAGNAME	USE
SLC 500	O:3/15	CONVEYOR	OUTPUT THAT TURNS THE CONVEYOR ON AND OFF (OUTPUT 15, SLOT 3)
SLC 500	I:7/8	PART-PRESENT	OVEN SENSOR TO CHECK FOR PART PRESENCE (INPUT 8, SLOT 7)
TEMPERATURE CONTROLLER	TEMPERATURE	TEMP_1	VARIABLE THAT HOLDS THE ACTUAL TEMPERATURE VALUE
TEMPERATURE CONTROLLER	SETPOINT	SET_1	USED TO CHANGE THE SETPOINT OF THE TEMPERATURE CONTROLLER
TEMPERATURE CONTROLLER	PROPORTIONAL	PROP_1	PROPORTIONAL GAIN OF THE TEMPERATURE CONTROLLER
TEMPERATURE CONTROLLER	INTEGRAL	INT_1	INTEGRAL GAIN OF THE TEMPERATURE CONTROLLER
TEMPERATURE CONTROLLER	DERIVATIVE	DER_1	DERIVATIVE GAIN OF THE TEMPERATURE CONTROLLER
TEMPERATURE CONTROLLER	STATUS	STAT_1	BIT IN THE TEMPERATURE CONTROLLER. IT IS A 1 IF COMMUNICATIONS ARE NORMAL AND A 0 IF THERE IS A COMMUNICATIONS ERROR

Figure 13–4 Information about the actual I/O and controllers used in the application. Tagnames have also been chosen for each I/O point.

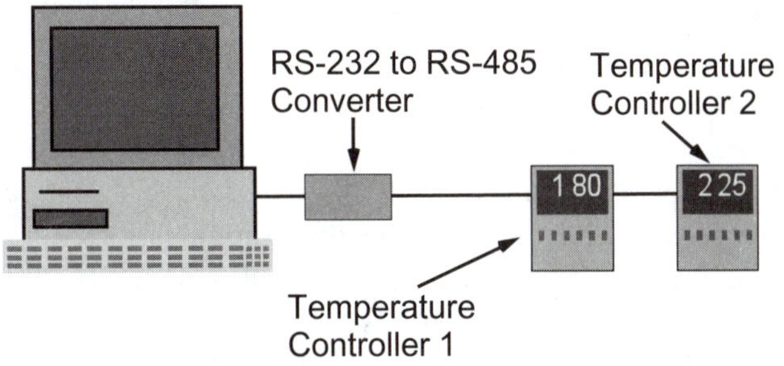

Figure 13–5 How the two temperature controllers are connected to the computer.

data to and from each other. It can establish a client-server relationship between two concurrently running applications. The server accepts requests for data from other applications and provides requested data to them. Applications that request data are called clients. This relationship can change. A client can become a server and vice versa as they share data with each other. DDE is a standard feature for most Windows applications that need data links to other applications. DDE-compliant applications include Microsoft Word, Excel, Access, InTouch, and many others. If we are putting data into a database program, the data can be shared with another software program running in the computer. For example, we might want to put investment data into a database every day and then graph the performance of our investments using a graph in a spreadsheet program. DDE can share the data between the applications so that we do not have to enter it twice.

DDE can also share data with other computers. It has the capability to communicate data over computer networks or over modems. In fact, NetDDE extends the DDE standard to make it possible to communicate over local area networks and through serial ports. DDE can be used to collect and distribute factory data. For example, we might want to get daily production data and put it into a spreadsheet, where it can be analyzed and graphed. It would be useful for a supervisor or management to see a graph that shows production over the last ten work days, for example. This data can be accessed automatically and shared via DDE with a spreadsheet. DDE can also be used to send production data to applications. Imagine a temperature control system that has several variables that change with the type of product manufactured. A spreadsheet can be set up that holds the variables needed for each product. When needed, the data can be sent via DDE to the controller.

OPC

OPC grew out of a different standard: object linking and embedding (OLE). OPC stands for "*o*bject linking and embedding for *p*rocess *c*ontrol." You will also see the phrase "OLE for process control." OLE has evolved from an object-oriented protocol to an object-based protocol. OLE is now called ActiveX. ActiveX is an open and integrated standard that allows developers to create portable applications, reduces the need for specialized communications drivers for control devices and equipment. It should also dramatically reduce the cost and confusion in communications integration of various devices.

OPC is an industry standard. The OPC Foundation manages the standard. The foundation has over 150 members from various industries. The standard was first released in 1996. The objective of OPC is to provide a plug-and-play standard that enables users to have a wider choice of solutions. OPC is designed to ensure that industrial automation systems can share information and integrate with other automation and business systems in the enterprise. OPC should be able to reduce dramatically the cost and time associated with integrating an automation system.

A computer is a good way to illustrate the goal of OPC. Have you ever added a drive, a new monitor, or other hardware to your computer? Most devices for a computer today are plug-and-play. You install the hardware and when Windows starts, it recognizes a device has been added. If it is plug-and-play-compatible, Windows accepts it. Plug-and-play in Windows makes it easy to connect one device with another automatically, without complex configuration and installation procedures. The goal of OPC is to make industrial devices just as easy to integrate. It is hoped that this will enable (or force) automation device suppliers' communications drivers to a standard form.

OPC enable OPC-compliant devices to work seamlessly with other OPC devices. This enables users to choose devices from a wide variety of suppliers and easily integrate them into systems. This should also reduce training and maintenance costs and reduce the need for custom development to integrate devices. This will also enable users to lose their fear of using more than one brand of controller or device, which should help lower costs of equipment.

Application Development

To develop an application, the programmer draws a picture of the objects that are needed to represent the application and then answers a few questions to describe the real-world I/O that each object represents.

Understanding the System Developing a Wonderware InTouch SCADA system is a relatively easy and straightforward task. Figure 13–6 shows a simple flow diagram of system development steps. The key to success, however, is understanding the manufacturing system. The application developer must know what types and brands of controllers are used in the system and what role they play.

First, we must examine and understand the system. There are three controllers in the simple system: two Omron temperature controllers and one Rockwell Automation SLC 5/04 PLC. Let's give each of the controllers a name. Let's call the SLC 5/04 - "SLC_1." We have two temperature controllers, so let's give each a descriptive name. Let's call the temperature controller for oven 1 Temp_Contrl_1. We can then call the second temperature controller Temp_Contrl_2. Note that almost any name can be used as long as naming conventions are used, but simple descriptive names are best. The names we just created are called DDE access names. They will be used by the application we create to tell InTouch which controller we are talking to. Figure 13–7 shows the controller, DDE access name, communications

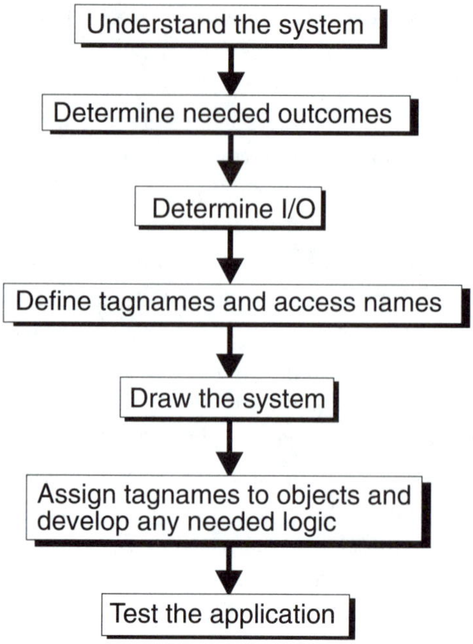

Figure 13–6 Steps in developing a SCADA application. Note that SCADA software is quite flexible. Experienced developers will develop their own preferences for the order in which they develop an application.

port used, and the purpose for each of the controllers. The PLC is not shown in this table because communications to it don't go through the serial port.

The application we will develop is shown in Figure 13–8. The figure shows what the application will look like in runtime. There are product selection buttons on the top of the application. These can be used by the operator to send parameters to the temperature controllers. For example, if the operator clicks on the first button, the parameters for product P134 are automatically sent to the temperature controllers. There are two indicator lights. One of the lights is a communications status light, and the other is used to indicate when product is present in the furnace.

The conveyor (long, thin rectangle on the bottom) turns red if the conveyor is on and green when it is off. The graph is used to show the setpoint and the current temperature of furnace 1. Text displays show the current temperature, the setpoint, and the product number being manufactured.

It will be helpful to your understanding to keep the final outcome in mind as the application is developed. Remember that this sample application development is not meant to teach you each specific step in application development. It is meant to show you that application development is a straightforward and relatively simple task. It does not take a programmer to develop most SCADA applications. A technician can easily develop applications.

CONTROLLER	DDE ACCESS NAME	COM PORT	USE
OMRON TEMPERATURE CONTROLLER	TEMP_CONTRL_1	SERIAL PORT 1	CONTROLS THE TEMPERATURE OF OVEN 1
OMRON TEMPERATURE CONTROLLER	TEMP_CONTRL_2	SERIAL PORT 1	CONTROLS THE TEMPERATURE OF OVEN 2

Figure 13–7 The controller, DDE access name, communications port used, and the purpose.

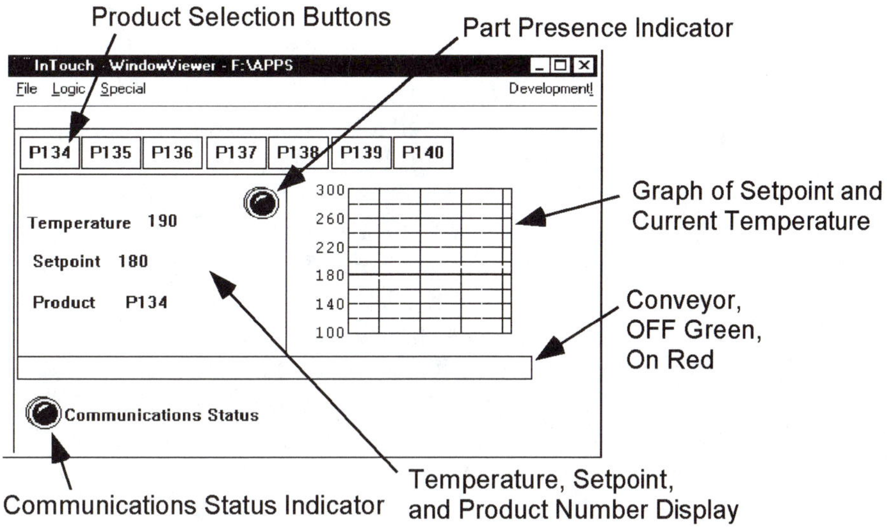

Figure 13–8 The runtime screen of the application that will be developed.

Communications Configuration for the Omron Controllers The first task we will do is create names for our controllers. InTouch calls these DDE access names. These will be used in our application to determine which controller we are talking to. Think of the DDE access name as containing all of the specific information needed to communicate with a particular controller. We will create the names for the Omron temperature controllers first. Remember that we decided to call them Temp_Contrl_1 and Temp_Contrl_2.

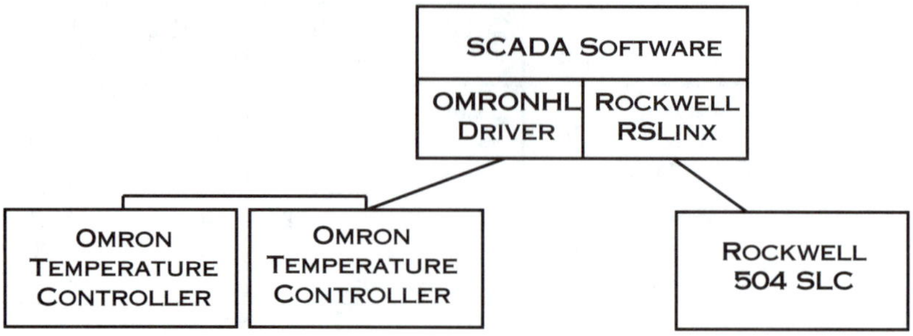

Figure 13–9 How SCADA software communicates with specific controllers. A software driver for each particular controller handles all communication with each controller. In this application, we need a driver for the Omron controllers (OMRONHL) and Rockwell RSLinx to talk to the Rockwell SLC.

DDE access names are created by running the driver software for each controller. Remember that driver software is like translator software for a particular controller (see Figure 13–9). Let's create the Temp_Contrl_1 DDE access name first. OmronHL is the name of the driver that talks to Omron PLCs and Omron temperature controllers. OmronHL stands for Omron host link. Remember that a driver is software that we would purchase for each brand of controller we need to talk to. Drivers are inexpensive translators that handle all communications protocol between the computer application we create and the controller we need to talk to. This makes communication with any controller a transparent, easy task.

The programmer started the OmronHL driver software and the screen shown in Figure 13–10 appeared. The user then chose Configure from the menu with the left mouse button and then Configure Topic.

The screen shown in Figure 13–11 then appears. Next, choose New to add a new topic. The Topic Definition screen shown in Figure 13–12 appears. This is the configuration screen for communications parameters. Note that there are three things that can be configured. The user chose COM port settings.

The programmer typed the topic name Temp_Contrl_1 in the Topic Name space (see Figure 13–12). Next COM1 was chosen. Then Temperature Controller was chosen. Note that other controllers are also shown. This driver can be used to talk to several different Omron products. It can be used with their PLCs and temperature controllers.

The correct model of temperature controller was then chosen from a drop-down menu. Next a unit address of 1 was entered. The unit number identifies which temperature controller we are talking to (see Figure 13–12). The programmer wanted to define Temp_Contrl_1, so 1 was entered. A unit address of 2 is entered for Temp_Contrl_2.

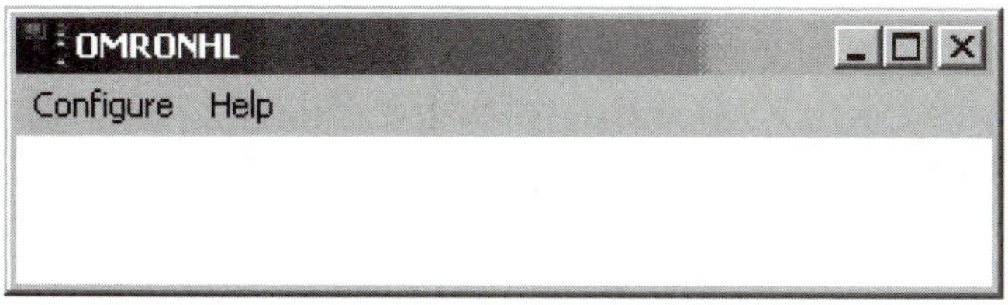

Figure 13–10 Configure screen for the OmronHL communications software driver.

Figure 13–11 Topic Definition screen. Note that any topics that have been created would appear as a list here. None have been created thus far.

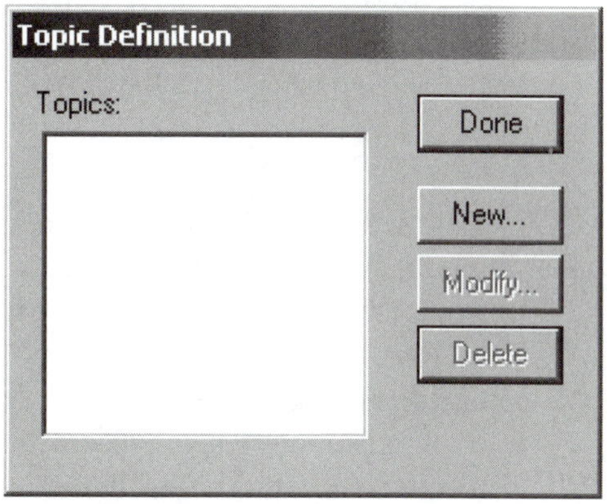

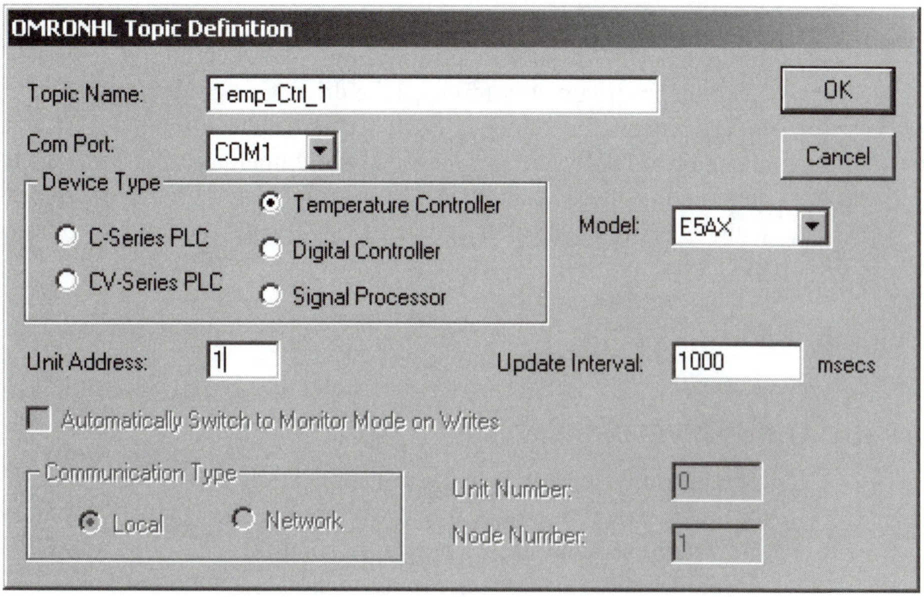

Figure 13–12 The actual port settings that were made.

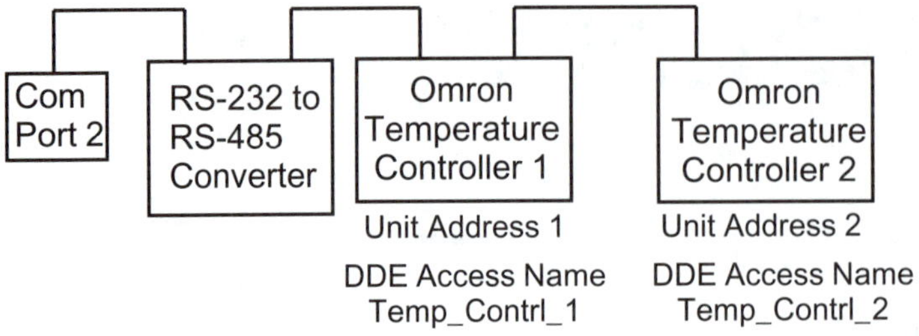

Figure 13–13 The connection of the two temperature controllers to COM1. The DDE access names and unit address are also shown for each. Remember that the unit addresses are set in the actual controller.

The actual Omron temperature controllers (hardware) need to be internally addressed. Each is given a unique address. The unit address for oven controller 1 was set to 1, and the unit address for oven controller 2 was set to 2. Note that this was done in the actual controller. COM1 will be used to talk to both temperature controllers (Figure 13–13).

An update interval can also be entered. The update interval determines how often values from this controller are to be updated. This completes the steps for configuring Temp_Contrl_1.

The same steps are followed to configure Temp_Contrl_2, except that Temp_Contrl_2 is substituted for Temp_Contrl_1, and the unit address is set to 2. Next the programmer configured the driver for the Rockwell SLC.

The process for setting up DDE access names is now complete. The programmer created two DDE access names: one for temperature controller 1 (Temp_Contrl_1) and one for the SLC 5/04 (SLC_1). Temp_Contrl_1 will be used to communicate with the Omron controller on oven 1. SLC_1 will be used to communicate with the SLC 5/04.

Application Development

Next, a simple temperature control application will be examined. We will examine the development of half of the whole temperature control system. We will develop the window (display) for temperature controller 1 as shown in Fig. 13–14.

InTouch refers to a screen of information as a window. Figure 13–15 shows a development screen with one window. We will develop the application in the development mode of InTouch. Development is easy. It involves drawing a picture of what

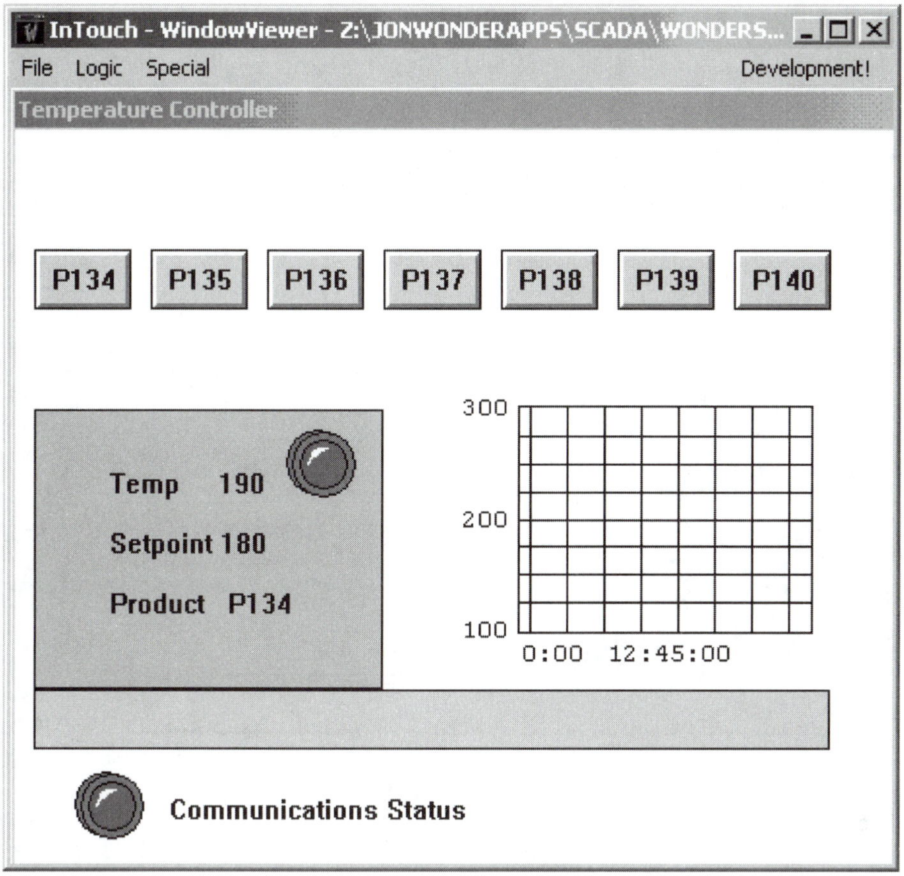

Figure 13–14 Configuring the topic for the Rockwell PLC.

the operator screen should look like and linking objects on the screen to I/O in the controllers so that pictures change on the screen as events change in the real world.

Study Figure 13–15. Note the tool palette on the right of the screen. The tool palette is similar to the tool palette in any drawing package. The tool palette is used to draw the application. The tools on the right side are: arrow (selection tool), rectangle, rounded rectangle, ellipse, line, and so on. Along the top of the drawing window are additional tools to help change font size, text line and fill color, and so on. We will use some of these tools as we develop this application.

The rectangle tool in Figure 13–15 is used to draw a rectangle that represents the oven. The rectangle tool is also used to draw a long, thin rectangle under the oven rectangle. This rectangle represents the conveyor. The text tool is used to add the

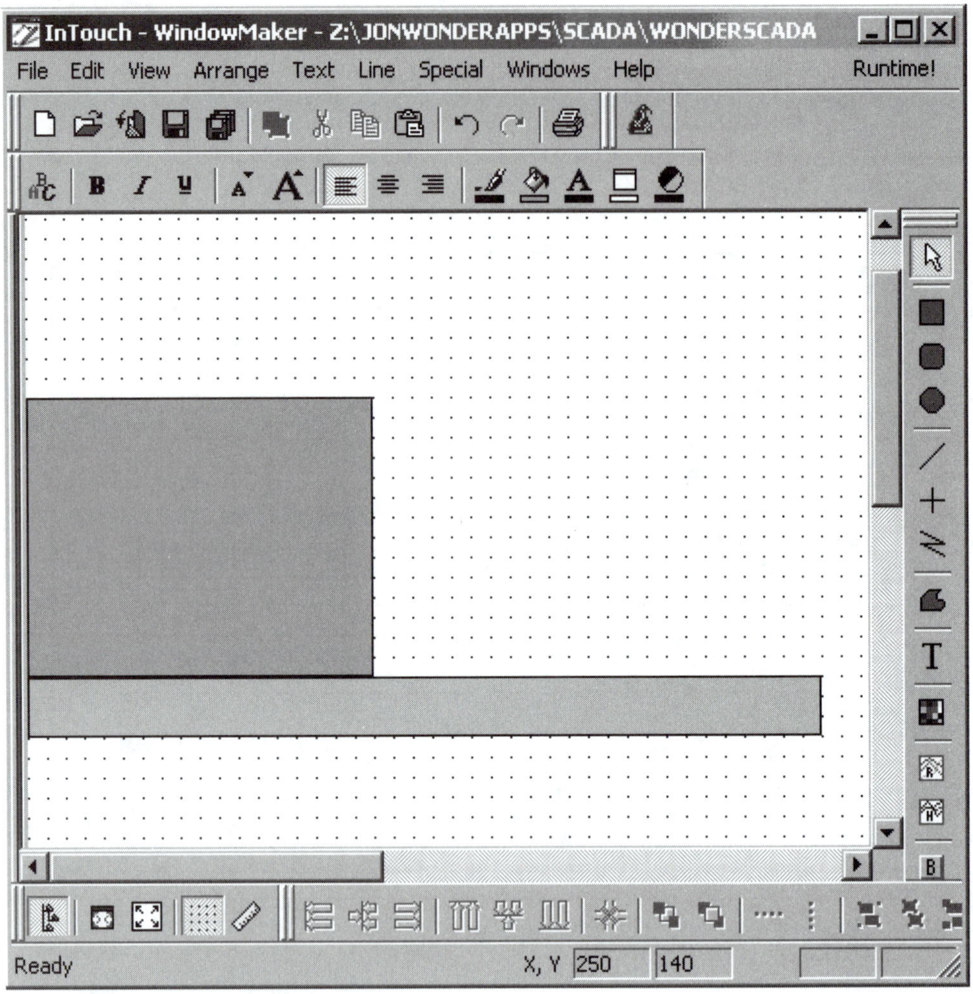

Figure 13–15 A typical development window in Wonderware InTouch. Note that two rectangles have been drawn. The tool palette is also shown in the window.

words "Communications Status" to the lower left portion of the window. The title bar above the working area shows the window name.

The conveyor is controlled by a Rockwell Automation SLC 504. The output number used is O:3/15 (output 15, slot 3). The programmer then links the conveyor rectangle and the real-world I/O in the PLC. A couple of methods are available for linking, but we will look at just one. The programmer double-clicks the left mouse

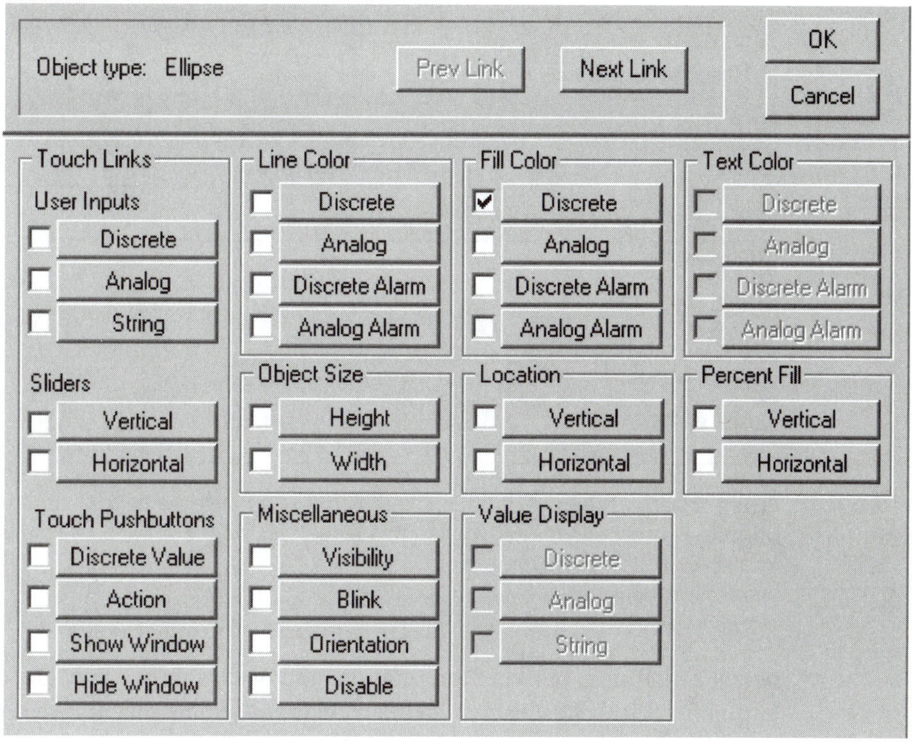

Figure 13–16 The screen that enables the user to decide how this object will behave. In this case, the rectangle represents the conveyor. Fill Color Discrete was chosen. Note the many other possible choices that could have been made. The choices are not exclusive. We can choose to have any combination of the object's attributes change depending on the real-world value.

button on the desired object (conveyor rectangle, in this case). A new screen appears that helps create a link between the chosen object and the real-world I/O in the controller. Figure 13–16 shows this input screen.

Note that this screen allows the programmer to choose how this object relates to the real-world I/O. The rectangle can be used for user inputs or to change the line color and/or fill color based on real-world I/O. The rectangle can be used as a slider to change the value of real-world I/O, or we can change the rectangle's size and/or location and/or fill level based on the value of the real-world I/O. The rectangle can also be used as a touch push button to input discrete values, change values, perform calculations, or to show or hide windows. Other miscellaneous choices control the

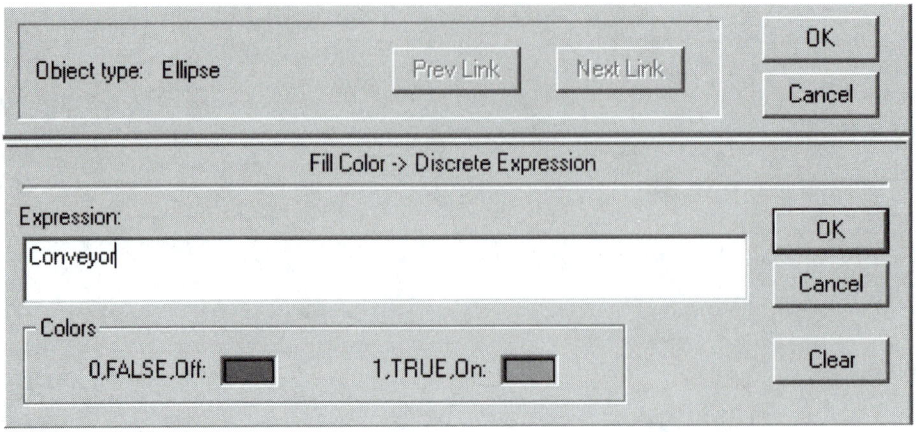

Figure 13–17 The configuration screen for the rectangle we called "conveyor." Fill Color Discrete was chosen for its type, so this screen allows the off and on colors to be chosen for the rectangle (conveyor).

object's visibility, blinking, orientation, and/or disable feature. The user chose the button for Fill Color Discrete, and a new input screen appears (see Figure 13–17). Note that multiple choices can be made. We can change an object's fill color, size, location, or other attributes if we need to. In this case the programmer wants the rectangle's color to change to red when the conveyor is on and green when the conveyor is off.

In this screen the programmer entered the tagname for the object in the Expression entry area. The programmer entered "Conveyor" for the tagname. Any name can be used for a tagname as long as naming rules are followed. The programmer clicks the mouse in the 0,FALSE,Off box. A color menu appears, and the programmer chooses the color for the OFF condition: green in this case. Then the programmer clicks on the 1,TRUE,On box and chooses the color red.

The programmer is done with data entry on the screen and chooses the OK button. The software realizes at this point that it doesn't know what the tagname conveyor is. A new screen appears asking the programmer if he or she wishes to define "Conveyor" (see Figure 13–18). The programmer chooses OK with the left mouse button.

A new input screen appears; it allows the programmer to define the "Conveyor" tagname (see Figure 13–19). The programmer chooses the type from a drop-down

Figure 13–18 This screen appears when the programmer uses a tagname that has not been defined yet. It allows the programmer to define the tagname at this point.

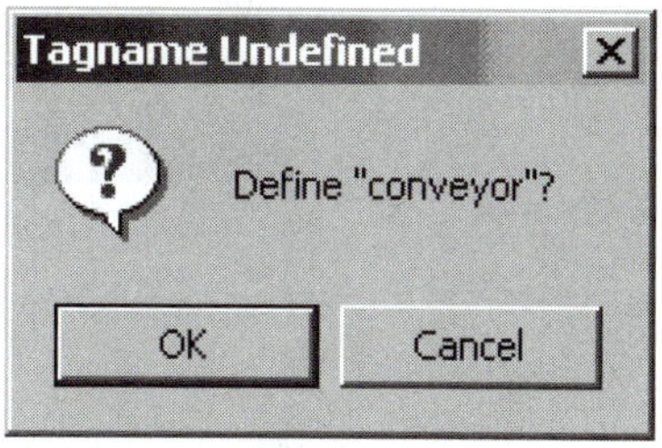

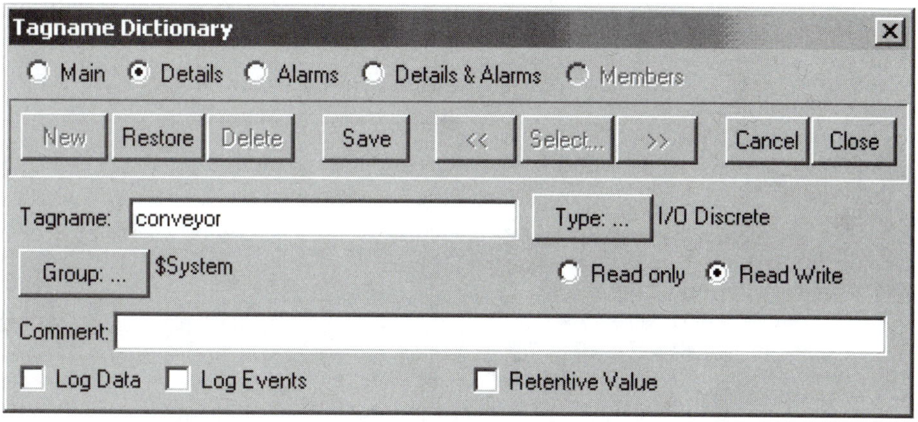

Figure 13–19 Tagname Dictionary screen for the "conveyor" tagname.

list. In this case the type is DDE discrete, which means that this real-world I/O point can have a value only of 0 or 1.

Next, the programmer chooses a DDE access name. One called SLC_1 was already created, so the programmer chooses it (see Figure 13–20). Remember that the DDE access name tells InTouch which controller the application needs to talk to. It associates the name of the controller and the I/O point (tagname) with each other so that when the application refers to the tagname "conveyor," InTouch knows that this tagname refers to the Rockwell Automation PLC that we need to talk to.

The programmer then enters O:3/15 for the item name (see Figure 13–21).

The programmer then adds a status light to indicate whether or not the computer is communicating with the Omron temperature controller. The programmer also added a part presence light to show when product is inside the oven. These were added using the Wizard tool.

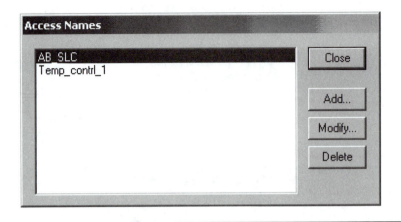

Figure 13–20 DDE Access Names Screen.

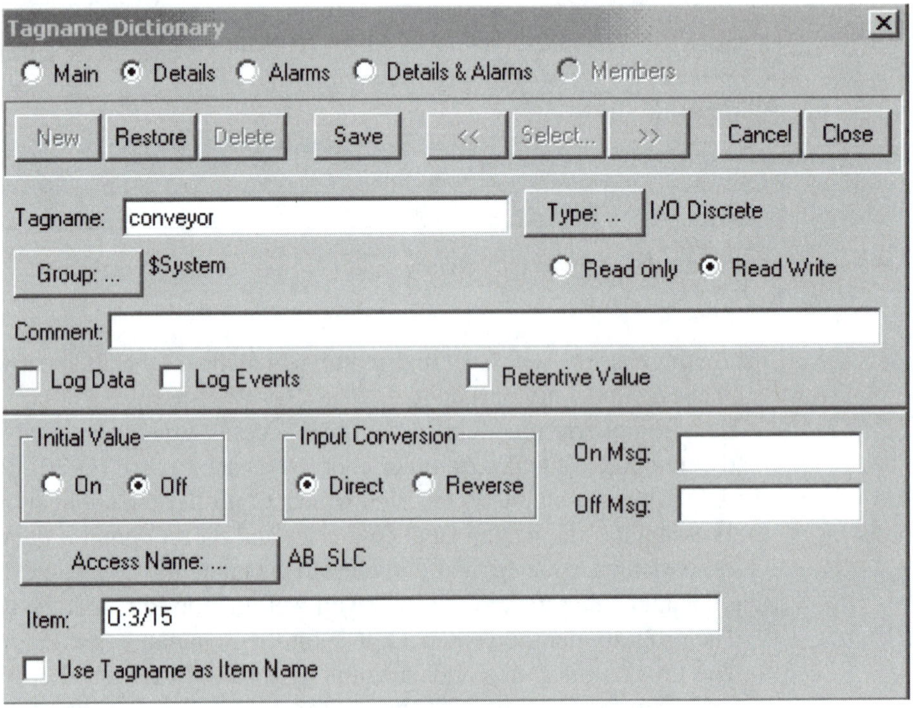

Figure 13–21 Tagname Dictionary screen.

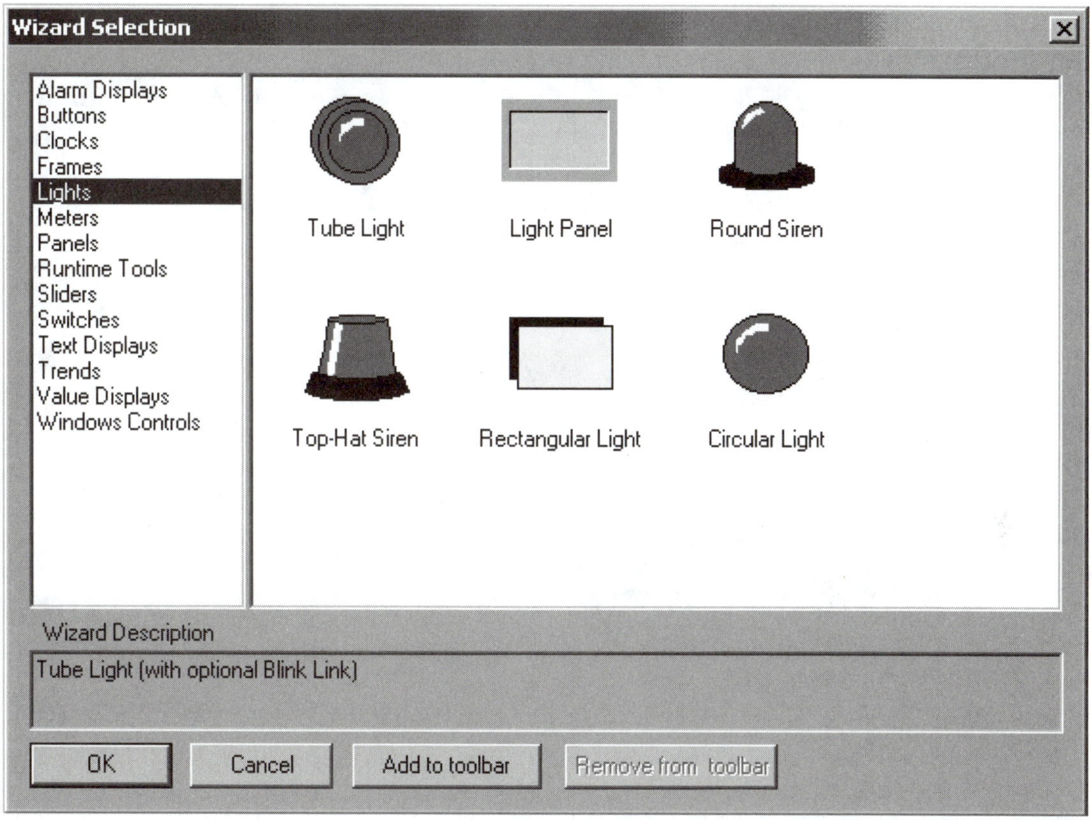

Figure 13–22 Wizard Selection screen.

Figure 13–22 shows a Wizard Selection screen. Note the list of types of Wizards that are available on the left side of the page. Lights is chosen, so the light Wizards are shown. The programmer chooses the first light and places it in the upper right corner of the oven (see Figure 13–23). The programmer then places another light on the lower left for a communications status indicator for the first temperature controller.

The programmer double-clicks on the light used for communications status and an input window appears (see Figure 13–24). The programmer enters a tagname (Status_1) for the status light that represents the temperature controller and then chooses red for the OFF state and green for the ON state. ON means that the computer and Omron temperature controller are communicating; red indicates that they are not communicating. Most control devices have a status bit in memory that indicates whether or not communications are normal. The programmer chooses OK

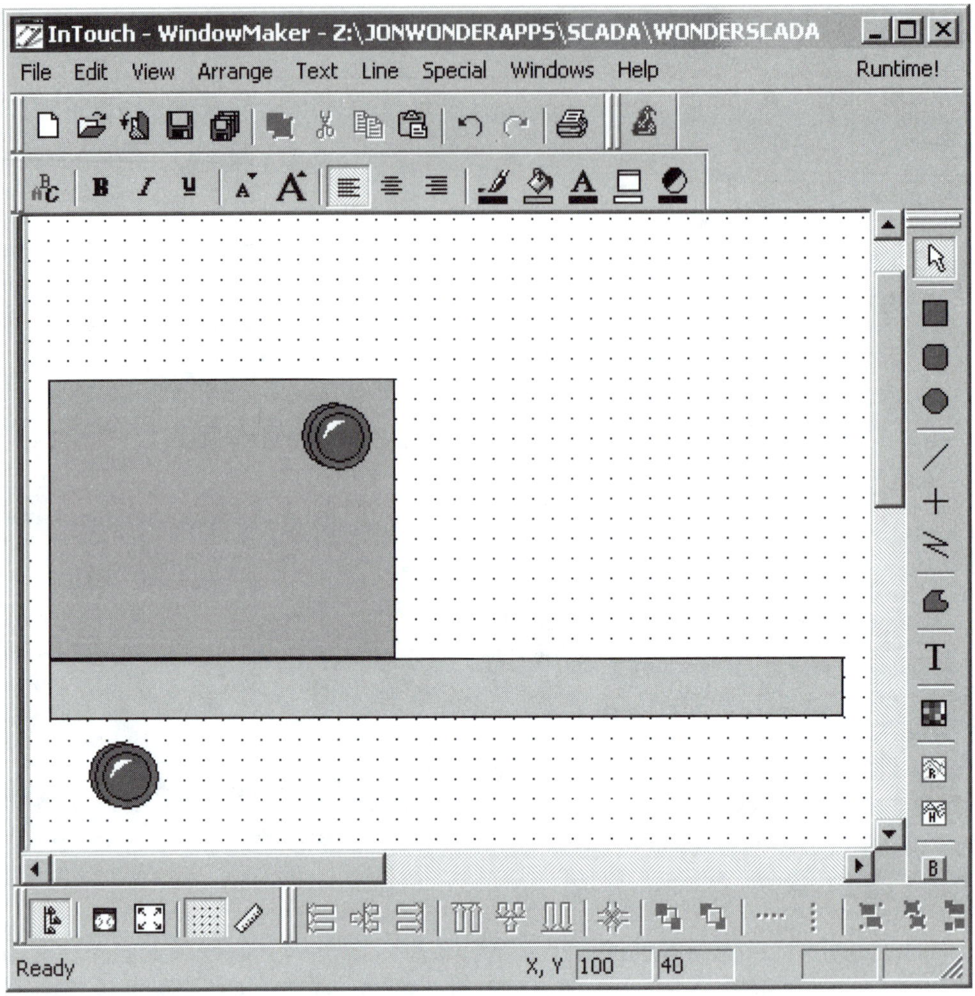

Figure 13–23 Application development screen showing the addition of a status light and product present light.

and the screen in Figure 13–25 appears asking the operator if he or she wants to define Status_1. The programmer chooses OK and the screen in Figure 13–26 appears. The programmer defines Status_1 (see Figure 13–26). The programmer chooses DDE discrete for the type. The programmer chooses Temp_Contrl_1 (remember that this is the DDE access name that we created earlier) for this controller's DDE access name.

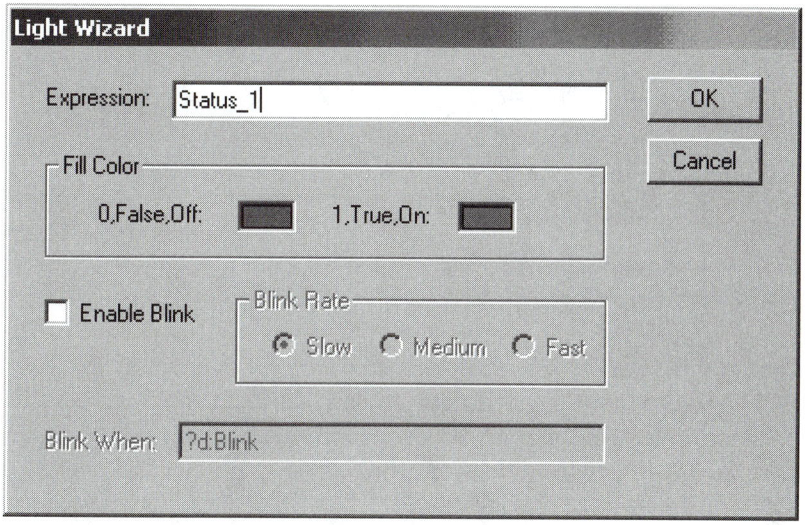

Figure 13–24 Light Wizard screen.

Figure 13–25 Tagname Undefined screen.

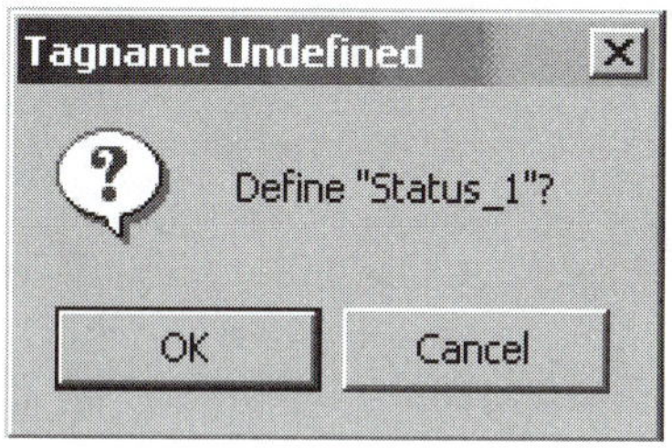

The programmer then enters the actual I/O item that the Omron temperature controller will understand. The driver for the Omron temperature controller understands the word "status." Status is entered as the item name. Remember that our tagname is Status_1 because the application has two temperature controllers and tagnames should clearly describe their purpose.

Next, the programmer adds a "real-time" graph to the display. The programmer needs to display a graph that shows how the oven has performed during the previous hour. The programmer decides to display the temperature and the setpoint on the graph. The trend-graph tool is chosen from the tool palette, and a graph is created just the way a rectangle would be drawn (see Figure 13–27).

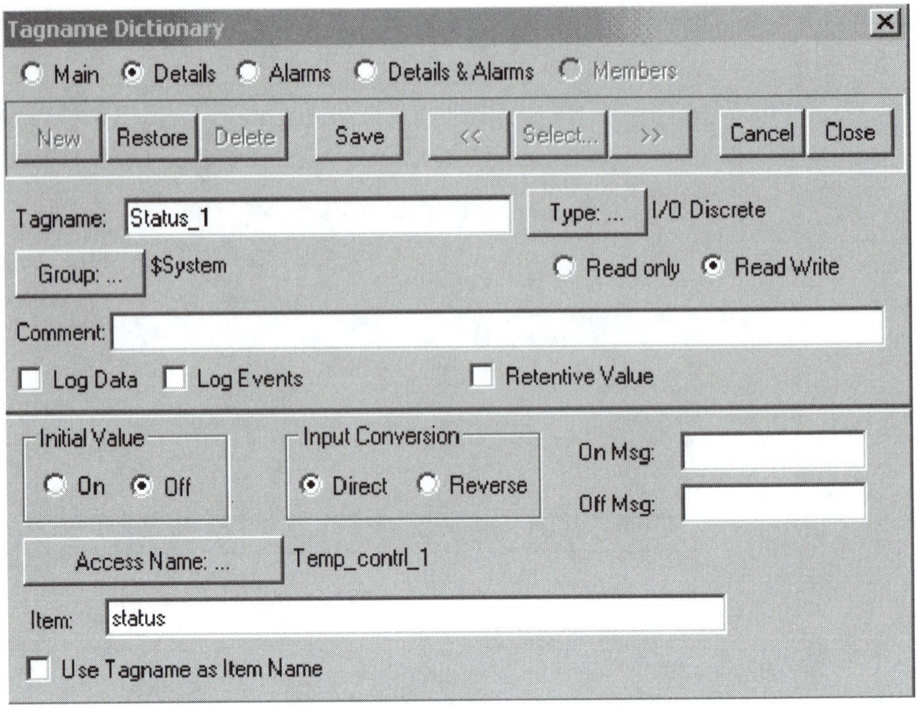

Figure 13–26 Status_1 definition screen.

Note that a historical trend graph can also be used to look at logged data. The historical graph can be used to look at logged data from past performance. The programmer adjusts the size of the graph by moving the side edges or the corners.

The programmer then double-clicks on the graph and an input window appears (see Figure 13–28). First, the time span is chosen for the graph. It was decided that the total time that the graph would show at one time would be 1 hour; 1 minute is chosen as the sample interval, which means that the graph will show from the present to the past 60 minutes, and it will sample and display new data every 1 minute.

Next, the programmer enters the tagnames of the items to be displayed on the graph. Temp_1 is chosen for the tagname of the temperature variable in the Omron temperature controller 1. Set_1 is chosen for the tagname of the setpoint variable in temperature controller 1. The programmer then changes the line width and color for each of the tagnames. The line width is set to 2 to make the lines more visible on the graph. Green is chosen for the temperature line (Temp_1) and blue is chosen for the setpoint line color (Set_1).

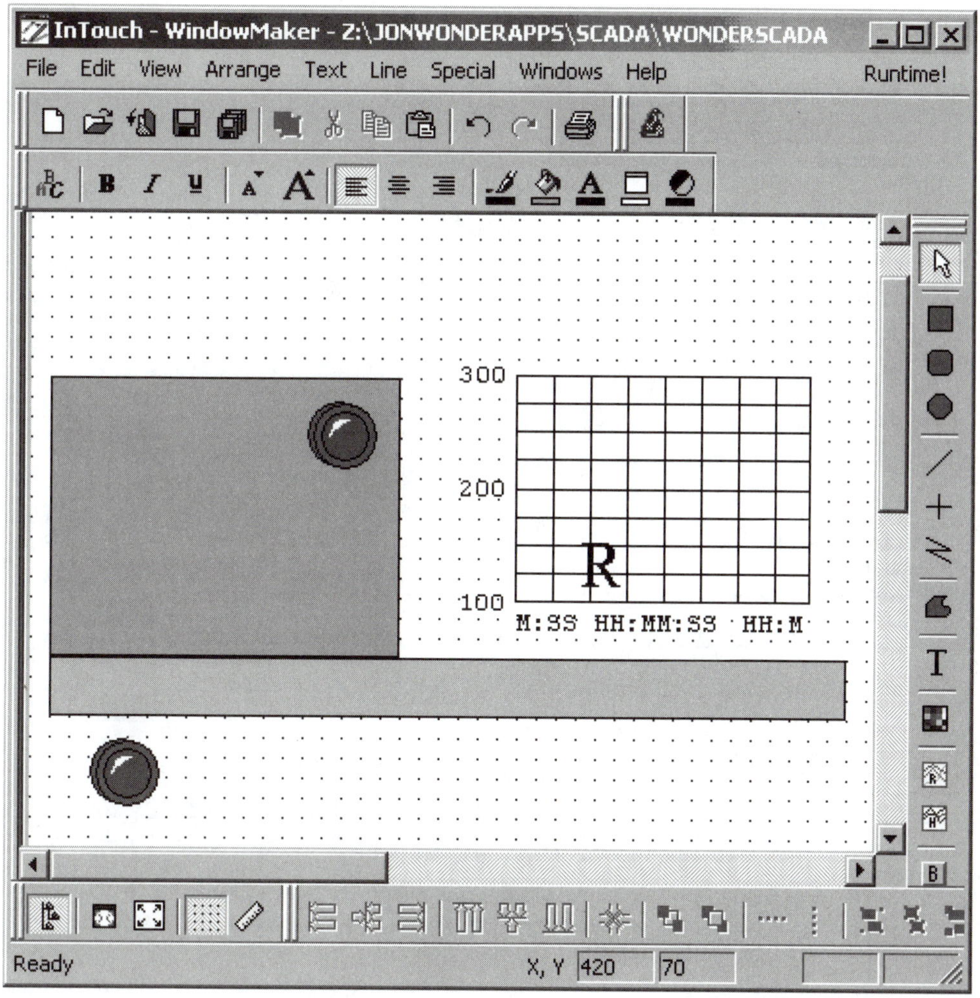

Figure 13–27 Application screen showing graph.

The programmer is done entering graph information at this point and clicks on the OK button. The programmer hasn't defined the Temp_1 tagname or the Set_1 tagname, so a window appears as shown in Figure 13–29. The programmer chooses OK, and a new window appears as shown in Figure 13–30.

The programmer chooses DDE integer for the type, Temp_Contrl_1 for the DDE access name, and temperature for the item name. Temperature is the name that the Omron driver uses to get the current temperature value from the controller.

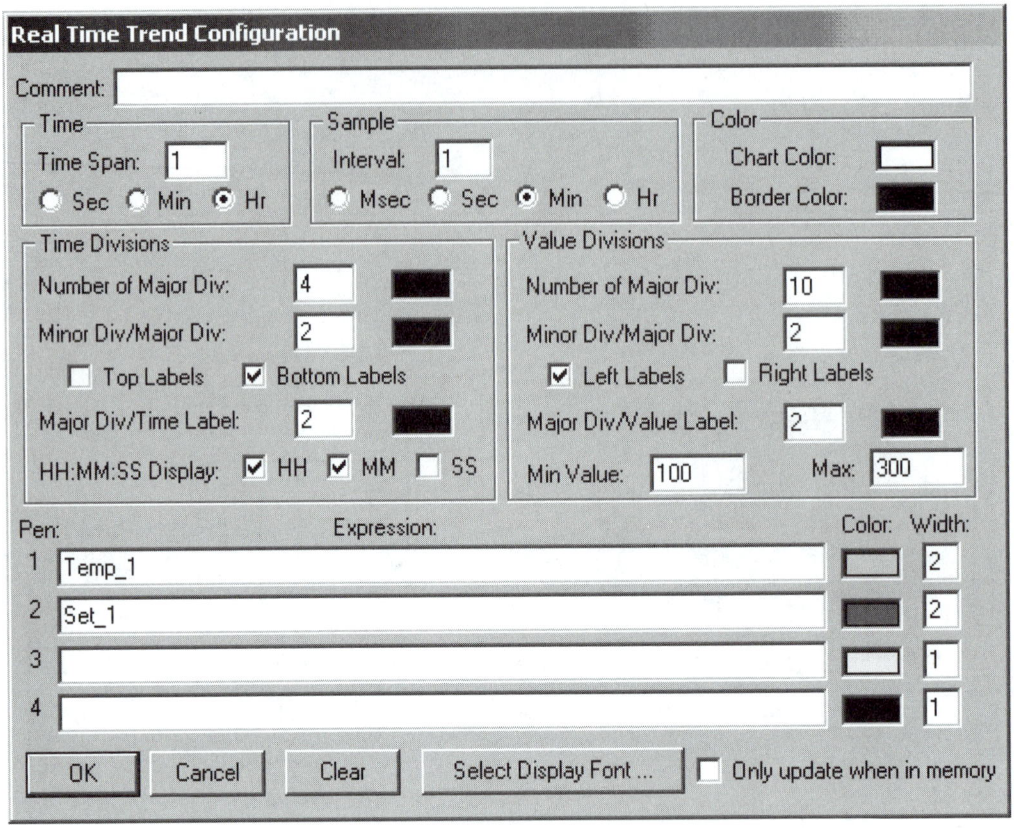

Figure 13–28 Graph configuration screen. Note that colors and time increments can be configured on this screen. The programmer can set the length of time that the graph will show on the screen. For example, the user might want to show the last hour's temperatures on the screen.

Figure 13–29 Tagname undefined screen.

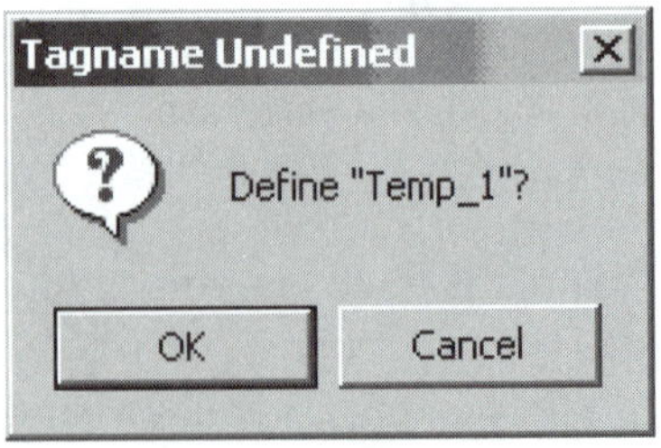

The programmer then chooses OK. The same procedure is followed to define tag name Set_1. The item name used was "setpoint." The Omron driver uses the name "setpoint" to send the setpoint value to the controller.

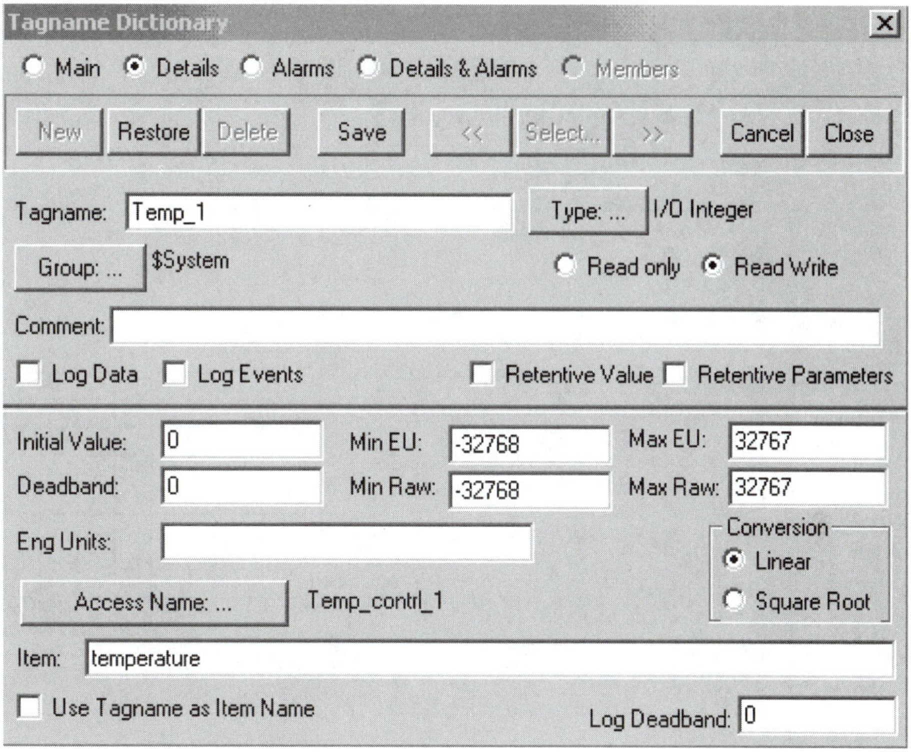

Figure 13–30 Tagname definition screen for Temp_1.

Next, the programmer adds a button so that the operator can automatically change all of the variables (proportional gain, integral gain, derivative gain, and the setpoint) in the temperature controller. The programmer uses the Button tool to create a button on the screen. Figure 13–31 shows the application with a button added for product P134. The programmer then substitutes the name P134 for the button name. The programmer also uses the Text tool to add three labels: Temperature, Setpoint, and Product. The Rectangle tool is used to draw a small rectangle to the right of the labels. The programmer will create links later so that the actual values can be displayed during runtime.

Examine Figure 13–32. This table shows the seven products manufactured in this process and the parameters for each product. Note that the setpoints are different for each controller.

The programmer then double-clicks on the button and a new input screen appears (see Figure 13–33). The Action Type button is chosen for this object. The programmer chooses OK and the screen shown in Figure 13–34 appears. This figure

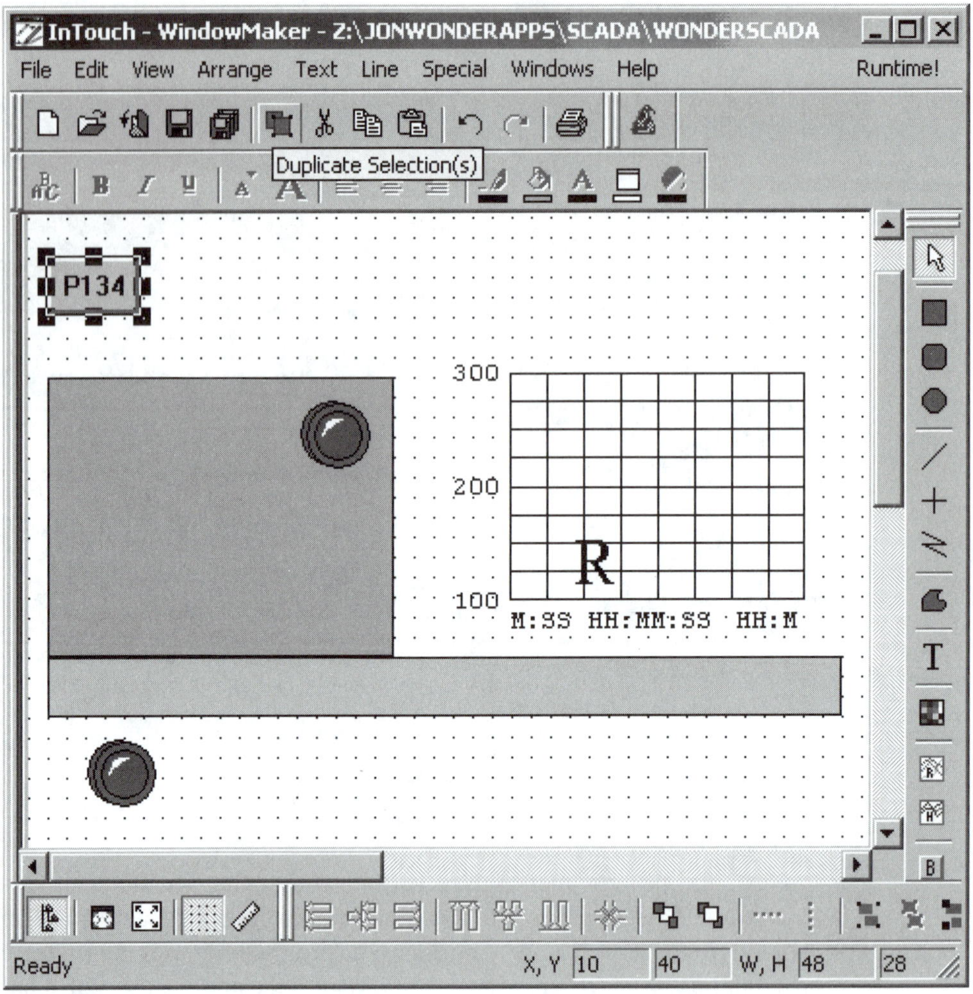

Figure 13–31 Application development screen showing graph, conveyor, and furnace and a product button.

shows the input screen for an action button. Note that four assignment statements were made in the input area. If the operator chooses this button while the actual application is running, these value assignments will be made. The values for this product will be sent to temperature controller 1 to set its parameters. The first statement assigns the value 180 to Set_1. Remember that Set_1 is the tagname for the temperature variable in the Omron temperature controller. The values for Prop_1,

Product	Setpoint Chamber 1	Setpoint Chamber 2	P	I	D
P134	180	225	90	10	30
P135	192	238	85	3	30
P136	163	207	97	12	32
P137	193	267	76	23	25
P138	215	237	85	12	24
P139	199	332	90	10	25
P140	137	183	87	8	22

Figure 13–32 Table showing temperature parameters for products.

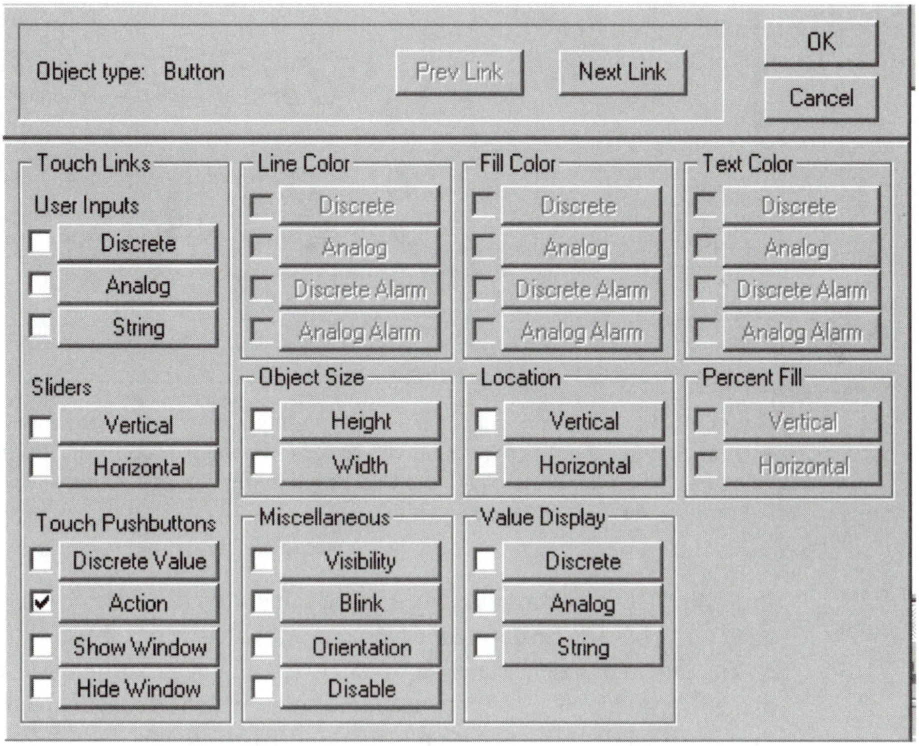

Figure 13–33 Link definition screen. Note that Action under Touch Pushbuttons was chosen.

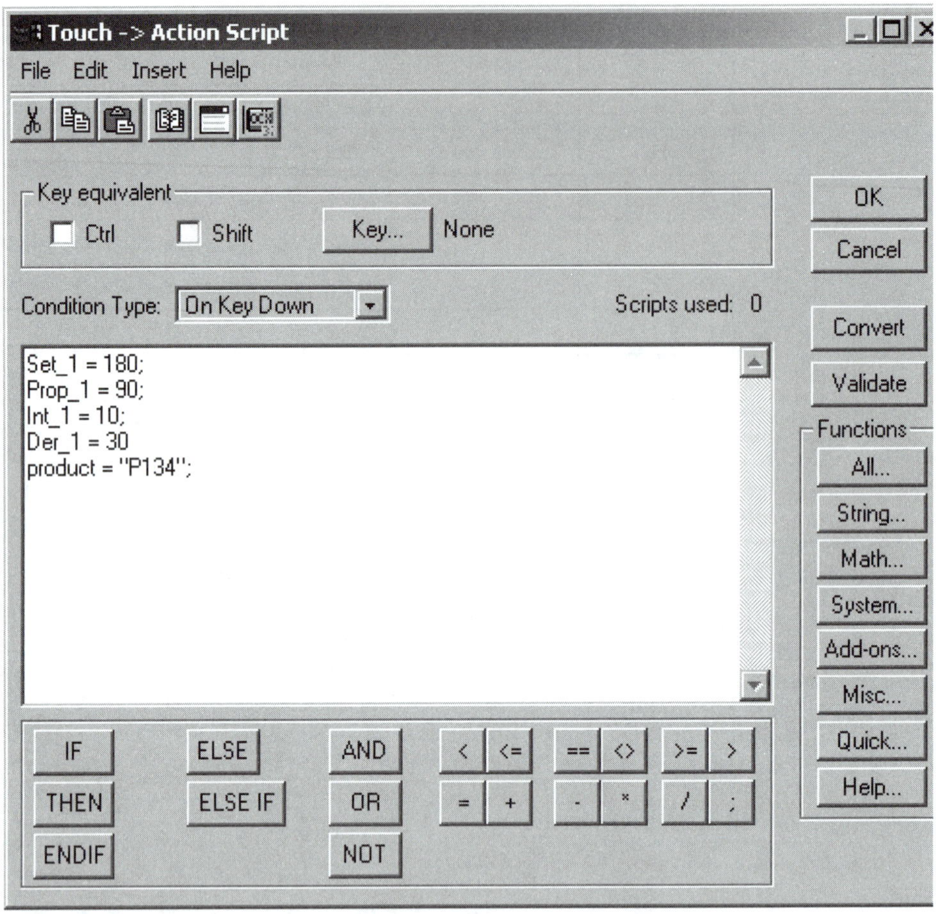

Figure 13–34 Script definition screen, where the user writes simple scripts to define actions. In this case, the parameter values for temperature controller 1, product 1 are being set. When the button linked to this action is clicked, this script executes and sends the new values to the controller.

Int_1, and Der_1 are also assigned in the same way. Note that Set_1, Prop_1, Int_1, and Der_1 have to be defined in the same way that Temp_1 was defined.

The last assignment statement assigns the character string P134 to the variable named "product." This will be used to display which product is being produced on the screen while the application is running. To create this display link for the product name, the programmer uses the Text tool and types a space in the rectangle labeled "product" in the window (see Figure 13–38. The programmer then

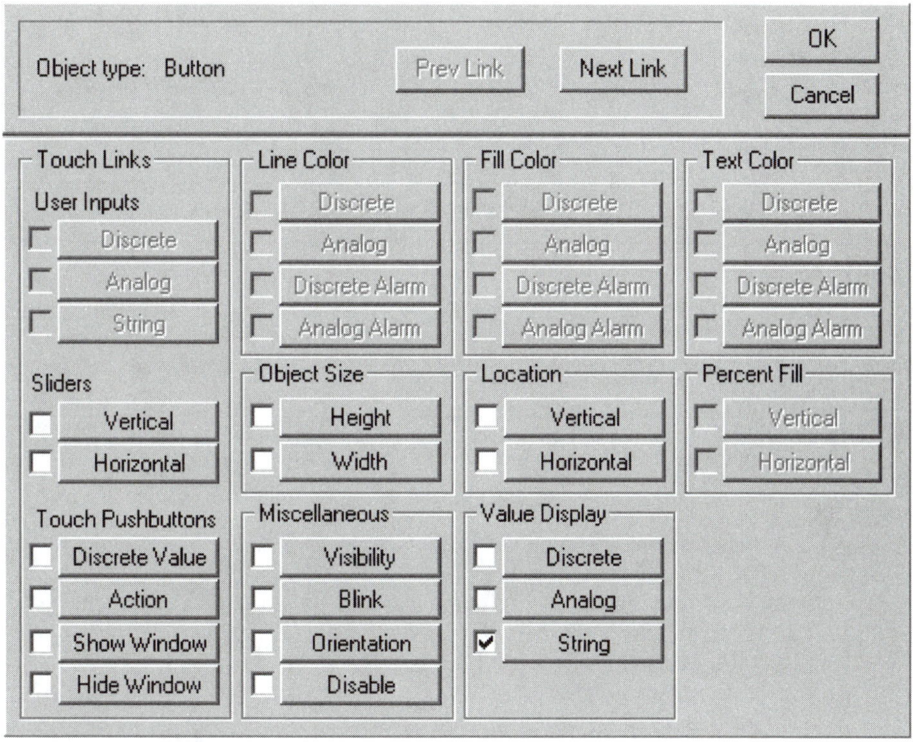

Figure 13–35 A string type value display was chosen.

double-clicks on the space and chooses String under Value Display (Figure 13–35). The programmer enters "product" for the tagname in Figure 13–36. Remember that "product" was entered as a variable name in the script for the product button. When the button is clicked, the script sets variable product equal to P134. The link that we are now developing will display it on the screen.

Tagname product was never defined, so the programmer has to define it. Figure 13–37 shows that the user chose Memory Message for the tagname type. This completes the display for product.

Next, the programmer would follow basically the same steps to create display links for the actual temperature and the setpoint. They have already been defined, so their addition to the application is easy. They use a DDE integer type because they both involve the actual controller. We will not add the links for temperature and setpoint.

The same process is followed to create buttons for each product (see Figure 13–38). The product variables and product name are changed for each product. Note that in a

Figure 13–36 Tagname definition screen.

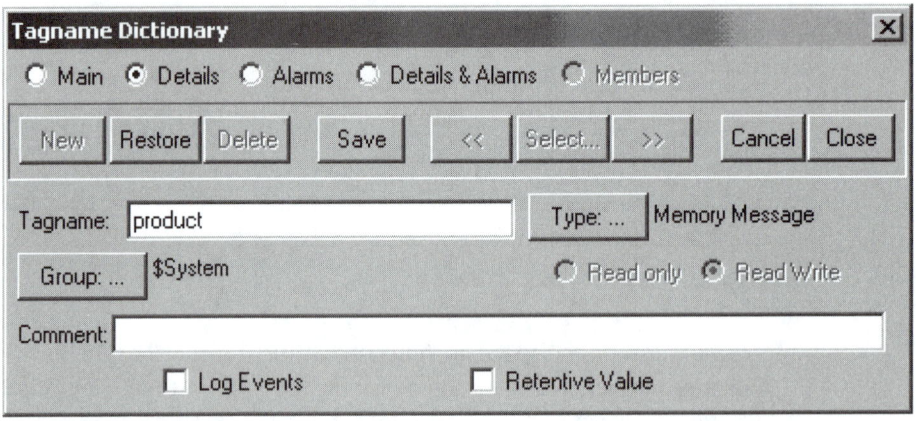

Figure 13–37 The programmer chose Memory Message for the type.

real application the buttons are used to change the parameters in both temperature controllers at once.

The programmer then creates a link to the part present indicator light on the furnace. The programmer double-clicks on the light on the upper right of the oven (see Figure 13–38). Part_present is entered for the tagname. Green is chosen for the ON color and red for the OFF color (see Figure 13–39). The programmer then enters OK and is asked if he or she wants to define Part_present. The programmer chooses OK, and the Tagname Dictionary screen appears (see Figure 13–40).

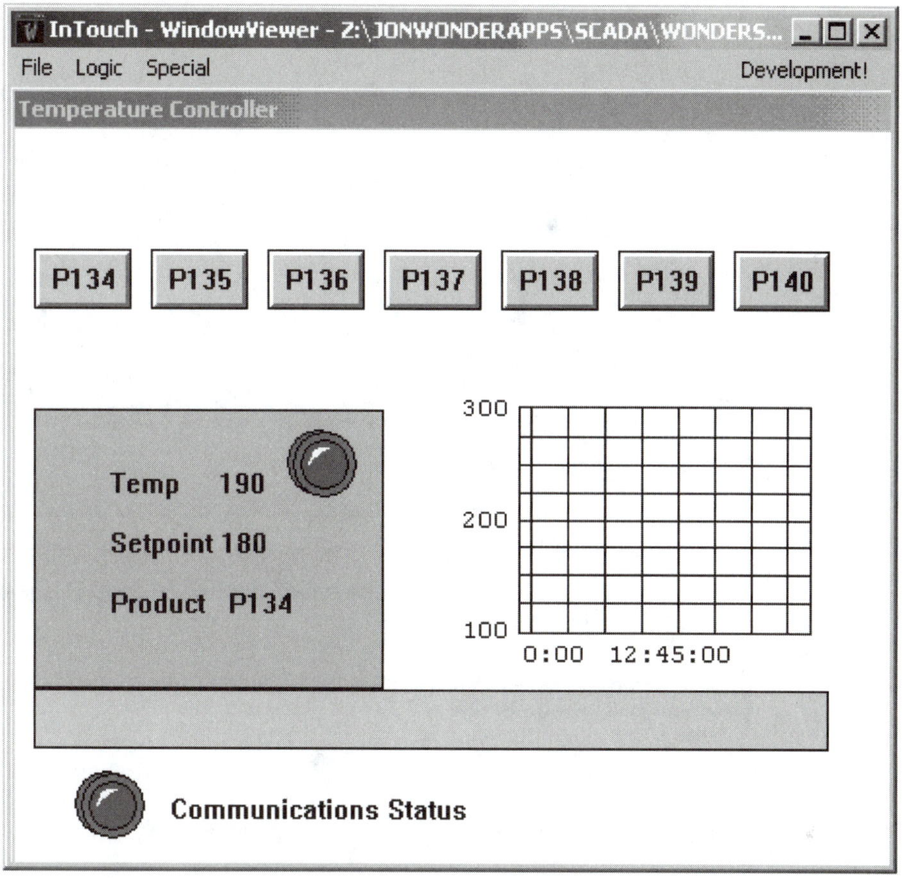

Figure 13–38 Application development screen showing that all product buttons have been added.

The programmer chooses DDE discrete for the I/O type (see Figure 13–40). SLC_1 is chosen for the DDE access name. Remember that this access name was created before to access the Rockwell Automation SLC 5/04 used in this application. I:7/8 is entered for the item name. This is the actual PLC input address for the presence sensor in the PLC.

The completed application is shown in Figure 13–41, which is what it would look like in the run mode. The operator monitors the process and uses the buttons to change to a different product. Note that the actual temperature, as well as the setpoint and the product being run, appear on the screen.

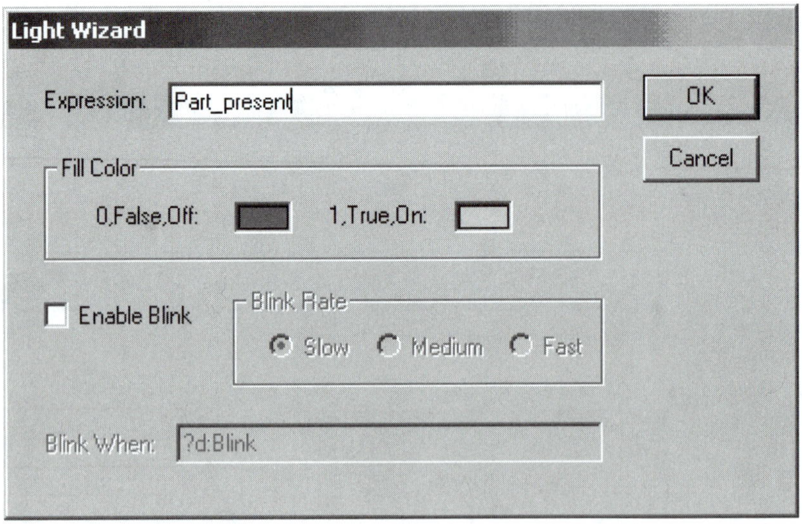

Figure 13–39 Tagname definition screen for Part_present. This is where the ON and OFF colors are chosen.

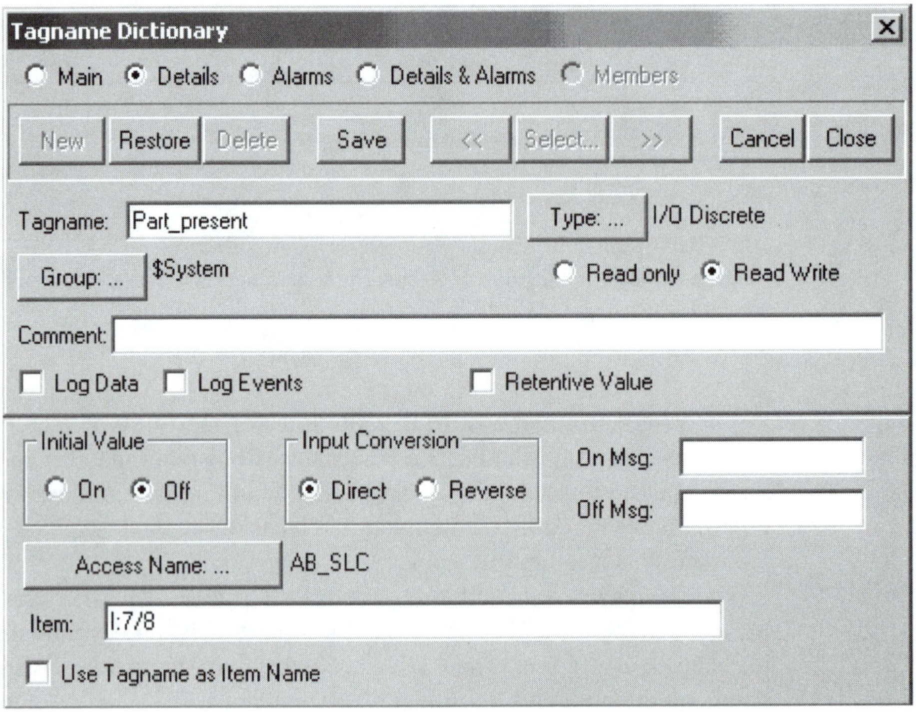

Figure 13–40 Tagname Dictionary screen for Part_present indicator light.

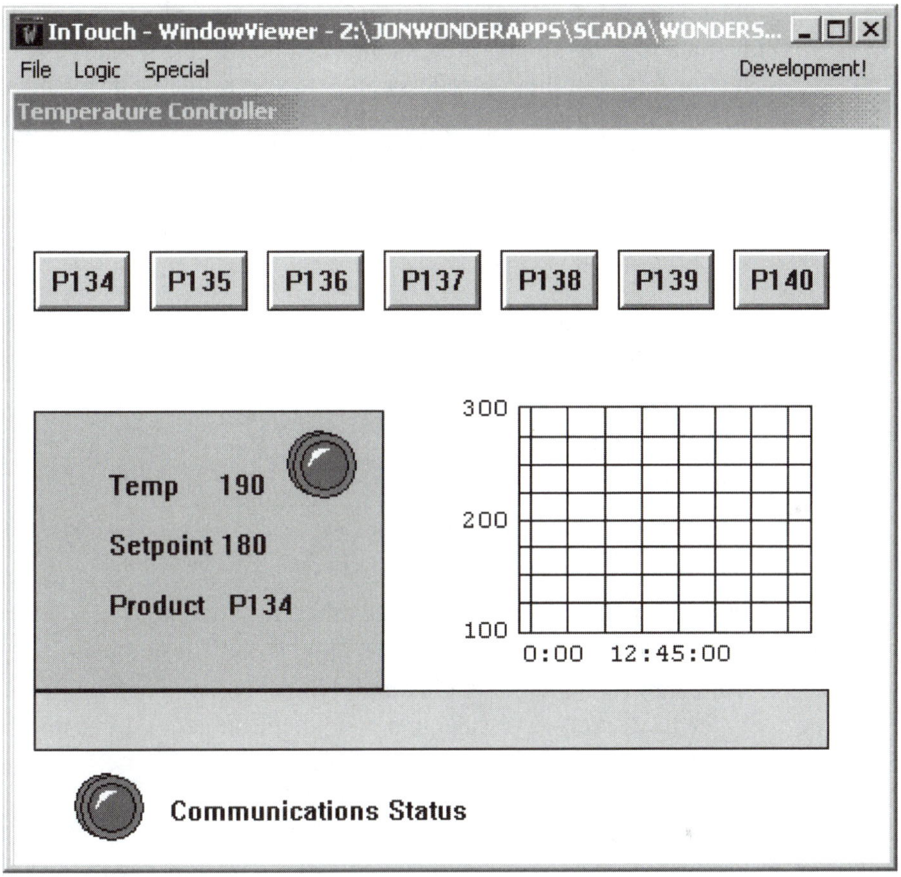

Figure 13–41 The completed application.

The steps for creating this temperature control window can be used to create a window for the second temperature controller. The window would be saved under a new name such as Temperature Controller 2. The programmer would change the tagnames and DDE access names so that the second controller would be used. A button can be added to each so that the operator can switch back and forth between screens. It would be easy to finish the application.

This simple application example gives you a feel for how SCADA applications are developed. We could have made this application much more complex and useful. For example, we could have added animation effects so that we can watch the product move down the conveyor line. Alarms could have been added in case something went wrong in the application. The alarm information could have been sent to another window, in the maintenance department, for example. We could have set up data logging so that process data can be saved for later use. In fact, any information in any of the controllers could now be available anywhere at any site worldwide. The information can be exchanged over the internal computer network, and the Internet can be used to link between worldwide sites.

Questions

1. What does SCADA stand for?
2. What does GUI stand for?
3. What does HMI stand for?
4. What does OPC stand for?
5. What is a topic?
6. What is an item?
7. What is a driver?
8. What is historical logging used for?
9. List and explain at least three different functional areas where performance might be improved by the use of a SCADA system.
10. Draw a diagram that illustrates typical communication in a SCADA system. Make sure you include two computers and at least two different industrial controllers.

chapter 14

Overview of Plant Floor Communication

If enterprises are to become more productive, they need to improve processes, and such improvement requires accurate data. Communications are vital. Production devices hold valuable data about their processes. In this chapter we will examine how these data can be acquired.

OBJECTIVES

Upon completion of this chapter, you will be able to:

1. Define terms such as *serial, synchronous, RS-232, RS-422, device, cell, area, host,* and *SCADA.*
2. Explain how computers can communicate with PLCs.
3. Indicate the opportunities available through factory communications.
4. Describe some of the industrial networks.

THE BASICS OF PLANT FLOOR COMMUNICATION

The programmable controller has revolutionized manufacturing. It has made automation flexible and affordable. PLCs control processes across the plant floor. In addition to producing product, PLCs also produce data, which can be more profitable than the product. This may not seem obvious; however, most processes are not as efficient as they could be. If we can use the data to improve processes, we can drastically improve profitability. The inefficiencies are not normally addressed because people are busy with other, more pressing problems. (A good friend of mine in a small manufacturing facility said it best: "It's hard to think about fire prevention when you're in the middle of a forest fire.") Manufacturing people are usually amazed to find that most of the data they would like to have about processes is already being produced in the PLCs or other smart devices in their processes. With very few changes, the data can be gathered and used to improve quality, productivity, and uptime. Huge gains are possible if these data are used.

To be used, the data must first be acquired. Many managers today say that they are already collecting much of the data. Why should they invest in electronic communications when they are already gathering data from the plant floor manually? The reasons are many. Often, data gathered manually are inaccurate. The data are not real-time either. The information must be written down by an operator, gathered by a supervisor, taken "upstairs," entered by a data process or person, printed into a report, and distributed, and all take time. It can often make data many days or weeks late. If mistakes in entry were made on the floor, it is often too late for corrections by the time they get to the office. The reports that are produced often contain too much extraneous information to be useful to anyone. The lateness and inaccuracy of the data gathering makes it almost counterproductive.

The other communication that is required is real-time information to the operator: accurate orders, accurate instructions, current specifications, and so on. This kind of communication is often lacking in industrial and service enterprises today, but it is quite easily achieved with the use of electronic communications. Many of the data required already exist in the smart devices on the factory floor. Much of the data that people write on forms in daily production already exist in the PLC and in the enterprise's computers. The improvements in computer hardware and software have made communication between control devices much easier. Many software communications packages make it easy to communicate with PLCs and other control devices.

Primitive Communications

Some devices do not have the capability to communicate. Some simple PLCs, for example, cannot communicate serially with other devices. In this case primitive methods are used. In the primitive mode, the devices essentially handshake with a few digital inputs and outputs. For example, a robot is programmed to wait until input 7 comes true before executing program number 13. It is also programmed to turn on output 1 after it completes the program. A PLC output can then be connected to input 1 of the robot, and output 1 from the robot can be connected to an input of the PLC. Now we have a simple one-device cell with primitive communications. The

PLC can command the robot to execute. When the program is complete, the robot notifies the PLC. Note that the process is a simple yes-or-no step (binary information).

Serial Communications

Many devices offer more than primitive communications capability. For example, we may need to upload and download programs or update variables. We cannot do that with primitive communications. Most machines offer serial communications capability using the asynchronous communications mode and have an RS-232 serial port available. You might think that any machine with an RS-232 port can communicate easily with any other device with an RS-232 port. This is definitely not true. Each machine may have its own protocol.

The RS-232 standard specifies a function for each of twenty-five pins. It does not say that all of the pins must be used, however. Some manufacturers use only three, as you can see in Figure 14–1. Some device manufacturers use more than three pins so some electrical handshaking can take place. In Figure 14–1 no handshaking is taking place. The first computer sends a message, whether or not another machine can accept it. The computer cable could be unplugged or the computer turned off; the sender would not know. Handshaking implies a cooperative operation. The first computer tells the second that it has a message it would like to transmit. It does this by setting pin 4 (the request to send pin) high. The second machine sees the request to send pin high, and if it is ready to receive, it sets the clear to send pin high. The first computer then knows that the cable is connected, the computer is on, and it is ready to receive. Some devices can be set up to handshake; others cannot.

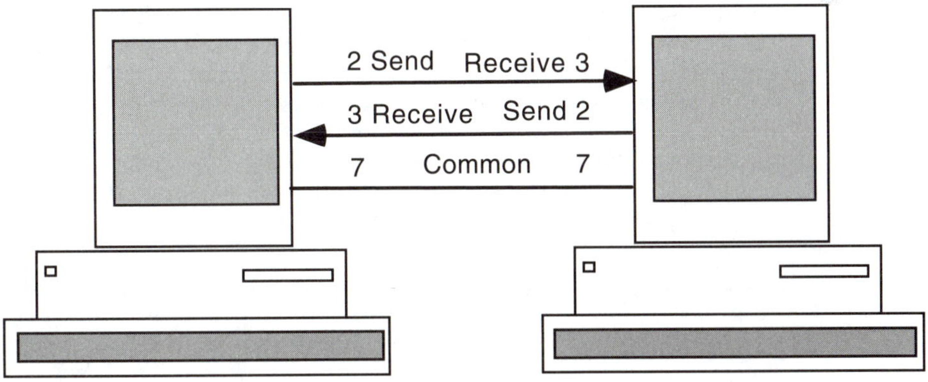

Figure 14–1 Simple RS-232 wiring scheme showing the simplest of RS-232 connections.

Fortunately when a machine is purchased, it is generally capable of communicating with a personal computer. The user usually opens the device manual to the section on communication to find a pinout for the proper cable. It is still difficult and expensive to communicate when a wide variety of devices are involved.

When a message is sent using asynchronous communications, the message is broken into individual characters and transmitted one bit at a time. The ASCII system is normally used. In ASCII every letter, number, and some special characters have a binary-coded equivalent. There is 7-bit ASCII and an 8-bit extended ASCII. In 7-bit ASCII 128 different letters, numbers, and special characters are possible. In 8-bit ASCII, 256 are possible.

Each character is sent as its ASCII equivalent. For example, the letter A is 1000001 in 7-bit ASCII (see Figure 14–2). It takes more than 7 bits to send a character in the asynchronous model, however. Other bits are used to make sure the receiving device knows a message is coming and that the message was not corrupted during transmission, and bits to let the receiver know that the character has been sent. The first bit sent is the start bit (see Figure 14–3), which lets the receiver know that a message is coming. The next 7 bits (8 if 8 bits are used) are the ASCII equiv-

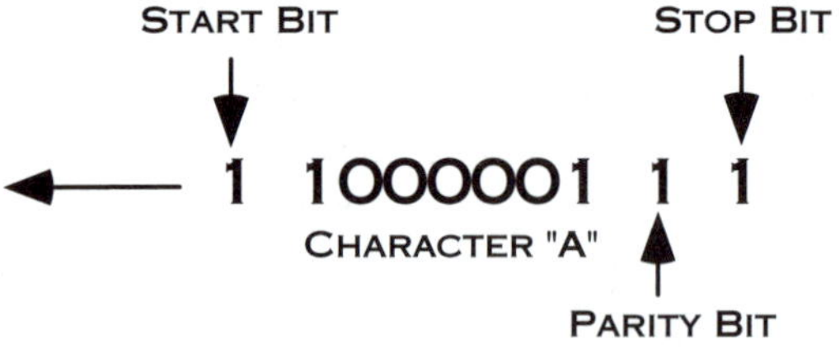

Figure 14–2 How the letter A is transmitted in the asynchronous serial mode of communications. This example assumes odd parity. There are an even number of 1s in the character A, so the parity bit is a 1 to make the total odd. If the character to be sent has an odd number of 1s, the parity bit would be a 0. The receiving device counts the number of 1s in the character and checks the parity bit. If they agree, the receiver assumes that the message was received accurately. This method is a rather crude way to check for errors. Note that two or more bits could change state and the parity bit could still be correct, but the message would be wrong.

START BIT	DATA BITS	PARITY BITS	STOP BIT
1 BIT	7 OR 8	1 BIT (ODD, EVEN, MARK, SPACE, OR NONE)	1, 1.5, OR 2 BITS

Figure 14–3 How a typical ASCII character is transmitted.

alent of the character. A bit is reserved for parity. Parity is used for error checking. The parity of most devices can be set up for odd or even, mark or space, or none.

Some new standards help integrate devices more easily. RS-422 and RS-423 were developed in 1977 to overcome some of the weaknesses of RS-232. The distance and speed of communications are drastically higher in RS-422 and RS-423.

RS-422 is called balanced serial. RS-232 has only one common. The transmit and receive line use the same common. This can lead to noise problems. RS-422 solves this problem by having separate commons for the transmit and receive lines, which makes each line immune to noise. The balanced mode of communications exhibits lower crosstalk between signals and is less susceptible to external interference. Crosstalk is the bleeding of one signal onto another, which reduces the potential speed and distance of communications. This is one reason that the distance and speed for RS-422 is much higher. RS-422 can be used at speeds of 10 megabits for distances of over 4000 feet, compared to 9600 baud and 50 feet for RS-232.

RS-423 is similar to RS-422 except that it is unbalanced. RS-423 has only one common, which the transmit and receive lines must share. RS-423 allows cable lengths exceeding 4000 feet. It is capable of speeds up to about 100,000 bps.

RS-449 was developed to specify the mechanical and electrical characteristics of the RS-422 and RS-423 specifications. The standard addressed some of the weaknesses of the RS-232 specification. The RS-449 standard specifies a thirty-seven-pin connector for the main and a nine-pin connector for the secondary. Remember that the RS-232 specification does not specify what type of connectors or how many pins must be used.

These standards are intended to replace RS-232 eventually. There are so many RS-232 devices, however, that it will take a long time. It is already occurring rapidly in industrial devices. Many PLCs' standard communications are done with RS-422. Adapters are cheap and readily available to convert RS-232 to RS-422, or vice versa (see Figure 14–4). Adapters can be used to advantage if a long cable length is needed for an RS-232 device (see Figure 14–5).

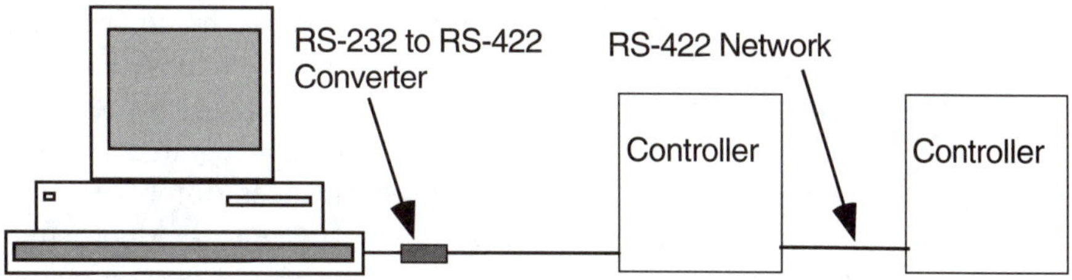

Figure 14–4 Use of a converter to change RS-232 communications to RS-422 communications. Note that the computer can now communicate with other devices on its network. The devices on the right are on an RS-422 network.

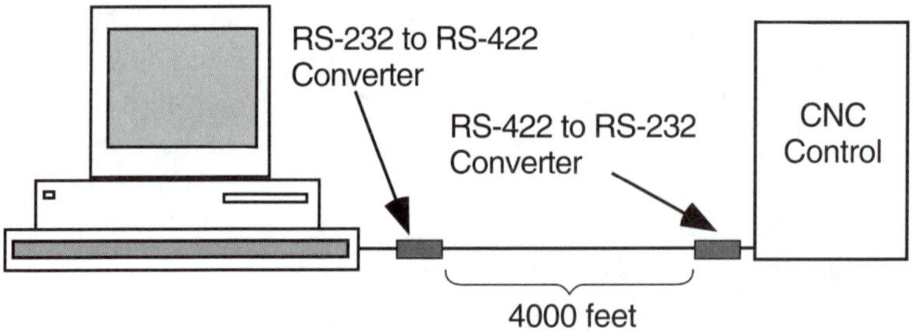

Figure 14–5 Use of two converters. In this case two are used to extend the cable length. Remember that RS-232 is reliable only to about 50 feet. The use of two converters allows 4000 feet to be covered by the RS-422 and then converted to RS-232 on each end. Note that the speed is limited by the RS-232.

RS-485 is a derivation of the RS-422 standard. The main difference is that it is a multidrop protocol, which means that many devices can be on the same line. This requires that the devices have some intelligence, however, because the devices must each have a name so that each knows when it is talked to. Many PLCs and other smart devices now utilize the RS-485 protocol. The standard specifies the electrical characteristics of receivers and transmitters connected to the network, and a differential signal between −7 to +12 volts. The standard limits the number of stations to thirty-two, which allows for up to thirty-two stations with transmission and reception capability, or one transmitter and up to thirty-one receiving stations.

This communication is usually accomplished by the use of supervisory control and data acquisition (SCADA) software. The concept is that software is run in a common microcomputer to enable communications to a wide variety of devices. The software is typically like a generic building block (see Figure 14–6).

The programmer writes the control application from menus or in some cases from graphic icons. The programmer then loads drivers for the specific devices in the application. Drivers are software. A driver is a specific package that was written to handle the communications with a specific brand and type of device. They are available for most common devices and are relatively inexpensive.

The main task of the software is to communicate easily with a wide range of devices. Most enterprises do not have the expertise required to write software drivers to communicate with devices. SCADA packages simplify the task. In addition to handling the communications, SCADA software makes it possible for applications personnel to write the control programs instead of programmers. This allows the people who best know the application to write it without learning complex programming languages. Drivers are available for all major brands of PLCs and other common manufacturing devices.

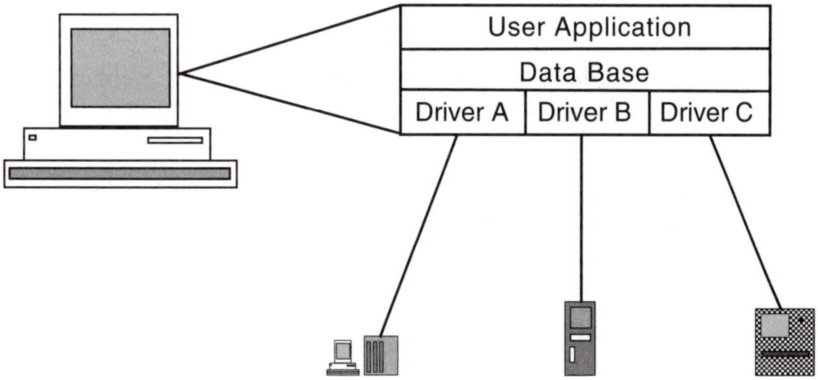

Figure 14–6 How a typical SCADA software package works. Note that the user application defines which variables from the devices must be communicated. These variables are collected through the drivers and stored in a database that is available to the application. Once the computer has the desired data, it is a relatively easy task to make the data available to other devices.

Device	Actual Number	Tagname
PLC 12	REG20	Temp_1
PLC 12	REG12	Cycletime_1
PLC 10	S19	Temp_2
PLC 07	N7:0	Quantity_1
Robot 1	R100	Quantity_2

Figure 14–7 What a tagname table might look like.

In general, an applications person writes the specific application using menu-driven software. The software is easy to use. Some are like spreadsheets and some use icons for programming. Instead of specific I/O numbers that the PLC uses, the programmer uses tagnames. For example, the application might involve temperature control. The actual temperature might be stored in register S20 in the PLC. The applications programmer would use a tagname instead of the actual number. The tagname might be Temp_1 (see Figure 14–7). This makes the programming transparent, which means that the application programmer does not have to worry about what brands of devices are in the application. A table is set up to assign specific PLC addresses to the tagnames (see Figure 14–7). In theory, if a different brand PLC were installed in the application, the only change required would be a change to the tagname table and the driver. Fortunately, more software is available daily to make the task of communications easier. The software is more user-friendly, faster, more flexible, and more graphics-oriented. The data gathered by SCADA packages can be used for statistical analysis, historical data collection, adjustment of the process, or graphical interface for the operator.

Local Area Networks (LANs)

Local area networks are the backbone of communications networks. The topic of LANs can be broken down into various methods of classification. We examine three: topology, cable types, and access method.

Topology

Topology refers to the physical layout of LANs. There are three main types of topology: star, bus, and ring.

Star Topology The star topology uses a hub to control all communications. All nodes are connected directly to the hub node (see Figure 14–8). All transmissions must be sent to the hub, which then sends them to the correct node. One problem with the star topology is that if the hub shuts down, the entire LAN shuts down.

Bus Topology The bus topology is a length of wire into which nodes can be tapped. Two types of transmission are available on a bus: baseband and broadband. Think of cable TV. All channels are transmitted at once on different frequencies. This is a broadband system. In a broadband system there is a head end (see Figure 14–9). The head end is an electronic box that performs several functions. The head end receives all communications, then remodulates (changes) the signal and sends it out to all nodes on another frequency. The head end changes the received signal frequency to another and sends it out for all nodes to hear. Only the nodes that are addressed pay attention to the message. The other end of the wire (bus) dissipates the signal. On a baseband bus, only one frequency is used so there is no need for a head end. Most industrial networks are based on baseband bus technology.

Ring Topology The ring topology has the appearance of a circle (see Figure 14–10). The output line (transmit line) from one computer goes to the input line (receive line) of the next computer, and so on. It is a straightforward topology. If a node wants to send a message, it sends it out on the transmit line. The message travels to the next

Figure 14–8 Star topology.

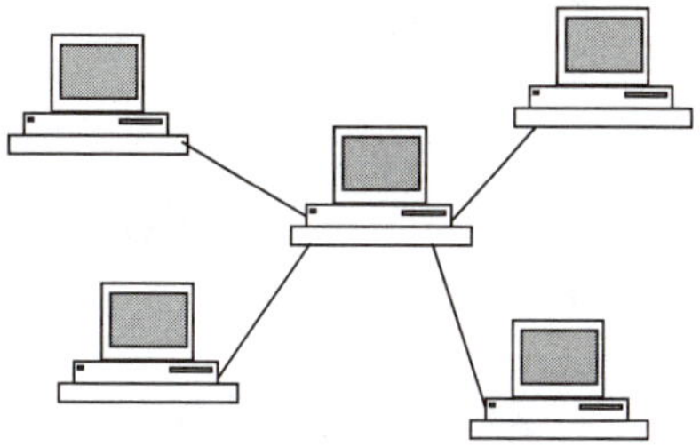

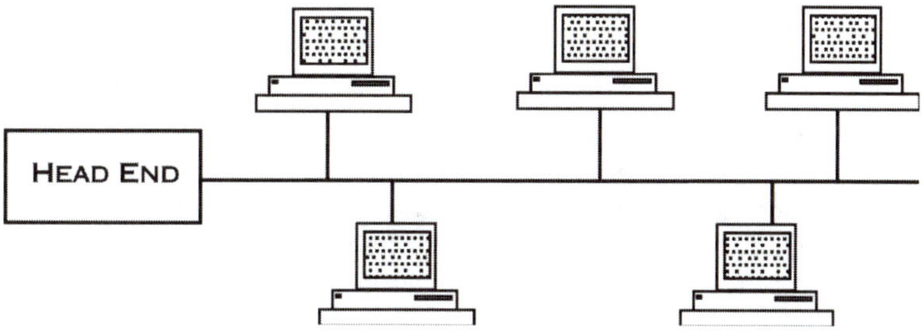

Figure 14–9 Typical bus topology. Each node (communication device) can speak on the bus. The message travels to the head end and is converted to a different frequency. It is then sent back out and every device receives the message. Only the device that the message was intended for pays attention to the message.

Figure 14–10 Ring topology.

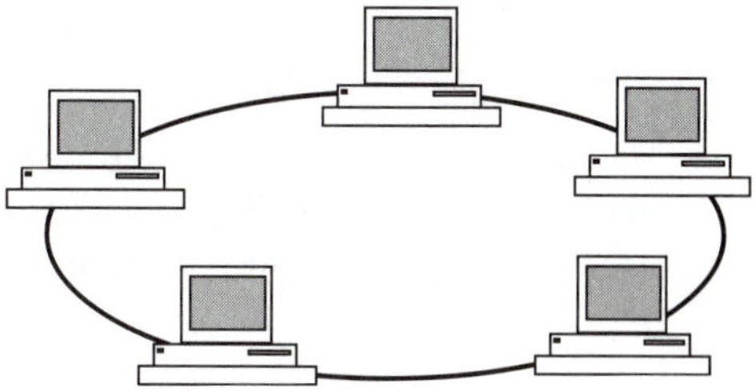

node. If the message is addressed to it, the node writes it down; if not, it passes it on until the correct node receives it.

The more likely configuration of a ring is shown in Figure 14–11. This style is still a ring, although it just does not appear to be one. This is the convenient way to wire a ring topology. The main ring (backbone) is run around the facility, and interface boxes are placed in line at convenient places around the building. These boxes are often placed in "wiring closets" close to where a group of computers will

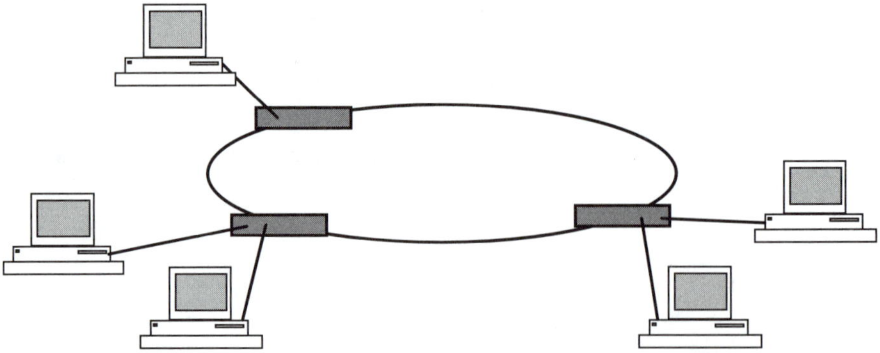

Figure 14–11 Typical ring topology. Note that it looks more like a star than a ring. It *is* a ring, however. The multiple station access units (MSAUs) allow multiple nodes to be connected to the ring. The MSAU looks like a set of electrical outlets. The computers are plugged into the MSAU and are then attached to the ring. The MSAU has relays for each port, which allows devices to be plugged in and removed without disrupting the ring.

be attached. These interface boxes are called multiple station access units (MSAUs). They are like electrical outlets. If we need to attach a computer, we just plug it into an outlet on the MSAU. The big advantage of the MSAU is that devices can be attached and detached without disrupting the ring. Communications are not disrupted at all.

Cable Types

There are four main types of transmission media: twisted pair, coaxial, fiber-optic, and radio frequency. Each has distinct advantages and disadvantages. The capabilities of the cable types are expanding continuously.

Twisted Pair

Twisted-pair wiring is, as its name implies, pairs of conductors (wires) twisted around each other along their entire length. The twisting of the wires helps make them more immune to noise. The telephone wires that enterprises have throughout their buildings are twisted pair. There are two types of twisted-pair wiring: shielded and unshielded. The shielded type has a shield around the outside of the twisted pair. This helps to make the wiring even more immune to noise. The wiring that is used for telephone wiring is typically unshielded. The newer types of unshielded cable are more immune to noise than in the past.

Coaxial Cable

Coaxial (or coax) is a common communication medium. Cable TV uses coaxial cable. Coax is broadband, which means that many channels can be transmitted simultaneously. Coax has excellent noise immunity because it is shielded (see Figure 14–12).

Broadband technology is more complex than *baseband* (single channel). With broadband technology there are two ends to the wire. One of the ends is called the head end. The head end receives all signals from devices that use the line. The head end then remodulates (changes to a different frequency) the signal and sends it back out on the line. All devices hear the transmission but pay attention only if it is intended for their address.

Frequency-division multiplexing is used in broadband technology. The transmission medium is divided into channels. Each channel has its own unique frequency. There are also buffer frequencies between each channel to help with noise immunity. Some channels are for transmission and some are for reception.

Time-division multiplexing is used in the baseband transmission method. This method is also called *time slicing*. Several devices may wish to talk on the line (see Figure 14–13). We cannot wait for one device to finish its transmission completely before another begins. They must share the line. One device takes a slice of time, then the next does, and so on.

Fiber-Optic Cable

Fiber-optic technology is also changing very rapidly. (See Figure 14–14 for the appearance of the cable.) The major arguments against fiber-optic are its complexity of installation and cost; however, the installation has become much easier and the

Figure 14–12 Coaxial cable. Note the shielding around the conductor.

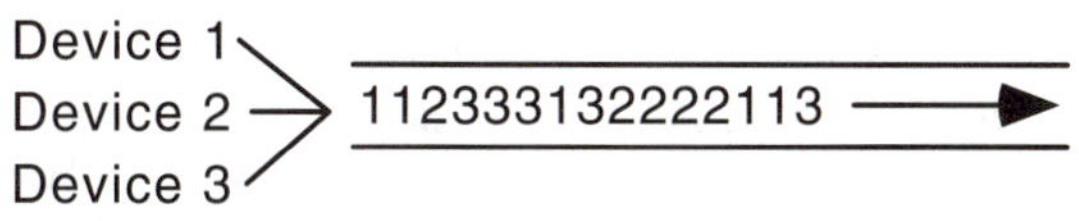

Figure 14–13 How time-division multiplexing works. Each device must share time on the line. Device 1 sends part of its message and then gives up the line so that another device can send, and so on.

Figure 14–14 Fiber-optic cable. Note the multiple fibers through one cable.

cost has fallen dramatically, to the point that when total cost is considered, fiber-optic is not much different for some installations than shielded twisted pair.

The advantages of fiber-optic are its perfect noise immunity, high security, low attenuation, and the high data transmission rates that are possible. Fiber-optic transmits with light, so it is unaffected by electrical noise. The security is good because fiber-optic does not create electrical fields that can be tapped like twisted pair or coaxial cable. The fiber must be physically cut to steal the signal, which makes it a much more secure system.

All transmission media attenuate signals, which means that the signal gets progressively weaker the farther it travels. Fiber-optic exhibits far less attenuation than other media. Fiber-optic can also handle far higher data transmission speeds than can other media. The fiber distributed data interchange (FDDI) standard was developed for fiber-optic. It calls for speeds of 100 megabits. This seemed very fast for a short period of time, but it is thought by many that it may be possible to get 100 megabits with twisted pair, so the speed standard for fiber-optic may be raised.

Radio Frequency

Radio frequency (RF) transmission has recently become very popular. The use of RF has exploded in the factory environment. The major makers of PLCs have RF modules available for their products. These modules use radio waves to transmit the data. The systems are immune to noise and perform well in industrial environments. RF is especially attractive because no wiring needs to be run. Wire is expensive to install and is susceptible to picking up electrical noise. Wire makes it harder and more expensive to move devices once they are installed. Wireless local area networks (WLANs) are finding a home everywhere from the factory floor to the office.

Wireless technology is becoming transparent. Wireless technology has become almost a black box. A wireless network operates exactly the same as a hardwired network. Devices are simply attached to a transceiver (combination transmitter/receiver). The transceiver does all of the translation and communication necessary to convert the electrical signals from the device to radio signals and then sends them via radio waves. The radio waves are received by another transceiver, which translates them back to network signals and puts them back on the network.

There are two parts to a wireless LAN: an access transceiver and remote client transceivers. The access transceiver is typically a stationary transceiver that attaches to the main hardwired LAN. The remote client transceivers link the remote parts of a LAN to the main LAN via radio waves. These transceivers can be used as *bridges* or *gateways*. A bridge is used to link two networks that utilize the same or similar protocols. A gateway is more of a translator. It is used to connect dissimilar network protocols so that two different kinds of networks can communicate.

Spread-spectrum technology is used for radio transmissions. Spread spectrum was developed by the U.S. military during World War II to prevent jamming of radio signals and also to make them hard to intercept. Spread-spectrum technology uses a wide frequency range. In spread-spectrum technology, the transmitted signal bandwidth is much wider than the information bandwidth. Radio stations utilize a narrow bandwidth for transmission. The transmitted signals (music and voice) utilize almost all of the bandwidth. In spread-spectrum the data being translated is modulated across the wider bandwidth. The transmission looks and sounds like noise to unauthorized receivers. It is very noise immune. A special pattern or code determines the actual transmitted bandwidth. Authorized receivers use the codes to pick out the data from the signal. The FCC has dedicated three frequency bands for commercial use: 900 MHz, 2.4 GHz, and 5.7 GHz. There is very little industrial electrical noise in these frequencies.

Token Passing Access Method

In the token-passing method, only one device can talk at a time. The device must have the token to be able to use the line. The token circulates among the devices until one of them wants to use the line (see Figure 14–15). The device then grabs the token and uses the line.

- The device that would like to talk waits for a free token.
- The sending station sets the token busy bit, adds an information field, adds the message it would like to send, and adds a trailer packet. The header packet contains the address of the station for which the message was intended. The entire message is then sent out on the line.
- Every station examines the header and checks the address to see if it is being talked to. If not, it ignores the message.
- The message arrives at the intended station and is copied. The receiving station sets bits in the trailer field to indicate that the message was received. It then regenerates the message and sends it back out on the line. The original station receives the message back and sees that the message was received. It then frees the token and sends it out for other stations to use.

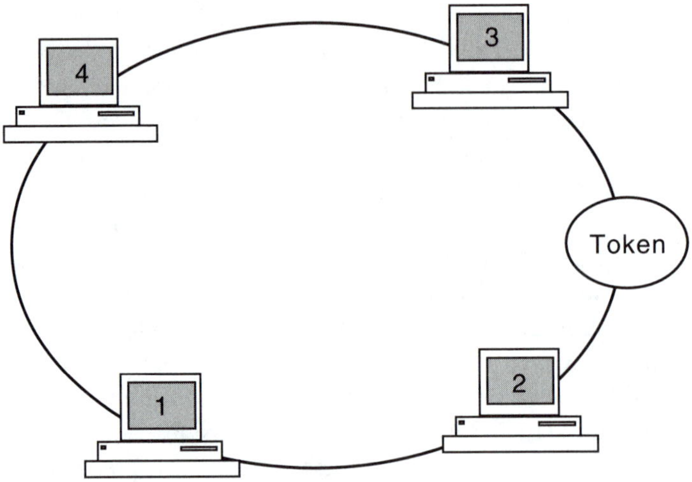

Figure 14–15 Token-passing method.

Token passing offers reliable performance. It also offers predictable access times, which can be important in manufacturing. Predictable access is often called deterministic because actual access times can be calculated based on the actual bus and nodes.

OVERVIEW OF INDUSTRIAL NETWORKS

Next we will examine the hierarchy of communications in a typical enterprise. Typical communications in an enterprise can be divided into three levels: device level, control level, and information or enterprise level. Rockwell Automation has a three-level model of communications (see Figure 14–16). The information layer, or enterprise level, is for plantwide data collection and maintenance.

The second level is the automation and control layer. This layer is used for real-time I/O control, messaging, and interlocking. The lowest level is the device level. The device level is used to integrate low-level devices into the overall enterprise cost effectively. Figure 14–16 shows an Ethernet network at the information (enterprise) level. ControlNet™ is used at the second level for automation and control. The device level utilizes DeviceNet to communicate.

This chapter will focus primarily on the device level, although the control level and enterprise level will be examined briefly. We will begin at the lowest level (device level) and move up in the communications hierarchy. It is important to note that the three levels are not always present or clearly defined.

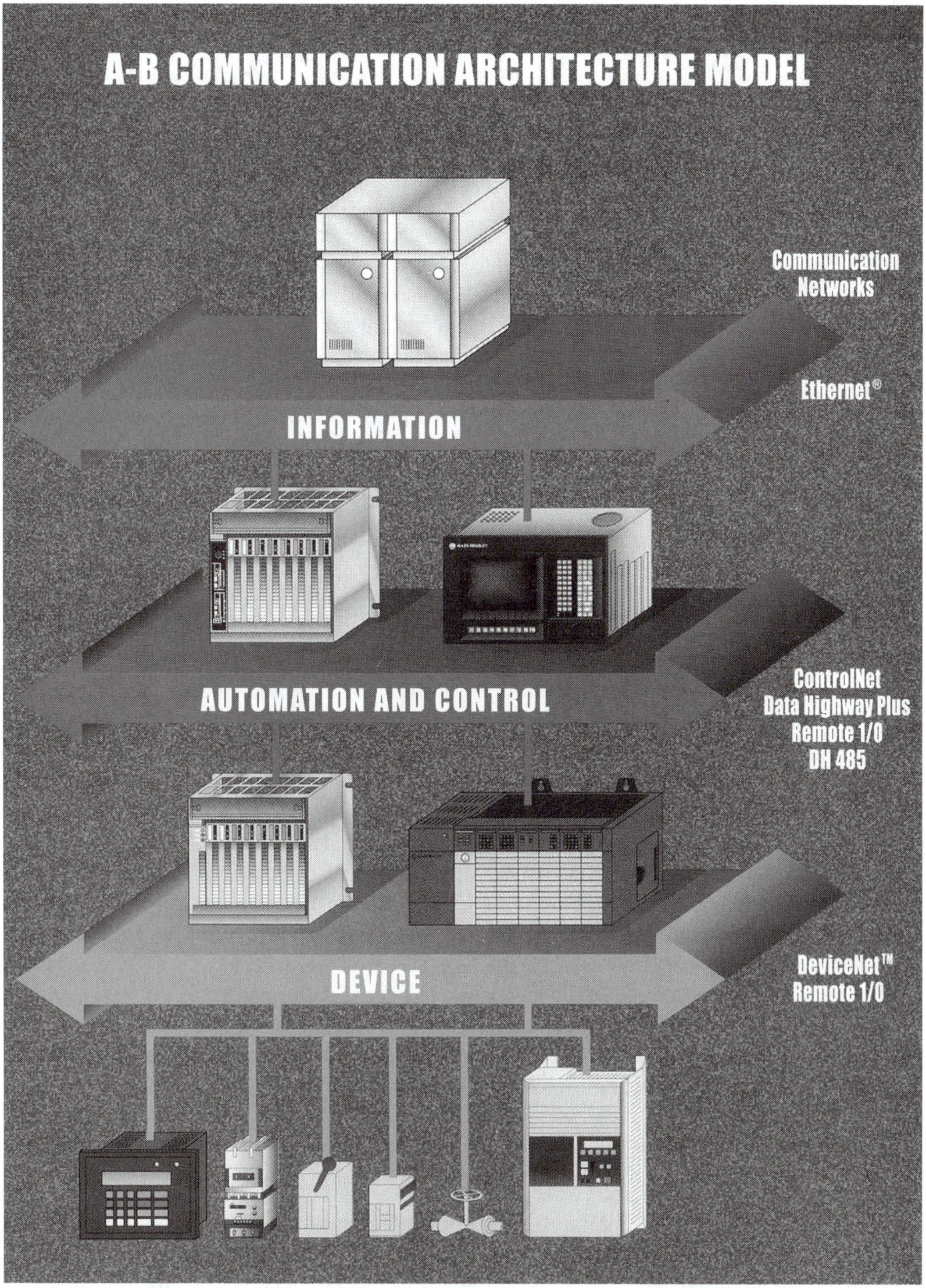

Figure 14–16 The Allen-Bradley communication architecture model. *(Courtesy Rockwell Automation Inc.)*

Device Level Networks

Device level networks are becoming more prevalent in industrial control. They are sometimes called field buses or industrial buses. These networks have many similarities to the conventional office computer network. Office networks allow many computers to communicate without being linked directly to each other. Each computer has only one connection to the network bus. Networks minimize the amount of wiring that needs to be done. Networks also allow devices such as printers to be shared. Industrial networks share some of the same advantages.

Imagine a complex automated machine that has hundreds of I/O devices and no device network. Now imagine the time and expense of connecting each and every I/O device back to the controller. This kind of connection can require hundreds of wires (or more) in a complex system and can involve long runs of wire. Conduit has to be bent and mounted for the vast number of wires that need to be run. Typical devices such as sensors and actuators require that two or more wires are connected at the point of use and then run to an I/O card at the control device. By contrast, in an industrial network, only a single twisted-pair wire bus needs to be run (see Figure 14–17). All devices can then connect directly to the bus. Multiple devices can even share one connection with the bus in some cases. Several manufacturers have developed I/O blocks that allow multiple I/O points to share one connection to the bus. One example is the valve manifold (see Figure 14–18). This particular unit can have up to thirty-two valves and up to thirty-two I/O with only two connections. Wiring and installation costs are drastically reduced when industrial networks are used.

The cost of field bus devices is higher than that of conventional devices. The field bus devices gain a cost advantage when one considers the labor cost of installation, maintenance, and troubleshooting. The cost benefit received from using distributed I/O is really in the labor saved during installation and start-up. There is also a material savings because wire does not have to be run. There is a tremendous savings in labor because only a fraction of the number of connections need to be made and only a fraction of the wire needs to be run. A field-bus system is also much easier to troubleshoot than a conventional system. If a problem exists, only one twisted-pair cable needs to be checked. A conventional system might require the technician to sort through hundreds of wires.

Field Devices

Sensors, valves, actuators, and starters are examples of I/O that are called field devices. Field devices can be digital or analog. Industrial applications often need analog information such as temperature, pressure, and so on. These devices typically convert the analog signal to a digital form and then need to communicate that to a controller. They may also be able to pass other information such as piece counts,

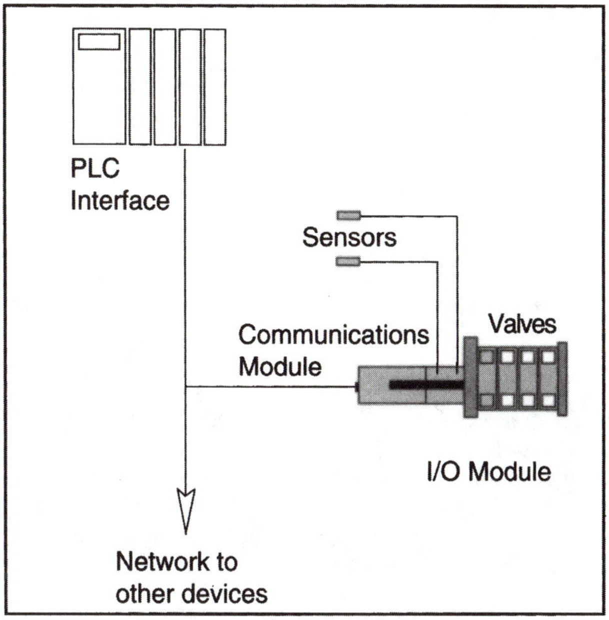

Figure 14–17 An example of a simple bus. Note the communications module and the I/O module. The communications module is DeviceNet in this example. *(Courtesy Parker Hannefin Corporation.)*

cycle times, error codes, and so on. Devices such as drives have also gained capability. An example of a smart device is a digital motor drive to which we can send parameters and commands serially. Parameters such as acceleration and other drive parameters can be sent by a computer or PLC to alter the way a drive operates. Every device manufacturer has had different communication protocols until now for communication.

The capabilities of field devices have increased rapidly, as has their ability to communicate. This has led to a need for a way to network them. Imagine what it would be like in your home if there were no electrical standard. Your television might use 100 volts at 50 cycles. Your refrigerator might use 220 volts at 50 cycles, and so on. Every device might use different plugs. It would be a nightmare to change any device to another brand because the wiring and outlets would need to be changed.

Industrial networks are an attempt to create standards to allow different brands of field devices to communicate and to be used interchangeably, no matter who made them. Figure 14–19 shows a simple industrial network with several field devices attached to it. Note the I/O block (I/O concentrator) and the valve manifold. Both are used to connect multiple devices to a network.

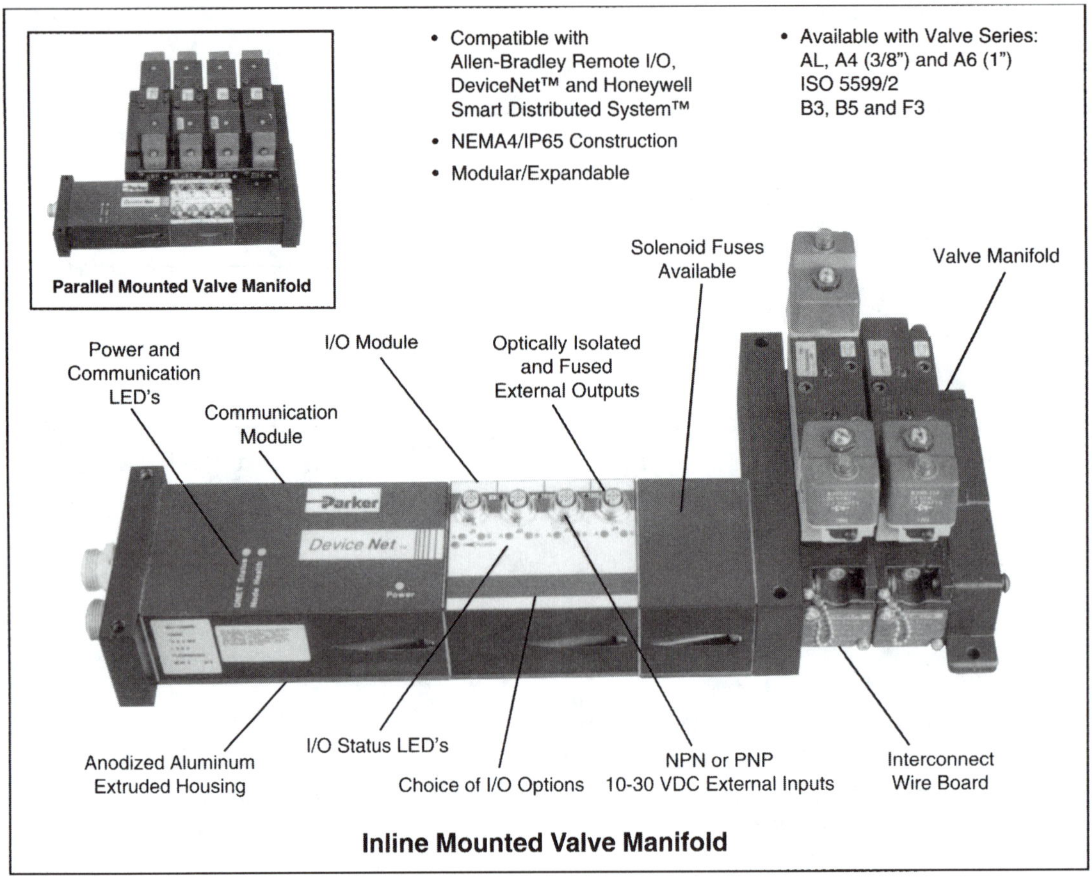

Figure 14–18 A DeviceNET valve manifold. *(Courtesy Parker Hannefin Corporation.)*

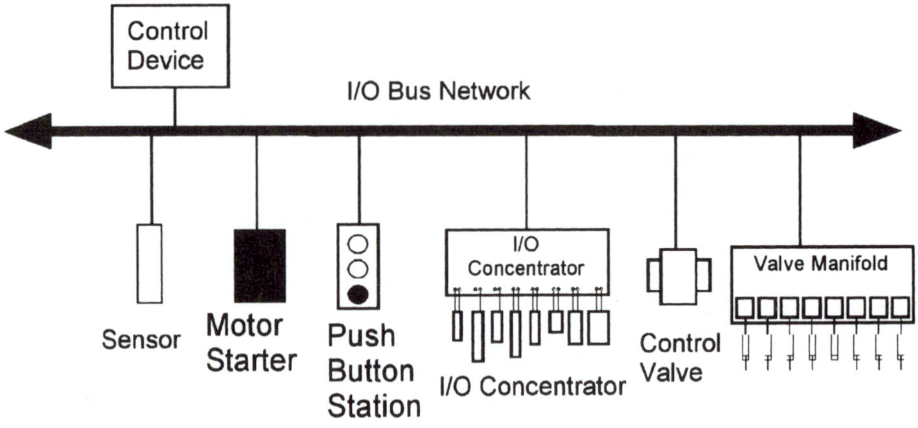

Figure 14–19 What a simple industrial network might look like.

Types of Device Level Buses

There are two basic categories that we could divide device bus networks into: process and device. Device-type buses are intended to handle the transmission of short messages, typically a few bytes in length. Process buses are capable of transmitting larger messages. Most devices in a device bus are discrete. They are devices such as sensors, push buttons, limit switches, etc. Many discrete buses can utilize some analog devices as well. They are typically devices that require only a few bytes of information transmission. Examples are some temperature controllers, some motor drives, thermocouples, etc. Since the transmission packets are small, device buses can transmit data packets from many devices in the same amount of time it would take to transmit one large packet of data on a process bus.

Industrial buses range from simple systems that can control discrete I/O (device-type bus) to buses that can be used to control large complex processes or a whole plant (process bus). Figure 14–20 shows a graph of the capabilities of several of the more common industrial buses. The left of the graph begins with simple discrete I/O control and goes up in capability to plant control. Block I/O are devices like manifolds that have several valves on one block, and only two to four wires are connected to control all of the I/O. With an industrial bus, only two to four wires have to be run instead of a few dozen.

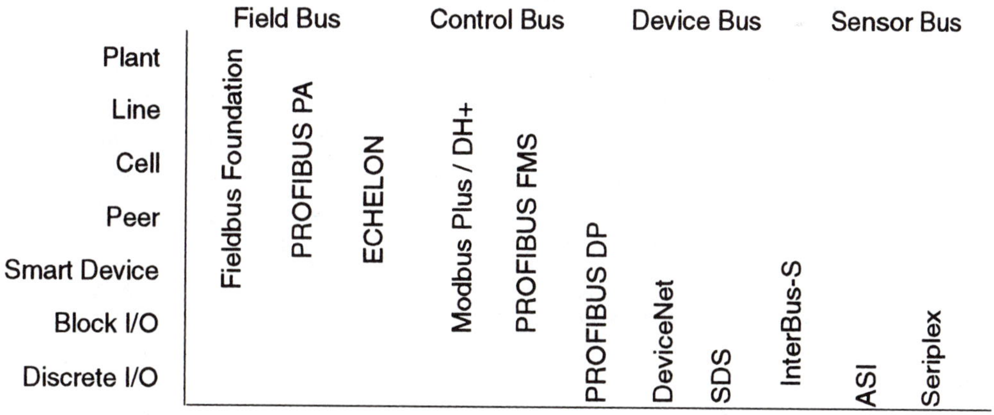

Figure 14–20 A comparison of some of the more common industrial buses.

Process Buses

Process buses are capable of communicating several hundred bytes of data per transmission. Process buses are slower because of their large data packet size. Most analog control devices do not require fast response times. Process controllers typically are smart devices. They are typically controlling analog types of variables such as flow, concentration, temperature, etc. These processes are typically slow to respond. Process buses are used to transmit process parameters to process controllers. Most devices in a process bus network are analog.

Process Bus Standards

There are two main organizations working on establishing process bus standards: the Fieldbus Foundation and the PROFIBUS (Process Field Bus) Trade Organization.

Fieldbus The Fieldbus Foundation's Fieldbus supports high-speed applications in discrete manufacturing such as motor starters, actuators, cell control, and remote I/O. Fieldbus devices can transmit and receive multivariable data and can also communicate directly with each other over the bus. System performance is enhanced due to the ability to communicate directly between two field devices rather than having to communicate through the control system.

Fieldbus is a new digital communications network that can be used in industry to replace the existing 4–20 mA analog signal. The network is a digital, bidirectional, multidrop, serial-bus communications network used to link isolated field devices such as controllers, transducers, actuators, and sensors.

Each field device has low-cost computing power installed in it, making each device a "smart" device. Each device can execute simple functions such as diagnostic, control, and maintenance functions on its own as well as providing bidirectional communication capabilities. With these devices, not only will the engineer be able to access the field devices, but they can also communicate with other field devices. In essence Fieldbus will replace centralized control networks with distributed-control networks.

Foundation Fieldbus is an open bus technology that is compatible with Instrument Society of America (ISA) SP50 standards. It is an interoperable, bidirectional communications protocol based on the International Standards Organization's Open System Interconnect (OSI/ISO) seven-layer communications model. The foundation protocol is not owned by any individual company or controlled by any nation or regulatory body. Foundation Fieldbus allows logic and control functions to be moved from host applications to field devices. The Foundation Fieldbus protocol was developed for critical applications in hazardous locations, with volatile

NETWORK SIZE	240 PER SEGMENT, 6500 SEGMENTS
NETWORK LENGTH	1900 M AT 31.25K OR 500M AT 2.5M
TOPOLOGY	MULTI-DROP WITH BUS POWERED DEVICES
PHYSICAL MEDIA	TWISTED PAIR
DATA TRANSFER SIZE	16.6 M OBJECTS/DEVICE
METHOD	CLIENT/SERVER, PUBLISHER/SUBSCRIBER
BUS ACCESS	CENTRALIZED SCHEDULER

Figure 14–21 Characteristics of a Fieldbus network.

processes, and with strict regulatory requirements. Each process cell requires only one wire to be run to the main cable, with a varying number of cells available. The Fieldbus protocol enables multiple devices to communicate over the same pair of wires. New devices can be added to the bus without disrupting control. Figure 14–21 shows a few of the characteristics of Fieldbus.

Fieldbus eases system debugging and maintenance by allowing online diagnostics to be carried out on individual field devices. Operators can monitor all of the devices included in the system and their interaction. Measurement and device values are available to all field and control devices in engineering units. This eliminates the need to convert raw data into the required units. This also frees the control system to perform other tasks. With Fieldbus technology, field instruments can be calibrated, initialized, and operated directly over the network. This reduces time for technicians, operators, and maintenance personnel.

PROFIBUS PROFIBUS specifies the functional and technical characteristics of a serial fieldbus. This bus interconnects digital field devices in the low (sensor/actuator level) up to the medium (cell level) performance range. The system contains master and slave devices. Masters are called active stations. Slave devices are simple devices such as valves, actuators, sensors, etc.

Slaves can respond only to received messages or, only if requested by the master, send messages. Master devices have the capability to control the bus. When the master has the right to access the bus, it may send messages without a remote request. Figure 14–22 shows an example of a PROFIBUS network with a PC card used as a PROFIBUS master. Figure 14–23 shows a GE Fanuc communicating with PROFIBUS I/O by using a PC card.

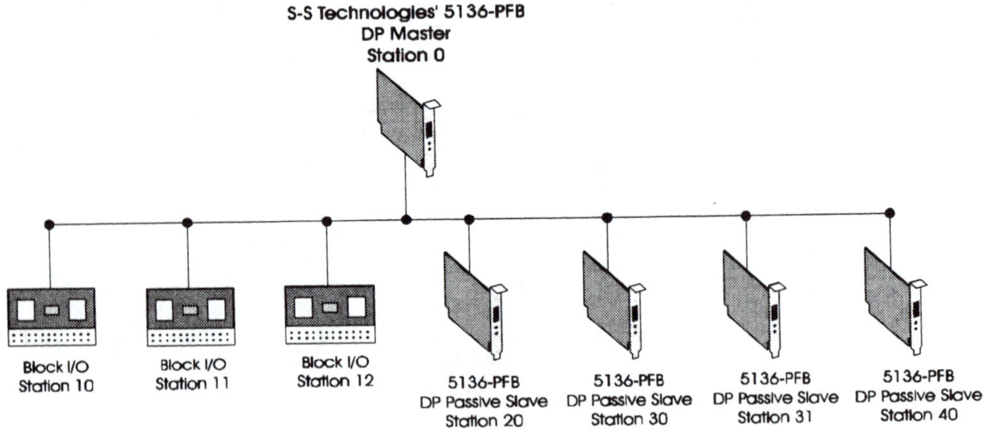

Figure 14–22 The use of an S-S Technologies PC card to act as a PROFIBUS DP master. *(Courtesy S-S Technologies Inc.)*

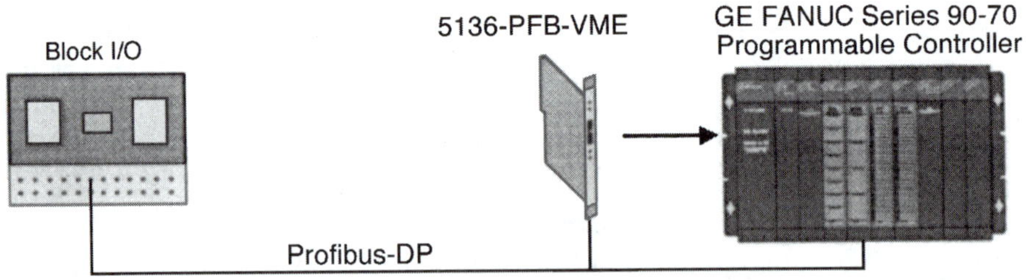

Figure 14–23 The use of an S-S Technologies PC card to enable a GE Fanuc PLC to communicate with and control PROFIBUS DP I/O. *(Courtesy S-S Technologies Inc.)*

The PROFIBUS access protocol includes the token-passing method for communication between complex stations (masters) and the master-slave method for communication between complex stations and simple peripheral devices (slaves). It is circulated in a maximum token rotation time among all masters. This maximum time is user configurable. PROFIBUS uses the token-passing method only among masters. Figure 14–24 shows some of the characteristics of a PROFIBUS network. The master-slave method allows the master that currently has the token to communicate with slave devices. Each master has the ability to transmit data to the slaves and receive data from the slaves. PROFIBUS allows the user a choice of access method:

- A pure master-slave system.
- A pure master-master system (token passing).
- A system with a combination of both methods, also called a hybrid system.

Network Size	Up to 127 nodes
Network Length	24 Km
Topology	Line, star, ring
Physical Media	Twisted pair or fiber
Data Transfer	244 bytes
Method	Master/slave, peer-to-peer
Bus Access	Token passing

Figure 14–24 Characteristics of a PROFIBUS network.

Device Buses

Device buses can be broken into two categories: bit-wide and byte-wide buses. Byte-wide buses can transfer 50 or more bytes of data at a time. Bit-wide buses typically transfer 1 to 8 bits of information to and from simple discrete devices. Byte-type systems are excellent for higher level communication, and bit-type systems are best for simple, physical-level I/O devices such as sensors and actuators. This chapter will concentrate on device-level buses, DeviceNet in particular.

Actuator Sensor Interface (ASI) Bus

AS Interface is a versatile, low-cost "smart" cabling solution designed specifically for use in low-level automation systems. It is designed to be easy to install, operate, maintain, and reconfigure. It was developed by a consortium of European companies, but the technology is now owned by AS International, an independent organization. It was specifically designed to ease installation and start-up and to have a low cost per node. Cost savings of up to 40 percent are possible in typical automation situations. AS Interface technology has been submitted for approval under the proposed IEC 947 standard. There are user groups in eight European countries. Figure 14–25 shows a few of the characteristics of ASI.

In an AS Interface network, data and power are carried over a single cable that links up to thirty-one slave devices to each master and up to 124 digital input and output nodes to a PLC or computer controller. Devices such as LEDs and indicators can be powered directly because the cable can handle up to eight amps. Higher power devices such as actuators are powered separately from a second cable. Worst-case cycle times for a full network are 5 ms per slave; faster times are possible when fewer devices are connected.

NETWORK SIZE	31 SLAVES
NETWORK LENGTH	100 METERS, 300 WITH A REPEATER
TOPOLOGY	BUS, MULTIDROP, STAR
PHYSICAL MEDIA	2-WIRE CABLE
DATA TRANSFER	MASTER AND SLAVES HAVE 4 BITS
BUS ACCESS	MASTER/SLAVE WITH CYCLIC POLLING

Figure 14–25 ASI bus characteristics table.

The cable is mechanically polarized and cannot be connected wrongly (it is foolproof). The cable is also self-healing, enabling I/O nodes to be disconnected and reconnected easily. The network utilizes cyclic polling of every network participant (sensor or actuator). With thirty-one slaves, the cycle time is 5 ms. Error detection is used to initiate a message repeat signal.

LonWorks

LonWorks has been used for a wide variety of applications: from simple control tasks to complex applications requiring thousands of devices. LonWorks networks can range in size from two to 32,000 devices. Figure 14–26 shows a few of the characteristics of LonWorks. LonWorks has been extensively used for building control. It has been successfully used for slot machines, home control, petroleum refining plants, aircraft, skyscraper building control, and so on.

LonWorks networks do not have a central control or master-slave arrangement. LonWorks uses a peer-to-peer architecture. Intelligent control devices called nodes have their own intelligence and can communicate with each other using a common protocol. Each contains intelligence that can implement the protocol and perform control functions. Nodes typically perform simple tasks. Devices such as switches, sensors, relays, drives, motion detectors and other home security devices, and so on, may all be nodes on the network. The overall network performs a complex control application, such as controlling a manufacturing line, securing your home, or automating a building.

Controller Area Network (CAN)

CAN was originally developed by Bosch for the European automobile network. It was designed to replace expensive wire harnesses with a low-cost network. There

NETWORK SIZE	32,000/DOMAIN
NETWORK LENGTH	2000 M AT 78 KBPS
TOPOLOGY	BUS, LOOP, STAR
PHYSICAL MEDIA	TWISTED PAIR, FIBER
DATA TRANSFER SIZE	228 BYTES
METHOD	MASTER/SLAVE PEER-TO-PEER

Figure 14–26 Characteristics of a LonWorks bus.

are many safety concerns with automobiles such as antilock brakes and airbags, so CAN was designed to be high-speed and very dependable. CAN is also found in commercial products. All of the major European car manufacturers are developing models using CAN to connect the various electronic components.

The Society of Automotive Engineers (SAE) developed a standard called SAE J1939. This standard is designed to give plug-and-play capabilities to car owners. The standard allocates CAN identifiers to different purposes. It uses extended (29-bit) identifiers.

CAN was designed to be a high-integrity serial data communications bus for real-time applications. Electronic chips were designed and built to implement the CAN standard. Demand for CAN chips has exploded. In 1994 more than 4 million CAN chips were shipped. This demand has led to the wide availability of reasonably priced CAN chips. CAN chips are typically 80 to 90 percent cheaper than chips for other networks. The CAN protocol is now being used in many other industrial automation and control applications

CAN is a broadcast bus-type system. The two-wire bus is usually a shielded or unshielded twisted pair. It can operate up to 1 Mbit/s. Many controllers are available for CAN systems. Messages are sent across the bus. The messages (or frames) can be variable in length: between 0 and 8 bytes in length. Each frame has an identifier. Each identifier must be unique. This means that two nodes may not send messages with the same identifier. Data messages transmitted from a node on a CAN bus do not contain addresses of either the transmitting node or of any intended receiving node. All nodes on the network receive each message and perform an acceptance test on the identifier to determine if the message is relevant to that particular node. If the message is relevant, it will be processed; otherwise it is ignored. This is known as multicast.

The node identifier also determines the priority of the message. The lower the numerical value of the identifier, the higher the priority. This allows arbitration if two (or more) nodes compete for access to the bus at the same time. The higher priority message is guaranteed bus access. Lower priority messages are retransmitted in the next bus cycle, or in a later bus cycle if there are other, higher priority messages waiting to be sent.

The ISO 11898 standard recommends that bus interface chips be designed so that communication can continue (but with reduced signal-to-noise ratio) even if:

- Either wire in the bus is broken.
- Either wire is shorted to power.
- Either wire is shorted to ground.

DeviceNET

DeviceNET is intended to be a low-cost method for connecting devices such as sensors, switches, valves, bar-code readers, drives, operator display panels, and so on, to a simple network. The DeviceNet standard is based on the CAN chip standard. By adhering to a standard, it is possible to interchange devices from different manufacturers (see Figures 14–27 and 14–28). This means that one could replace a device from one manufacturer with a similar device from a different manufacturer. DeviceNET allows simple devices to be interchanged and makes interconnectivity of more complex devices possible. In addition to reading the state of discrete devices, DeviceNET can report temperatures, read the load current in a motor starter, change the parameters of drives, and so on.

DeviceNET is an open network standard. The standard is not proprietary; it is open to any manufacturer. Any manufacturer can participate in the Open DeviceNET Vendor Association (ODVA), Inc. ODVA is an independent organization that manages the DeviceNET specification and supports the worldwide growth of DeviceNET. Figure 14–29 shows a few of the characteristics of DeviceNET. ODVA works with vendors and provides assistance through developer tools, developer training, compliance testing, and marketing activities. ODVA publishes a DeviceNET product catalog.

DeviceNET is based on the CAN standard. DeviceNET is a broadcast-based communications protocol. It uses the CAN chip, which helps keep costs low. DeviceNET supports strobed, polled, cyclic, change-of-state, and application-triggered data movement. The user can choose master/slave, multimaster, or peer-to-peer or a combination configuration depending on device capability and application requirements.

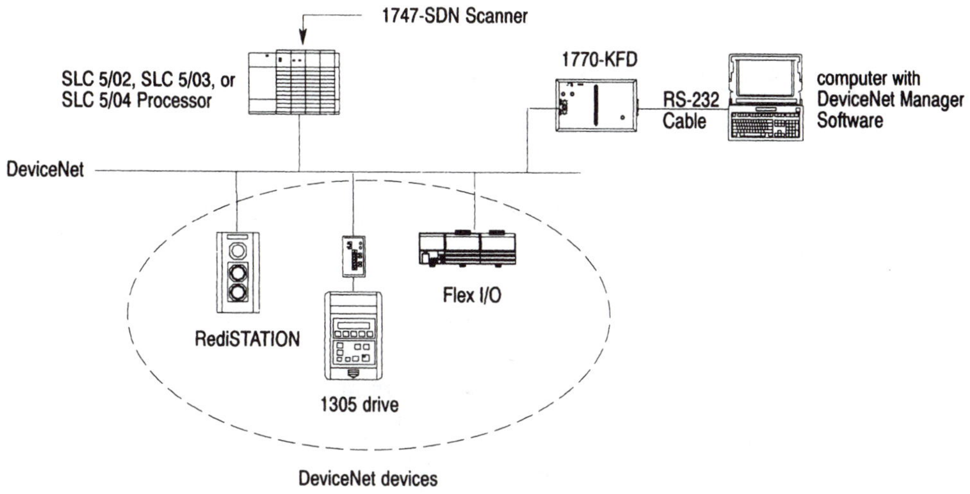

Figure 14–27 Example of a DeviceNET network. *(Courtesy Rockwell Automation Inc.)*

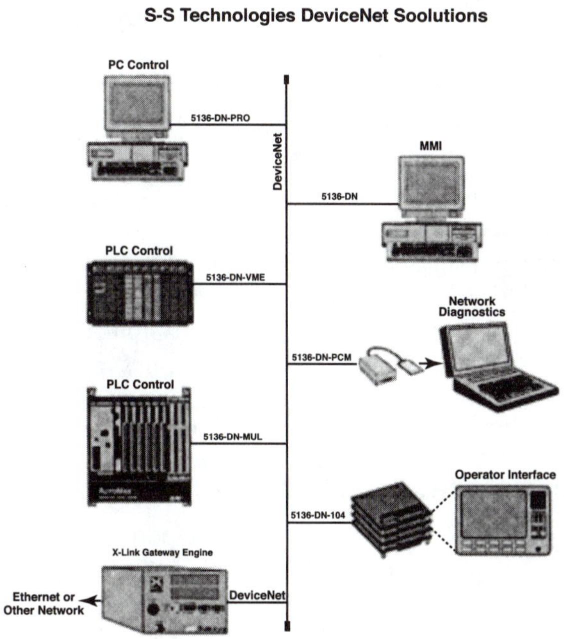

Figure 14–28 Example of a DeviceNET network. *(Courtesy S-S Technologies Inc.)*

NETWORK SIZE	UP TO 64 NODES
NETWORK LENGTH	500 METERS AT 125 KBPS 250 METERS AT 250 KBPS 100 METERS AT 500 KBPS
TOPOLOGY	LINEAR (TRUNKLINE/DROPLINE). POWER AND SIGNAL ON THE SAME CABLE.
PHYSICAL MEDIA	TWISTED PAIR FOR SIGNAL AND POWER
DATA PACKETS	0 TO 8 BYTES
METHOD	PEER-TO-PEER WITH MULTICAST (ONE-TO-MANY): MASTER/SLAVE SPECIAL CASE AND MULTI-MASTER: POLLED OR CHANGE-OF-STATE.
BUS ACCESS	CARRIER SENSE/MULTIPLE ACCESS

Figure 14–29 DeviceNET characteristics.

Higher priority data gets the right of way. This provides inherent peer-to-peer capability. If two or more nodes try to access the network simultaneously, a bit-wise nondestructive arbitration mechanism resolves the potential conflict with no loss of data or bandwidth. By comparison, Ethernet uses collision detectors, which result in loss of data and bandwidth because both nodes have to back off and re-send their data.

Peer-to-peer data exchange means that any DeviceNET product can produce and consume messages. Master/slave operation is defined as a proper subset of peer-to-peer. A DeviceNET product may behave as a client or a server or both. A DeviceNET network may have up to sixty-four media access control identifiers or MAC IDs (node addresses). Each node can support an infinite number of I/O. Typical I/O counts for pneumatic valve actuators are sixteen or thirty-two.

Change-of-State and Cyclic Transmission

Change-of-state means that a device reports its data only when the data changes. To be sure the consuming device knows that the producer is still alive and active, DeviceNET provides an adjustable, background heartbeat rate. Devices send data whenever their data changes or the heartbeat timer expires. This keeps the connection alive and lets the consumer know that the data source is still alive and active. The heartbeat timing prevents talkative nodes from dominating the network. The device generates the heartbeat, which frees the controller from having to send a nuisance request periodically to make sure that the device is still there.

The cyclic transmission method can reduce unnecessary traffic on a network. For example, instead of a slow-changing temperature being scanned many times every second, devices can be set up to report their data on a regular basis that is adequate to monitor their change.

Device Profiles

The DeviceNET specification promotes interoperability of devices by specifying standard device models. All devices of the same model must support common identity and communication status data. Device-specific data is contained in device profiles that are defined for various types of devices. This ensures that devices such as push buttons, valves, starters, and so on, from multiple manufacturers that comply with the device-type profile are interchangeable. Manufacturers can offer extended capabilities for their devices, but the base functionality must be the same.

Control-Level Communications

Communications at the control level take place between control-type devices. Control devices include PLCs, robots, CNC controllers, and so on.

Control-Level Networking

Standardization of control-level communications is necessary so that control equipment from various manufacturers can communicate. One of the emerging standards for control-level networking is ControlNet™.

ControlNet™

ControlNet™ is a high-speed, deterministic network developed by Allen-Bradley for the transmission of critical automation and control information. It has very high throughput (5 Mbit/s for I/O), PLC interlocking, and peer-to-peer messaging and programming. This network can perform multiple functions. Multiple PLCs, human/machine interfaces, network access by PCs for programming and troubleshooting from any node can all be performed on the network. The capability of ControlNet™ to perform all these tasks can reduce the need for multiple networks for integration. Figure 14–30 shows an example of integrating various networks.

ControlNet™ is compatible with Rockwell Automation PLCs, I/O, and software. ControlNet™ supports bus, star, or tree topologies. It utilizes RG6 cable, which is identical to cable television cable. This means that taps, cable, and connectors are all easy to obtain and very reasonable in price. ControlNet™ also has a dual-media option (see Figure 14–31). This means that two separate cables can be installed to guard against failures such as cut cables, loose connectors, or noise.

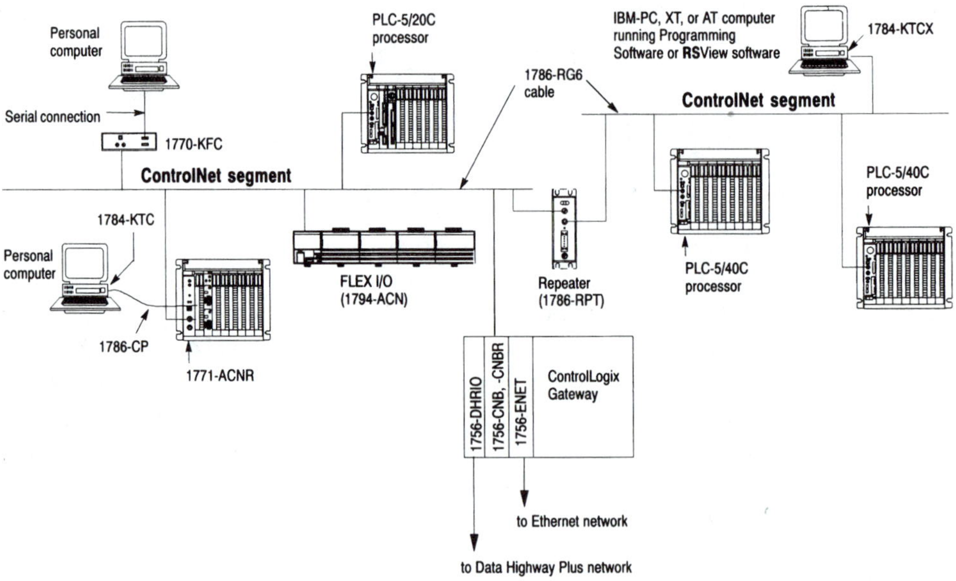

Figure 14–30 Example of integrating various networks. *(Courtesy Rockwell Automation Inc.)*

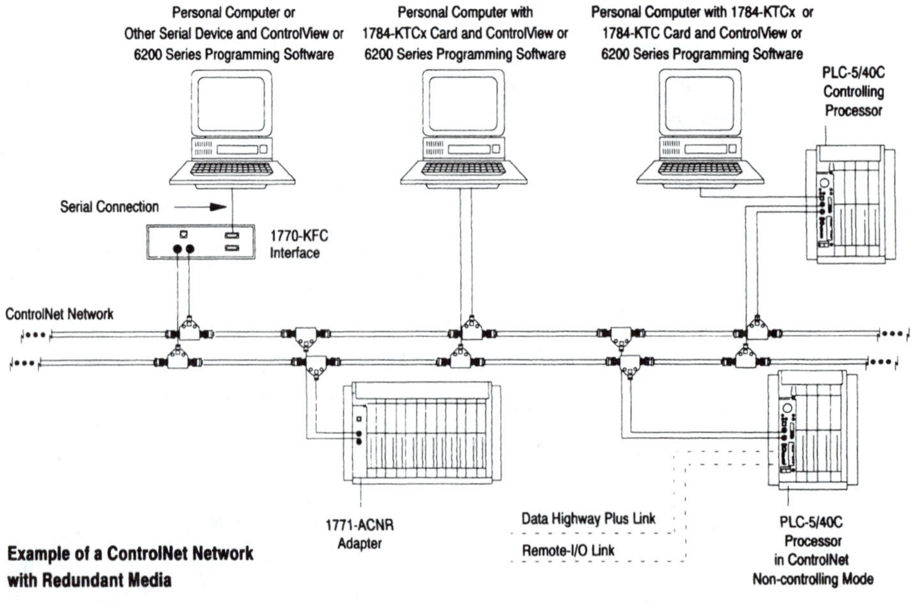

Figure 14–31 Redundant cabling. *(Courtesy Rockwell Automation Inc.)*

The figure also shows some of the types of computers and controllers that can be integrated, as well as the wiring.

Ethernet is also making inroads in control-level networking. Ethernet is the most prevalent personal computer networking standard. It is based on CSMA/CD access methods and bus structure. Increased speeds in Ethernet may make it an even more popular choice. Ethernet is already widely used for computer networking and it is relatively inexpensive to implement. Most controllers can now be purchased with Ethernet capability.

Enterprise Level Communications

The enterpise level seems to have settled on Ethernet for its standard; it meets the needs for the enterprise level. Ethernet will be covered in the next chapter, but robots, PLCs, and many other devices are available with Ethernet capability, which means that they can be plugged into the plant's Ethernet network for communication.

The rapid advancements in computers and networking are creating vast changes in industrial communication hierarchies. Figure 14–32 shows an example of what is occurring in networking. Note that some of the field devices are connected to a bus-type network in Figure 14–33.

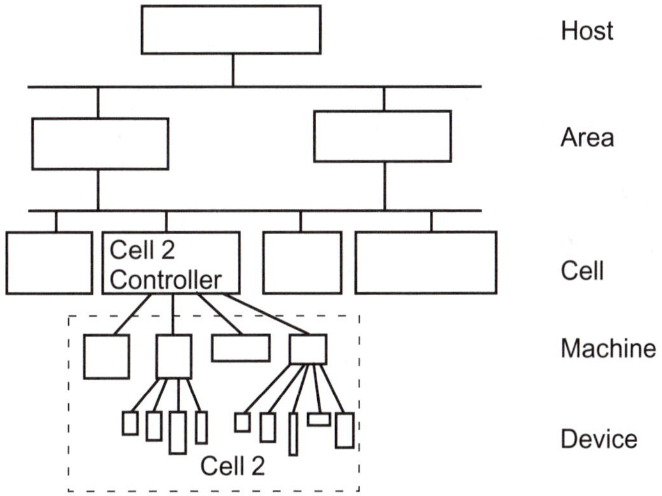

Figure 14–32 Typical host-level control. Also called the enterprise level.

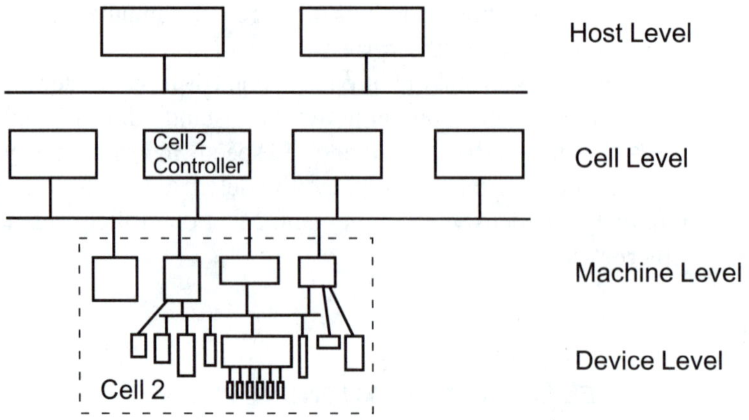

Figure 14–33 Example of networking.

QUESTIONS

1. Describe the term *serial communications.*
2. Describe the term *asynchronous.*
3. Explain the term *SCADA.*
4. Complete the following table.

	Speed	Length Limit	Balanced or Unbalanced	Special Characteristics
RS-232				
RS-422				
RS-423				
RS-485				

5. Explain industrial networks and why they are becoming more prevalent.
6. What are some of the differences between a device bus and a process bus?
7. What is a field device?
8. Describe how the capabilities of field devices are changing.
9. Where are device buses appropriate?
10. Where are process buses appropriate?
11. Why is the CAN system so important?
12. Describe the DeviceNET standard.
13. Describe the purpose of Fieldbus.
14. Describe the purpose of ControlNet™.

chapter 15

Communications, ControlLogix, and DeviceNet

This chapter will examine some of the newer technologies in communications and control. Topics include the three levels of industrial communications (information networks, control networks, and device networks), ControlLogix, and DeviceNet.

OBJECTIVES

Upon completion of this chapter, you will be able to:

1. Describe the three levels of industrial communications.
2. Describe the purpose and capabilities of ControlLogix.
3. Describe the purpose and capabilities of DeviceNet.
4. Describe appropriate uses for ControlLogix and DeviceNet.

INDUSTRIAL COMMUNICATIONS

The three main types of industrial networks are information networks, control networks, and device networks. Figure 15–1 shows the hierarchy for the three typical networks.

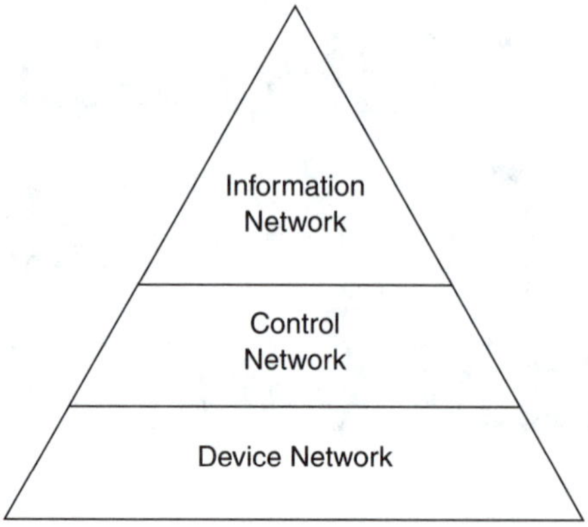

Figure 15–1 Industrial communications hierarchy.

Information Level Network

The top level is the information network layer. This is the enterprise's computer network. It typically consists of Ethernet local area networks (LANs) that can exchange information among the enterprise's computers. It also serves as a link between the plant floor and the manufacturing software systems.

Ethernet

Ethernet is the most widely used protocol at the information systems level of communications. Computer networking has become increasingly important as enterprises have integrated their computers and systems. Computers are linked to share enterprise data, e-mail, and other resources such as printers and data storage. These computers are connected through LANs.

Ethernet is popular because it is fast, inexpensive, and easy to install. The Institute for Electrical and Electronic Engineers (IEEE) defined a standard for Ethernet. It is known as IEEE Standard 802.3. Ethernet is the most prevalent personal computer networking standard. It is based on the carrier sense, multiple access/collision detect (CSMA/CD) access method. A computer must have some way to gain access to a network when it needs to communicate. If CSMA/CD is used, the computer "listens" to the access line (carrier) to see if it is being used or available (carrier sense). If available, the computer can talk. This, by the way, is similar to how telephone service once operated. Neighbors, especially in country areas, would share a line, called a party line. If you wanted to use the phone, you would pick it up and see if anyone else was talking on it. You were supposed to hang up if it was being used. If no one was on the line, you could call someone. (Party lines also provided entertainment for nosy neighbors.) CSMA/CD is similar. If the line is available, the computer can talk. Sometimes, however, the line is available, but two or more computers may try to talk at the same time, which can lead to a data collision. This explains why collision detection (CD) is necessary. The computers that had the collision of data would sense the collision and then wait to talk again. The wait is controlled by algorithms so that they do not attempt to talk again at exactly the same time. These collisions are minor and do not affect speed unless the line utilization is very high.

Ethernet also employs a bus structure, which makes for easy connections in a manufacturing or office environment. It is also easy to add additional length to a bus network as a company grows or expands a network.

Fast Ethernet A faster standard has also been established for Ethernet. IEEE has established the Fast Ethernet standard (IEEE 802.3u). This standard raises the Ethernet speed limit from 10 Megabits per second (Mbps) to 100 Mbps with only minor changes to the existing cable structure. There are three types of Fast Ethernet: 100BASE-TX for use with level 5 UTP cable; 100BASE-FX for use with fiber-optic cable; and 100BASE-T4, which utilizes an extra two wires for use with level 3 UTP cable. The 100BASE-TX standard has become the most popular due to its close compatibility with the 10BASE-T Ethernet standard. An enterprise's communications networks will have a combination of Ethernet and Fast Ethernet. The future promises even faster Ethernet standards.

LANs A network is any collection of computers that communicate with one another over a shared network. LANs can be small and link a few computers, or they can be huge and link hundreds of computers. An enterprise will often have many smaller LANs that are also interconnected. The standardization of protocols is what enables enterprises to make good use of LANs. This has certainly been the case with Ethernet. As more and more enterprises decide to use Ethernet, the prices of hardware and software are dramatically reduced. Many midsize and large organizations have sites that are not in close physical proximity. These enterprises often utilize wide area networks (WANs), so that the LANs in each site can communicate with the whole organization.

WANs WANs can integrate LANs that are geographically separate. They connect the different LANs using dedicated leased phone lines, dial-up phone lines, data carrier services, or even satellite links. Wide-area networking can be as simple as a modem and remote-access server into which employees can dial, or it can be hundreds of branches linked over any geographic area.

Control Level Network

The control level network is the level at which the controllers reside. This level includes the PLCs, personal computers, operator I/O devices, drives, motion controllers, robot controllers, and vision systems. Communication at this level is mainly between the devices; for example, we can share data among PLCs. Communication at this level must be dependable. Rockwell offers several protocols for this level. Data Highway+ (DH+) is one standard for connecting controllers at this level. ControlNet is another popular protocol at this level for communicating between controllers.

ControlNet

ControlNet was developed to provide a real-time, control-layer network to provide high-speed transport of both time-critical I/O data and messaging data, including upload/download of programming and configuration data and peer-to-peer messaging, on a single physical media link. ControlNet's access method is deterministic. Remember that Ethernet employs CSMA/CD, which is not deterministic because of collisions. ControlNet's access method is deterministic and repeatable. ControlNet's 5 Mbps control and data capabilities enhance I/O performance and peer-to-peer communications. ControlNet is highly deterministic and repeatable—critical requirements for manufacturing control and communication. Determinism is the ability to predict reliably when data will be delivered; repeatability ensures that transmit times are constant and unaffected by devices connecting to, or leaving, the network.

ControlNet's access method utilizes the producer/consumer model. ControlNet allows multiple controllers to control I/O on the same wire. Many other networks allow only one master controller on the wire. ControlNet also allows multicast of both inputs and peer-to-peer data, thus reducing traffic on the wire and increasing system performance.

ControlNet Characteristics

High bandwidth for I/O, peer-to-peer messaging, and programming.

Network access from any node.

Deterministic, repeatable performance for both discrete and process applications.

Multiple controllers controlling I/O on the same link.

Simple installation requiring no special tools to install or tune the network.

Network access from any node.

Flexibility in topology options (bus, tree, star) and media types (coaxial, fiber, other).

Device Level Network

The device level network is the lowest level. It consists of plant-floor devices such as sensors, valves, and drives. Many device-type networks are available. One of the more common is DeviceNet. It is like a LAN for plant-floor devices. A device level network can connect many types of plant-floor devices.

Device level networks are becoming prevalent in industrial control. They are sometimes called field buses or industrial buses. These networks have many similarities to the conventional office computer network. Office networks allow many computers to communicate without being connected directly to each other. Each computer has only one connection to the network. Imagine if every computer had to be connected physically to every other computer. Networks therefore minimize the amount of wiring that needs to be done. Networks also allow devices such as printers to be shared. Industrial networks share some of the same advantages.

Imagine a complex automated machine that has hundreds of I/O devices. Now imagine the time and expense of connecting each and every I/O device back to the controller. Conduit has to be bent and mounted, and hundreds of wires need to be run. This can require hundreds of wires (or more) in a complex system and can involve long runs of wire. Typical devices such as sensors and actuators now require that two or more wires be connected at the point of use and then run to an I/O card at the control device. Industrial networks eliminate this need so only a single twisted-pair wire bus (generally two to four wires) needs to be used for communication. All devices can then connect directly to the bus. This could also be called distributed I/O.

Figure 15–2 shows an example of two valve manifolds. The top one has a DeviceNet module. This module can be plugged into a DeviceNet bus. The bottom one is not a bus-type and requires many connections through a 25-pin connector. Figure 15–3 shows two communications modules that can be used with a valve manifold.

Multiple devices can even share one connection with the bus in some cases. Several manufacturers have developed I/O blocks that allow multiple I/O points to share one connection to the bus. One example is a valve manifold. This particular unit could have up to thirty-two valves and up to thirty-two I/O with only two connections. Wiring and installation costs are drastically reduced when industrial networks are used.

The cost of field bus devices is higher than that of conventional devices. The field bus devices gain a cost advantage when one considers the labor cost of installation, maintenance, and troubleshooting. The cost benefit received from using distributed I/O is really in the labor saved during installation and start-up. There is also a material savings when one considers the wire that does not have to be run—there is a tremendous savings in labor when only a fraction of the number of connections

Figure 15–2 Two valve manifolds. The one on top is DeviceNet capable. *(Courtesy Parker Hannefin Inc.)*

need to be made and only a fraction of the wire needs to be run. A field bus system is also much easier to troubleshoot than a conventional system. If a problem exists, only one twisted-pair cable needs to be checked. A conventional system might require the technician to sort out hundreds of wires.

Field Devices

Sensors, valves, actuators, and starters are examples of I/O called field devices. The capabilities of field devices, as well as their ability to communicate, have increased rapidly. This has led to a need for a way to network them. Imagine what it would be like in your home if there was no electrical standard. Your television might use 100 volts at 50 cycles. Your refrigerator might use 220 volts at 60 cycles, and so on. Every device might use a different plug. It would be a nightmare to replace any de-

P2S-EA162BP16A
Head & Tail Set with
Profibus-DP®

P2S-EA162BD16A
Head & Tail Set with
DeviceNet"

Figure 15–3 Two communications modules that can be used with a valve manifold. The one on the left is a Profibus module and the one on the right is a DeviceNet module. *(Courtesy Parker Hannefin Inc.)*

vice with another brand because the wiring and outlets would need to be changed. This has been the case in terms of communications between industrial devices.

Simple industrial digital devices such as sensors can be interchanged. They typically operate on 24 volts or 110 volts, and their outputs are digital, so any brand can be used. In other words, we can replace a simple digital photo sensor from one manufacturer with a different brand because it is only an on/off signal.

Figure 15–4 shows a simple industrial network with several field devices attached. Note the I/O block (I/O concentrator) and the valve manifold. Both are used to connect multiple devices to a network.

Industrial applications often need analog information such as temperature and pressure. These devices typically convert the analog signal to a digital form and then need to communicate that information to a controller. They may also be able to pass other information such as piece counts, cycle times, and error codes. Devices such as drives have also gained capability. Parameters such as acceleration and other drive parameters can be sent by a computer or PLC to alter the way a drive operates. Every device manufacturer has had different communication protocols, until now, for that communication. This situation required a separate network for each brand of device.

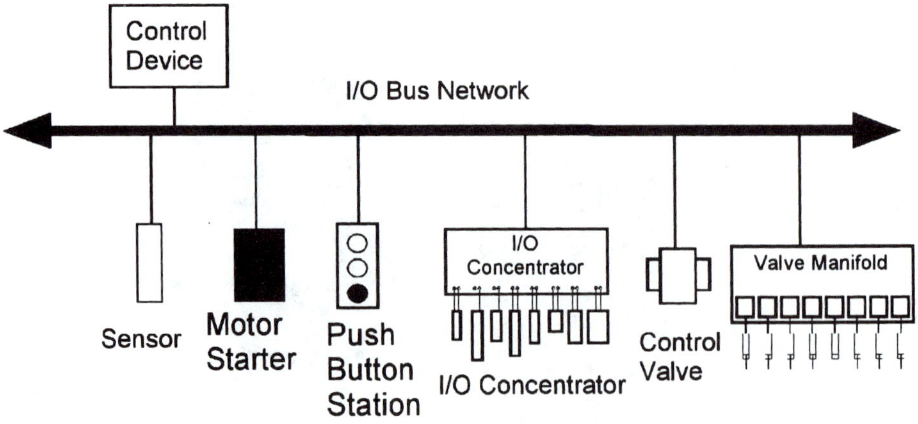

Figure 15–4 A simple industrial network.

Types of Industrial Buses

The two basic categories of industrial bus networks are device and process. Device-type buses are intended to handle the transmission of short messages, typically a few bytes in length. Most devices in a device bus are discrete, such as sensors, push buttons, and limit switches. Many discrete buses can also utilize some analog devices. They would typically be devices that require only a few bytes of information transmission, for example, temperature controllers, some motor drives, and thermocouples. Since the transmission packets are small, device buses can transmit data packets from many devices in the same amount of time as it would take to transmit one large packet of data on a process bus.

Industrial buses range from simple systems that can control discrete I/O (device-type bus) to buses that can be used to control a whole plant (process bus). Figure 15–5 shows the capabilities of several common industrial buses. The left of the graph begins with simple discrete I/O control and increases in capability to plant control. Block I/O would be devices like manifolds that would have several valves on one block, and only two to four wires would need to be connected to control all of the I/O. By using an industrial bus, only two to four wires would have to be run instead of a few dozen.

An example of a smart device would be a digital motor drive to which we could send parameters and commands serially. The peer level would be the capability for peer controllers to communicate with each other. The next levels are cell control, control of a line of cells, and overall plant control.

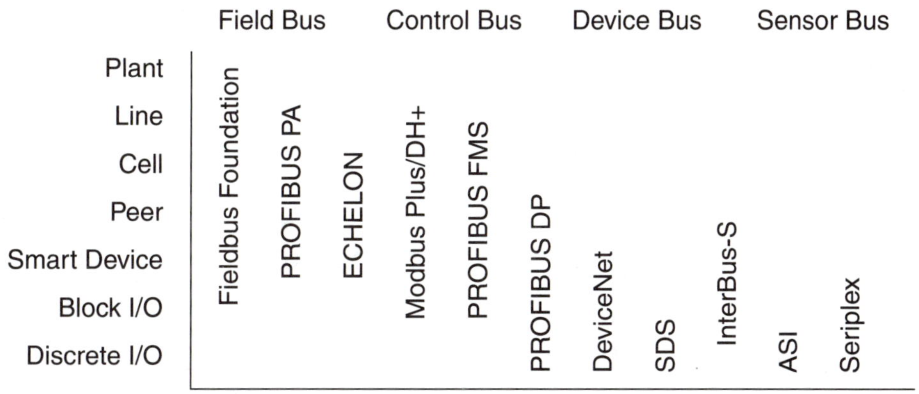

Figure 15–5 A comparison of common industrial buses.

Device Buses

Device buses have two categories: bit wide and byte wide. Byte-wide buses can transfer 50 or more bytes of data at a time. Bit-wide buses typically transfer 1 to 8 bits of information to and from simple discrete devices. Byte-type systems are excellent for higher level communication, and bit-type systems are best for simple, physical-level I/O devices such as sensors and actuators.

Process Buses

Process buses are capable of communicating several hundred bytes of data per transmission. Process buses are slower because of their large data packet size. Most analog control devices do not require fast response times. Process controllers typically are smart devices that control analog types of variables such as flow, concentration, and temperature. These processes are typically slow to respond. Process buses are used to transmit process parameters to process controllers. Most devices in a process bus network are analog.

This chapter will focus on the device bus type. DeviceNet will be covered in detail later in the chapter.

Example Communications System

Figure 15–6 shows an example of an industrial communications hierarchy, with three networks. There is the typical Ethernet LAN for the computers to communicate at the information systems level. Ethernet is the overwhelming choice at this level. The computers are networked with Ethernet. Rockwell Automation RSLinx

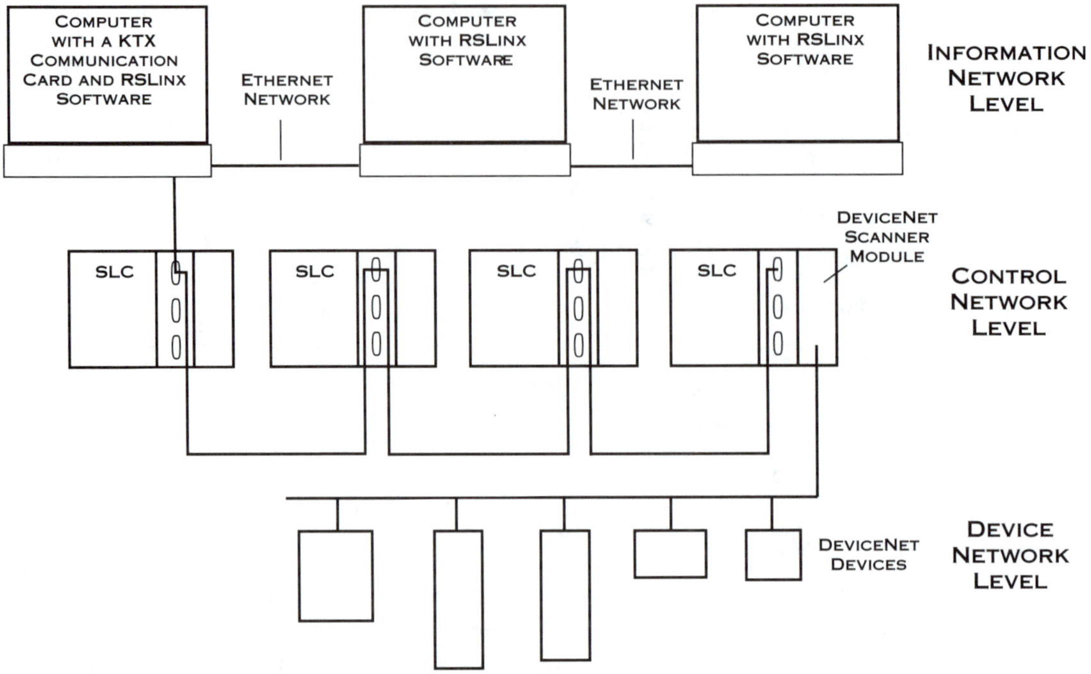

Figure 15–6 Example communications network.

software is running on each of the computers, so that they can communicate through the KTX card to any of the PLCs. If we have programming software on the computers, any of them can be used to program, upload/download, or monitor any PLC on the DH+ network. The KTX card in the one computer acts as a gateway so that the computer network can talk to the DH+ network. The use of RSLinx and the KTX card as a bridge between the Ethernet and DH+ networks allows any of the PLCs and computers to share information. For example, supervisory control and data acquisition (SCADA) software such as RSView32 or Wonderware can be used to gather information from floor-level devices and controllers and share it with the rest of the enterprise network.

At the controller level are several communications protocols available for communicating between controllers. Data Highway Plus (DH+) is one of them. DH+ can be used to program the controllers and to communicate between them. DH+ is easy to implement. It is simply a three-wire conductor that is daisy-chained between controllers. Daisy chaining means that the wire comes into a controller and back out to the next. Figure 15–6 shows an example of PLCs that are daisy chained.

Our example has a DH+ network for the PLCs to communicate with each other at the control level. DH+ enables data to be exchanged between controllers and even enables one controller to control another's I/O.

Note that ControlNet or Ethernet can be used instead of DH+ at this level to communicate between controllers. SLC 5/05s are Ethernet capable. They can be plugged into the Ethernet computer network as well. If we want to use ControlNet, we need to add a ControlNet module in each SLC. ControlNet is a controller communications standard developed by Rockwell Automation and then given to a user group to be an open standard.

DeviceNet was used at the device level. There is a DeviceNet scanner module in the fourth PLC. The DeviceNet network would network floor-level devices into the fourth PLC.

Next we will consider ControlLogix, another alternative to linking these levels and different protocols.

ControlLogix

Rockwell Automation ControlLogix (CL) is a new technology that can be used for control, communication, and integrating various types of networks.

Imagine a device that can be used for controlling any type of system or can be used as a gateway so that many different types of communications networks can communicate. Until now it has been relatively difficult to communicate between different protocols and systems. A device that can link different types of communications protocols so all can communicate with each other is called a gateway. Imagine a device that can do control or act as a communications gateway, or both, simultaneously.

Rockwell Automation developed a family of products to accomplish this task. The backbone of the system is a chassis that can house control modules, I/O modules, and communications modules.

A chassis with a control module and some I/O modules becomes a controller. A chassis with communications modules to integrate various protocols becomes a communications gateway to link various networks. A chassis with a control module and communications modules can act as a controller and a communications gateway.

ControlLogix is a modular platform for multiple types of control and communications. As a controller, the user can utilize a CL system for sequential, motion control, and process control in any combination. As a communications gateway, CL also enables multiple computers, PLCs, networks, and I/O communications to be integrated.

Figure 15–7 A typical CL system. Note that a CL system is modular so the user can build it to meet the application needs for communication and control. *(Courtesy Rockwell Automation Inc.)*

Control Choices

CL can integrate various types of control systems and communications systems into a seamless network of control (Figure 15–7). To make the task easier, there is a wide variety of software and hardware available.

A ControlLogix system can be a simple stand-alone controller. Programming is provided through RSLogix 5000. It is IEC 1131-3 compliant. IEC 1131-3 is an international standard for programming languages for PLCs. It allows the user to program in ladder logic, function block, and even symbolic programming with structures and arrays.

Figure 15–8 shows an example of a simple system. In this system a CL controller has been utilized with CL I/O modules. I/O would be wired directly to the modules. This system would look like a typical PLC system.

One or more controllers can be used in a chassis. Chassis sizes range from four to seventeen slots. Multiple controllers can read input values from all inputs. The controllers can address a large number of I/O points (128,000 digital maximum/4000 analog maximum).

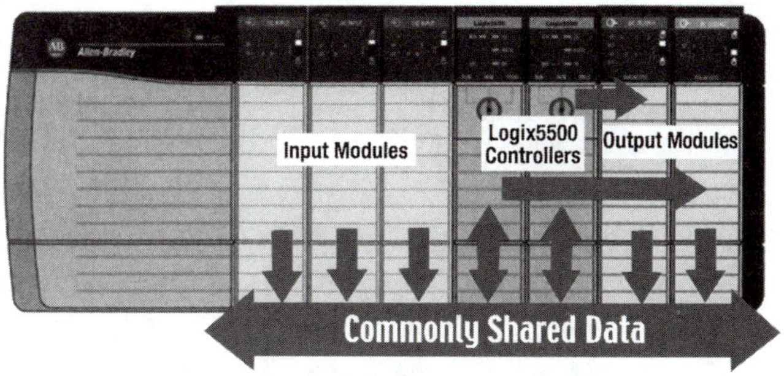

Figure 15–8 A simple CL system. *(Courtesy Rockwell Automation Inc.)*

I/O Modules for a CL Controller

A wide variety of I/O modules is available. AC and DC input and output modules feature advanced diagnostic, fusing, and isolation capability. I/O modules with diagnostics can provide diagnostics to the point level on the module. Modules with electronic fusing have internal fusing to prevent too much current from flowing through the module. Modules are also available with individually isolated I/O.

Another special-purpose I/O module is a configurable flow-meter module. It can be used to control metering applications for process control, and it can be used for high-speed frequency measurements for speed or rate control. The module has two outputs that can be triggered based on flow, frequency, or other states.

A high-speed counter module is also available; it can be used with devices that provide a pulsed output such as sensors or encoders.

A programmable limit switch module is also available. This type of module is common in packaging operations. The module takes its input from a resolver. It can provide control of packaging operations up to 1500 parts per minute. It is based on controlling outputs (as based on the rotation of the resolver) and can detect to within 1.08 degrees of rotation at 1800 RPM.

A variety of wiring options is available for the I/O modules. Removable terminal blocks, similar to typical PLC terminal blocks, for wiring directly to the module are available in screw-clamp or spring-clamp types.

Interface modules can also be used to separate the terminations from the actual I/O module. They are mounted on DIN rails and connected to the actual I/O module by a cable.

If one needs to create a distributed I/O system, FlexLogix can be used to control and integrate Flex I/O modules.

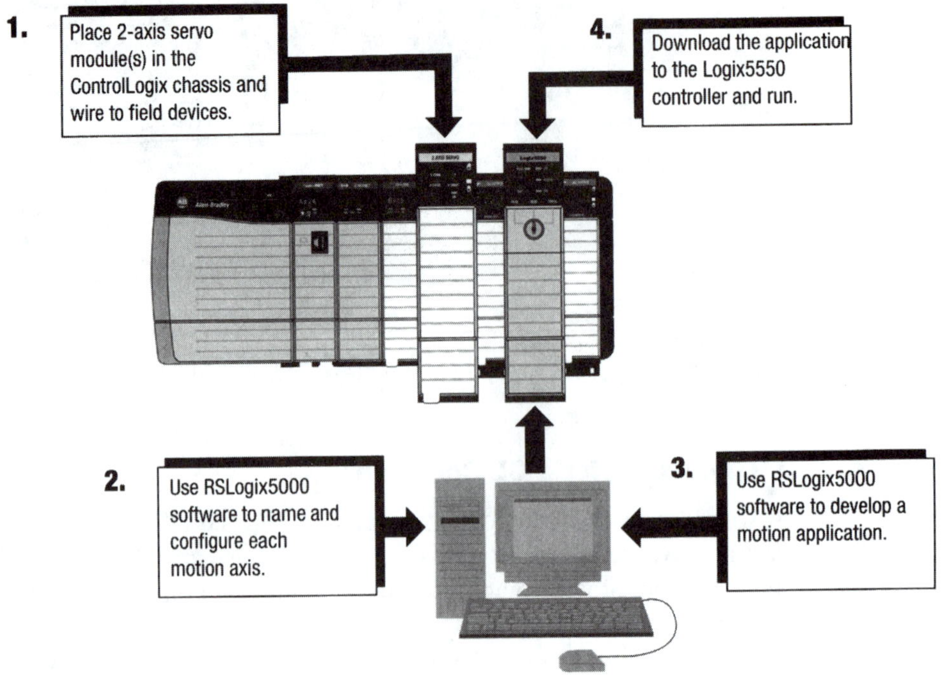

1.
Place 2-axis servo module(s) in the ControlLogix chassis and wire to field devices.
4.
Download the application to the Logix5550 controller and run.
2.
Use RSLogix5000 software to name and configure each motion axis.
3.
Use RSLogix5000 software to develop a motion application.

Figure 15–9 A typical CL motion-control system. *(Courtesy Rockwell Automation Inc.)*

Motion Control

Figure 15–9 shows how a typical motion-control system is implemented in a CL system. Modules are chosen and installed in the chassis. RSLogix 5000 software is then used to configure each axis of motion. RSLogix 5000 software is used to develop the motion application; the application is downloaded to the controller and can be run.

Figure 15–10 diagrams the development and operation of a motion application to control two axes. In the first panel, RSLogix 5000 is used to configure and program the application. The second panel shows the ControlLogix control module running the program logic and communicating with the servo module. The third panel shows the servo module that actually controls the motor drives. In this case, two drives are being controlled.

There is an international open standard for multi-axis motion-synchronized motion control, called the Serial Real-time Communication System (SERCOS). It is designed to be a protocol over a fiber-optic medium. Modules that employ the SERCOS standard are also available. Figure 15–11 shows an example of a SERCOS.

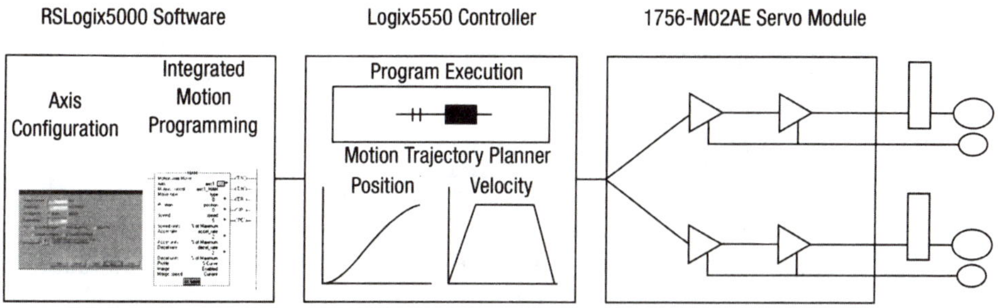

Figure 15–10 The development and operation of a motion application. *(Courtesy Rockwell Automation Inc.)*

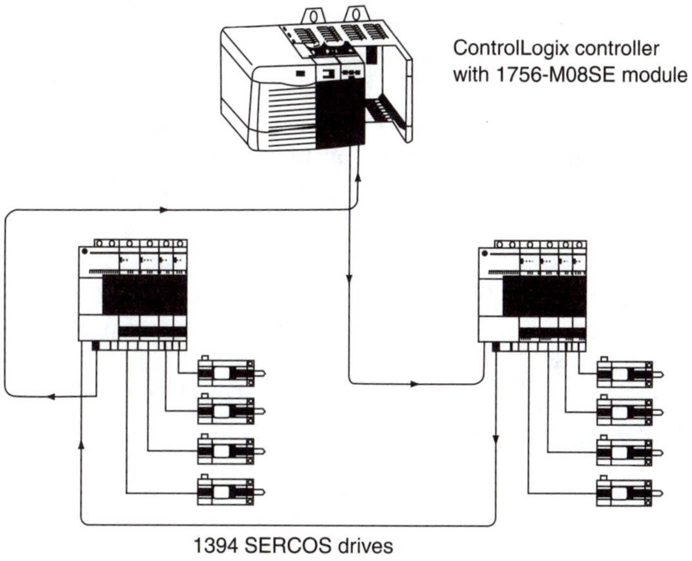

Figure 15–11 A SERCOS drive system. Note that the communications network is a ring. *(Courtesy Rockwell Automation Inc.)*

The SERCOS interface uses a ring topology with one master and multiple slaves (axes). The fiber-optic ring begins and ends at the master.

Analog modules are also available.

A CL system can also be used as a communications gateway. The user adds the communications modules to connect to the required networks, which allows the CL to communicate with other PLC-based networks in multiple protocols. If the system is used for communications, the controller is not required. Protocols include Ethernet, ControlNet, DeviceNet, and others.

Process Control

CL can be used for process control by utilizing ProcessLogix, a distributed control system (DCS) for process control. It can be used to control various types of process controllers and can utilize the FieldBus communications standard.

DeviceNet

We now take an in-depth look at DeviceNet. Device level networks must be installed and configured. It is likely that the technician will need to be actively involved with any device level network implementation.

Industrial networks are increasing in acceptance and use. DeviceNet is a communications link to connect industrial devices (such as limit switches, photo sensors, motor starters, drives, valves and valve manifolds, push buttons, bar-code readers, panel displays, and operator interfaces) to a network, and at the same time eliminate costly and time-consuming hand wiring.

The basis of many device level networks is the controller area network (CAN) chip. The CAN chip was developed for the European automotive industry. Figure 15–12 shows a conceptual automobile network.

Note that the network has a bus configuration. Imagine one communication wire that runs from the front of the car to the back. Imagine that each device is plugged into the bus. At the back of the car the taillights are plugged in. The door locks and power windows are also plugged in. The CAN chip handles the communications to other devices on the bus. The CAN chip is a smart communications front-end for

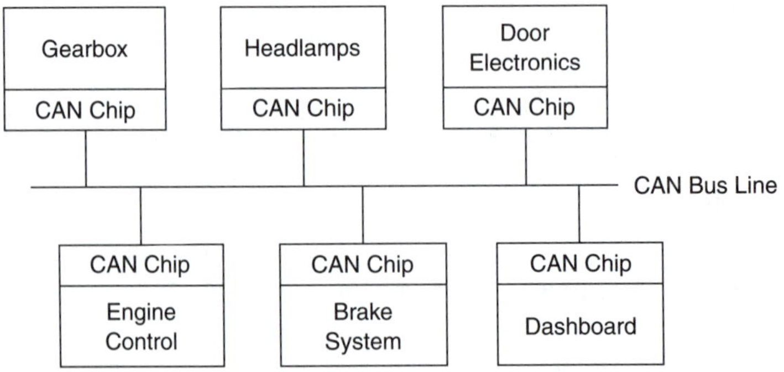

Figure 15–12 Automobile CAN network. *(Courtesy Rockwell Automation Inc.)*

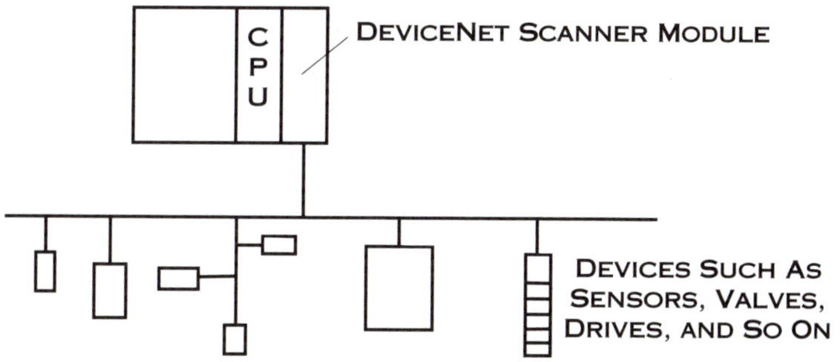

Figure 15–13 DeviceNet system. Note that only one cable is running to the scanner for all of the I/Os.

each device. The CAN chip is a receiver/transmitter. If the driver hits the brakes, the computer senses the brake pressure and sends a message over the bus to turn on the brakes and brake lights.

This technology offers several advantages. First, it simplifies the car wiring. The large and complex wiring harness is no longer needed. Each device simply plugs into the bus. Second, the chips are quite inexpensive—over 40 million CAN chips are sold annually, at less than $3 each. A network can also simplify troubleshooting. CAN chip systems can have powerful diagnostic capabilities. A bus system also simplifies installation of the wiring.

Rockwell Automation used the CAN chip when it developed DeviceNet. DeviceNet is an open standard invented by Rockwell Automation in 1993. The network specifications and protocol are open, which means that vendors are not required to purchase hardware, software, or licensing rights to connect devices to a system. The DeviceNet standard is maintained by the Open DeviceNet Vendors Association (ODVA). You can find them on the Web at www.odva.org. It is open to any manufacturer who wants to make and sell devices that are DeviceNet compatible. You can buy DeviceNet devices from any manufacturer and they will work in your DeviceNet system.

Figure 15–13 shows an example of a DeviceNet networked system. Note that there is only one cable running to the PLC, and that all devices plug into the trunk line. The PLC has a scanning module that acts as the brains of the network.

DeviceNet Components

Figure 15–14 shows the typical components for a DeviceNet cable system. The trunk line is the backbone of the system. It is the one cable to which all devices attach. In DeviceNet terminology, a device that has an address is called a node. There

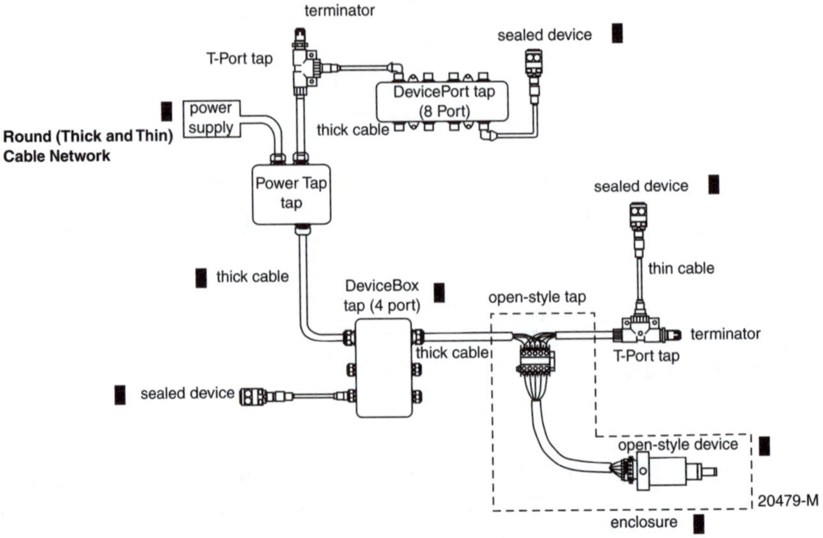

Figure 15–14 Typical components in a DeviceNet system. *(Courtesy Rockwell Automation Inc.)*

are two types of wire: thick and thin. Thick wire is used for the trunk line. The wire that connects devices to the trunk lines is called a drop line. Drop lines usually utilize thin cable.

The most important device in a DeviceNet network is a scanner (Figure 15–15). For our sample system, the scanner is a module installed in the PLC rack. The scanner acts as an interface between the PLC CPU and the inputs and outputs in our system. The scanner reads inputs and writes to outputs in the system. The scanner also monitors the devices for faults and downloads configuration data to each device. The scanner may also be equipped with a readout to help with troubleshooting the network. A Rockwell Automation scanner will flash a number that represents the error code and the number of the node that has the problem.

Of the devices, many types of sensors are available. Let's look at a Rockwell Series 9000 photo sensor (Figure 15–16). This sensor has several important features. Normally we would have to stock light-on and dark-on sensors for various applications. This sensor can be configured to be light-on or dark-on, so that we need to stock only one type. It also features some troubleshooting capability. The sensor has a margin output. Imagine that the sensor transmits light to a reflector and then back to the sensor where it is sensed. The margin output from the sensor will turn

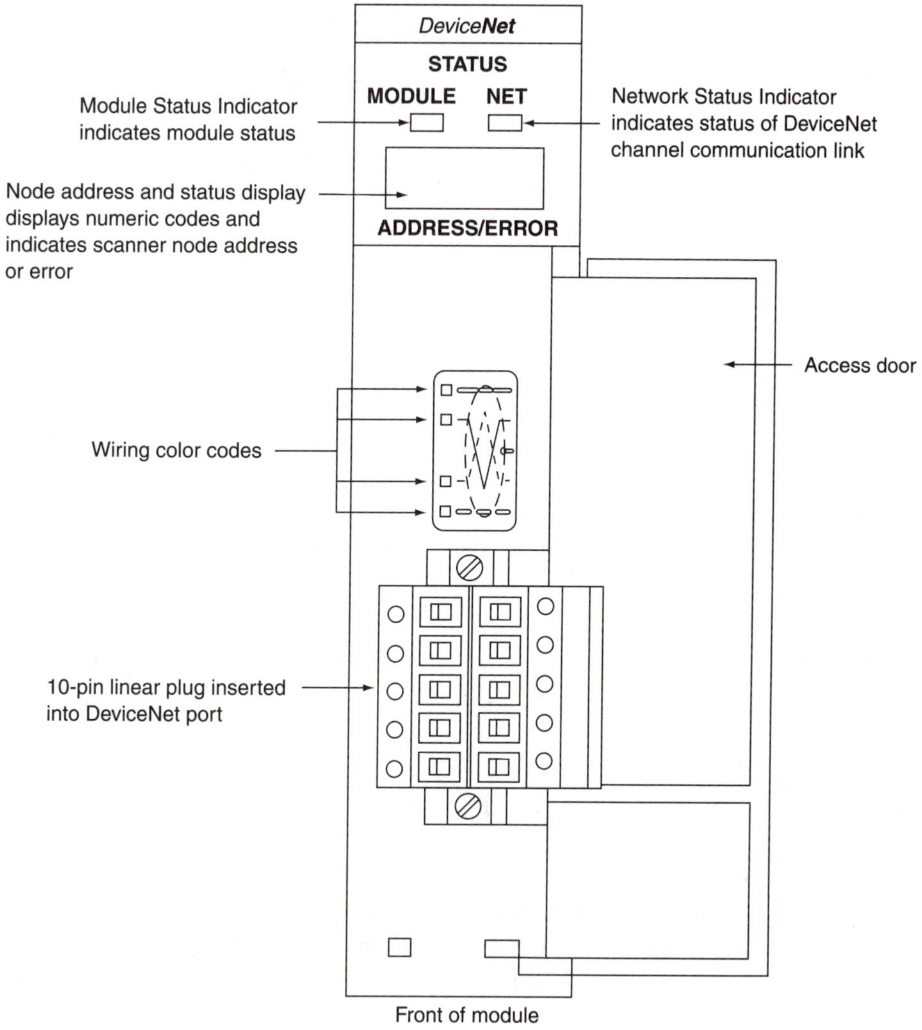

Figure 15–15 DeviceNet scanner. *(Courtesy Rockwell Automation Inc.)*

on if the reflector gets dirty and does not reflect the amount of light the sensor expects. It would also turn on if the reflector or sensor is bumped out of alignment. We can use the margin output in our ladder logic for troubleshooting if so desired. The device must be configured with a node address. DeviceNet can have sixty-four nodes, of which one is the scanner, which leaves sixty-three other potential node addresses. New devices come configured as node number 63. The sensor can be

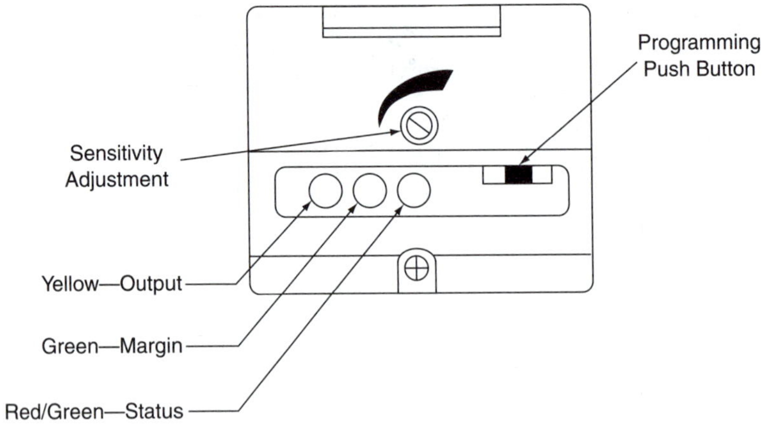

Figure 15–16 A Rockwell Series 9000 DeviceNet photo sensor. *(Courtesy Rockwell Automation Inc.)*

configured by switches on the sensor or by utilizing DeviceNet software and a PC. The sensor has LEDs on it for troubleshooting. Many manufacturers make DeviceNet sensors.

Another example of a DeviceNet device is a Rockwell Automation RediSTATION (Figure 15–17). This device features a start button, a stop button, and a red light. It requires one node address, but has two inputs (start and stop push buttons) and an output (red light).

Rockwell Automation also makes a device named DeviceLink for devices in a system that are not DeviceNet compatible. A DeviceLink can be used to connect a non-DeviceNet digital device to a DeviceNet network. The DeviceLink is connected to the network and given an address. The device (a limit switch, for example) is connected to the DeviceLink, which can transmit its state (on or off) to the scanner. Each DeviceLink requires one node address.

A motor drive is another example of a DeviceNet device. Figure 15–17 shows a Rockwell Automation 160 AC drive with DeviceNet capability. It is possible with this setup for the scanner to send speed, accell/decell, direction, and so on, to the drive. It is also possible for the scanner to read operating information from the drive.

Thus far we have looked at devices that basically have one device per node address. Let us examine a few that can have multiple devices with only one node address. Rockwell Automation Flex I/O is a way to connect many I/O devices and use

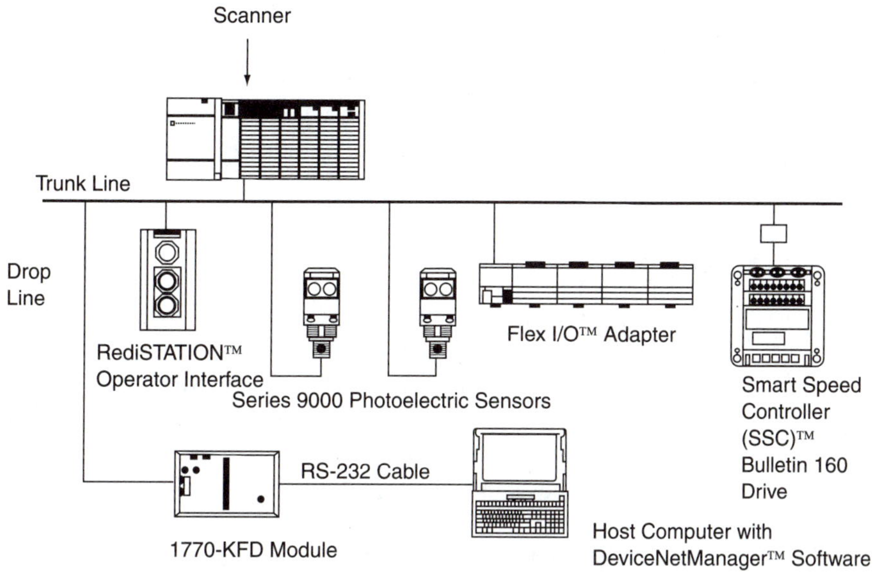

Figure 15–17 A simple conceptual DeviceNet network. *(Courtesy Rockwell Automation Inc.)*

only one node address. The I/O that is connected to the modules does not have to be DeviceNet capable. Any digital or analog I/O can be connected to the modules. Flex I/O modules are available for multiple inputs or outputs in digital or analog. Up to eight modules can be plugged together, meaning that up to 128 discrete devices or sixty-four analog channels can be connected to a DeviceNet network using a FLEX I/O system. All these devices can be connected and use only one node address. The modules are then attached to a DeviceNet communications module that is connected to the network. Flex I/O is appropriate for connecting non-DeviceNet devices to a DeviceNet network and also when a large number of devices may be concentrated in one area of a machine away from the controller.

Pneumatic components are also available for DeviceNet. Several manufacturers make multiple valve arrangements with a DeviceNet communications module. Thirty-two or more valves can be utilized with only one node address.

We also need a way to connect to a DeviceNet network with a computer. Rockwell utilizes a communications module between the computer and the network. A Rockwell 1770-KFD module plugs into the computer serial port. The other side of the KFD module plugs into a DeviceNet device for configuration or to the actual network. DeviceNet configuration software is run on the computer. Rockwell calls its software DeviceNet Manager. Other DeviceNet manufacturers have their own versions of software. DeviceNet Manager software can be used online or offline to build a network. The software can also be used to configure individual devices; for example, we can plug a Rockwell 9000 series photo sensor into the KFD module and utilize the computer and software to set the sensor's node address (remember that new devices come with a default address of 63). We can also set operation parameters such as light-on or dark-on.

Wiring

Let us take a closer look at wiring a DeviceNet system. Figure 15–18 shows a simple, generic network. Note that the trunk line is thick cable, which is capable of handling 8 A, although National Electric Code (NEC) only allows 4 A. Thin cable is normally used for drop lines. Thin cable can handle up to 3 A of power. Both thick and thin cable have power and communication lines, which is important because many devices can be powered directly from the DeviceNet thick or thin cable. Two wires supply 24 volts to the network; two other wires are used for communication. The ends of the trunk line must be terminated with 121 ohm resistors. Terminating resistors should never be used on drop lines.

DeviceNet is quite flexible in wiring layout. Figure 15–18 shows several possible topologies for wiring. Devices are shown as circles. The leftmost example shows a tree-type structure. Next we see a single device attached through a drop line to the trunk line, a daisy-chain topology, and then a bus-type topology.

There is a maximum allowable drop-line length of 20 feet (Figure 15–19), and a maximum cumulative drop-line length, which varies by the network speed. The slower the speed, the longer the line length allowed; the faster the network, the less drop-line length allowed. The maximum drop-line length is based on the cumula-

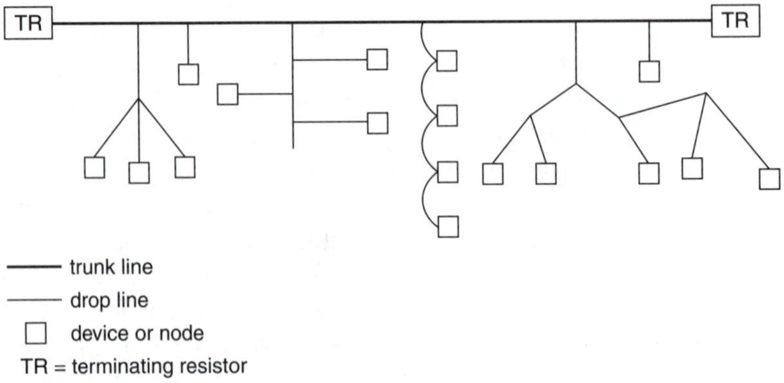

Figure 15–18 Wiring topologies. *(Courtesy Rockwell Automation Inc.)*

Network Speed	Maximum Drop Line Length	Maximum Cumulative Drop-Line Length
125 K baud	20 feet	512 feet
250 K baud		256 feet
500 K baud		128 feet

Figure 15–19 Allowable line lengths for DeviceNet network speeds.

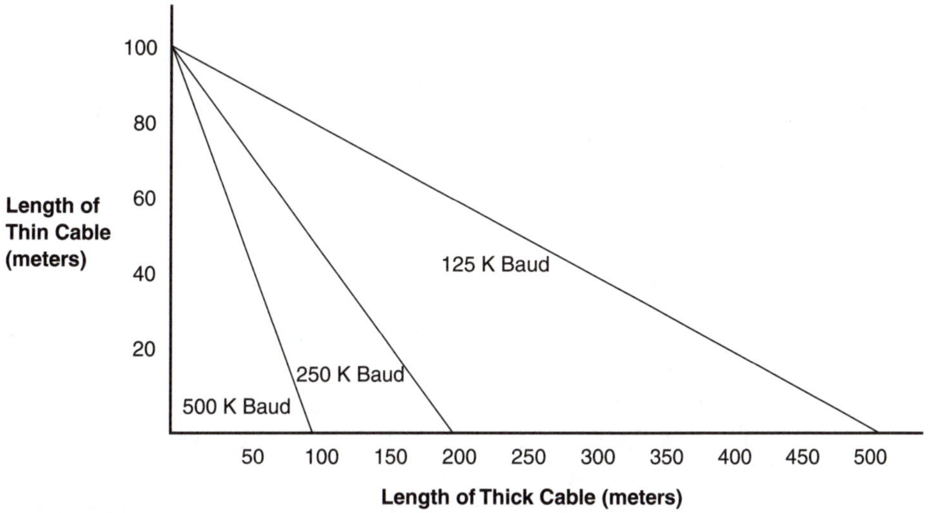

Figure 15–20 Graph of allowable line length. *(Courtesy Rockwell Automation Inc.)*

tive drop-line length. Figure 15–19 shows the maximum cumulative line length for the three network speeds allowed in a DeviceNet system.

Figure 15–20 shows approximate line lengths for combinations of thick and thin cable at the three speeds. For example, if we assume a speed of 125 K baud we could have a line length of only 300 feet if all thin cable were used. If all thick cable were used, we could have a line length of approximately 1500 feet; or we could choose any point along the graph of the 125 K speed. For example, assume we will use thick cable for the trunk line. The trunk needs to be 300 feet long. Look for the 300-foot point on the horizontal line of the graph. Look directly above the 300-foot point and you will see that approximately 270 feet of thin cable could still be used.

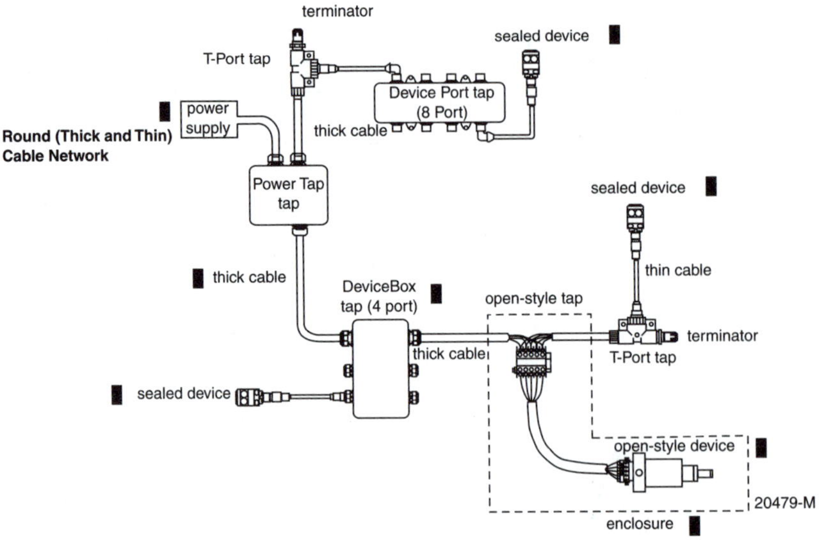

Figure 15–21 DeviceNet connectors. *(Courtesy Rockwell Automation Inc.)*

Connections

Figure 15–21 shows examples of connection devices that can be used. The trunk line is attached to the left and right sides of the T-port tap. The drop line would be connected to the third port of the T-port tap (Figure 15–22). These are screw-type connections. A DevicePort tap (Figure 15–23) can be used to connect up to seven devices to the trunk line. The trunk line is connected to one port on the DevicePort and up to six other devices can be connected to the other ports. These are also screw-type connections. A DeviceBox tap is very similar except these are compression-type fittings. The top is removed from the box and the wires are individually connected to terminals (Figure 15–24). The trunk line (thick cable) is attached to the larger compression fittings on the top left and top right, and up to four devices are connected through the four compression-type fittings (Figure 15–25). The compression fittings grip the cord and also seal the cable entry. DeviceBox taps are available in two-port, four-port, and eight-port configurations.

The third type of connector (Figure 15–26) is an open-style connector. Three lines can be connected with this device.

A box called a PowerTap is used to connect 24 volt DC power to the network. Two lines are used for the trunk line (compression style) and the third is used to run a power line into the box. There are two fuses in the PowerTap—one for the left side of the network and the other for the right side of the network. One can be re-

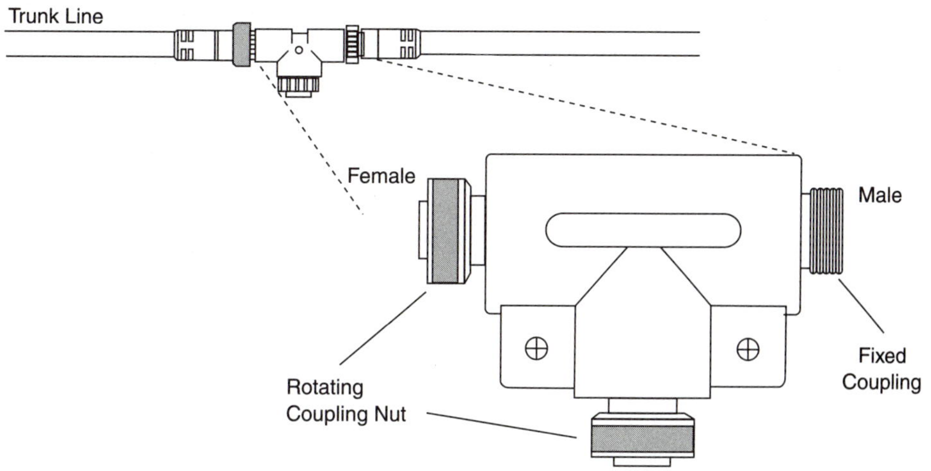

Figure 15–22 A T-port tap. *(Courtesy Rockwell Automation Inc.)*

Figure 15–23 A DevicePort tap. *(Courtesy Rockwell Automation Inc.)*

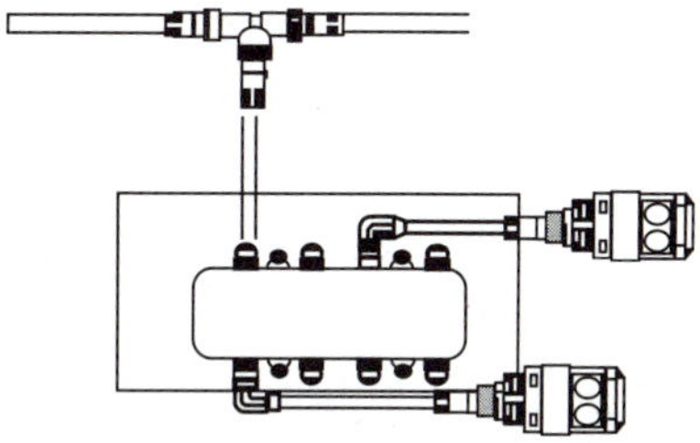

moved so that the power supply powers only one side of the network (you will see why this might be done later). Even if the fuse is removed, the communications lines run to both sides of the network. The fuse is just used to protect the 24-volt power supply.

Attaching a Computer to a DeviceNet Network

Figure 15–27 shows a typical network with a computer attached. Note the Rockwell Automation KFD module. The purpose of this module is to act as an interface between the serial port of a computer and the DeviceNet protocol of the network. Note that power must be supplied to the module through the power supply connector on the side or directly from a DeviceNet network. The switch on the side is used to choose the proper source of power.

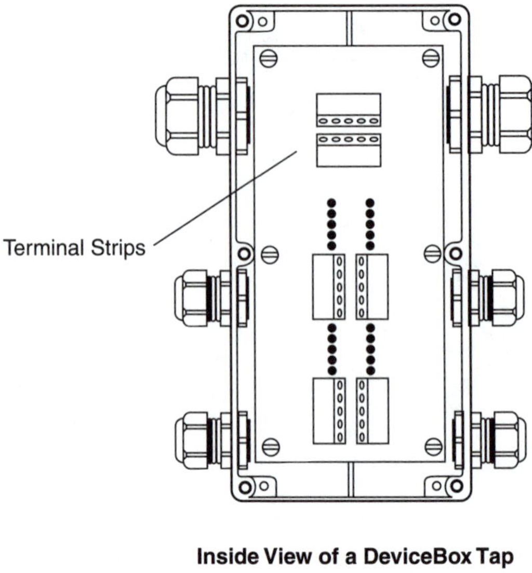

Figure 15–24 An inside view of a DeviceBox tap. *(Courtesy Rockwell Automation Inc.)*

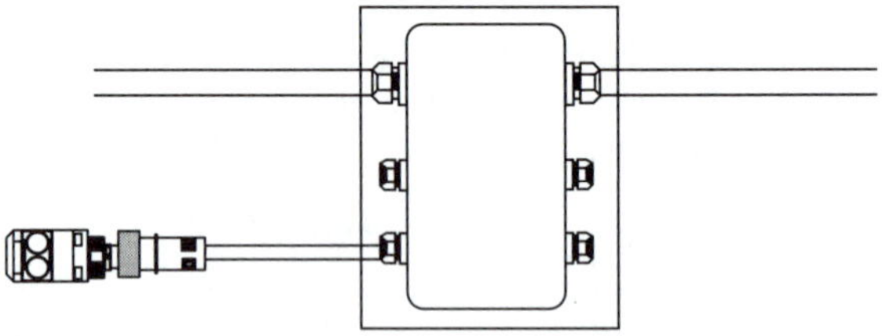

Figure 15–25 A DeviceBox tap. *(Courtesy Rockwell Automation Inc.)*

Figure 15–26 An open-style connector. *(Courtesy Rockwell Automation Inc.)*

KFD modules have three LEDs for troubleshooting (Figure 15–28). The RS232 status LED is used to troubleshoot the serial connection between the module and the computer. The module status LED is used to troubleshoot the module. The network status indicator is used to troubleshoot the DeviceNet network. Figures 15–29 through 15–31 show how the LEDs can be used for troubleshooting.

Supplying Power to the Network

Remember that a DeviceNet network handles communications, but it also provides the 24-volt DC power to operate many devices. It is very important to design the system so that current limits are not exceeded. One important consideration is that

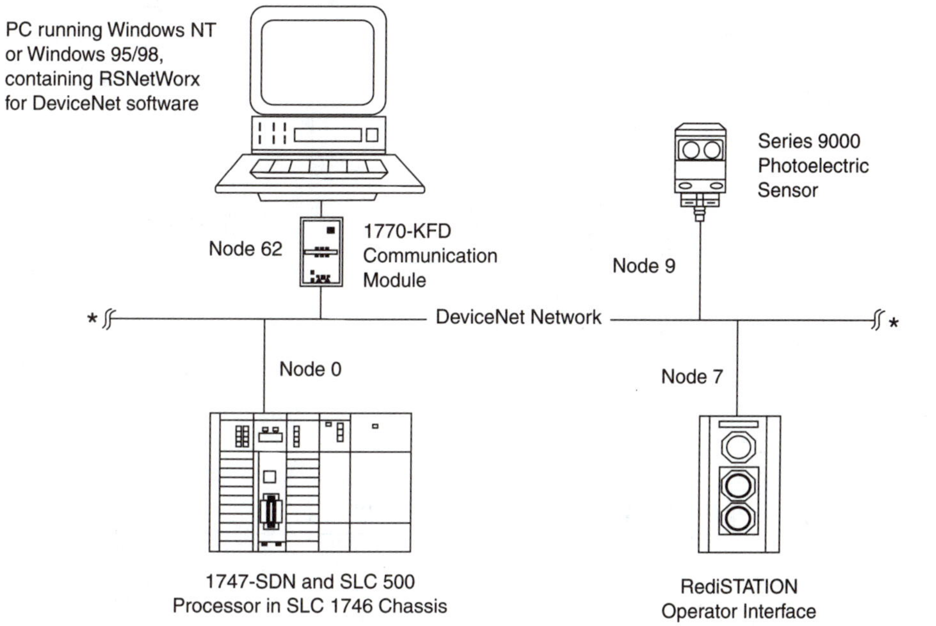

Figure 15–27 Typical computer connection. *(Courtesy Rockwell Automation Inc.)*

Figure 15–28 KFD module and troubleshooting status indicators. *(Courtesy Rockwell Automation Inc.)*

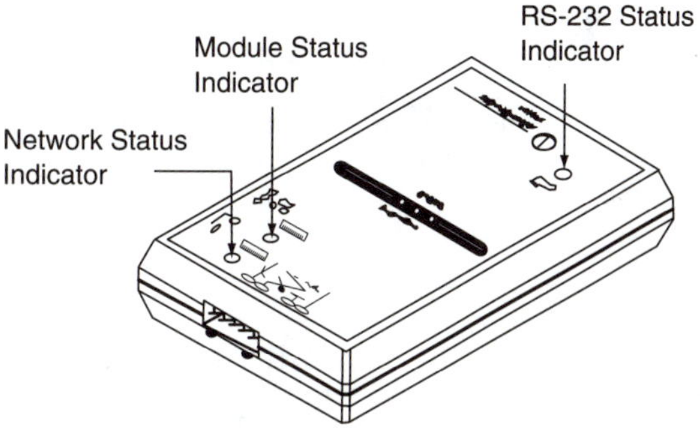

NEC does not allow over 4 A on the trunk line. Drop lines also limit the amount of allowable current. Figure 15–32 shows the current limits for various drop-line lengths. Note that the longer the drop line, the lower the current limit.

Figure 15–33 is a chart that can be used to determine maximum currents for given network lengths. For example, if your network is 525 feet long, the current limit would be approximately 1.9 A. Note that any current on the chart above 4 A is not allowed under NEC guidelines.

If the RS232 Status Indicator Is:	Then:
Off	No activity—link OK
Flickering green	Activity—link OK
Solid red	Link failed—critical fault
Flashing red	Link failed—noncritical fault

Figure 15–29 RS232 status indicator conditions.

If the Module Status Indicator Is:	Then:
Off	No power
Solid green	Device OK
Flashing green	Not configured
Solid red	Critical fault
Flashing red	Noncritical fault

Figure 15–30 KFD module status indicator conditions.

If the Network Status Indicator Is:	Then:
Off	Offline
Solid green	Online—communicating
Flashing green	Online
Solid red	Link failed—critical fault

Figure 15–31 Network status indicators.

Drop-Line Length	Maximum Allowable Current
5 feet	3 A
6.6 feet	2 A
10 feet	1.5 A
15 feet	1 A
20 feet	0.75 A

Figure 15–32 Allowable drop-line current for various lengths.

Let us look at a few examples of power supply calculation. Figure 15–34 shows a simple system. The network is 450 feet long. If we add the device currents, we find that the total current draw is 2.0 A. Our network length is 450 feet. (If we go to the chart in Figure 15–33, we see that we are allowed approximately 2.1 A.) Since our total draw is 2.0 A, we are within the limits.

Note that device current is based on how much current the device will draw from the DeviceNet network. Some devices require external power. A drive, for example, will require an external supply and may draw very little current from the network. Base your current calculations on how much the devices draw from the network.

Let us consider another example (Figure 15–35). In this case, the power supply has been placed in the middle of the network. There is a distinct advantage to putting the power supply somewhere in the middle instead of on the end of the network. In this case we have 2.5 A total current on the left leg of the network and 1.5 A of current on the right. Consider the left side first. The length of the left side of the network is 450 feet. If we look at Figure 15–33, we see that for 325 feet we are allowed approximately 2.9 A. Our total draw for this side of the network is 2.5 A. The actual draw is less than the allowable draw, so the left side is within limits. The total current draw for the system is under 4 A, so this system is within NEC guidelines.

The right side of the network is 550 feet long. (The allowable current for a 550-foot length from Figure 15–33 is approximately 1.8 A.) The actual current is 1.25 A, so the current draw for each side is within the limits. The total draw for the system is 2.5 A + 1.25 A, or 3.75 A. The total of 3.75 A does not exceed the code of 4 A, so the whole network is within specifications.

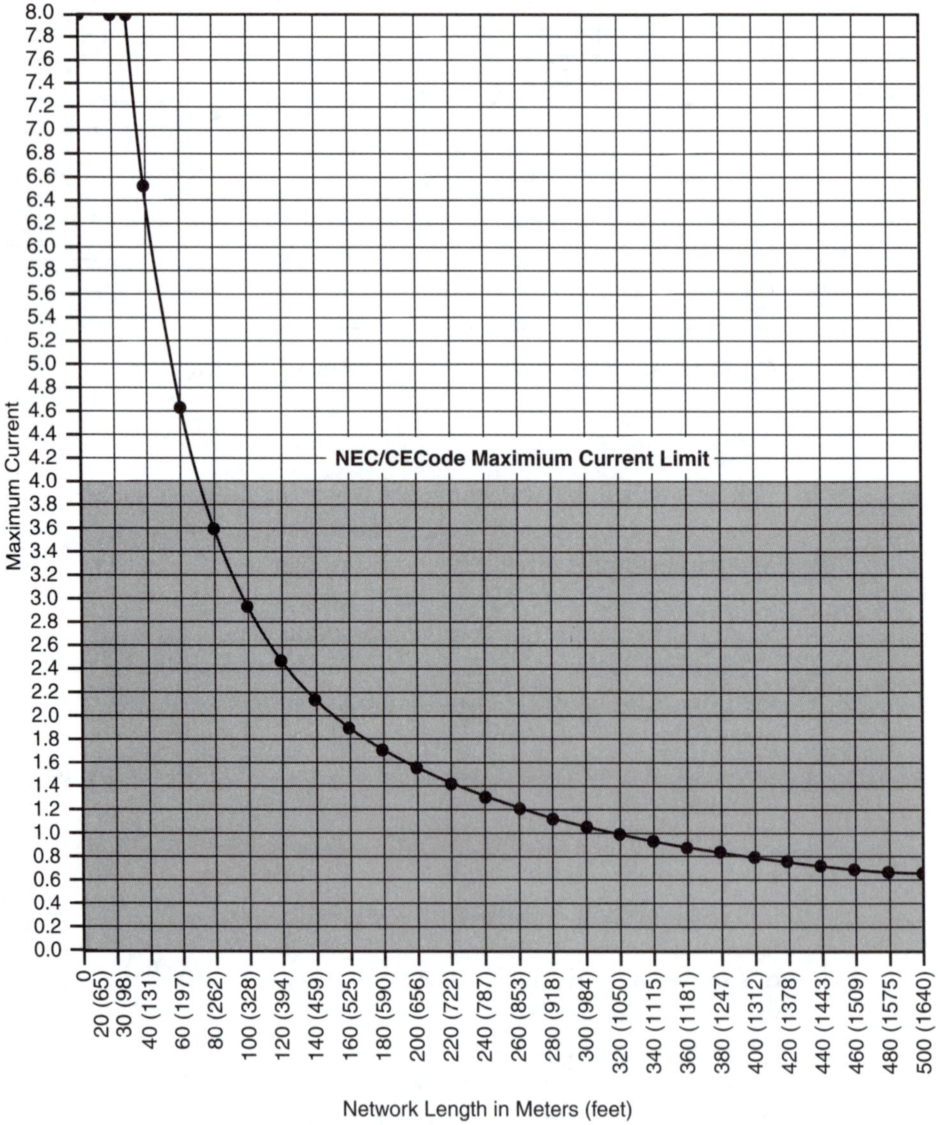

Figure 15–33 Maximum current chart. *(Courtesy Rockwell Automation Inc.)*

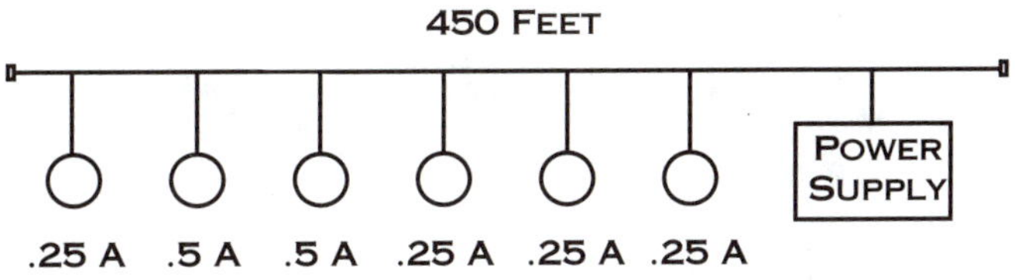

Figure 15–34 Simple DeviceNet System.

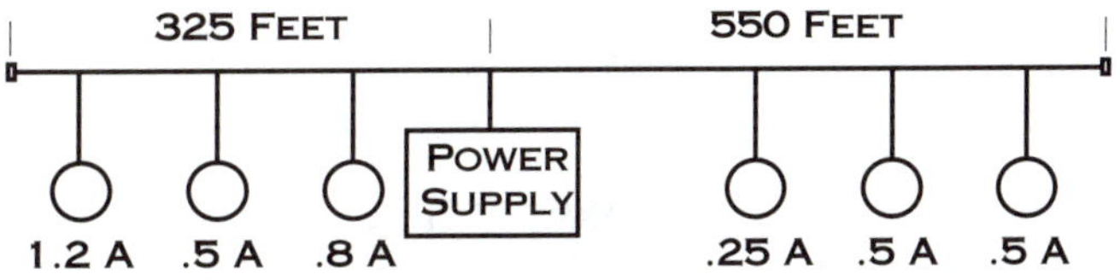

Figure 15–35 Network with power supply in the middle of the network.

Figure 15–36 shows an example in which the total current draw exceeds the NEC 4 A limit. If we add the current draws, we get a total of 6.25 A. We can construct this network if we split the network into two segments and utilize two power supplies. Note that the dotted lines between the segments represent the communications lines. They are connected over the whole network. The power lines are not, however. If you recall, the PowerTap allowed us to power one side or both sides of the network. In this case, the first power supply powers only the left side of the network. The second power supply powers the right segment. Remember, however, that the communications lines are connected over the whole length. The left side draws only 2.5 A and the right side draws 3.75 A. Both segments are under 4 A, so this network is within the limits.

Communication Flow in a DeviceNet Network

The scanner card (or master) coordinates and controls all communication in a DeviceNet system. Figure 15–37 shows a typical system. The information that a scanner receives or transmits is stored in its memory. The CPU (processor) of the SLC can utilize this information in ladder logic. The CPU can receive or transmit information in two modes. One of the modes is I/O. I/O messaging is used for time-critical, control-oriented data exchange. The second mode is explicit messaging, which enables the CPU to transmit and receive a minimum of six words, up to a maximum of twenty-six words.

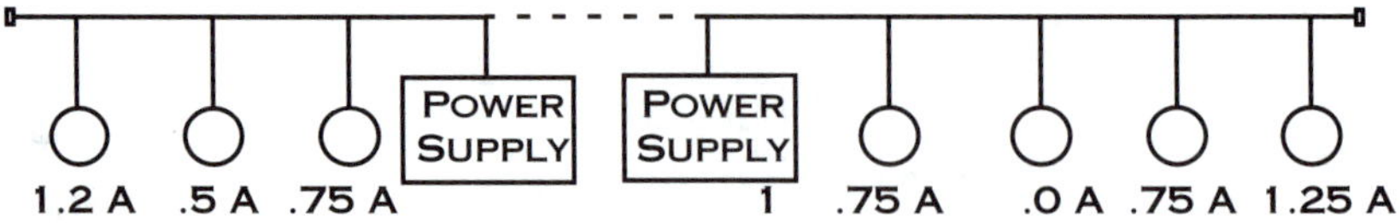

Figure 15–36 Note that two power supplies are used to keep power under 4 A on each side of the network. Note also that the communications line connects both sides; only the power lines are separate for each side.

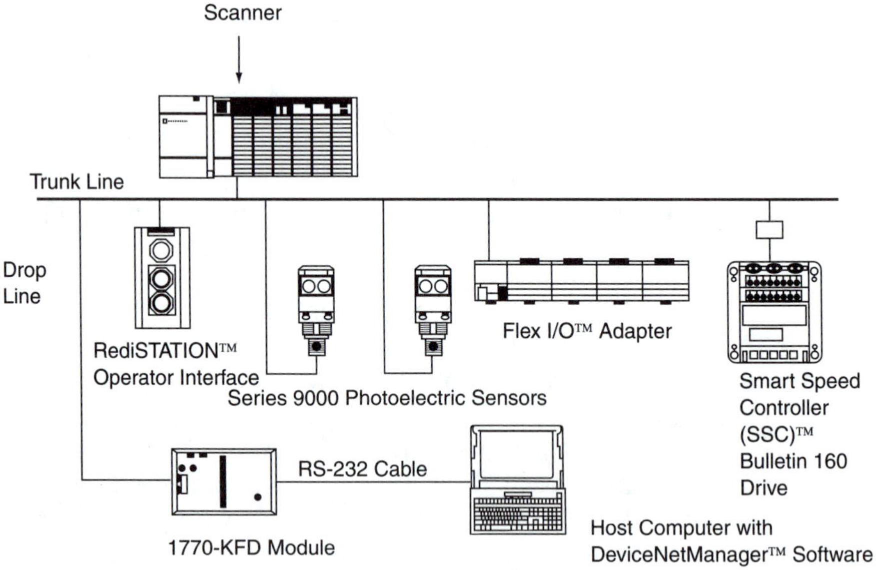

Figure 15–37 Typical DeviceNet system. *(Courtesy Rockwell Automation Inc.)*

Explicit messaging can be used to transmit and receive information for a device like a drive. The ladder diagram in the CPU can determine speeds, acceleration/deceleration, and so on. It would then use an explicit message to the scanner module (master). The scanner would then transmit the message to the drive as a block of information. The CPU can also request drive information from the scanner. The scanner would then request the information from the drive and make it available to the CPU. As you can see, the scanner handles all communications with the devices and makes the information available to the CPU. In a Rockwell system, explicit messaging is initiated by an instruction in the ladder. A write from the processor to the scanner module would utilize the M0 file in the scanner. A write from the processor utilizes the M1 file of the scanner. These reads and writes are performed only when they are called by the CPU. The programmer should be careful to make these calls only when they are needed. They should not be called every scan because they are time intensive.

I/O messaging is almost transparent to the user. The user can access I/O like regular I/O in the rung, utilizing the slot and bit number of the particular I/O in the scanner.

Configuring the Scanner

Let us assume for this example that the DeviceNet scanner is in slot 1 of the SLC. Remember that this is the physical address (slot 1) for the scanner. The scanner will also have a node address for DeviceNet communications.

Scanners need to be configured. Configuration is done with a computer and software attached to the network. Rockwell's software is called RSNetworx. First, node addresses would need to be set on individual devices. Remember that new devices come with an address of 63. Some devices can be configured manually, whereas others might have to be connected individually to the computer and RSNetworx software used for configuring the device and setting its node address.

Consider an imaginary application. We will go through the steps to create a DeviceNet network. Note that this example will not include every step, but it is intended to give an overview of the process. There is one photo sensor, a drive, one limit switch, and the scanner. First, we connect to the scanner module and set the node address to 0. It is desirable to have the scanner at the lowest address. The rest of the devices are connected and the addresses are set as shown in Figure 15–38.

Device	Node Address
Scanner	0
Drive	1
Photo Sensor	2
Limit Switch Through a DeviceLink	3
RediSTATION	4
KFD Module for Computer Link	62

Figure 15–38 Node addresses.

Note that this is just one possibility. Any node numbers could have been chosen. It is a good idea, however, to set the scanner to node and to leave node 63 open so that new devices can be connected and then set to a new address.

For our example we will utilize Rockwell Automation RSNetworx software (DeviceNet Manager software can also be used) and Rockwell products. Remember that any DeviceNet compatible software or hardware can be used.

Figure 15–39 shows the physical setup needed to set up our devices and network. Note the computer, KFD module, and photo sensor in this example.

The user first needs to change the node addresses of the devices in our application. Remember that new devices come with a default address of 63. The user must commission each device to the correct node number.

Next the user constructs the actual network. The user can then connect all of the devices together for the network. The KFD module is also connected into the network. RSNetworx is now able to communicate with all the devices and build the network on the screen. Next the user must build a scan list so that the scanner knows which devices it should be scanning. The devices must then be mapped to memory. The scanner can do this automatically or the user can do it manually.

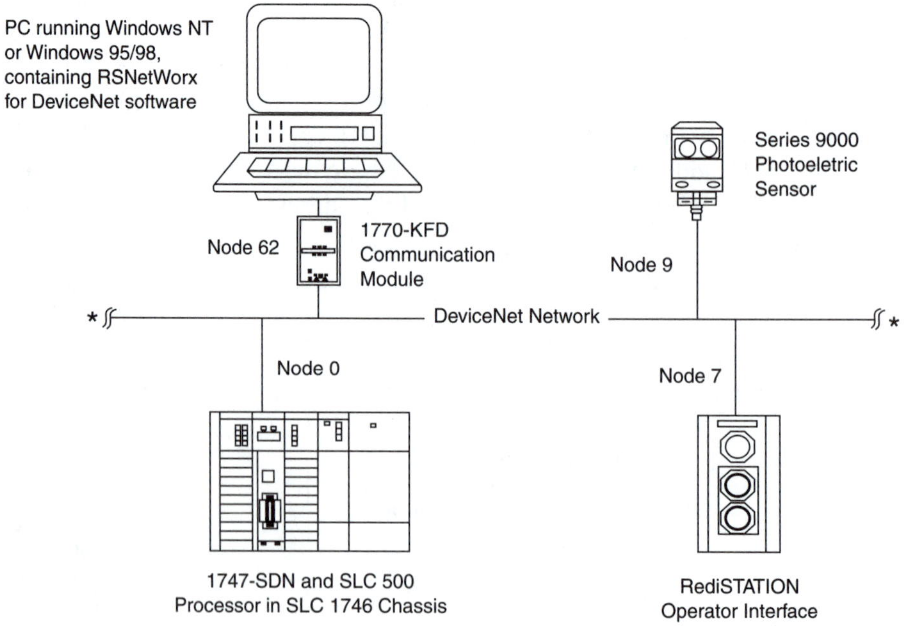

Figure 15–39 Computer, KFD communications module, and a series 9000 Rockwell Automation photo sensor. *(Courtesy Rockwell Automation Inc.)*

Drive Example

A drive is an example of a device that uses multiple words in memory. For this example, we will use a Rockwell Automation 160 drive. A DeviceNet communications module can be attached to the front of the drive. The drive will be one node address in a DeviceNet system.

The address and baud rate are set with dip switches or configured through software. The front panel of the DeviceNet drive module has troubleshooting LEDs for communications, fault, and ready.

Figures 15–40 and 15–41 show the input and output words available to control and monitor the drive using DeviceNet. The user chooses from several choices of input and output selections for these tables. In this case, a reversing speed control drive format was chosen. The terms *output* and *input* are in relation to the scanner. So the output information would be used to send information to the drive such as a speed command (reference).

Byte	Bit 7	Bit 6	Bit 5	Bit 4	Bit 3	Bit 2	Bit 1	Bit 0
0		Net Reference	Net Control			Fault Reset	Run Reverse	Run Forward
1								
2	Speed Reference (RPM) Low Byte							
3	Speed Reference (RPM) High Byte							

Figure 15–40 Output data position in memory.

Byte	Bit 7	Bit 6	Bit 5	Bit 4	Bit 3	Bit 2	Bit 1	Bit 0
0	At Speed	Ref From Net	CTRL from Net	Ready	Running Reverse	Running Forward		Faulted
1								
2	Actual Speed (RPM) Low Byte							
3	Actual Speed (RPM) High Byte							

Figure 15–41 Input data position in memory.

Figure 15–41 shows the input information available from the drive. Remember that the information is in relation to the scanner so that input information for the scanner is output data from the drive. As you can see from Figure 15–41, bit 7 would be a 1 if the drive has achieved the commanded speed. Bit 2 would be a 1 if the drive is running forward. Bit 0 would be a 1 if the drive has faulted. Any of these bits can be used in ladder logic for logic or troubleshooting.

The drive outputs (input to scanner) the actual speed of the drive. That information is in bytes 2 and 3.

Revisit Figures 15–40 and 15–41. Note that there are 4 bytes of input information available and 4 bytes of output information available. These must be mapped to the DeviceNet scanner's memory. The user can choose where to map the drive's data in input and output memory or the scanner can automatically map the memory. Then the user would look at the starting memory location and determine where all of the drive parameters and information are in memory.

Once the scan list and memory mapping are complete and the scan list has been downloaded to the scanner, the DeviceNet system is ready to run. At this point, ladder logic in the SLC can control the devices.

Figure 15–42 shows an example portion of ladder logic used to control a drive. Note that in this case, the drive is mapped to discrete memory. Note also that the drive is a polled device. Rung 0 turns on the poll bit in the scanner so that the scanner polls all polled devices.

Remember that the scanner handles all communication with the network so that the CPU can worry about running the program (ladder logic). We can directly access the discrete information that we mapped to the scanner just like any other I/O. DeviceNet Manager enables us to configure the drive's parameters during setup if we wish.

The drive parameters can also be changed or monitored "on the fly." These can be accessed in a different manner, however, called explicit messaging. Explicit messaging utilizes M0 and M1 files. We would use copy (COP) instructions in the CPU to copy the M1 (input information from the scanner) files to integer files (N7) in the CPU. COP instructions would also be used to copy information (output information to the scanner—parameters in this case) from integer files (N7) to scanner memory. Scan time is optimized if we copy M0 and M1 files only when necessary. For example, if we want to change the acceleration parameter in the drive, we would copy the new value during one scan to M0 memory in the scanner and the scanner would write it to the drive.

RSLogix can also be used to examine the state of the devices. If a user wanted to see the states of DeviceNet inputs, the user would bring up the input file for slot 1. The scanner in this sample application was mounted in slot 1. The user can bring up the input or output files in the run mode and monitor DeviceNet I/O.

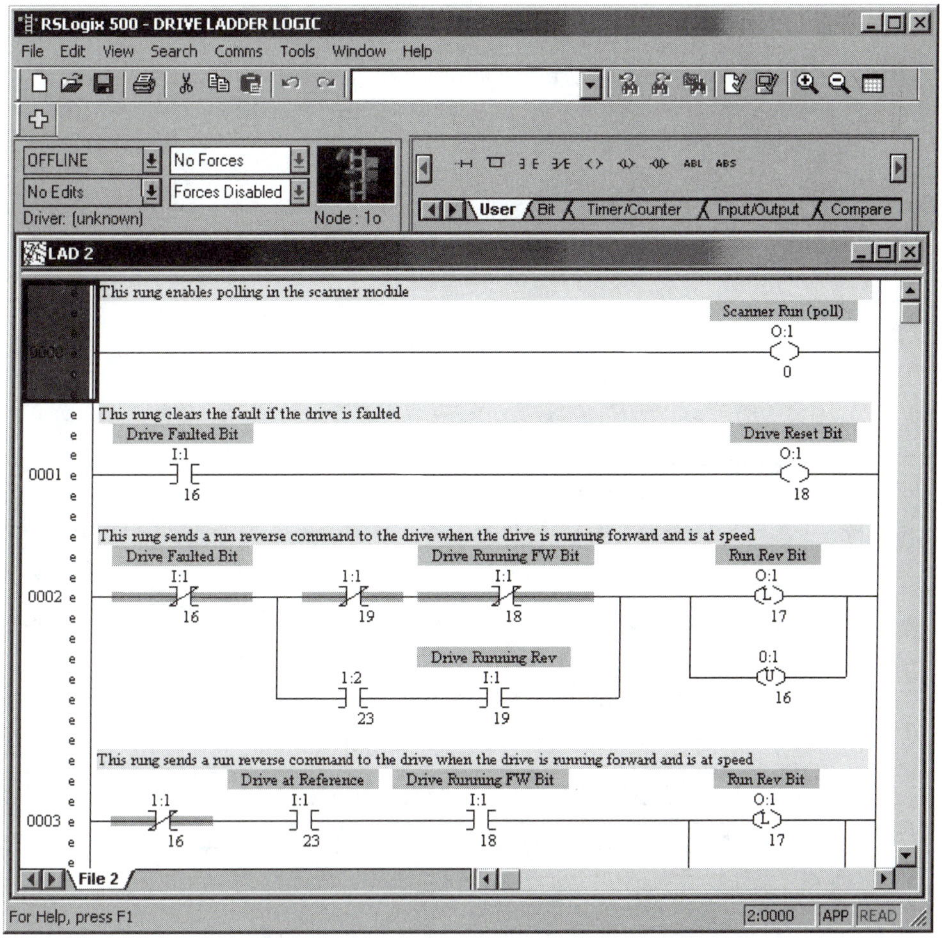

Figure 15–42 Note that the drive is mapped to discrete memory.

Troubleshooting

One of the greatest strengths of DeviceNet is the troubleshooting that is possible. There are two LEDs on the front of the scanner and a two-digit numeric display (Figure 15–43).

Figures 15–44 and 15–45 show the net and module status indicator conditions and troubleshooting help. Figure 15–46 shows some of the two-digit codes for troubleshooting the DeviceNet network. These two-digit codes are helpful in pinpointing the problem and correcting it. For example, if a node goes bad or a line gets cut, the scanner's readout will continually flash the error code and the nodes that are not

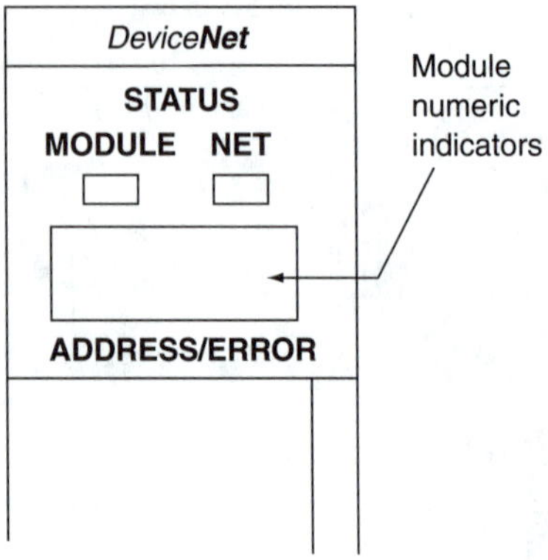

Figure 15–43 Top of scanner module showing status LEDs and numeric display. *(Courtesy Rockwell Automation Inc.)*

If the Module Indicator Is:	Then:	Required Action:
Off	No power to the module	Apply power
Green	Module operating in normal mode	None
Flashing green	Module not configured	Configure the module
Flashing red	Invalid configuration	Check configuration setup
Red	Unrecoverable fault in the module	Replace the module

Figure 15–44 Scanner module status LED indicators.

communicating. Assume that the drop line to the photo eye was cut. The scanner readout would continually flash 72 and then 2. The 72 means that a node has stopped communicating and the 2 is the node that is missing. If 3 was also missing, it would flash 72-2-3. If there is a power problem, the scanner will flash 92.

Devices also offer additional capability. For example, the Rockwell Automation photo sensor has a margin output bit that can be used in the user's logic. If the sensor is not receiving the amount of light it should be, it turns on the margin output. This could mean that the reflector for the sensor is dirty or misaligned. The user can use the margin bit as an input in the ladder logic to notify an operator or maintenance worker that this sensor has a margin problem. In this manner the sensor can be repaired or replaced before it failed.

If the Net Indicator Is:	Then:	Which Indicates:	Required Action
Off	The device has no power or the channel is disabled for communication due to a bus off condition, or loss of network power, or it has been intentionally disabled.	The channel is disabled for DeviceNet communication.	Power-up the module, provide network power to the channel, and be sure that the channel is enabled in both the module configuration table and the module command word.
Green	Normal operation.	All slave devices in the scan list are communicating normally with the module.	None.
Flashing green	The two-digit numeric display for the channel indicates an error code that provides more information about the condition of the channel.	The channel is enabled but no communication is occurring.	Configure the scan list table for the channel to add devices.
Flashing red	The two-digit numeric display on the scanner displays an error code that provides information about the cause.	At least one of the slave devices has failed to communicate with the module.	Examine the failed device and the scan list for accuracy.
Red	The communications channel has failed. The two-digit numeric display for the scanner displays an error code that provides more information about the condition of the channel.	The module may be defective.	Reset the module. If failures continue, replace the module.

Figure 15–45 Network status LED indicators.

Code	Description	Required Action
0–63	Under normal operation the display will show the network address of the scanner.	None.
70	Module failed the duplicate node address check.	Change the module address to another available one.
72	Slave device stopped communicating. The node number of the device will flash alternately with the 72.	Inspect the field devices and verify connections.
75	No scan list is active in the module.	Enter a scan list.
78	Slave device in the scan list does not exist. The node number will alternately flash with the 78.	Add the device to the network or delete the device from the scan list.
80	The scanner is in idle mode.	None.
81	The scanner is in fault mode.	None.
92	No network power is detected on the communications port.	Provide network power. Make sure that the DC power is present, clean, and sufficient.

Figure 15–46 A partial list of the troubleshooting codes.

DeviceNet Summary

DeviceNet is a simple, open networking solution that reduces the cost and time to wire and install automation. DeviceNet enables devices from any DeviceNet compatible vendors to be interchanged.

The troubleshooting it provides can also dramatically reduce downtime. The scanner module error codes can direct you to the malfunction and even the problem node. The user can also utilize ladder logic and the device's intelligence to pinpoint problems. An example would be using the light margin bit in the photo sensors in ladder logic to alert the operator when the sensor may need to be aligned, cleaned, or replaced.

QUESTIONS

1. Name and describe the three levels of communications.
2. What is daisy chaining?
3. What is DH+ and what is its purpose?
4. What is ControlNet used for?
5. What is DeviceNet used for?
6. Which of the following can a ControlLogix system be used for?
   a. Stand-alone controller
   b. Process controller
   c. Motion controller
   d. Communications gateway
   e. Only a, c, and d above
   f. All of the above
7. What is SERCOS?
8. How many nodes can a DeviceNet network have?
9. What is a scanner and what does it do?
10. What is a scanlist?

Chapter 16
PC-Based Control

The use of computers for industrial control is expanding rapidly. This chapter will examine the use of PCs for industrial control. Different approaches to PC-based control will also be examined.

OBJECTIVES

Upon completion of this chapter, you will be able to:

1. Describe the types of applications that are most appropriate for industrial PC-based control.
2. Describe the advantages of PC-based control.
3. Compare and contrast the use of PC-based control versus PLC-based control.
4. Describe flowchart programming and its advantages.
5. List some alternative approaches to PC-based control.

OVERVIEW OF PC-BASED CONTROL

PLCs were first sold as simple control devices that could replace relays. Computers at this time were not user-friendly, and users had to be software programmers to develop an application. There was almost no automation software available. Industrial personnel were unfamiliar and uncomfortable with computers. There were no fancy graphics or user-friendly help screens. Everything was text-based. This made PLCs very attractive to manufacturers who wanted easier, quicker, and more flexible controllers than relay control methods.

PLCs were often called relay replacers. They were intended to be easy for plant electricians to use and program. PLCs were proprietary. Many manufacturers began producing and selling PLCs. They all used ladder logic for programming but it varied substantially between PLC brands. Remember that a PLC is a special-purpose microcomputer. The simple functions that the first PLCs could perform soon were not enough. People demanded more capability. Users wanted math functions and analog capability. Users soon wanted to control more complex systems with PID functions. Manufacturers were more than happy to add and sell additional capability to users, competition among PLC manufacturers increased.

Users also began to demand networking capability. PLC manufacturers offered proprietary solutions again. The networks, for the most part, were intended to be used with the manufacturer's own brand of PLCs. They would not allow other brands of PLCs on their network.

PLCs are proprietary in software and hardware. You cannot buy I/O modules from one manufacturer and a CPU from another and expect them to work together. When you choose a PLC brand for an application, you are forced to buy all related equipment from that manufacturer.

Personal computers started this way also. Many companies tried to sell proprietary computers. Most have fallen by the wayside. The IBM PC became a standard. DOS and then Windows became the standard operating system. Today there are many manufacturers of PC microprocessor motherboards, memory, hard drives, monitors, and so on. This has caused extreme price competition and real value for consumers. In fact, today almost anyone can start a computer business by buying standard components, assembling them into a computer, putting a name on them, and then selling them. Apple was not so open with their system and has not flourished. Apple is very strong in niche markets like graphics and the printing industry.

This has not occurred with PLCs. PLC manufacturers have kept their products proprietary. Ladder diagramming was usually done on a dedicated programming terminal purchased from the PLC manufacturer. These were costly and cumbersome for programming. If you had multiple brands of PLCs, you had to have multiple dedicated programming terminals, one for each brand of PLC.

Computers, meanwhile, were becoming more user-friendly and commonplace. People were becoming more competent at using computers. Third-party software companies began to sell ladder diagram programming software that could be used on a standard microcomputer. The software was more user-friendly than the manufacturers' dedicated systems, and cheaper because the computer could be used for other purposes. Instead of buying multiple programming terminals, users could purchase programming software for each brand of PLC and utilize one computer to program all their PLCs. This was a more economical solution. It also helped make microcomputers popular in industry.

PLCs and industrial computers are just microprocessor-based systems. Both use a microprocessor for logic and control functions. People are now realizing that the PLC is really a special-purpose microcomputer. Users are now willing to consider the microcomputer as a control device in industry.

Hardware improvements were even more incredible. Processing power and speeds doubled every 18 months. Memory prices as well as the price of all related hardware steadily decreased. More and more people could afford computers on their desks and in their homes. Computers have become an ever-present and widely used tool in industry and our daily lives.

Advantages of Computers for Industrial Control

The interface between user and an automated system is crucial. Computers have a distinct advantage in operator interface because of their graphics capability. A whole computer system is cheaper than purchasing one high-quality color display panel for a PLC. The software available for the computer is often easier to use in the development of these interface screens. It is also possible to utilize the computer for other tasks, if the application allows. The user may be able to use other software in Windows to accomplish tasks. Data from the process might be used in spreadsheets, word processors, or other analysis software. Microcomputers are excellent for data logging also; they can easily store vast amounts of historical process data.

The ease of use and understandability of the Microsoft Windows environment makes all software similar and makes training easier and less extensive. It is also easier to connect the computer to a standard LAN such as Ethernet and share manufacturing data throughout the organization. This ease of networking is a major advantage. The networking cost is also low. Network cards are available at a fraction of the cost of a PLC network card. The network capability allows integration of other systems for data exchange, tracking, maintenance, production planning, quality control, recipe downloading, order tracking, etc. Another advantage is that the same hardware can be used for programming and control. The hardware can be used to develop the program and then to operate and troubleshoot the system. Microcomputer hardware is standardized and can survive several generations of product change. This is not always the case with PLCs.

An endless array of peripheral devices is available for microcomputers: printers, bar code scanners, bar-code printers, sound cards, multimedia cards, and so on. The user can purchase hardware from any manufacturer. The prices of hardware are continually falling. Several years ago someone observed that if luxury cars had made the same improvements in power and speed and decreases in price that computers had made, a luxury car today would cost about 50 cents. The improvements in computers have been phenomenal.

The drive toward a common programming standard (IEC 1131-3) has also made computers more attractive. Various programming methods are available for PLCs and microcomputers: ladder, flow diagram, function block, statement list, state logic, C, BASIC, etc.

Microcomputer Concerns

Anyone who has ever had a computer lock up for no apparent reason knows that this situation cannot be allowed to happen when industrial systems are involved, especially when safety might be a concern. A manufacturer cannot tolerate a control system that locks up. PLCs are designed to be very rugged in an industrial environment. They are very noise immune. They can operate in extremes of temperature, shock, and vibration and in dirty environments.

Several companies have devised microcomputer solutions to address these concerns. There is a wide array of equipment and software to utilize computers for industrial control. Next we will consider a few of the alternative approaches to computer control.

There are two types of PC-based control systems: soft logic and hard real-time control (see Figures 16–1 and 16–2). Most soft-logic systems run the control as a high-priority real-time task under Windows. Real-time tasks can be interrupted by deferred procedure calls used to service Windows' system functions such as disk access, network communications, and so on. Soft logic does not provide the same level of deterministic control as a PLC because higher-priority functions can interrupt and delay real-time control systems. If the application requires high-speed control, the lack of determinism may be unacceptable or, worse yet, unsafe.

Hard real-time control is used by PLC systems. In a hard real-time system, the real-time operating system is loaded first. The control engine runs as the highest priority task. All Windows functions run as the lowest priority task within the operating system. The logic control engine always has priority and is completely protected from Windows. Windows is not allowed to preempt real-time control.

Remember that a PLC is really a microprocessor with a real-time operating system (RTOS). The RTOS is the code that controls all tasks and operations that run on the microprocessor. The RTOS controls the PLC scan and logic and gives the PLC its deterministic response.

Next we will look at a few different approaches to PC-based control.

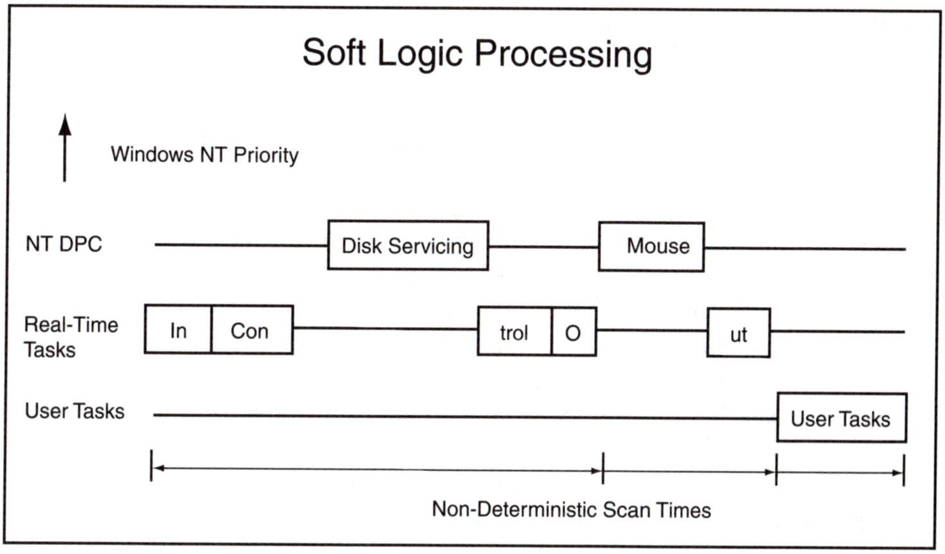

Figure 16–1 Soft logic processing. *(Courtesy Entivity, Inc.)*

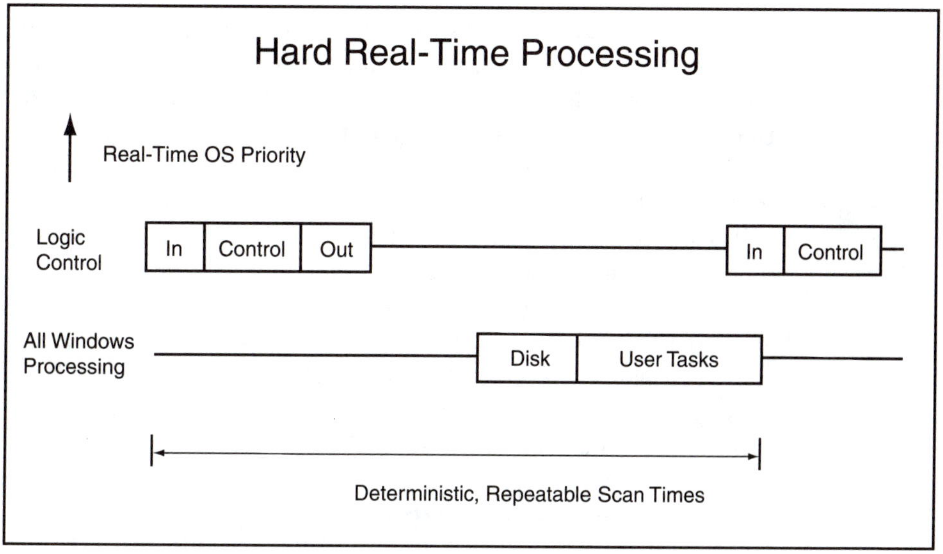

Figure 16–2 Hard real-time processing. *(Courtesy Entivity, Inc.)*

Flowchart Programming

Flowchart programming is becoming more popular in industrial control. There are many reasons for its growth as a programming language. Steeplechase Software, Inc., has a PC-based control system that utilizes flowchart programming as its fundamental programming method. The user can also program in IEC 61131-3–based ladder logic. Flowcharts allow a system to be developed as a simple, intuitive, graphical description of a process that everyone can understand. Flowchart programming allows engineers, operators, and plant-floor technicians alike to interpret and understand the step-by-step process used to program, operate, and troubleshoot machines. The same cannot be said about ladder logic. Steeplechase calls their product the Visual Logic Controller (VLC). Their system was the first Windows-based system on the market to offer integrated flowchart programming, control, and simulation and a man-machine interface (MMI) on a simple PC.

Steeplechase's approach marries the benefits of hard real-time control and Windows. It is compatible with Windows 2000/NT/XT. The Windows environment provides many benefits over the traditional PLC. Windows-based systems can be used to provide network communications and graphical user interface and can utilize Windows-based software to process information. There is also a substantial cost savings in some systems. In systems where communication and graphics are important, a PC-based controller can be more cost effective (see Figure 16–3).

Steeplechase loads Windows as the lowest priority task in the hard real-time operating system. All the control functions are run as higher-priority tasks in the real-time operating system and are isolated in memory from Windows applications and drivers. Steeplechase uses the memory protection functions inside the Intel processors to prohibit Windows from accessing any of the memory or CPU cycles that are dedicated to the real-time engine.

This approach yields several benefits. Windows can crash without affecting system control. This allows the control program to continue to operate as normal or execute an orderly shutdown to put the machine into a safe state. The system can survive a hard disk crash. The hard real-time operating system is loaded into the PC, so dependence on the hard disk is eliminated. The entire RTOS is loaded and active in memory, so a failure of the hard disk, regardless of its impact on Windows, does not affect the control activities.

Benefits of Flowchart Programming

Flowchart programming, unlike ladder diagramming, allows programs to be easily broken down into logical steps. Control system design can be reduced 50 to 70 percent using flowchart programming compared to ladder logic. Remember that a large part of every ladder diagram is logic to make sure outputs are not

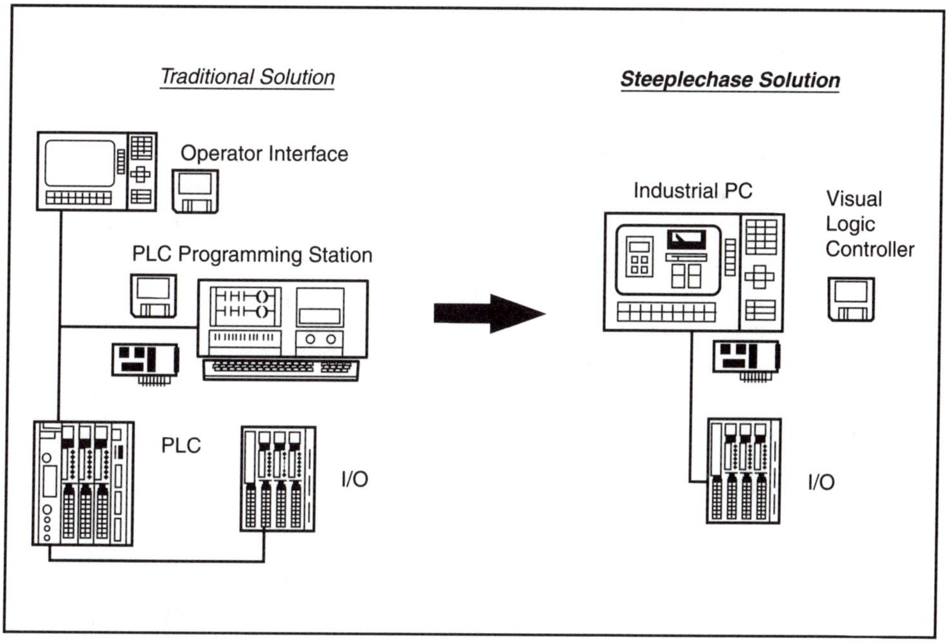

Figure 16–3 A comparison of a PLC system to a system controlled by a PC-based controller. *(Courtesy Entivity, Inc.)*

on when they should not be. Flowcharts break the process into logical steps that avoid that problem. Flowcharts allow the programmer to draw the desired machine control sequence as a simple set of flowchart steps. Boxes are actions where outputs are turned on and off and math or logic functions are performed. Diamonds are decisions that direct the control sequencing based on input states or logic conditions. The operation of any process can be easily described by a set of flowcharts. Flowcharting is similar to the way in which a system would be designed anyway. The engineers, technicians, and operators would get together to develop and understand the system. They would discuss and write down the process as a series of steps. Flowchart programming makes good use of that effort and the flowchart becomes the program. During runtime, troubleshooting is made much easier using flowcharts. In fact, operators can often find the problem by watching the control program flowchart as it highlights every step of the process. This information then helps maintenance technicians pinpoint and repair the problem quickly. Flowchart programming makes it easy for operators and technicians to understand the control process and contribute to debugging a process.

The simplicity and understandability of flowchart programs allows systems to be modified and improved easily. Ladder diagrams are typically understood only by the one person who wrote them. Ladder diagrams typically have hundreds of rungs of logic. The interrelationship and interdependence of rungs make wholesale changes to logic difficult after a system is running. Flowcharts, however, break even a complex system into small understandable blocks. This ability allows modifications to be made easily without undue worry about the change's effect on other logic in the program. Process improvement can thus be made easily. Flowcharting is also self-documenting. A flowchart is understood easily by all plant personnel. The flowchart acts as program and documentation. Comments can be added to explain the process even further.

Application Development

There are three steps in application development:

- Planning the control sequences and entering the flowcharts into the PC.
- Setting up the operator control panel screens (MMI).
- Simulating the control system and when ready assigning the real-world I/O to tagnames and running the system.

Planning Control Sequences and Entering the Flowcharts

The first step is to plan the machine sequences. Let's consider a simple tank-fill application. The sequence is simple. First, the pump must be turned on, then the level must be monitored so that the pump can be turned off when the tank is full.

Programming Each flowchart begins with a start element (see Figure 16–4). The elements (start, action, decision, subprogram, and stop) are simply chosen from the flow tool palette and placed on the screen. This is done by clicking the mouse on the desired element in the tool palette and then moving the cursor to the desired position on the screen and clicking the mouse button again. The programmer simply places all the desired elements on the screen in the desired location. The programmer then uses the flow tool to connect the elements together.

In our example (see Figure 16–4) the first element was a start element. The start was connected to an action element with a flow line. This action element is used to turn the pump motor on. Action elements can also be used to initiate functions such as turning outputs on and off and performing mathematical computations, logic, or any combination of these. If the programmer double clicks on any element, a new screen appears to allow the programer to configure the element. Figure 16–5 shows

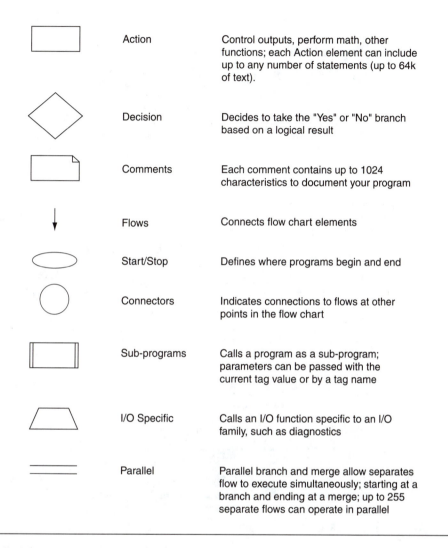

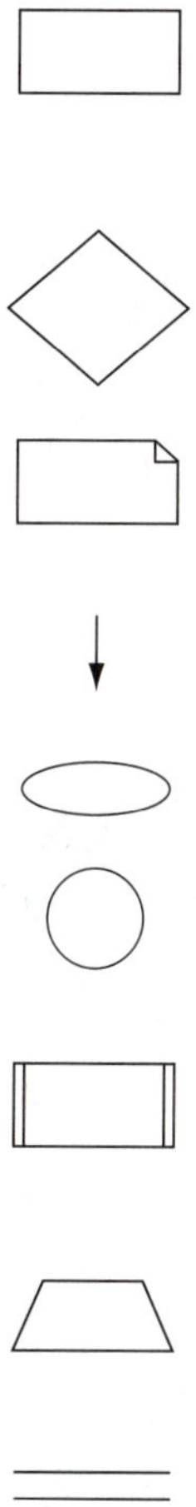

Action	Control outputs, perform math, other functions; each Action element can include up to any number of statements (up to 64k of text).
Decision	Decides to take the "Yes" or "No" branch based on a logical result
Comments	Each comment contains up to 1024 characteristics to document your program
Flows	Connects flow chart elements
Start/Stop	Defines where programs begin and end
Connectors	Indicates connections to flows at other points in the flow chart
Sub-programs	Calls a program as a sub-program; parameters can be passed with the current tag value or by a tag name
I/O Specific	Calls an I/O function specific to an I/O family, such as diagnostics
Parallel	Parallel branch and merge allow separates flow to execute simultaneously; starting at a branch and ending at a merge; up to 255 separate flows can operate in parallel

Figure 16–4 The elements that are used to create a flow diagram program. *(Diagram courtesy of Entivity, Inc.)*

a typical input screen for an action element. Flowchart enhancements loop commands (IF-THEN-ELSE and WHILE) are also available. In addition, multidimensional arrays have been incorporated to aid in the generation and processing of complex algorithms traditionally used in various material-handling and high-speed sorting applications.

If the programmer chooses edit tag from the screen shown in Figure 16–5, the screen shown in Figure 16–6 appears. The programmer then enters the specific

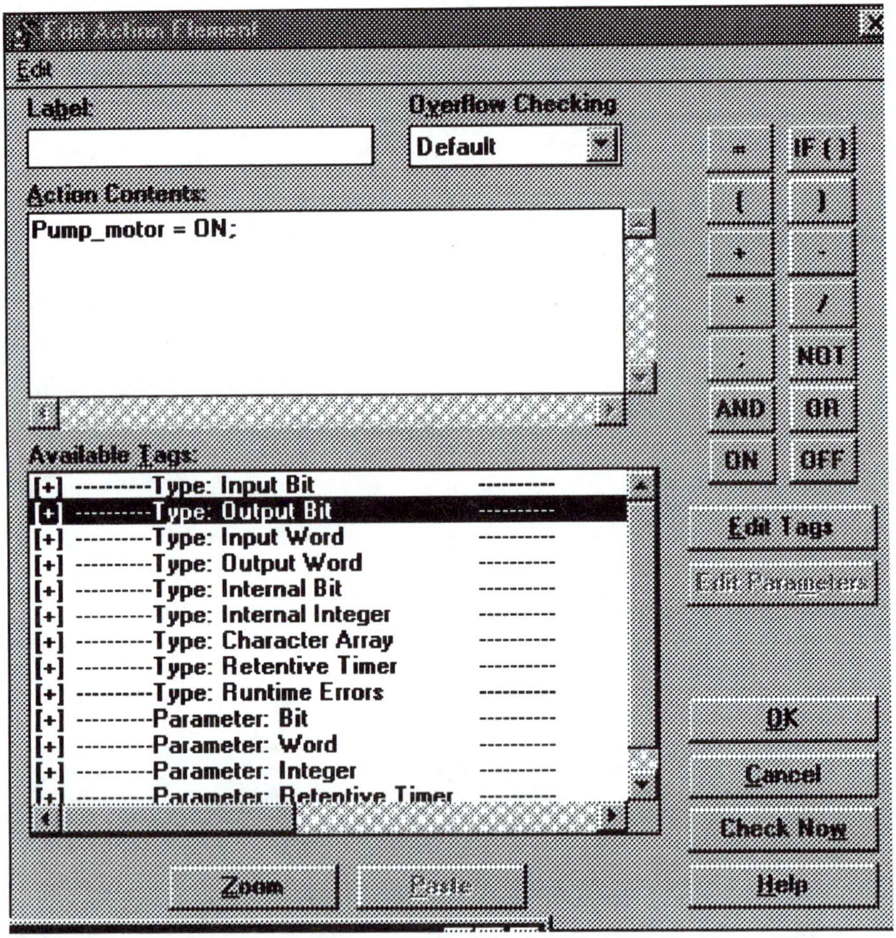

Figure 16–5 A configuration screen for a tag.

information for this tag. The name of the tag will be Pump_motor. The device is then chosen from a drop-down list. The specific point is entered. The point is the actual address of the I/O point in the device that was chosen. This is an important point. With this type of control (industrial PC), multiple devices and/or I/O devices can be attached to the computer. They will all respond as if they are one system. The programmer just chooses the device from the list for each particular tag. The tag configuration can be done at the time the element is placed on the screen or after all graphic elements have been placed. This same process would be followed for all elements.

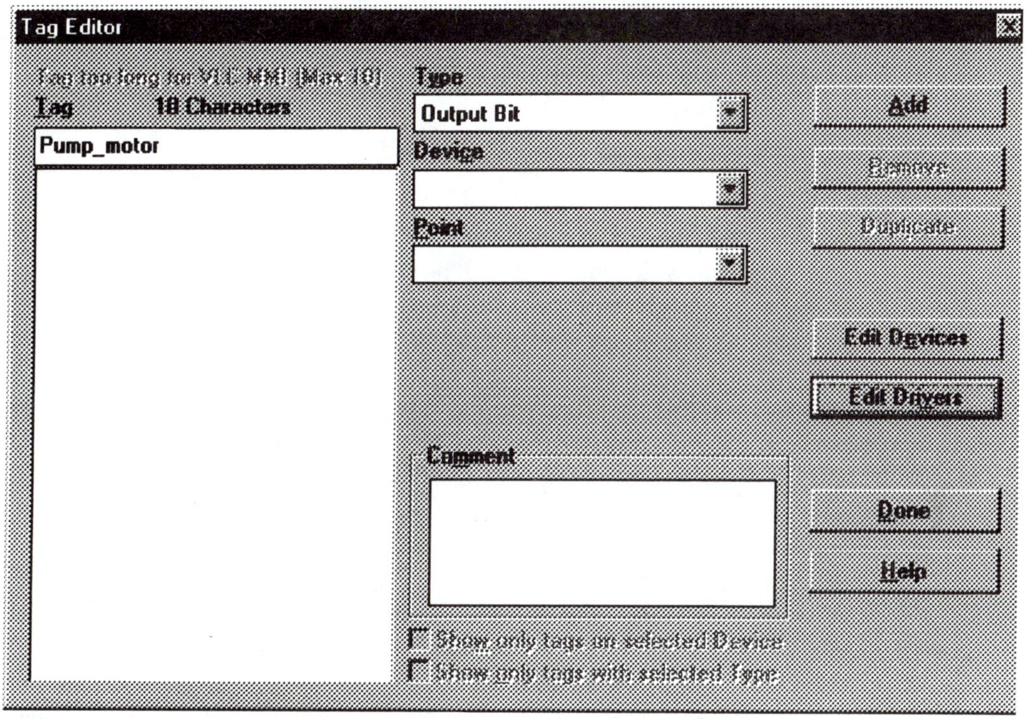

Figure 16–6 A tag definition screen.

This feature is quite an advantage over a PLC because devices and I/O from different manufacturers can be integrated easily. The software supports I/O modules from Allen Bradley, GE Fanuc, APC Seriplex, DeviceNet, Modicon Remote I/O, Interbus-S, PID Controller, RS-232/RS-422 Drivers, Smart Distributed System (SDS), and many others.

The next element is a decision element. This element is used to decide if the tank is full. Note that two paths flow out from a decision block: a yes and a no path. Each of the paths can connect to other elements. In our system the yes flow line is connected to the next action element. The no flow line is connected back to the input flow line to the decision block. This decision element is used to determine if the tank is full. If the tank is full, the yes flow line is followed to the next element. If the tank is not full, the no flow line is followed back to the decision element again. It will continue to repeat this process until the tank is full. When the tank is full, processing then moves to the next action element. Note that the programmer entered Pump_motor = ON in the action contents box. The programmer also chose the output type bit for the tag. This action element is used to turn the pump motor off. After the pump is turned off processing moves to the next element. The next (and last) element is the stop element.

Setting Up Operator Control Panel Screens

Next, the operator panel is created. Figure 16–7 shows what the completed operator panel looks like. There is a meter shown in the upper right of the panel. The meter is used to show the pump speed. The slider control is used by the operator to set the pump speed. The operator can use the mouse to move the slider. The pump speed can also be controlled by using the up and down arrow keys. The meter, slider, and up and down arrows are all linked to the actual analog register in the controller that controls the pump speed. All of the controls on this screen were created by selecting them from a toolbox (see Figure 16–8) and placing them on the screen. Figure 16–9 shows an example of the types of sliders available. The programmer then double-clicks on each and assigns a tagname, device, and the actual I/O point address.

The graph is used to show tank level versus time. Note that multiple variables can be graphed at one time so the operator can get instant, real-time feedback on actual system operation.

The programmer can also draw or import text objects and animate them. By double-clicking on any drawn object, the programmer can link the graphic to a tag

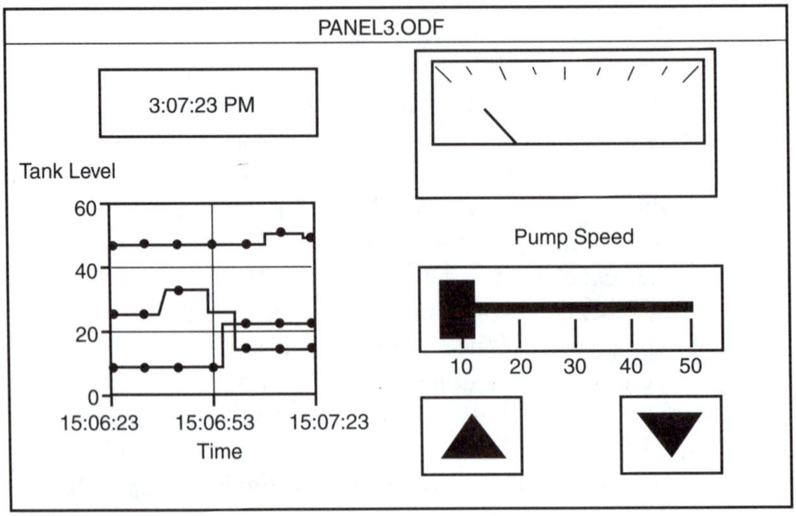

Figure 16–7 The completed operator panel. *(Diagram courtesy Entivity, Inc.)*

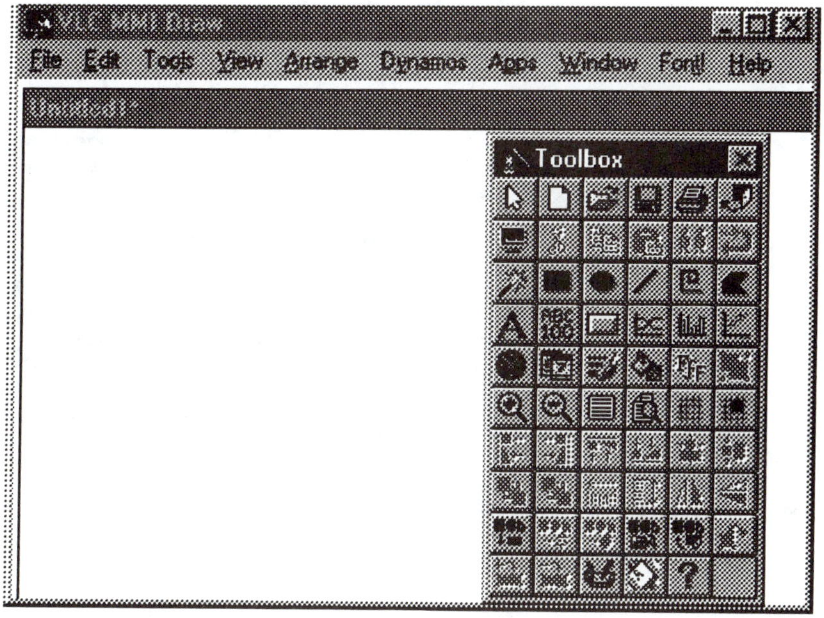

Figure 16–8 The MMI development screen. Note the tool palette.

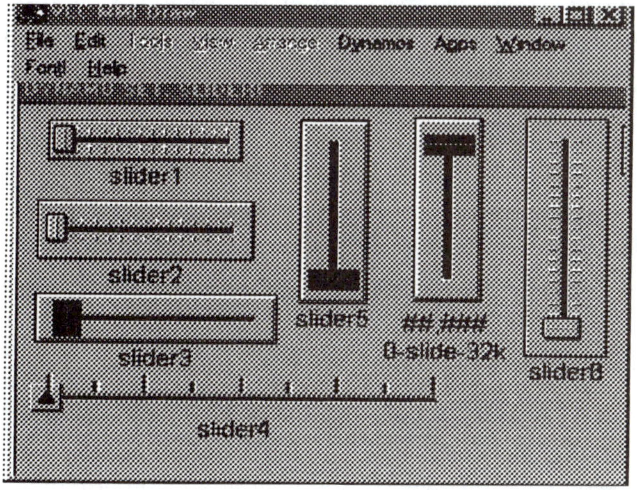

Figure 16–9 Some of the slider controls that are available in the library.

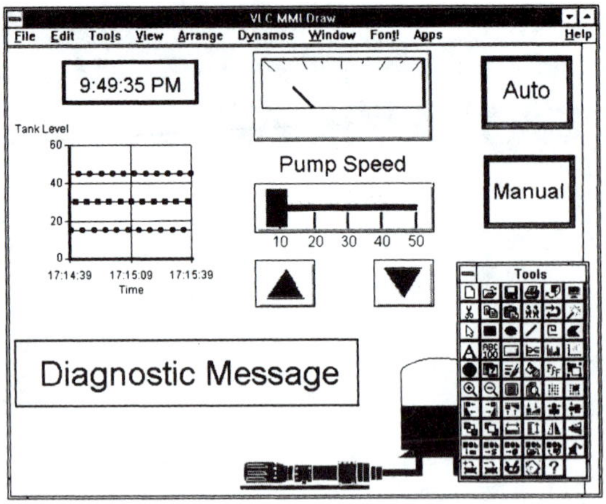

Figure 16–10 The operator panel for the system during its development. *(Diagram courtesy of Entivity, Inc.)*

and control its color, position, size, and fill. Figure 16–10 shows an example of the screen during development.

Simulating the Control System and Assigning I/O

After the flowchart and operator panel have been developed, the system can be simulated to be sure that everything works correctly. Figure 16–11 shows the way the screen might appear for our system during simulation. Note that the active step is highlighted. Note also that the operator panel shows the tank about half full. The level changes as the real-world level changes. The tank-level display is linked to an analog register in the controller. The programmer can utilize any of the controls on the screen for test purposes.

At this point, if it has not been done already, the programmer can assign real-world I/O to the tags and run the machine. Note that this example is simple and does not show the real power or ease of use of this software.

Concurrent processing can also take place. Figure 16–12 shows an example. In this case three different flows were started from one point. These are called parallel branches. The double lines at the top and bottom can be used to create a parallel branch or parallel merge.

Another type of powerful element is a subprogram. Subprograms can be used to simplify the view of the process. A subprogram element is actually used to call another flowchart. They can hide portions of programs that have already been tested

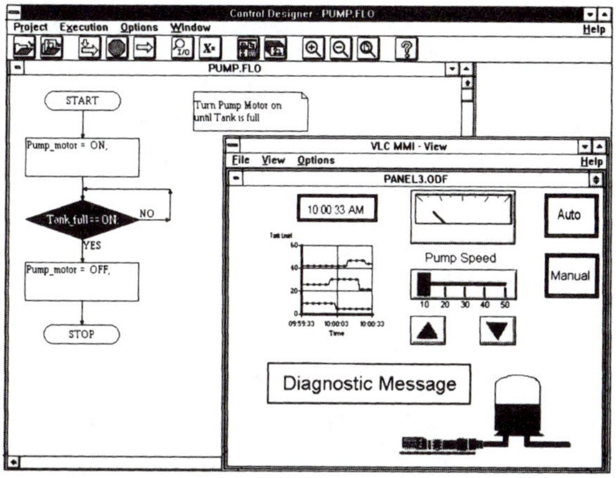

Figure 16–11 The system in simulation mode. *(Diagram courtesy of Entivity, Inc.)*

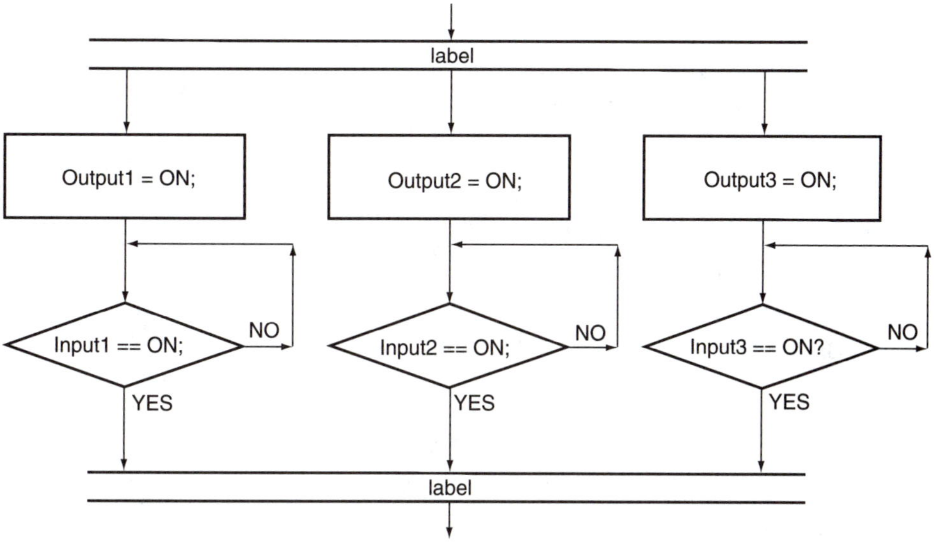

Figure 16–12 An example of parallel branch elements. *(Diagram courtesy of Entivity, Inc.)*

or organize programs in hierarchies so that each view remains simple and compact. The number and nesting of subprograms is almost unlimited.

Ladder Logic Editor Ladder logic is still the most commonly used control language on the factory floor. Steeplechase Software allows users to write programs in traditional flowcharts or a ladder logic editor based on the IEC-61131-3 specification. To increase the power and usability of ladder logic, powerful flowchart elements can be added, allowing users to create more powerful program routines than with pure ladder logic. The incorporation of self-documenting flowchart elements in a ladder logic program reduces the time required to develop a control program and greatly simplifies troubleshooting.

Scan Time The programmer can set the desired scan time. The scan time can be set in 5-millisecond increments from 5 to 500 milliseconds. During each scan, the VLC reads inputs, executes the program, and writes outputs. In the time remaining, it executes DOS and Windows applications (see Figure 16–13). If the control programs exceed the target scan time during any cycle, the DOS and Windows applications do not execute in that cycle. The control program can also detect the scan overrun error and report it.

As you can see, flowcharting is a very powerful and yet easy to understand programming language.

SoftPLC

A different approach to PC-based control was taken by SoftPLC Corporation. They decided to utilize ladder logic as their programming language. SoftPLC is software technology that turns a standard industrial computer (PC) into a full-function PLC-like process controller. SoftPLC combines PID discrete and analog I/O control with the data handling, computational, and networking capabilities of computers. A multitasking control kernel, SoftPLC provides a powerful instruction set, fast and deterministic scan time, reliable operation, and an open architecture for connection to a wide variety of I/O systems, other devices, and networks.

When added to supporting computer hardware components, SoftPLC creates a control system with throughput, performance, and programming capabilities exceeding those of conventional programmable controllers. SoftPLC is not a DOS application; it is a real-time, multitasking operating system, or kernel. Once SoftPLC is loaded into memory, it is in control of the computer CPU at all times. SoftPLC

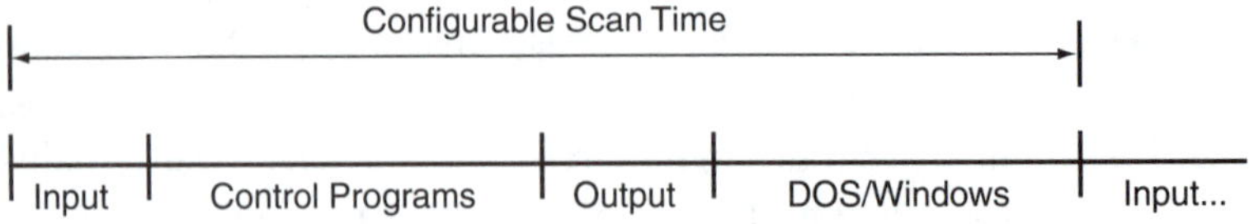

Figure 16–13 An example of scan time. *(Diagram courtesy of Entivity, Inc.)*

loads from DOS, then runs in computer RAM memory. SoftPLC can be loaded from any defined drive available to the computer, including EPROM or flash memory, a local hard/floppy drive, a network drive, etc. DOS is never called. This gives SoftPLC the reliability and characteristics of proprietary, dedicated hardware PLCs, such as deterministic scan time, protection from bugs in the operating system or other software applications, etc. In general, SoftPLC's instruction execution times are two to ten times faster than traditional PLCs on a 486 system. Online run mode program changes can be made. SoftPLC logic can be changed while the machine is running. There is no need to recompile the application or to stop the machine control to download the new program.

SoftPLC provides several features that ensure consistent and reliable operation of a computer-based control system. SoftPLC manages its own application program and data files directly for automatic backup and restore (upload and download) purposes. Advance power loss detection logic combined with an uninterruptible power supply (UPS) or battery-backed RAM may be used to enable SoftPLC to save the ladder program and data table to a disk file (which may reside on a network file server). SoftPLC's status file includes fault bits to enable easy detection of the reason for a runtime error and to clear and correct the fault. (Examples of these faults include divide by 0 and jump to a nonexistent label.) The keyboard lockout feature can be used to protect the operation of the system. Even [Ctrl]-[Alt]-[Del] won't shut down the system.

SoftPLC Features

There are four user-configurable communication channels (network, data paths) for data or program logic access from other computer applications or PLCs programmed with TOPDOC. It runs imported and/or converted Allen-Bradley PLC, PLC-2, PLC-5, or SLC-500 programs.

Instruction Set Programming SoftPLC's ladder logic execution capabilities are similar to those of an Allen-Bradley PLC-5. In fact, they are similar enough that an existing PLC-2, PLC-5, or SLC-500 program can be easily converted, then loaded into SoftPLC. Standard ladder instructions include: contacts, coils, timers/counters, data comparisons/moves, math and logical operations, shift registers/sequencers, branches, jumps and subroutines, and specials (PID, message, diagnostics). SoftPLC's standard instruction set also includes several loadable functions such as COMGENIE, which is used for general-purpose ASCII communications to and from RS-232, RS-422, or RS-485 devices through COM ports. TOPDOC also provides a PID loop display and a PID auto-tuning utility, as well as advanced math instructions such as statistics, trigonometric functions, etc. Users can create their own instructions with C, C++, or Java.

A-B Conversion Utilities Software utilities are available to import and convert PLC, PLC-2, PLC-5, or SLC-500 programs and documentation into SoftPLC/TOPDOC format. These utilities thus provide an easy upgrade path from existing Allen-Bradley PLC systems to SoftPLC.

I/O Capabilities SoftPLC connects to I/O systems through the use of loadable software I/O drivers. The user can select whichever I/O vendor hardware is best for the specific application. The drivers for some of the more popular I/O systems are included with SoftPLC. Other drivers have been developed and are available from I/O vendors or third parties. SoftPLC I/O drivers are called loadable modules (or TLMs).

SoftPLC works with various I/O types. The hardware is available from several I/O vendors. SoftPLC Corporation also distributes some I/O interface cards. I/O can reside on the backplane of the computer or can be remote. Some I/O cards fit directly into the computer, in which case the I/O driver talks to the I/O over a PC bus or backplane. Remote or fieldbus I/O systems typically utilize an I/O scanner or interface card that fits into the PC. In these cases, the SoftPLC I/O driver communicates to RAM memory on the card, which stores the I/O status information from inputs or sends it to outputs.

SoftPLC's maximum I/O capacity is 8192 I/O and includes support for analog, digital, and special I/O. I/O forcing is independently controllable for inputs and outputs. Motion applications can also be controlled. The user adds a motion control card to the PC. The combination of SoftPLC and a computer resident motion control card makes a powerful, tightly integrated system. Tuning, scaling, trajectory, and velocity data can come from SoftPLC using simple ladder logic commands. SoftPLC can then send data to each axis on an event-driven or timed basis.

HARDWARE

Major changes are coming in control hardware. The line between PLCs and PC-based controllers will certainly blur. One of the approaches will achieve dual use from the equipment. Rockwell Automation has a product called the 1747 open controller. This controller was their first entry in a family of PC-based controllers.

The heart of this controller is the 1747-OC CPU module. This is the processor module. It can be used in the standard SLC PLC chassis. The CPU is capable of using any existing Allen-Bradley hardware, I/O chassis, and power supply. The controller module has dual CPUs. One is used as an I/O scanner and the other for the operating system (see Figure 16–14). This setup ensures rapid deterministic response times and reliable operation.

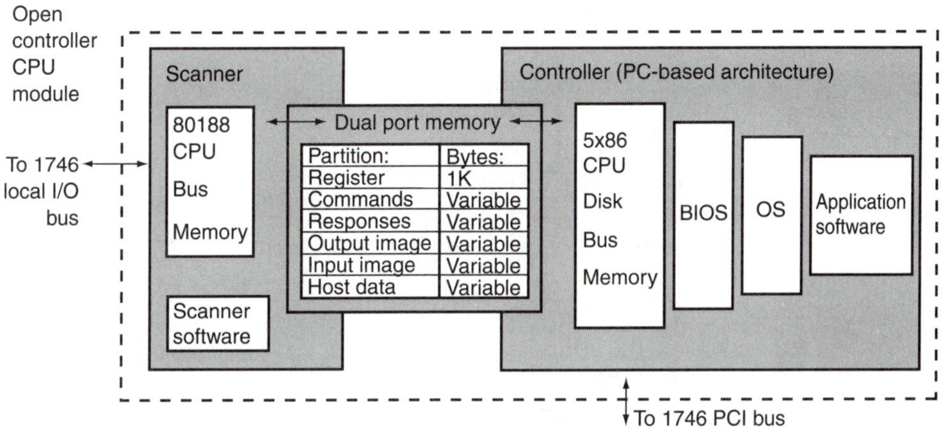

Figure 16–14 Diagram of the open controller architecture. *(Courtesy of Rockwell Automation Inc.)*

Users can choose the operating system and software they want to control the system (see Figure 16–15). Any PC-based commercial software control package can be used with the controller and standard Rockwell Automation I/O. The system can also be expanded to utilize other I/O.

The CPU utilizes SRAM, DRAM, and Flash Drive™ technology so that it is not dependent on a hard drive. There are many communications options because it is a PC-based system. Ethernet interfaces, Allen-Bradley communications cards, SCSI adapters, modems, and so on, are all available to expand the capability of the system. Figure 16–16 shows examples of the use of modules to connect to networks. The system is capable of booting without a monitor or keyboard. It can also be used with a keyboard and monitor for operator interface.

Rockwell Automation also has a product line called ControlLogix that combines the benefits of the PLC and computer.

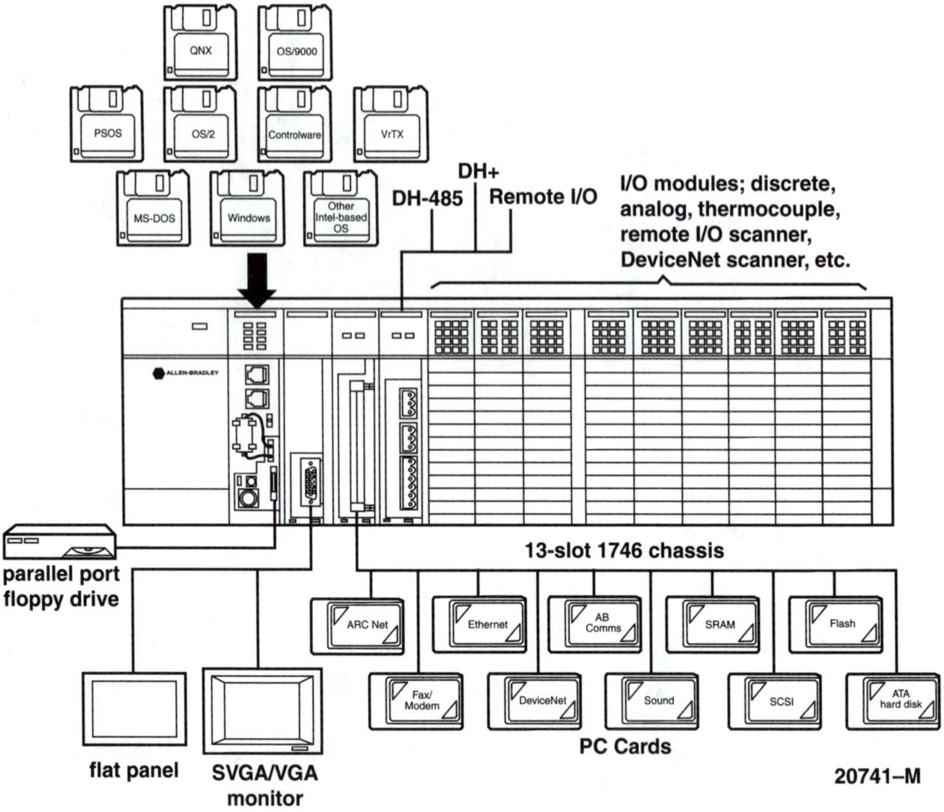

Figure 16–15 1747 Open Controller System and options. *(Courtesy of Rockwell Automation Inc.)*

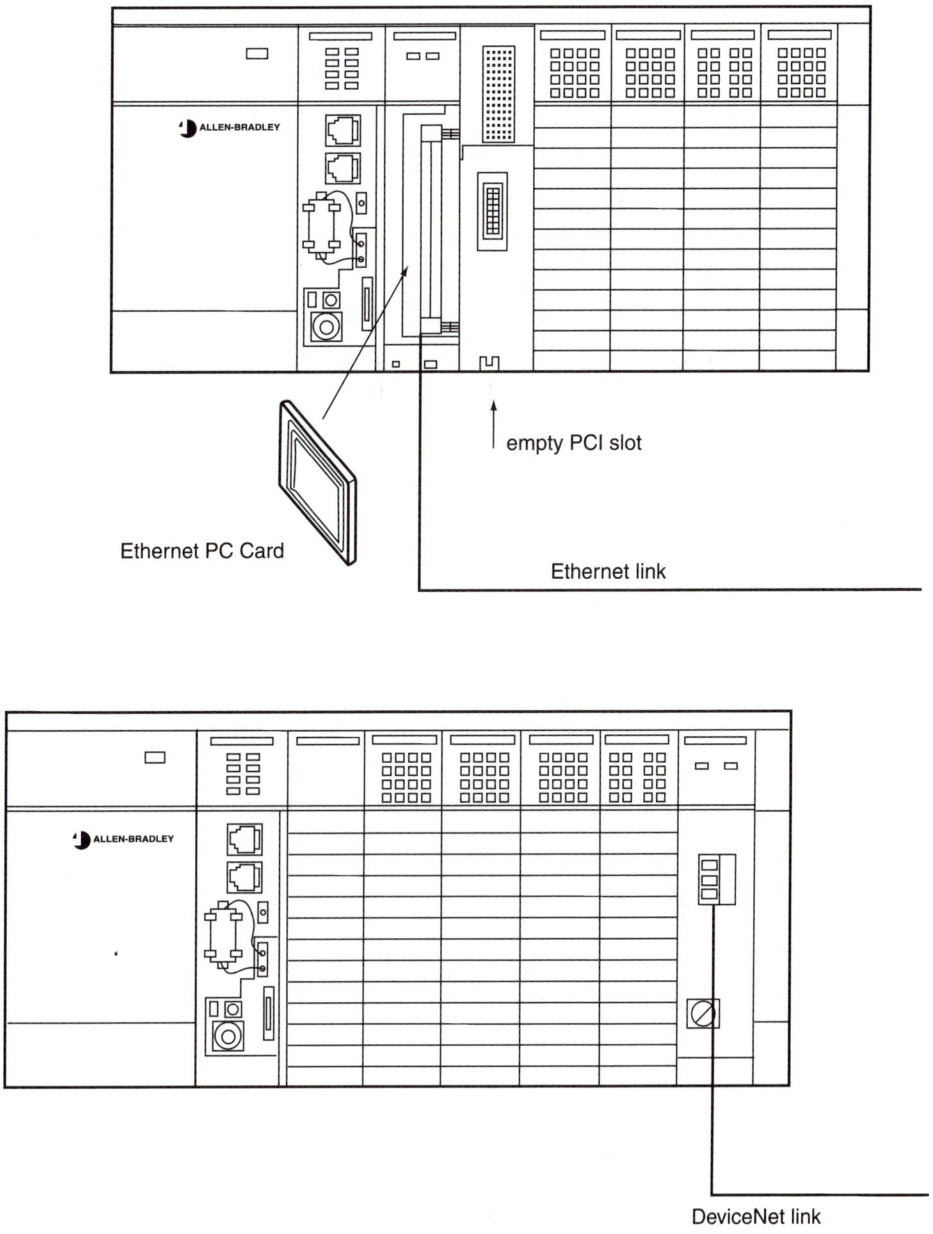

Figure 16–16 Examples of connecting to communications networks. *(Courtesy of Rockwell Automation Inc.)*

QUESTIONS

1. List at least five advantages of PC-based control systems.
2. List at least three potential disadvantages of PC-based control.
3. What is flowchart programming?
4. Describe at least three advantages of flowchart programming.
5. Describe at least three advantages of ladder logic programming for PC-based control.
6. Explain object-oriented programming.
7. Compare and contrast the methods and advantages of each of the four methods of PC-based control discussed in this chapter.

Chapter 17

Fundamentals of Process Control

This chapter covers the fundamentals of process control. Open- and closed-loop systems will be examined. Proportional integral and derivative control will all be covered.

OBJECTIVES

Upon completion of this chapter, you will be able to:

1. Define these terms: *open loop, closed loop, feedback, summing junction, command, error, disturbances,* and *damping.*
2. Describe the principle of on/off control.
3. Describe the principle of proportional, integral, and derivative (PID) control.
4. Describe tuning a PID system.

CONTROL SYSTEMS

Control system performance can be evaluated in many ways. The most important characteristic is stability, which indicates that a system can control smoothly without undue oscillation or overcorrection. Another characteristic is how closely the system can control to the setpoint, also called the *setting* or *command.* The difference between the setpoint and the actual value of the variable is called *error.* The smaller the error, the better the control system. The objective is zero error, but this is not actually possible because the system must always respond to disturbances to try to make the actual value equal the setpoint.

A third characteristic is *steady state.* A system must be designed with an allowable error. For example, an engineer might be asked to design a temperature control system to regulate temperature to a value of 375 degrees plus or minus 2 degrees. This means that the allowable deviation in the actual system is 373 degrees to 377 degrees. When the system is operating at a given setpoint under normal operating conditions, it is called *steady state.*

Another important characteristic of a control system is how quickly it can respond to an error and correct it. In many systems, speed of response is more important than very close control to setpoint.

All these factors are interrelated. An adjustment that attempts to improve the system's accuracy reduces its stability. An adjustment to make a system more responsive to an error reduces the stability. In process control, as in life, system adjustment often seeks to achieve a happy medium.

Open- Versus Closed-Loop Control Systems

There are two basic types of process control: open and closed loop. Open loop is the simplest, so we examine it first. Figure 17–1 shows a level control system that has a tank with an output pipe and a valve that allows input to the tank. Figure 17–2 shows a block diagram of an open-loop system. Block diagrams are often used to illustrate control systems. The open-loop system in Figure 17–2 has no feedback because it has no sensor to sense the fluid level. Open-loop systems have no sensor, so they can provide no feedback.

Let's examine the simple level control application in more detail (see Figure 17–1). First imagine that the valve has only two positions: completely open or completely

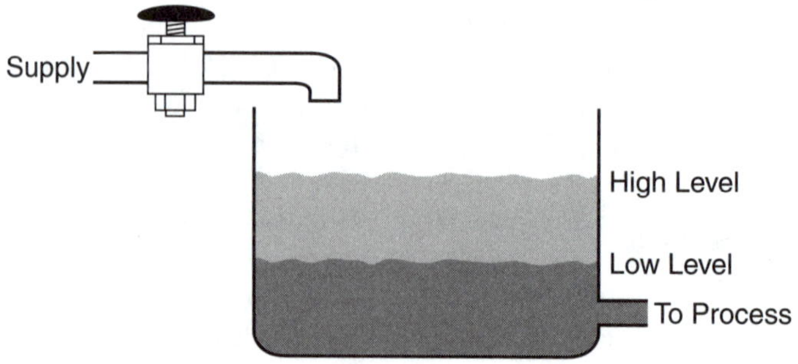

Figure 17–1 Level control system.

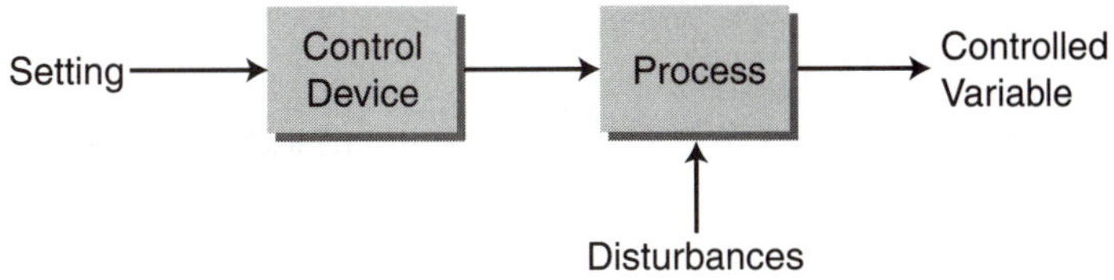

Figure 17–2 Block diagram of an open-loop control system.

closed. Note that two lines have been drawn on the tank: a high-level indicator and a low-level indicator. If the level in the tank goes above the high-indication line, the operator should close the valve. If the level gets below the low-level indication line, the operator should open the valve. This is obviously very crude control but it works if the operator stays awake and is attentive to the process. This is simple on/off control. The operator closes the loop; if the operator's attention is diverted, the level system is open loop and the tank will certainly overflow or empty itself.

Closed-Loop Control System

Closed-loop is a system that has feedback and can self-correct for various changes. We can improve this system if we make it capable of self-correction. Let's use the typical home sump pump for our example (see Figure 17–3).

The float was added to provide feedback and to activate a switch that turns on the pump. The float can be adjusted up or down on the rod to change the setpoint. In our home sump pump, the setpoint is the level of water in the sump hole.

As with other on/off systems, the level is never held right at setpoint. Under normal conditions, the water gradually rises until the float on the rod forces the pump switch on, and the level falls rapidly until the float rod forces the switch to turn off. The water then gradually rises again. Under abnormal conditions, such as during a large spring thunderstorm, the water rises very quickly in the sump hole, and the float rises and forces the pump switch on. If it is raining hard enough, the level might actually rise above the setpoint if the pump cannot keep up, but the pump runs until the water level subsides. While the on/off home sump pump does not control setpoint (one exact level), it certainly does a more than adequate job of controlling the level of water.

Another example of a closed-loop on/off control system is the home furnace (see Figure 17–4). Note in the drawing that the home heating system is relatively simple. Its thermostat located on the first floor of the house measures the actual temperature in the house. The furnace is often found in the basement. If the temperature drops below the setpoint, the thermostat transmits a signal to the furnace, which

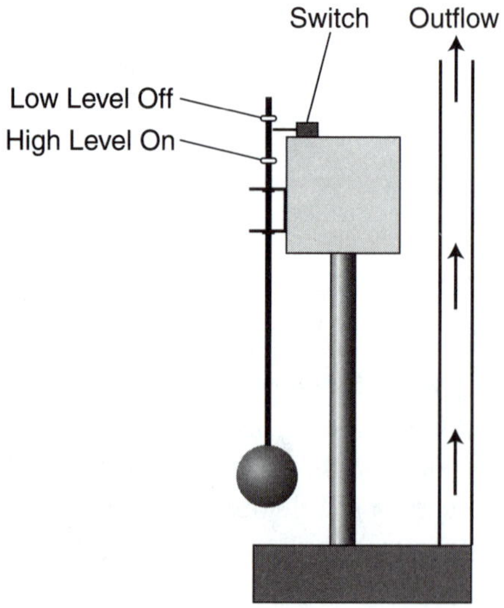

Figure 17–3 Typical home sump pump.

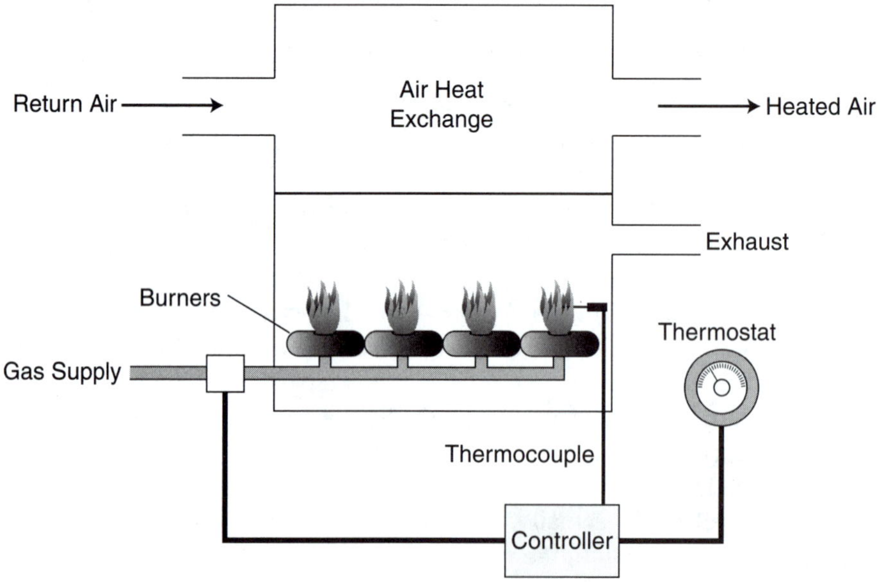

Figure 17–4 Closed-loop on/off control system for a furnace.

opens the gas valve completely and turns on the fan to circulate heated air from a heat exchanger through the house. When the temperature rises above the setpoint, the thermostat opens, and the furnace valve is closed completely. The setpoint in this system is the desired operation point, perhaps 78 degrees.

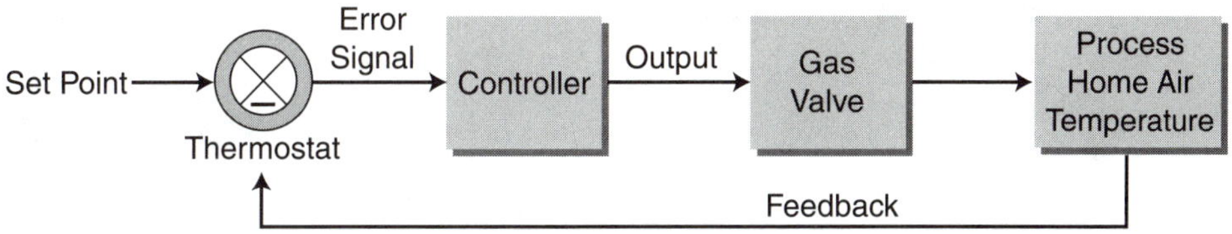

Figure 17–5 Block diagram of a home temperature control system.

Figure 17–5 is a block diagram of the home temperature control system. The process block always determines how the other blocks function. A feedback device must be chosen to measure the process variable associated with the process (the thermostat in this example). The adjusting device (furnace) must be chosen to adjust the manipulated variable (heated air) used to change the process variable (house temperature).

The homeowner sets the setpoint on the thermostat. The thermostat acts as a comparator, also called a *comparer* or *summing junction* (a descriptive name for the comparer that accurately describes its function), that compares the setpoint and the air temperature in the room. The summing junction compares the setpoint and the feedback from the system and generates an error. The thermostat compares the setpoint and feedback (air temperature in the room) and generates an error signal to the furnace. In effect, the thermostat closes a contact to the furnace if the temperature is too low and opens the contact if the temperature is too high.

The summing junction has two inputs: the command (setpoint) and the feedback signal. The summing junction's output is an error signal used to control the final correction device. Note in Figure 17–5 the negative sign on the feedback connection. This means that the negative feedback principle is used. The polarity of the feedback signal is opposite polarity and is then summed with the command to generate an error signal.

Feedback *Feedback* provides current information about a variable we are trying to control, such as level. Many types of sensors (often called *transducers*) can provide feedback. Transducers convert one form of energy to another form. In the usual home furnace, the thermostat's ambient air temperature is converted to a mechanical change in a bimetallic strip in the thermostat that closes or opens an electrical contact (see Figure 17–6). Most transducers for process control systems convert the measured variable (in the furnace, the measured variable is pressure) into an electrical signal that can be measured. In this case, the feedback device is the float, which changes level into mechanical movement to a change in valve position. In other cases, thermocouples convert temperature to proportional voltage and tachometers convert velocity (motion) to proportional voltage.

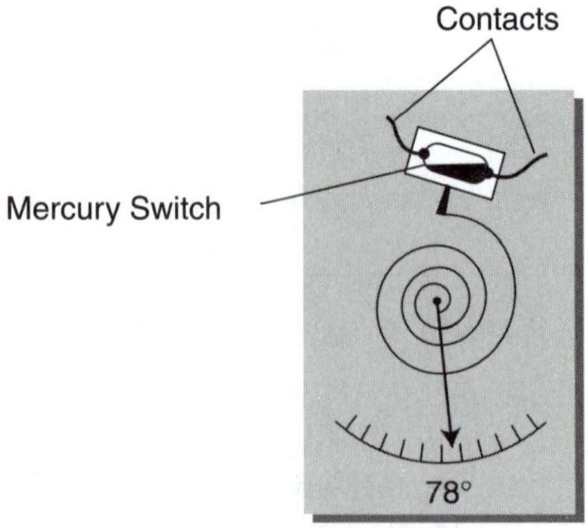

Figure 17–6 Principle of a thermostat.

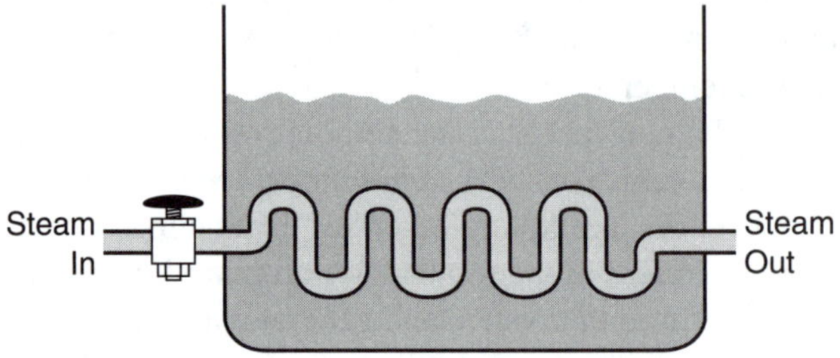

Figure 17–7 Tank with a steam line for heating.

Controlled Variable The *controlled variable* is what is actually manipulated to control the *process variable.* It is not always possible to alter the process variable directly. For example, consider a tank that has a steam line running through it (see Figure 17–7). An operator can control the temperature of fluid in the tank only by controlling how much steam flows through the steam line. In this example, the controlled variable is the *flow rate of steam.* The process variable is the *temperature of the fluid in the tank.*

The operator can improve the system's operation by installing a variable valve at the input to the tank. The valve can then be set so that the flow into the tank from the valve approximately equals the tank's flow out (outflow). If the outflow is constant, the operator can get the input valve position quite close to perfect, and the level will stay pretty much in the middle between the upper limit and lower limit lines. If the level falls, the operator opens the valve slightly; if it rises, the operator closes the valve slightly.

Imagine that the operator set the valve perfectly so that the inflow exactly equaled the outflow. The operator observed the system for 30 minutes and was sure it was operating properly. The operator decided to take a long break. It would not take very long before the tank is empty or overflows. The system is open-loop the minute the operator leaves, meaning that there is no feedback. The operator is not there to watch the process. The next feedback would probably be the boss's comments to the worker when the tank overflowed or ran empty.

Disturbances The operator thought the valve was set perfectly but the tank overflowed the minute he left for several reasons. The tank outflow cannot be perfectly stable; it varies. The inflow cannot be stable either. Think about the water pressure in your house. The pressure in the system is constantly varying. The washer might be running, someone may be taking a shower, someone could be watering the lawn, or the neighbor is washing the car. These are all disturbances to our system; every system always has disturbances.

A *disturbance* is anything that disturbs or disrupts the process. Consider the tank level example. There are many possible disturbances. Imagine that the tank just contains water. The inlet pipe comes directly from the city water department. The water pressure varies continuously, depending on current usage throughout the city. (A running story maintains that city water pressure is always the lowest during halftime of a Super Bowl. Why would that be?)

Low water pressure is an immediate disturbance to the system. Less fluid is flowing in, which causes the level to drop in the tank example. When the second half of the Super Bowl begins, no one is using the water, and the water pressure goes up. Now too much water enters, and the level increases. So water pressure is continuously varying. It is a constant disturbance to the system.

The outflow also varies depending on the process using the water from the tank. Temperature can also have an effect on the viscosity of fluids, which affects flow rates. Imagine a thick fluid, like molasses. Cold temperatures make the fluid thicker and slow its flow rate dramatically. If the molasses is hot, the flow rate increases dramatically. Another reason could be that over time the inside of the pipes have corroded or built up scale, which impedes the flow and disturbs the system.

The home furnace also has many disturbances, including the air temperature outside; the wind; leakage through windows, walls, and doors; and someone opening a door. Systems experience many disturbances, and they affect the process. Some have a dramatic effect on the process, and some have only a small effect. A good control system can correct for all disturbances.

Consider the home furnace. The homeowner chooses a setpoint of 78 degrees. The furnace attempts to keep the temperature in the house at 78 degrees. The home furnace has an on/off controller. If the temperature is above 78, the furnace is shut off. If the temperature is below 78 degrees, the furnace is turned on (see Figure 17–8). The gas valve in a home furnace is either completely open (100 percent open) or completely closed (0 percent open).

Figure 17–8 Furnace on/off versus setpoint.

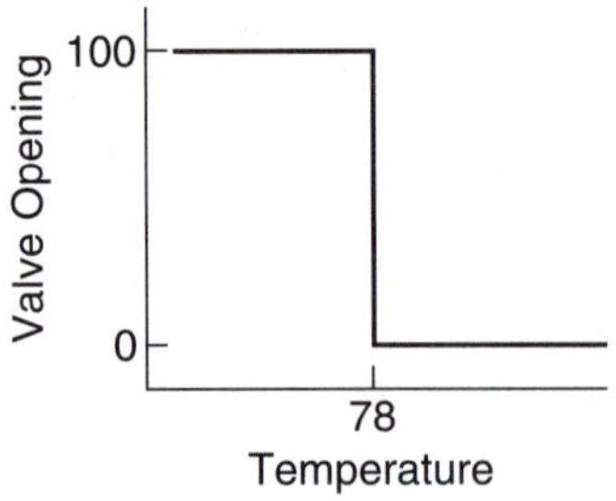

A graph of the temperature in the home would be a sine wave (see Figure 17–9). When the furnace is on, the temperature graph is rising; when the furnace is off, the temperature graph is falling. The graph of the furnace response is a square wave, however, because it is either on or off.

In reality, the furnace cannot change states at exactly one temperature. If the furnace had to change from on to off, or off to on, at exactly 78 degrees, it could not operate. There needs to be a narrow temperature band where the furnace does not change states. This is typically called *deadband* or *hysteresis.*

To help illustrate the concept of deadband, consider a railroad crossing. Imagine two sensors, one on the left of the railroad crossing and one to the right of the railroad crossing. If a train crosses either of these sensors, the crossing light should flash its warning and the gates should close after a delay. The gates should stay closed until after the train completely leaves the crossing. Could the gates open after the first sensor does not sense the train anymore? No! The train may have left the first sensor but stopped in the crossing. Could the gates wait to open until after the first sensor and the second sensor no longer sense a train? This would not be safe either. It would work for a train that is longer than the distance between the two sensors, but what would happen with a short train? This situation would not be safe either because the first sensor would turn on and then off as the train passed by it, but the train is too short and the second sensor would not be on, so the gates open. Logic in this case becomes very important. It ensures that the first sensor is on before the gates are closed. The logic then watches the second sensor to be sure that it turned on and off before the gates were opened (note that this logic is not perfect either). This example does help explain deadband. We do not want any change in the gate position while the train is between the sensors. We do not want the gates to reopen until the train has gone completely past the two sensors.

In the simple temperature control system, we do not want the furnace to change state unless the measured temperature changes completely through the deadband. Study Figure 17–10. Notice the 4-degree deadband. If the temperature is below 76 degrees, the furnace will turn on and stay on until the temperature exceeds 80 degrees. The furnace will then turn off when the temperature exceeds 80 and will re-

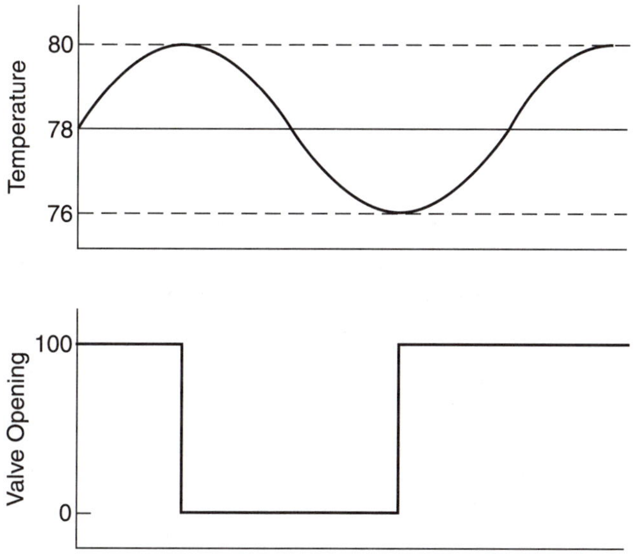

Figure 17–9 Graph comparing the output to the response for a simplified on/off controller.

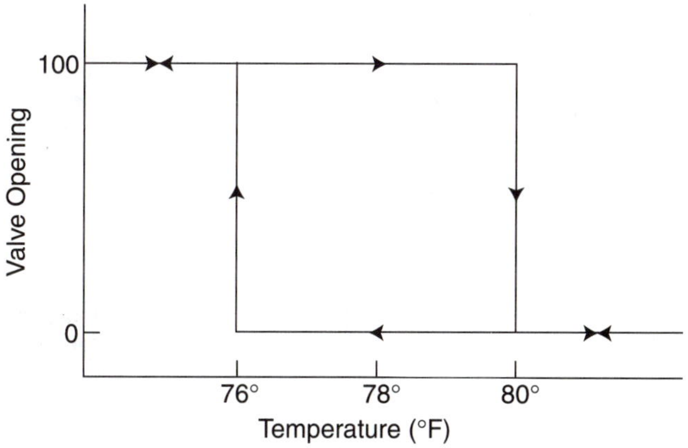

Figure 17–10 Comparison of the output to the response for an on/off controller with hysteresis.

main off until the temperature falls below 76 degrees. Note that the furnace never changes state in the deadband. It changes state only if the temperature is above or below the deadband. As a matter of fact, the valve will never change state unless the measured value (temperature) goes completely through the deadband. In other words, if the temperature is below 76, the furnace turns on and the temperature begins to rise. The temperature moves into the deadband area, and the furnace remains on. It stays on until the temperature rises above the deadband (80 degrees). Now the furnace will remain off until the temperature falls all of the way through the deadband and below 76 degrees.

Note that an on/off controller can never maintain the actual temperature at exactly the setpoint. It continually oscillates above or below the setpoint. The oscillation frequency changes depending on process conditions. Consider your house on a nice fall day with the warm sun shining through the windows. The furnace does not run very often. The temperature oscillation in the house is very slow. Maybe the furnace has to run only once every hour. Now imagine a cold day in February with the wind blowing hard outside. The furnace is running every five minutes to try to keep the temperature at 78 degrees. In this case, it operates at a much higher frequency (see Figure 17–11). Note that the amplitude is approximately the same but the temperature oscillates at a much higher frequency.

Imagine a normal December day in Wisconsin (20 degrees or so). Imagine the furnace is running about once every 30 minutes. What would happen if we narrow the deadband to 2 degrees instead of 4? The furnace would turn on and off more

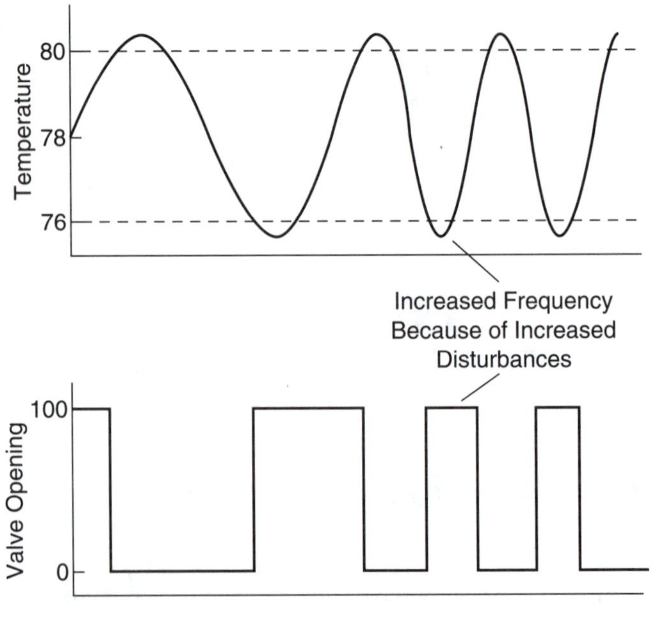

Figure 17–11 Graphs of on/off control.

frequently to maintain the closer temperature. What if we widened the deadband to 8 degrees? The furnace would not have to run very often but the temperature amplitude (variation in temperature) would dramatically increase, and our comfort level would decrease.

Thus, there are limits to increasing the deadband. If we widen it too much, the furnace will not run very often but will run longer and we will not be comfortable. There is, however, a limit to how much we can narrow the deadband because at some point, the system will not be able to keep up with the rapid on and off commands.

Advantages and Disadvantages of On/Off (Closed-Loop) Control On/off control is inexpensive and simple to implement and maintain. It is more than adequate for many control applications.

An on/off controller cannot keep the temperature at the exact setpoint. On/off controllers have permanent oscillation. The temperature always oscillates above and below the setpoint. An on/off controller also needs to have a deadband, which means that on/off controllers cannot hold tight control of the variable (temperature in this example).

Proportional (Closed-Loop) Control

Figure 17–12 shows a closed-loop level control system that is similar to the typical basement sump pump and that has feedback. The block diagram for it is shown in Figure 17–13. The float measures the height of the fluid in the tank and provides the feedback. The float is attached to a rod that automatically opens or closes the

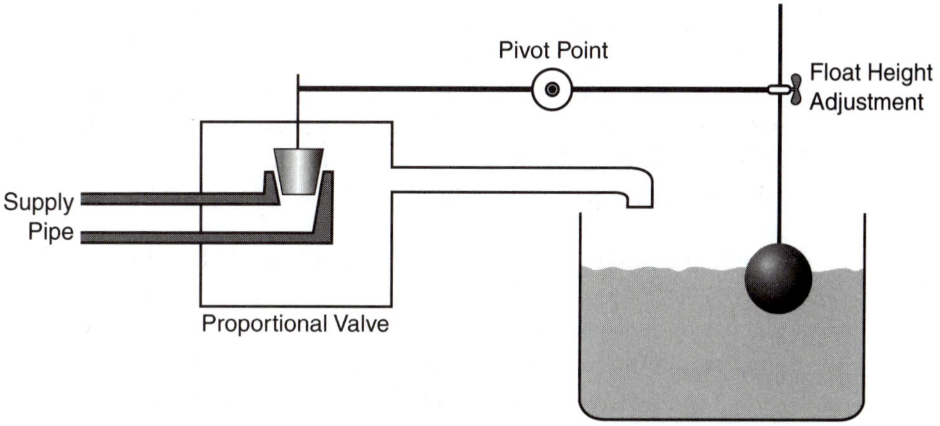

Figure 17–12 Closed-loop level control system.

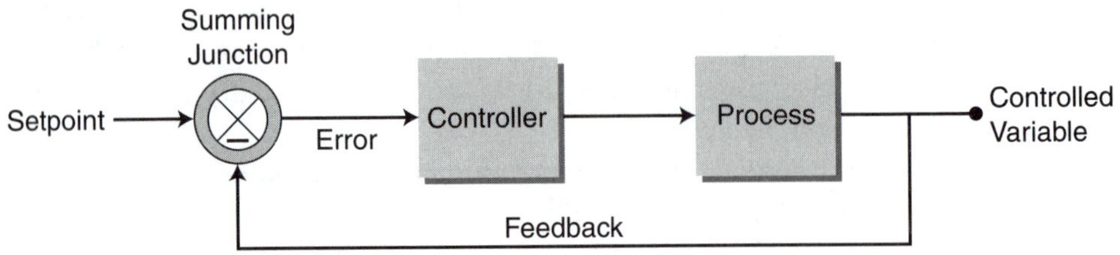

Figure 17–13 Block diagram of a closed-loop control system.

input flow valve. The valve can adjust between 0 percent and 100 percent open. Study the pivot point for the float rod; it can be adjusted to the right or left. This controls how much effect the float has on the valve position. If the pivot is moved to the right, the float has a large effect on valve position. Even a small movement of the float dramatically changes the valve position. If the pivot point is adjusted to the left, the system is less responsive. A change in the float height has less effect on valve position. The fluid level (setpoint) is adjusted by moving the float up or down the vertical rod.

Consider the operation of this system. If the outflow from the tank is relatively constant, we should find a point for the pivot that allows the system to operate quite nicely. The float would cause the valve to open just enough to allow approximately the same amount of fluid into the tank as is flowing out. If we could find the perfect point for locating the pivot, the inflow would always equal the outflow. In reality, there is no perfect pivot point.

If the level varies, it must be corrected. For example, consider the tank level system. If the commanded level in the tank is 5 feet of fluid and the feedback from the float shows 4 feet, there is a *one-foot error* (5 feet − 4 feet). The *summing junction* generates an error and then changes the valve position (generates an error signal) to correct for the 1-foot error. In this case it causes the valve to open more and enable more flow into the tank. This correction continues until the setting (commanded level) equals the feedback and the summing junction generates a zero error.

Overadjustment Now imagine the pivot moved from the "equilibrium" point to a point farther to the right. Now any movement in the float would cause a dramatic change in the valve position. Assume that the fluid level decreased slightly. The float would move down a small amount but cause a drastic change in valve position. The valve in this case would open a great deal and allow too much fluid to flow into the tank. The level would increase rapidly and the float would move up. The float would cause the valve to close too much and the inflow would be too small to keep up with the outflow. With the valve too far to the right, the response to any error would be too great and the system would begin to oscillate. In other words, it would start to fight itself to try and make rapid changes to keep up with its overcorrection. The control in this case is poor and the level would fluctuate rapidly above and below the setpoint.

Now imagine that we move the pivot point too far to the left. A change in the float would cause very little change in the valve opening. Assume that the outflow increases. The float begins to drop, which opens the valve slightly. The valve does not open enough so the level continues to drop, which drops the float and opens the valve more but not enough. The level continues to fall until the float opens the valve enough to make inflow equal to outflow. With the pivot in this position, the response is very slow, and the height of the fluid in the tank varies a great deal. It is very slow to correct for a change in level.

The proper size valve to control the flow is important. If a chosen valve is too small, the valve might not be able to correct for certain errors. For example, the tank eventually empties if the outflow is greater than the inflow and the valve cannot allow enough flow to keep up. If the valve is too large, it could have a tendency to overcorrect and oscillate on and off more. The size of the valve and the setting (pivot position in this simple application) are crucial to proper control of a system. A valve should generally be chosen so that it operates between 30 percent and 70 percent open during normal operation.

Stability *Stability* is key in any system. A system can be very stable if the pivot point (proportional band) is set correctly. If we set the proportional band too small, the system overcorrects for any error and becomes unstable. The level varies a great deal, and the valve opens and closes erratically to try to maintain the level.

In a stable system, the proportional band is set so that the valve makes appropriate adjustments and the level remains relatively constant. Stability is one of the most important characteristics in any system. To be stable, the system components must be appropriately sized for the application and the system must be correctly adjusted (tuned).

The proportional gain portion of *proportional, integral,* and *derivative* (PID) control looks at the magnitude of the error. The proportional response to an error has the largest effect on the system. Proportional control reacts proportionally to errors. A large error receives a large response. (In the case of a large temperature error, the fuel valve is opened a great deal.) A small error receives a small response.

Imagine a furnace that can be heated to 1500 degrees. There is a portion of this range of temperature at which the response of the system is proportional. For example, let's assume that between 1000 and 1500 degrees, the system adjusts the valve opening in proportion to the error (see Figure 17–14). Below 1000 degrees, the valve is open 100 percent. Above 1500 degrees, the valve is open 0 percent. The

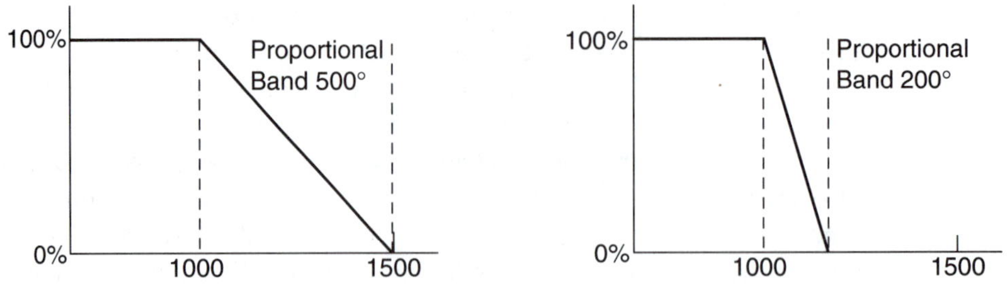

Figure 17–14 Graphs of proportional band between 1000 and 1500 degrees.

$\frac{500}{1440} * 100 = 34.7\%$
.2% Valve Response to 1° Error

$\frac{200}{1440} * 100 = 13.9\%$
.5% Valve Response to 1° Error

Figure 17–15 Proportional band example.

proportional band in this case is 500 degrees. Proportional band is normally given as a percentage (see Figure 17–15). The percentage is calculated by dividing the proportional band in degrees by the full controller range and multiplying by 100. The full controller range is simply the range of temperatures that the furnace can control. In this case let's assume that the full controller range is 1440 (1500 − 60). Are you wondering why 60 was subtracted? Assume that room temperature is 60 degrees. The furnace cannot control below room temperature, so that is excluded in our calculation. In this case the proportional band, in degrees, equals (500/1440) * 100, or 34.7 percent. The technician can adjust the width of the proportional band to make the system more or less responsive to an error. The narrower the proportional band, the greater the response to a given error (see Figure 17–16).

Proportional Control Gain The gain of a system can be calculated as shown in Figure 17–17. For example, in the previous example, the proportional gain was 34.7 percent. If we divide 100 percent by 34.7 percent, we get a gain of approximately 3.

Figure 17–18 is an example of a disturbance in a system. The system's response to three different proportional bands is shown. The setpoint and the disturbance were the same in each case. The narrowest proportional band (20 percent) has the

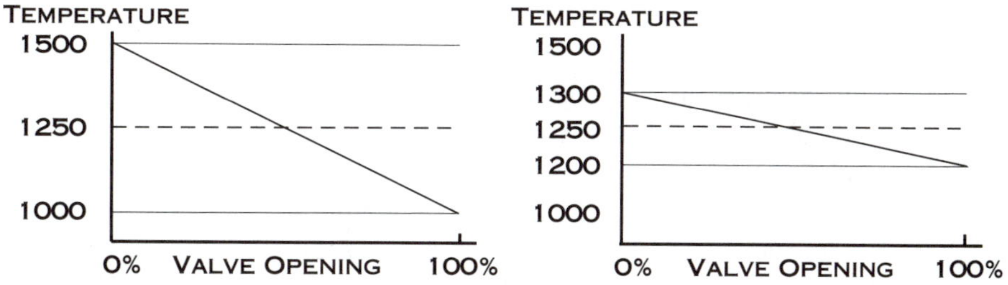

Figure 17–16 The effects of a wide and narrow proportional band.

Figure 17–17 Proportional gain calculation.

$$\text{Proportional Gain} = \frac{100\%}{\text{Proportional Band}}$$

$$\text{Gain} \sim 3 = \frac{100\%}{34.7\%}$$

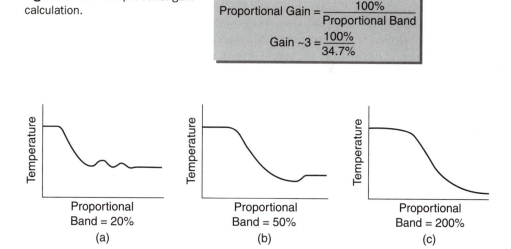

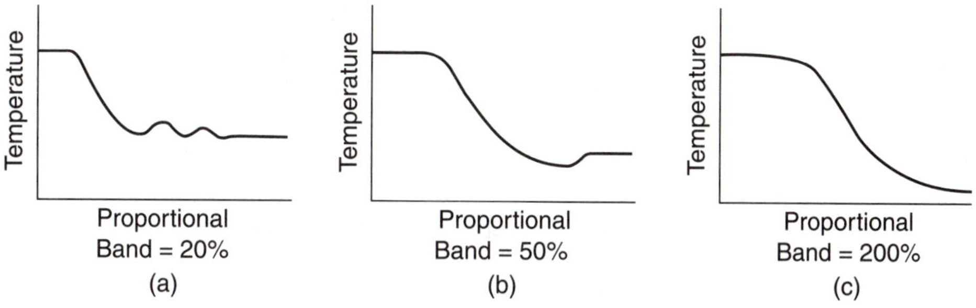

Figure 17–18 Effects of width of proportional band on response to a disturbance.

greatest response to the disturbance (error). The 50 percent proportional band has a smaller response to the same disturbance. The 200 percent proportional band has the least response to the disturbance. This simple illustration presents several important concepts. Note that none of the examples returned the actual temperature to the setpoint, but each did return the temperature to an equilibrium, or steady state. Note also that the narrower the proportional band (most response to an error), the closer the system returns the temperature to the setpoint. The wider the proportional band (least response to an error), the farther from the setpoint is the new equilibrium. Let's examine the reason for this. You may have to read the following paragraph several times to understand this concept.

Assume that the furnace had been controlling the furnace on setpoint (assume 1000 degrees) at 60 percent valve opening. In other words, a 60 percent valve opening keeps the furnace's actual temperature at 1000 degrees. Now imagine that a disturbance occurs. The furnace operator does not completely close the furnace door, and the temperature drops, which increases the error. The controller opens the valve to 70 percent to bring the actual temperature back to the setpoint. The error begins to decrease as the temperature approaches the setpoint. But if the temperature gets back to the setpoint, the error is zero, the valve returns to a 60 percent valve opening, and the temperature falls again. A proportional system can never return to setpoint. Some small error must be maintained to keep the valve at a slightly different position than the setpoint calls for.

Refer again to Figure 17–18. Note that the narrow proportional band recovers more fully from a disturbance because a narrow proportional band causes a larger response to an error. In other words, a 20 percent proportional band responds 10 times as much to the same error as our 200 percent proportional band.

Note the difference in the graphs again. The narrow proportional band response shows some oscillation around the setpoint until the system settles. The second shows a smoother response and just a slight overshoot. The third example takes the longest to correct but is the smoothest response. Which is best? It depends on the application. If we need very quick correction for errors, the narrow band is best. If smooth response is crucial, the second or third band might be best.

Thus far, we have assumed that a proportional controller can control on setpoint if there are no disturbances. Unfortunately, this is not true. In reality and assuming no disturbances, there is only one point at which a proportional controller can control at exactly the setpoint.

Offset If a proportional-only control system has its setpoint set exactly at the 50 percent valve opening needed to maintain the temperature, it can control right on setpoint, assuming no load changes or disturbances. This, however, is never the case. Proportional-only controllers can never control right on setpoint. The difference between the setpoint and the actual measured value is called *offset* (see Figure 17–19). The offset may cause the measured value to be below or above the setpoint. The amount of offset varies by how close the setpoint is to the 50 percent valve opening (ideal setpoint). The farther the setpoint is from the ideal setpoint, the larger the offset (see Figure 17–20). Offset can also be caused by load changes. Remember that there is one ideal temperature assuming a given load. The valve would be 50 percent open to maintain that value. Now assume a larger load but the same setpoint. A 50 percent valve opening cannot maintain the setpoint anymore because of the larger load. So the actual value cannot be held at the exact setpoint because if there were a zero error, the valve would return to 50 percent open and the actual

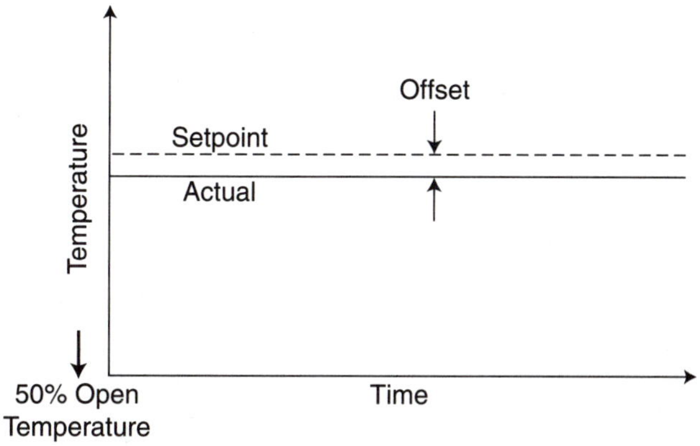

Figure 17–19 Offset in a proportional system.

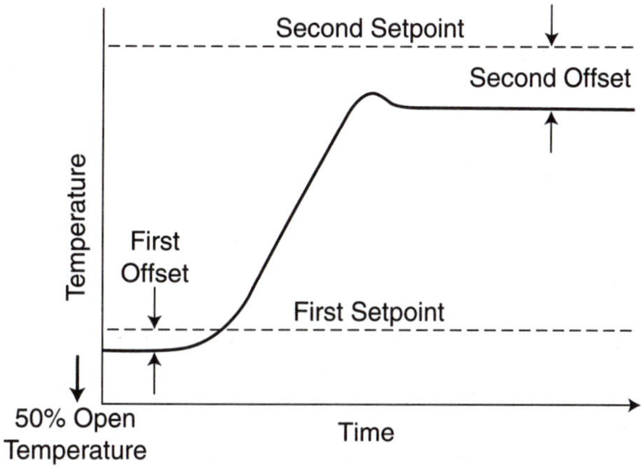

Figure 17–20 The farther the actual setpoint is from the ideal setpoint, the larger the offset.

value would drop again. Offset can be reduced by narrowing the proportional band. Remember that a narrow proportional band responds more to even a small error. Remember also that if the proportional band is narrowed too much, the system becomes unstable. To illustrate, imagine a proportional band of zero. It would be an on/off controller without hysteresis.

Effect of Load The load on a system affects how the system responds. A heavier load requires more valve opening to keep the system on setpoint. Figure 17–21

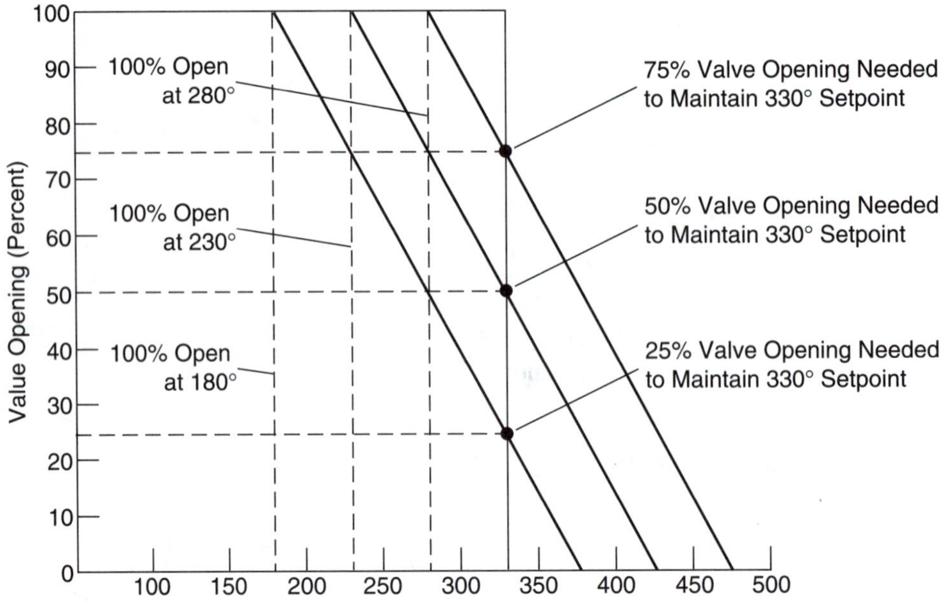

Figure 17–21 Effect of load on a system.

shows a system with three different loads. Imagine a furnace in which a small metal part needs to be brought up to temperature. Then imagine how the same furnace would have to respond to a large piece of metal.

Let's examine proportional gain first. The proportional factor responds to the magnitude of an error: large error, large response; small error, small response. Proportional control means we make the response to an error proportional to the error. Proportional control can help eliminate some of the shortcomings of on/off control.

Consider the simple temperature control system in Figure 17–22. This system represents an industrial furnace with a proportional valve to control the flow of gas to the burner. The valve can be controlled between 0 percent and 100 percent open. To keep the numbers simple, assume the furnace can control the temperature between 70 and 1070 degrees, or 1000 degrees. This means that the valve is 100 percent open at 70 and 0 percent open at 1070. Thus 100 percent of the valve opening is spread over 1000 degrees. The graph in Figure 17–23 shows the relationship between valve opening and temperature for this system. For this graph, assume that the 50 percent valve opening represents the setpoint. If the temperature rises above 570, the valve must close proportionally to bring the temperature back to 570 degrees. For any 10 degree temperature change, the valve changes 1 percent (see

Figure 17–22 Industrial furnace.

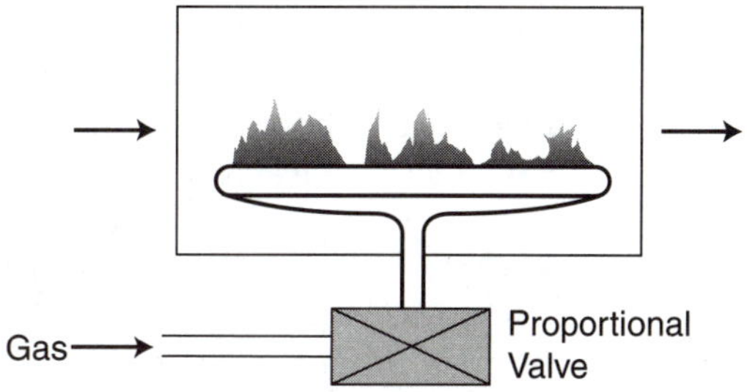

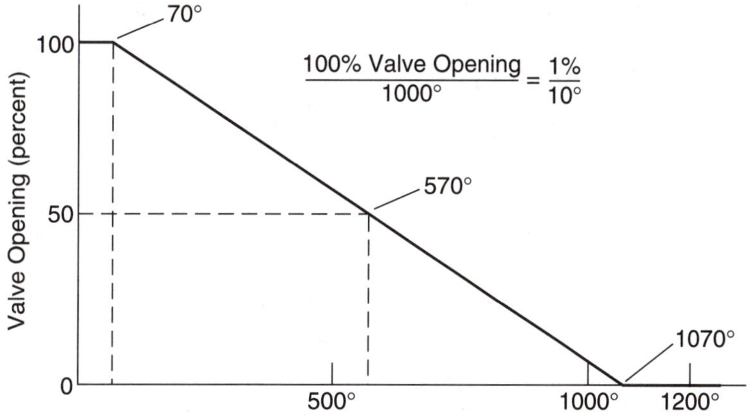

Figure 17–23 Relationship between valve opening and temperature.

Figure 17–23). Thus, if the temperature rises to 580 degrees, the valve closes 1 percent more to bring the temperature back to 570 degrees. If the temperature falls to 560 degrees, the valve opens 1 percent more. This system would have very low response to an error. For example, if the operator changes the setpoint from 300 to 350, the valve changes only 5 percent.

Imagine that it takes five minutes for the furnace to get to the new temperature. The highest temperature we need for our process is 450. What if we could change our system so that the valve closes at 70 degrees and is 100 percent open at 570 degrees? This would improve the system response. In other words, the system would correct errors more quickly. This change can be done by revising our proportional band over which the valve changes from 0 to 100 percent open.

In the previous example, it took five minutes to change the temperature 50 degrees when the setpoint was changed. There was a 1 percent change in valve position for every 10 degree error. With the new smaller proportional band, the system has twice the response to a given error. If the setpoint is changed from 300 to 350 degrees, there is a 50 degree error. The system now responds with a 2 percent valve change, twice the response. It can correct for a 50-degree error in approximately 2.5 minutes—twice as fast as the previous system.

Summary of Proportional Control Proportional control acts on the magnitude of the error. It can eliminate the permanent oscillation that exists with an on/off controller. Proportional control also comes closer to setpoint than an on/off controller. A few problems are associated with proportional control, however. They cannot control exactly on setpoint because of permanent error and offset, and they are appropriate for most slowly changing systems.

Integral Gain Control

Proportional control cannot solve all problems. Imagine the cruise control system in a car. We set the setpoint at 65 miles per hour. The cruise control sets the fuel flow rate. If there are no disturbances, the car moves at 65 miles per hour. There are many disturbances, however, such as going up a gradual incline. The speed falls slightly based on the proportional fuel flow rate.

Proportional cannot correct for very small errors, called *offset* or *steady-state errors.* An example of a steady-state error might be driving your car on a level road with the cruise control on. The car does not maintain the exact speed that you chose. The proportional control does not need to adjust, however. The proportional control cannot correct small variations, or steady-state error.

The integral gain portion corrects for the small error (offset) that proportional cannot. Integral gain control looks at the error over time. It increases the importance of even a small error over time. Integral is determined by multiplying the error by the time the error has persisted. A small error at time zero has zero importance. A small error at time 10 has an importance of $10 \times$ error. In this way integral gain increases the response of the system to a given error over time until the problem is corrected. Integral can also be adjusted. The integral adjustment is called *reset rate,* which is a time factor. The shorter the reset rate, the quicker the correction of an error (see Figure 17–24). In hardware-based systems, a *potentiometer* makes the adjustment. The potentiometer essentially adjusts the time constant of a resistor–capacitor (*RC*) circuit. Too short a reset rate can cause erratic performance. In software-based systems, the user sets a parameter that is based on reports/minute, or minutes/repeat.

Proportional versus Integral Gain Control Proportional considers the magnitude of the error, but it cannot control exactly on setpoint. There is always a small permanent error and offset in a proportional controller. Integral gain control corrects

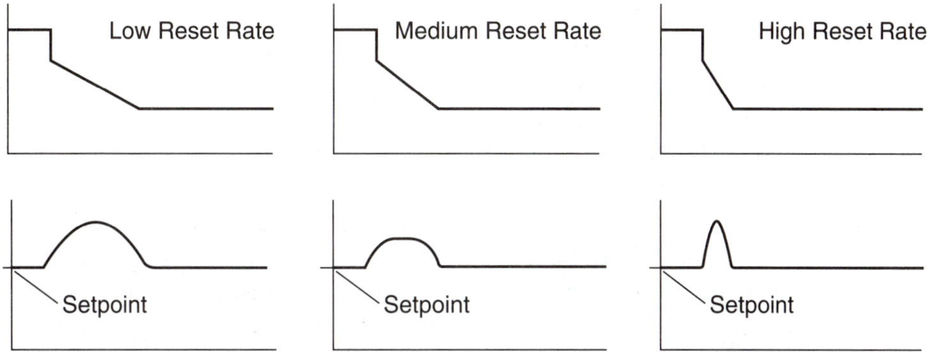

Figure 17–24 Effects of short, medium, and long reset rate. The vertical line in the three top graphs shows the proportional response to the disturbance. The proportional response to the error was the same in each example.

these small errors over time. The addition of integral control enables a proportional system to control on setpoint and can eliminate permanent error and offset.

One More Analogy of Proportional versus Integral Control Imagine that you are driving your car at 45 miles per hour (mph). You begin to drive up a long hill and your speed begins to drop. You respond by giving the car more gas until your speed stops declining. Let's assume that the car is going at a steady speed of 42 mph now, which is the proportional response.

Now you give the car a little more gas and check the speedometer. Then you give the car more gas and check the speedometer. You repeat this process until the speed is back at 45 mph. This process is the integral response.

Derivative Gain Control

Both proportional and integral have problems. Neither considers the rate of change in the error. A system should react differently to a rapidly changing error than a slowly changing error. Derivative gain control, sometimes called *rate time,* considers the rate of change in the error and attempts to look ahead and prevent overshoot and undershoot of a proportional controller. This is often called *damping*; it allows the gain (or proportional factor) to be set higher than would be possible without some derivative.

Consider the cruise control again. The car is going along at 65 miles an hour and all of a sudden begins going down a very steep hill. The proportional and integral controls were keeping the speed constant until now. The rate for change in the error is large now (the speed is rapidly increasing). Derivative control acts and makes a large change in fuel flow. This brings the speed back close to setpoint again so that integral may again act if a small error persists over time.

Let's consider another automobile example. A major problem with proportional control is that it cannot adjust its output based on the rate of change in the error (remember that the proportional term looks at magnitude of error, not rate of change in the error). We are driving on the highway and the brake lights on the car ahead of us come on. We react normally with what we think is appropriate brake pressure, perhaps 50 percent. This is the proportional response. Then we watch the gap between our car and the car ahead of us. If the gap remains constant (constant error), our proportional response was correct and no additional response (derivative) is needed. If, however, the gap is decreasing rapidly (rapid rate of change in the error signal), we quickly apply additional brake pressure—a derivative response. Likewise, if the gap is increasing rapidly (large rate of change in the error signal), we reduce pressure on the brake pedal.

In other words, proportional responds first and then the derivative monitors the rate of change in the error. The derivative acts when a change in the rate of error occurs. We can control how quickly derivative responds to a change in the rate of error. Derivative causes a greater system response to a rapid rate of change than to a small rate of change. Think of it this way: if a system's error continues to increase, the control device must not be responding with enough correction. Derivative control senses this rate of change in the error and causes a greater response.

The derivative gain helps "damp" a system. If the goal is to have a system that is responsive and quickly correct errors, it is necessary to have a relatively high proportional gain. A high proportional gain causes the system to overshoot. The derivative gain can help damp the overshoot. When a system overshoots the commanded value (position, velocity, or temperature), the error rapidly increases.

The derivative gain responds to this rapid rate of change in the error and damps the proportional response. The derivative term provides a much higher proportional gain and a stable system. The derivative term also adds "stiffness" to a system. If we were to turn the shaft on a small motor with proportional but no derivative term, it would move relatively easily. When we add derivative gain, the shaft becomes much more difficult to move. The system fights more to maintain position or zero velocity.

Proportional, Integral, and Derivative Control Proportional gain control considers the magnitude of the error. Integral control considers small errors over time. Derivative control considers the rate of change in the error. It can help damp a system. The addition of derivative allows the proportional gain to be higher and improves the response and stability of a system. Derivative is more common in slower changing systems, such as those involving temperature. It is less common in flow control.

PID Control Systems

Proportional, integral, and derivative controls are used to control processes. PLC manufacturers have approached PID control in two ways: some offer special-purpose processing modules, and some utilize standard analog/digital I/O with PID software in the CPU (see Figure 17–25). PID can be used to control physical variables such as temperature, pressure, concentration, and moisture content. It is widely used in industrial control to achieve accurate control under a wide variety of process conditions. Although PID often seems complex, it need not be. It is essentially an equation that the controller uses to evaluate the controlled variable (Figure 17–26). The controlled variable (temperature, for example) is measured and feedback is generated. The control device receives this feedback. The control device compares the feedback to the setpoint and generates an error signal. The error is examined by proportional, integral, and derivative methodology. Each of the three factors can be thought of as a *gain.* Each can affect the amount of response to a given error. We can control how much of an effect each has. The controller then uses these gains (proportional gain, integral gain, and derivative gain) to calculate a command (output signal) to correct for any measured error.

Figure 17–27 shows how a PLC implements the PID algorithm. Note the loop table in the diagram; it holds the user parameters for each gain.

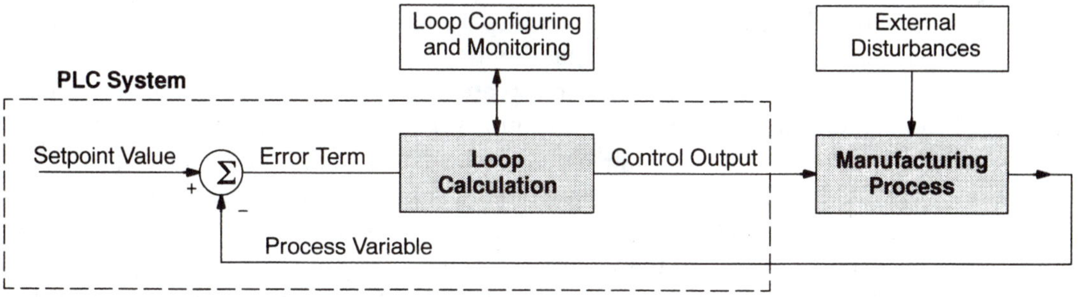

Figure 17–25 Block diagram of a process controlled by a PID controller. The system is a tank that controls temperature. A PLC controls the system. The CPU takes input from the analog input module and performs the PID calculation. The CPU generates an error signal and sends it to the output module (digital or analog). The output controls the fuel valve. *(Courtesy AutomationDirect.)*

$$Output = K_C\left[(E) + \frac{1}{T_1}\int(E)dt + T_D\frac{D(PV)}{df}\right] + Feed\ Forward/Bias$$

Figure 17–26 PID equation.

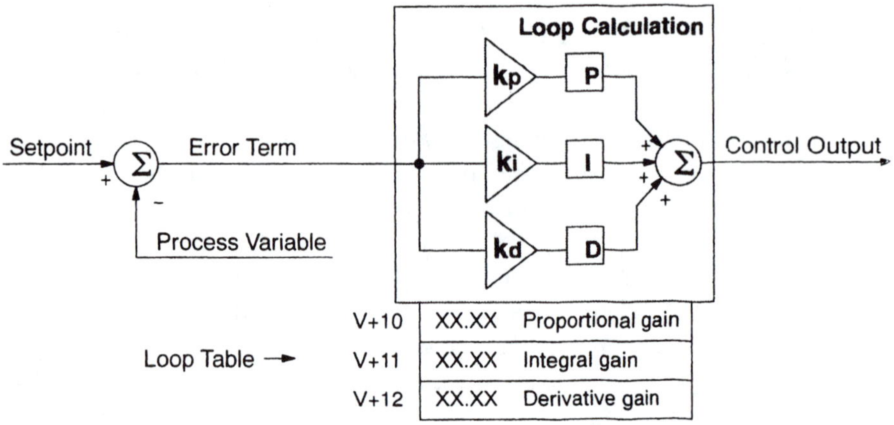

Figure 17–27 Block diagram of the PID loop calculation. *(Courtesy AutomationDirect.)*

Rockwell Automation PID Instruction Rockwell Automation has a PID instruction that can control a closed loop using inputs from an analog input module and providing an output to an analog output module. The instruction allows the user to convert an analog output to a time proportioning on/off output for driving a heater or cooling unit. A Rockwell Automation PID instruction is shown in Figure 17–28.

The control block is a file that stores the data required to operate the instruction. The file length is fixed at twenty-three words. The user enters the address of the first integer word to be used. For example, if the user enters N7:10, the instruction allocates words N7:10 through N7:22.

The process variable (PV) is an element that stores the process input value. This address normally is the address of the analog input word. This value can also be the address of an integer value if the user chooses to prescale the input value to the range of 0–16383.

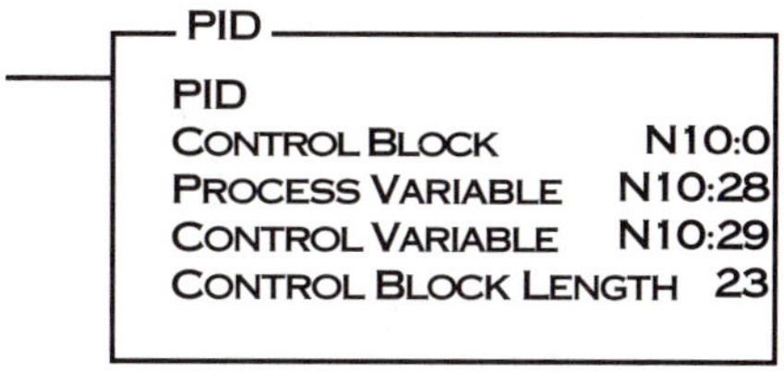

Figure 17–28 Rockwell Automation PID instruction.

The control variable (CV) is the address that stores the output of the PID instruction. The output ranges from 0–16383; 16,383 is 100 percent of the value. The user can use the address of the analog output word or an integer that can then be scaled to the particular range needed.

After these addresses have been filled in, the user double-clicks on the instruction and configures the parameters that determine how the instruction will perform for the application.

Most larger PLCs offer PID capability through the use of software and analog I/O modules. Some offer the capability through special PID modules. The DirectLogic 250 CPU offers extensive PID capability. Figure 17–29 is an example of the use of a PLC to control a tank process using PID control.

Note that there is an analog input module so that the process can be monitored. The output can be either analog or digital. In this application, a PLC Direct 250 processor was used. The 250 DirectSOFT programming software allows the user to use dialog boxes to create a formlike editor to set up the loops. DirectSOFT's PID Trend View can then be used to view and tune each PID loop.

Figure 17–30 is a diagram of the system. The process in this case is the level of fluid in the tank. Note the external disturbances that affect the process, which are always present. Temperature changes can cause the viscosity of the fluid to change, the pressure of the supplied fluid into the tank can vary, and so on. There are many disturbances, and their individual effects are constantly changing. The PID control system must overcome these disturbances.

Cascade Control Better control in some industrial systems can be achieved by utilizing two control systems. Figure 17–31 shows a simple level control system. A controller monitors the level in the tank. If there is an error, it opens or closes the control valve to correct the error and brings the tank back to the correct level. This setup would be adequate for most applications.

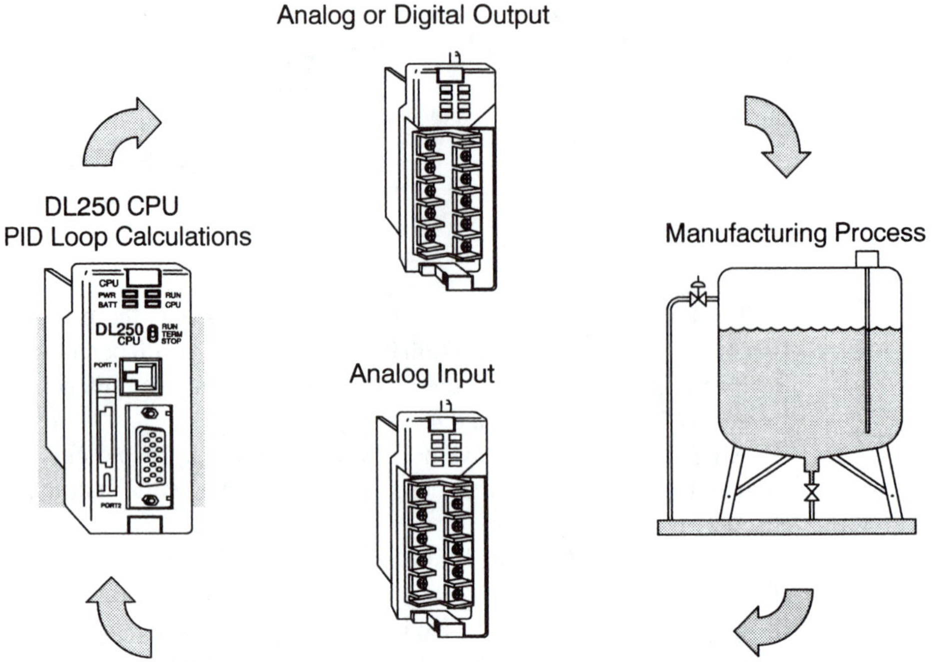

Figure 17–29 PID-controlled tank process. *(Courtesy AutomationDirect.)*

One weakness in this system is that it responds only to a level change. Many disturbances can occur upstream that will change the flow rate. The flow rate can vary widely, but the change is seen only when the level changes. The level change is a second-stage effect. This system cannot respond to a flow-rate change until the level changes.

Figure 17–32 shows a level control system in which a second controller and a second feedback device has been added. In this system, the output (error) of the second controller becomes the setpoint for the flow controller. This is called cascading. Any change in the outer loop (level) generates an error signal that becomes a new setpoint for the flow controller. Even though the measured value of the inner loop (flow) has not changed, the new setpoint causes the flow controller to generate an error signal and change the flow. Figure 17–33 shows a block diagram of a cascade level system. Cascade control can provide better control of the outer loop variable (level) than can be accomplished in a single variable system.

Feedforward Feedforward is an enhancement to PID control that improves system response when a predictable error occurs from changing setpoints or commands. For

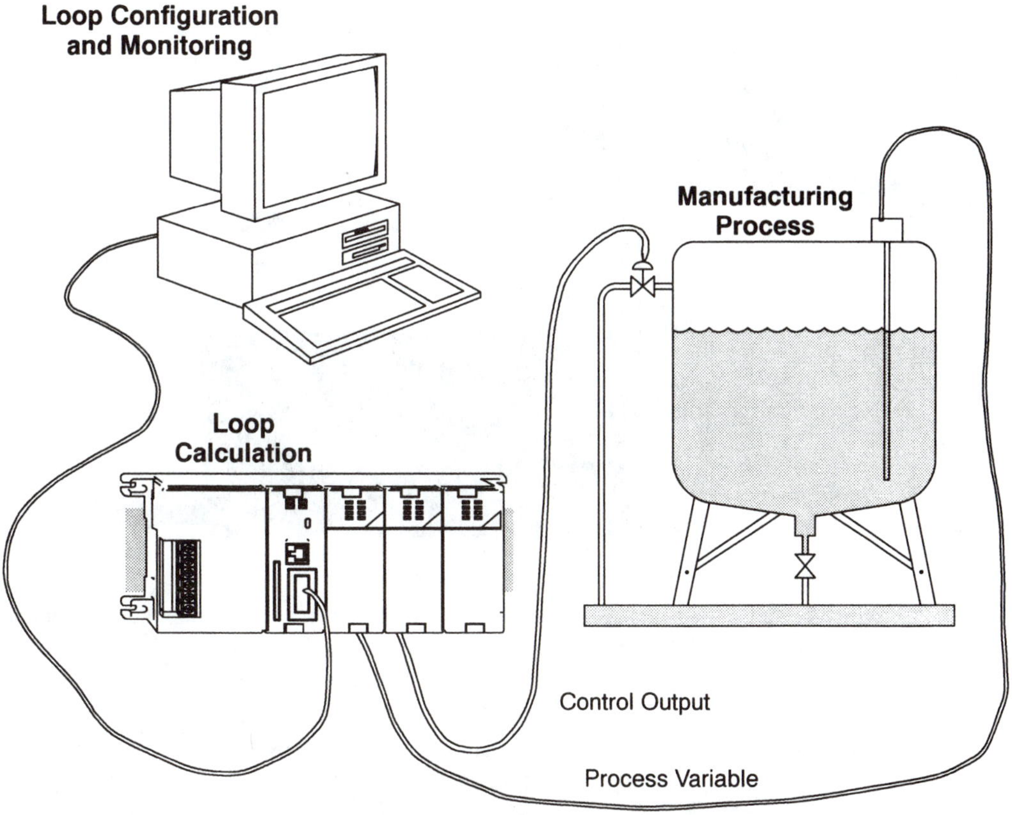

Figure 17–30 PLC system to control the tank process. *(Courtesy AutomationDirect.)*

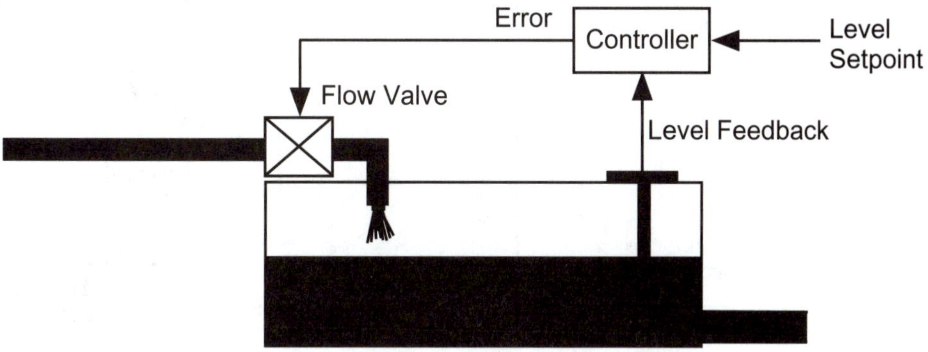

Figure 17–31 A simple level control system.

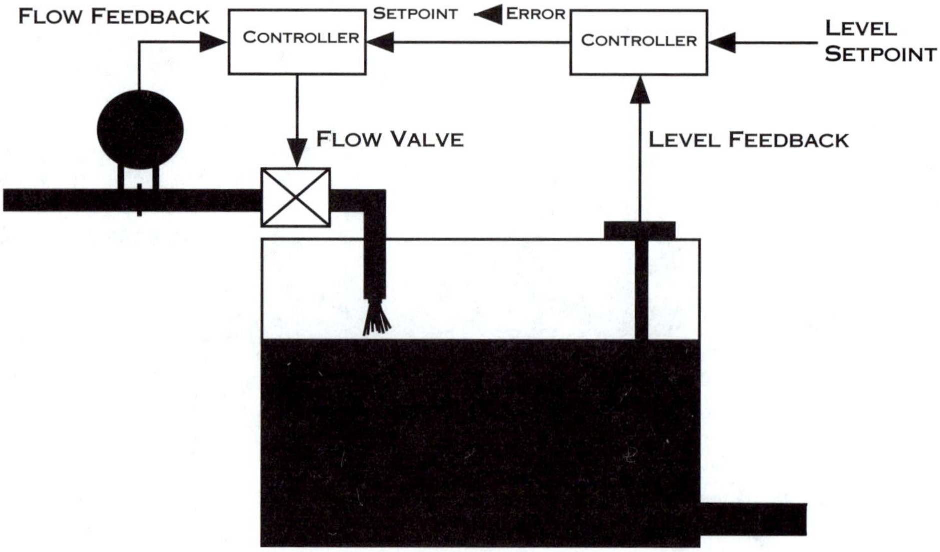

Figure 17–32 A level control system in which a second controller and a second feedback device has been added.

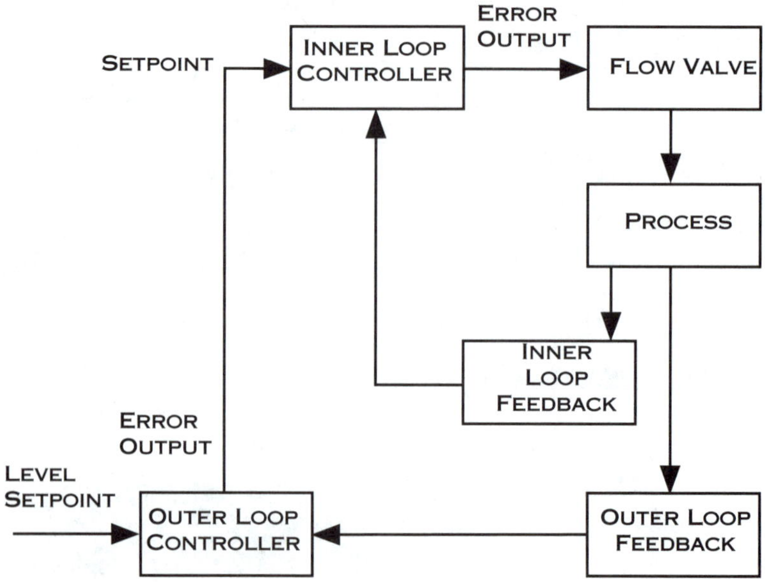

Figure 17–33 Block diagram of a cascade control loop for a level control system.

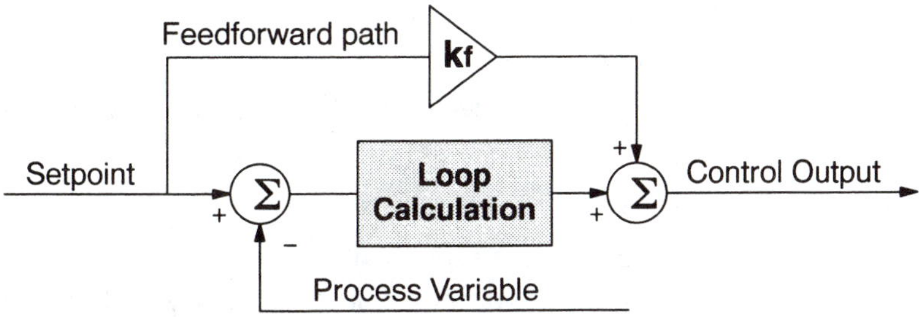

Figure 17–34 Feedforward compensation. *(Courtesy AutomationDirect.)*

example, after adjusting the PID parameters as well as is possible, a predictable following error exists when a new position command is sent. Feedforward can be used to compensate for these errors. Figure 17–34 shows a block diagram of *feedforward* or *bias*. The feedforward term is "fed forward" around the PID equation and summed with its output. Without feedforward when a new command is issued, the loop does not know what the new operating point is. The loop essentially must increment and/or decrement its way until the error disappears. When the error disappears, the loop has found the new operating point. If the error is somewhat predictable (known from previous testing) when a new command is issued, we can change the output directly using feedforward. This term can be added in many controllers to help improve system response. If used correctly, it can also help reduce integral gain and improve system stability.

Tuning a PID System Tuning a PID system involves adjusting potentiometers or software parameters in the controller. Many systems now employ software-based PID control allowing the technician to change a software parameter value for each factor. The goal of tuning the system is to adjust the gains so that the loop has optimal performance under dynamic conditions. Tuning is often a trial-and-error process, but formulas can be used to develop approximate PID settings. One of these will be examined later in this chapter.

The technician should follow the procedures in the technical manual for the system being adjusted and must take care whenever making changes to a system. The following example of tuning a drive is based on a system in which the technician can watch the actual error terms while the system is operating. In general, the tuning procedure is as follows:

1. Turn all gains to zero.
2. Adjust the proportional gain until the system begins to oscillate.

3. Reduce the proportional gain until the oscillation stops and then reduce it by about 20 percent more.
4. Increase the derivative term to improve the system stability.
5. Next raise the integral term until the system reaches the point of instability and reduce it slightly.

An oscilloscope can be used to view system performance during tuning. In general, a command signal, often a square wave, is issued to the drive. By comparing the square wave command on the oscilloscope screen to the system response, the technician can make adjustments to the PID parameters. The system response can be traced by checking the tachometer or the current feedback.

Zigler-Nichols PID Tuning A well-tuned control loop should exhibit the following characteristics: stability, acceptable performance at steady state, and adequate response to setpoint changes and disturbances.

One of the oldest and most widely used PID tuning techniques is called the *Zigler-Nichols method.* The user makes adjustments to a closed-loop system until steady oscillations begin to occur. The user must find the proportional gain (critical gain) that causes a steady-state oscillation to occur and then measure the period (critical period) of the oscillations. The user then calculates settings for proportional, integral, and derivative settings based on the two critical gains and the critical period that caused the steady oscillation.

The Zigler-Nichols method begins by allowing only the proportional factor to influence the system. The integral and derivative are set so that they have zero effect on the loop. Next the user gradually increases the proportional gain to establish the point at which the system begins to oscillate steadily around the setpoint. This is the critical gain. The user must also create small disturbances during this phase of the tuning to force the system into oscillation. The user should then note the critical gain K_C when the measured variable first begins to oscillate around the setpoint. At this point the user must record the critical period (T_C) of the oscillations in minutes.

For a controller that utilizes only a proportional gain, the proportional gain is calculated using the following formula:

$$K_p = 0.5 * K_C$$

The values for a proportional plus integral controller can be calculated as follows:

$$K_p = 0.45 * K_C$$
$$T_I = T_C/1.2$$

If the quarter-amplitude criteria is desired, use this formula:

$$T_I = T_C$$

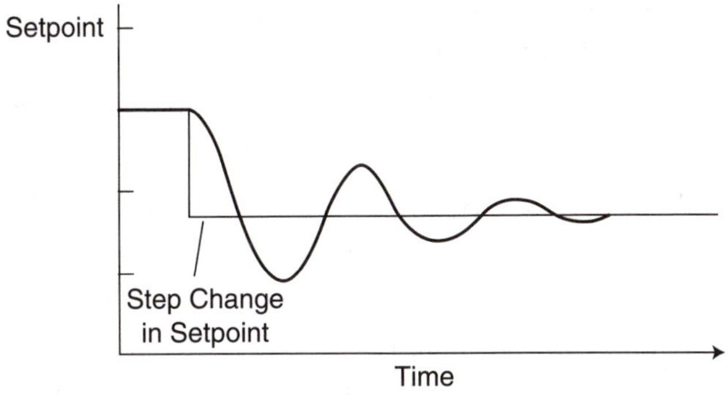

Figure 17–35 A step change in input and a quarter-amplitude response.

Note that the proportional gain must then be adjusted to achieve the quarter-amplitude response. The values for a full proportional, integral, and derivative control system are calculated using the following formulas:

$$K_p = 0.6 * K_C$$
$$T_I = T_C/2.0$$
$$T_D = T_C/8$$

Note that for a proportional-only controller, the gain is found by multiplying the critical period by 0.5. When the derivative is added, the proportional gain is calculated by multiplying the critical period by 0.6. This means that the proportional gain can be set higher (approximately 20 percent) when a derivative is added.

If quarter-amplitude response is desired, the formulas are:

$$T_i = T_C/1.5$$
$$T_D = T_C/6$$

Quarter-amplitude tuning (see Figure 17–35) implies that the amplitude of each peak of a cyclic response must be one quarter of the preceding peak. The proportional gain must then be adjusted to achieve a quarter-amplitude response. Many combinations of PID values can achieve acceptable system performance.

Tuning Problem A technician needs to tune a temperature control loop for a new application. The technician begins adjusting the proportional band down from 100 percent while applying regular disturbances for the system to react to. When a 40 percent proportional band is reached, the system starts to oscillate. The technician measures the time period and finds it to be 8.5 minutes.

Tuning Problem Solution The critical gain can be calculated from the proportional band using the following formula:

$$K_C \text{ (critical gain)} = 100/\text{proportional band}$$
$$2.5 = 100/40$$

The critical gain (K_C) is 2.5. Next the proportional gain is calculated:

$$K_p \text{ (proportional gain)} = 0.6K_C$$
$$1.5 = 0.6 * 2.5$$

The proportional gain is 1.5.

The setting for the integral is found using the following formula:

$$T_I = T_C/1.5$$
$$5.7 = 8.5/1.5$$
$$T_I = 5.7 \text{ minutes}$$

The setting for the derivative is found as follows:

$$T_D = T_C/8$$
$$1.06 = 8/8.5$$
$$T_D = 1.06 \text{ minutes}$$

Industrial Process Reaction Delay

Industrial processes do not respond fully to a change in input instantly. Imagine a tank of water that needs to be heated to a given temperature. The tank has a capacity to hold heat just like a capacitor has a capacity to hold a charge. The concept of time constants applies to both. Study Figure 17–36. It took 200 seconds for the fluid

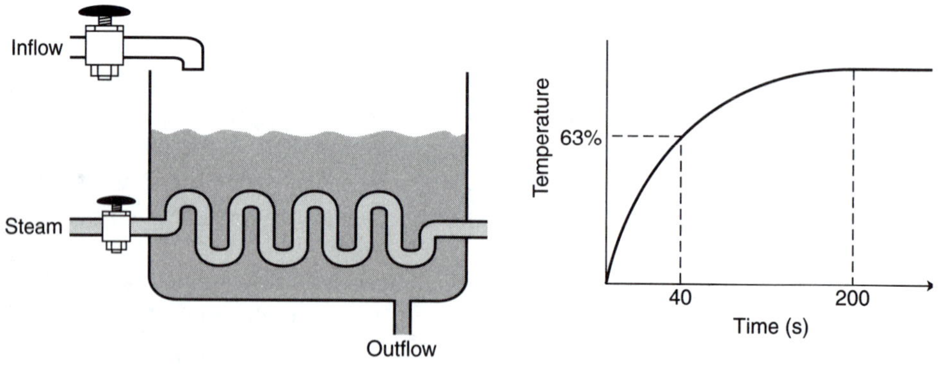

Figure 17–36 Time constant principle.

in this system to get to the new temperature. As you can see, it took about one time constant to get to 63 percent of the new setpoint. It takes five time constants to get to the new temperature. Remember that a capacitor charges to 63 percent of its capacity in one time constant also.

The tank has thermal capacity but also thermal resistance. All materials have a natural reluctance to conduct heat. In this example the steam heats the walls of the steam pipe, not the water directly, and the pipe heats the water. The thermal resistance depends on the thermal conductivity of the pipe wall, its thickness, and the surface area of the coils. The thermal capacity depends on the specific heat of the liquid and the size of the tank. The thermal time constant is based on the combination of the thermal resistance and the thermal capacity.

Almost all industrial processes exhibit process reaction delay (time constant delay). The length of the delay varies widely from very short times to hours, but almost all processes experience a delay between taking corrective action and seeing the final result.

Transfer Lag Many processes have more than one resistance-capacity combination (see Figure 17–37). For example, gas is used in a furnace to create heat in the heat chamber. The heat must move through the wall of the heat chamber into the air in the chamber. The air then heats the parts in the furnace. If the parts could be heated directly instead of using this indirect method, the curve is a normal *RC* time constant curve. But the air must be heated first and then the parts, resulting in two capacities: the capacity of the air and the capacity of the part, as well as two resistances (the resistance of the wall of the heating chamber to transmit heat and the resistance of the part to transfer heat). The curve looks more like the one in Figure 17–38. The significance of transfer lag is that it makes control more difficult.

Figure 17–37 Furnace with transfer lag.

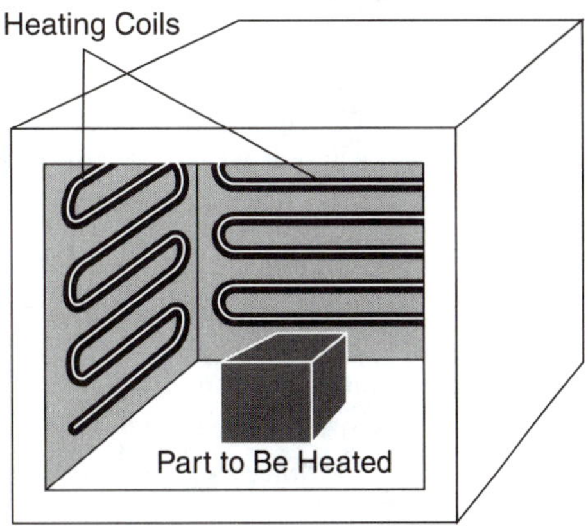

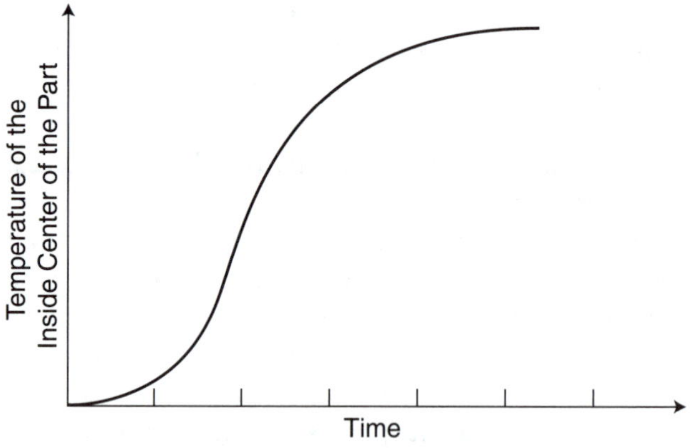

Figure 17–38 Graphed temperature versus time for a two-capacity process.

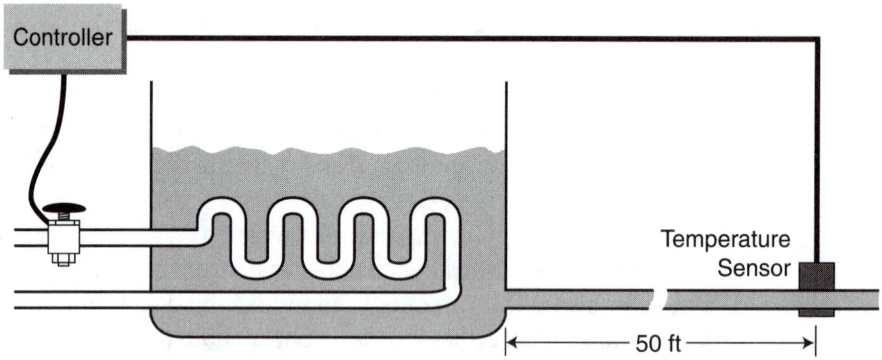

Figure 17–39 Transportation lag example.

Transportation Lag Figure 17–39 is an example of transportation lag. The process involves heating a fluid in a tank to a given temperature and then sending it to the process. The temperature is measured at the process rather than in the tank. Opening the steam valve produces an immediate response in the tank. Even though it might be small, an immediate change can be measured. Assume that the fluid takes five seconds to travel down the pipe to the process. This means that even though the change is immediate in the tank, it will be 5 seconds before the temperature sensor senses the change. This period between making a change and sensing it is called *deadtime.* Here, it is five seconds. To put this in a more meaningful situation, imag-

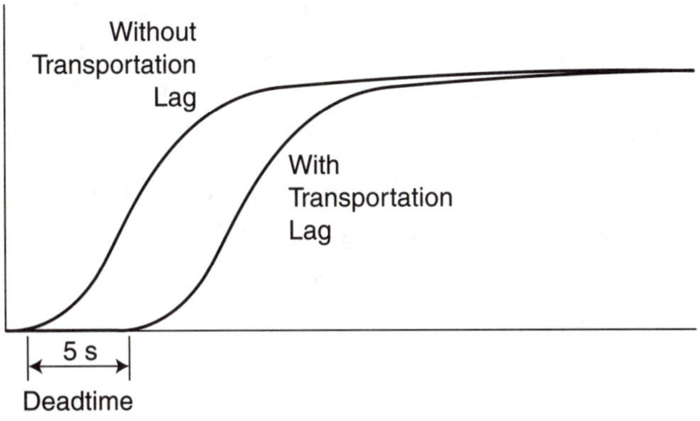

Figure 17–40 Response with and without transportation lag.

ine what would happen if the human brain could not respond to a change for 5 seconds. It would be impossible, for example, for a person to drive a car if it took 5 seconds to see the effect of moving the steering wheel. The graph in Figure 17–40 shows the response of the system with and without transportation lag.

Questions

1. List four characteristics of an effective control system.
2. What are the major differences between open- and closed-loop control?
3. Is the typical home sump pump an open- or closed-loop system?
4. Define the term *controlled variable.*
5. Describe the term *deadband.*
6. How does narrowing the deadband affect a system?
7. Describe proportional control.
8. Describe the term *proportional band.*
9. What is offset? What causes it?
10. A furnace can be heated between room temperature and 1060° F. Assume that there is a proportional response between 800 and 1000° F. Calculate the proportional band percentage.
11. A furnace can be heated between room temperature and 1560° F. Assume that there is a proportional response between 800 and 1200° F. Calculate the proportional band percentage.

12. A furnace can be heated between room temperature and 500° F. Assume that there is a proportional response between 300 and 500° F. Calculate the proportional band percentage.
13. What is the proportional gain of the system in Question 10?
14. What is the proportional gain of the system in Question 11?
15. What is the proportional gain of the system in Question 12?
16. What effect does load have on a system?
17. Describe integral control.
18. Describe derivative control.
19. Describe PID tuning utilizing the Zigler-Nichols method.
20. What is transfer lag? What effect does it have on a control system?
21. What is transportation lag? What effect does it have on a control system?

chapter 18
Instrumentation

This chapter examines various types of process control systems and instrumentation.

OBJECTIVES

Upon completion of this chapter, you will be able to:

1. Describe various flow, level, density, and temperature measurement devices.
2. Describe the types of signals used in process measurement.
3. Describe common methods of measurement and control for flow, level, density, and temperature measurement systems.
4. Explain the purpose and importance of calibration.

FLOW CONTROL PROCESS

The flow rate of a liquid, gas, or solid (particles) often must be controlled in industry. Flow control is accomplished by measuring the force of the flow and then using a variable valve. Some devices measure the actual flow; others measure changes in pressure to approximate the flow rate. The velocity of fluid flow depends on the amount of pressure forcing the fluid through the pipe; it can be determined as follows:

$$\text{Flow} = \text{cross-sectional area of the pipe} \times \text{average velocity of the flow}$$

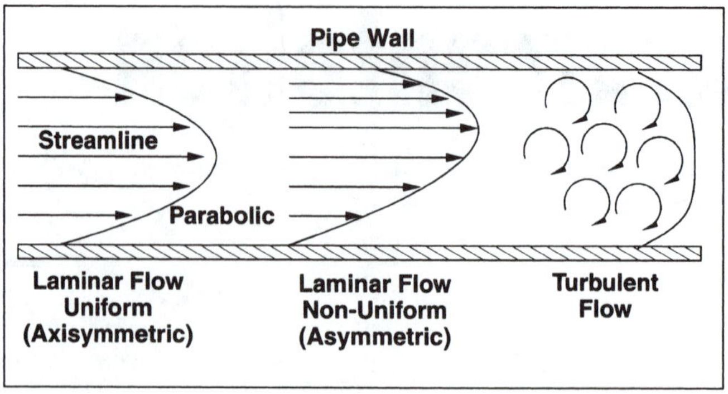

Figure 18–1 Three types of flow in a pipe. *(Figure published in the Rosemount Product Catalog, Publication No. 00805-0100-1025 © 1998, Rosemount, Inc. Used with permission.)*

Fluid Flow Through Pipes

The three main types of fluid flow in a pipe are laminar uniform, laminar nonuniform, and turbulent (see Figure 18–1). Most fluid flow applications involve turbulent flow, which occurs when velocities are high or viscosities are low. A value called a *Reynolds number* is used to represent the turbulence of flow. The density and viscosity of fluids as well as the friction created by the flow over pipe walls affect the flow rate. The pipe diameter and the viscosity of the fluid act as drag forces on the fluid.

In a given application, the specific gravity of the fluid and the pipe diameter remain constant, but the velocity and the viscosity vary. Note the parabolic shape and the flow in smooth layers in Figure 18–1. The highest velocity is at the center of the pipe and the slowest is at the pipe wall because viscous properties of the fluid against the wall restrain it.

Flowmeters

Mass, positive displacement, differential pressure, and velocity are the four main types of meters available to measure flow. Selection of the proper flowmeter is crucial to the success of the application.

Mass Flowmeter One of the more interesting transducers used to measure mass rate of flow directly rather than the volume of the flow and changes in density is the Coriolis meter, which is one of the more commonly used measuring instru-

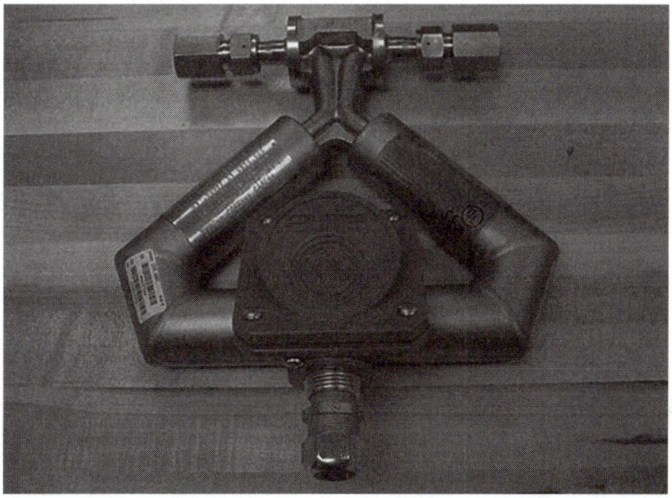

Figure 18–2 Coriolis meter.

ments. Its meters are very linear. It uses u-shaped tubes (see Figure 18–2). An electromechanical coil vibrates the tubes. When fluid passes through a u-shaped tube that is vibrated up and down, the flow causes the tube to twist. Motion sensors mounted on each side of the tube sense how much twist exists by measuring the differences in timing as the tube passes the sensor on each side. The Coriolis principle can be used to measure mass flow or density (Figure 18–3). The output from a Coriolis meter is a 4–20 mA signal that is proportional to the flow or the density of the fluid. You can find an excellent tutorial on Coriolis meters at www.micromotion.com.

In process control 4–20 mA is a very common signal, and almost all devices (measurement and control) are available in 4–20 mA. There are several advantages in using 4–20 mA signals. A voltage signal is subject to losses as it travels through the wire; the longer the wire, the larger the loss. A current signal is not subject to this loss. Another advantage of a 4–20 mA signal is in tuning the system (calibration will be covered later). The "zero" adjustment for the system is adjusted until 4.0 mA is reached. If you read 3.98 mA, you adjust the "zero" back to 4.0 mA. If 0 is used instead of 4 mA, you can never be sure that it has been adjusted to zero because it could be adjusted too far below zero and you would not be able to see it. So the 4–20 mA signal is a good choice for process control systems, and almost all devices (measurement and control) are available in 4–20 mA.

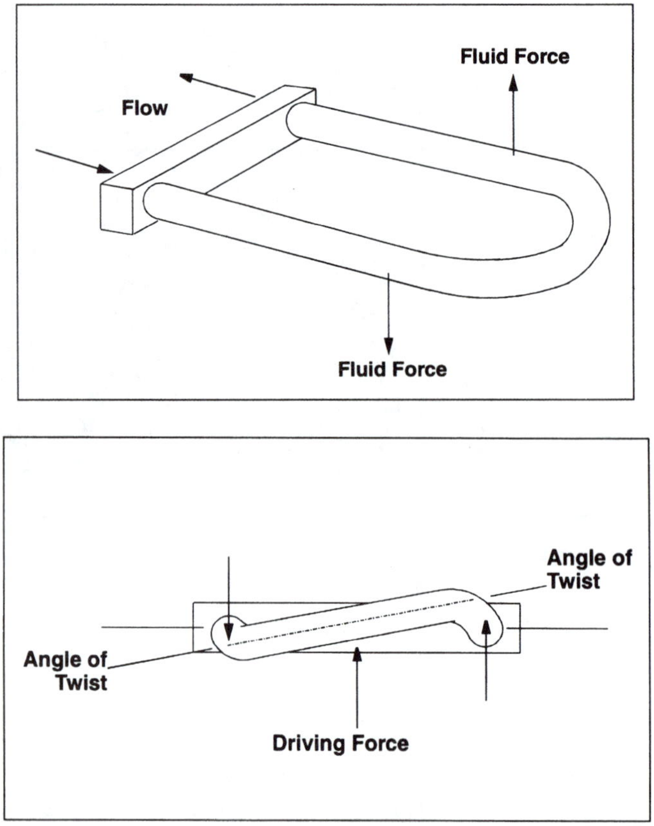

Figure 18–3 Principle of a Coriolis meter. *(Figure published in the Rosemount Product Catalog, Publication No. 00805-0100-1025 © 1998, Rosemount, Inc. Used with permission.)*

Differential Pressure Meter The most common type of flow measurement is a differential pressure meter or transducer (see Figure 18–4). It measures the pressure difference between the inlet and the outlet; therefore, a pressure drop must be created in a pipe. The basic principle of a differential pressure meter is that the pressure drop across the meter is proportional to the square of the flow rate.

Flow Obstruction Devices Numerous devices are used to obstruct the flow and create a pressure drop. An orifice plate is widely used in gas, liquid, and steam flow control applications. It is usually a thin circular plate with a tab for handling it. It is normally inserted into a pipe between two flanges, and the pressure drop is measured to determine the fluid's flow rate.

Figure 18–4 Differential pressure transducer. *(Figure published in the Rosemount Product Catalog, Publication No. 00805-0100-1025 © 1998, Rosemount, Inc. Used with permission.)*

Another device to obstruct flow is a flow nozzle, which is a derivation of an orifice plate but has a larger flow capacity. See Figure 18–5.

A venturi tube also creates a pressure drop (see Figure 18–5). It has a tapered entrance to restrict the flow. The straight section that follows increases the velocity of the fluid, creating a pressure differential between the inlet and outlet that can be measured by a differential pressure meter. A venturi tube can pass 25 to 50 percent more flow than an orifice plate with the same pressure drop. It requires less straight pipe than an orifice plate and is more immune to buildup of scale and corrosion. Because it is expensive, it is normally used for more demanding or difficult flow control applications.

A *pitot tube* installed in couplings in the pipe creates a differential pressure; it is one of the simplest flow sensors. Bent at a right angle to the flow direction (see Figure 18–5), it measures air flow in pipes and ducts and fluid flow in pipes and open channels. Pitot tubes are simple and reliable and inexpensive, but they are susceptible to clogging.

Positive Displacement Meter A positive displacement meter essentially breaks the fluid into discrete amounts and passes them on (see Figure 18–6). The volume is indicated by the number or rotations of the meter.

A rotary-vane meter is one type of positive displacement meter. It contains an equally divided rotating impeller in the meter's housing that contacts the walls of

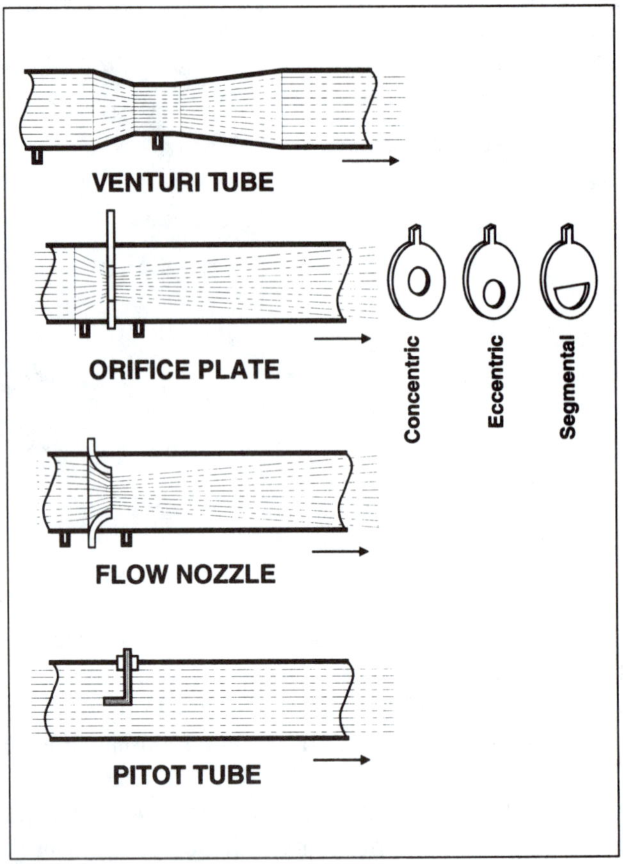

Figure 18–5 Four types of flow restrictors. *(Figure published in the Rosemount Product Catalog, Publication No. 00805-0100-1025 © 1998, Rosemount, Inc. Used with permission.)*

Figure 18–6 Positive displacement meter.

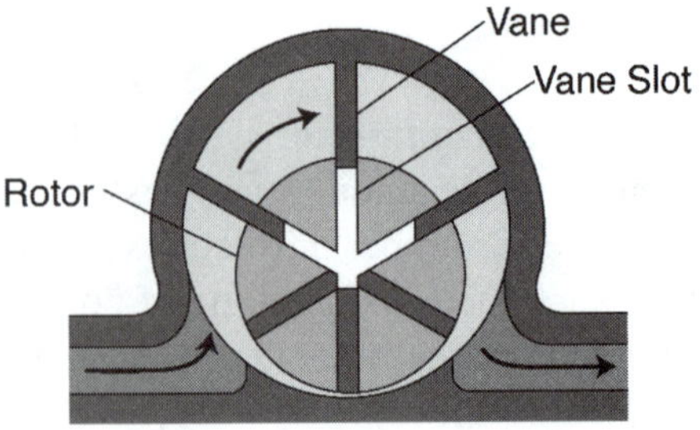

the housing. The fluid flow causes the impeller to turn. Each segment of the impeller releases a fixed volume of fluid to the outlet. The revolutions of the impeller can then be counted and a flow rate can be calculated.

Velocity Meters

Turbine Meter A turbine meter has a bladed rotor mounted perpendicular to the flow that causes the blades to rotate. The speed of rotation is directly proportional to the flow rate. The rotational speed can be sensed by a magnetic pickup that counts the pulses to establish the flow rate. A turbine meter is a good choice for accurate liquid measurement applications, but it is susceptible to bearing wear.

Vortex Meter A vortex meter works on the principle of *vortex shedding* (see Figure 18–7). A sensor in the meter senses the presence of vortices and generates pulses. The meter amplifies and conditions the signal and provides an output that is proportional to the flow rate. Figure 18–8 shows a typical vortex flowmeter.

Target Meter A *target meter* measures the force exerted on it by the fluid. Imagine a strain gage attached to a target that is placed in the pipe (see Figure 18–9). The higher the flow, the higher the force that is exerted on the target.

Magnetic Flowmeter A magnetic flowmeter induces an AC field through the flow (see Figure 18–10). The measured AC frequency varies proportionally to the

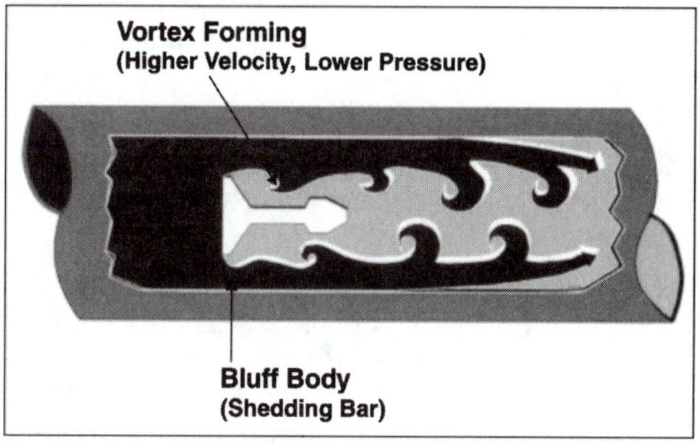

Figure 18–7 Principle of a vortex flowmeter. *(Figure published in the Rosemount Product Catalog, Publication No. 00805-0100-1025 © 1998, Rosemount, Inc. Used with permission.)*

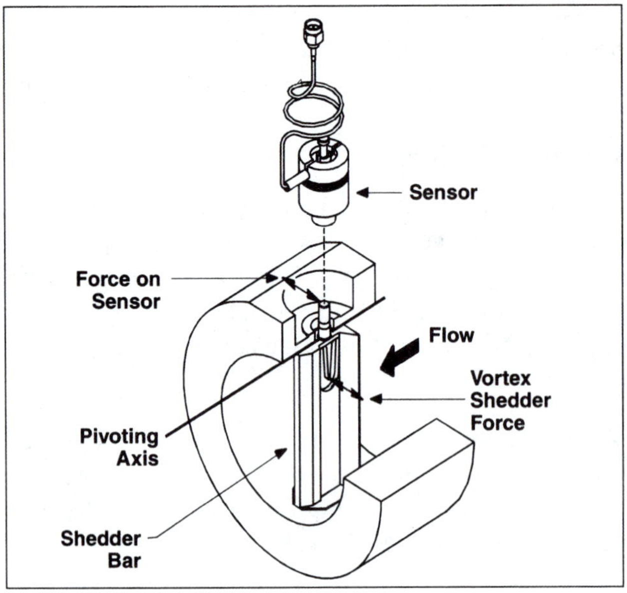

Figure 18–8 Vortex flowmeter. *(Figure published in the Rosemount Product Catalog, Publication No. 00805-0100-1025 © 1998, Rosemount, Inc. Used with permission.)*

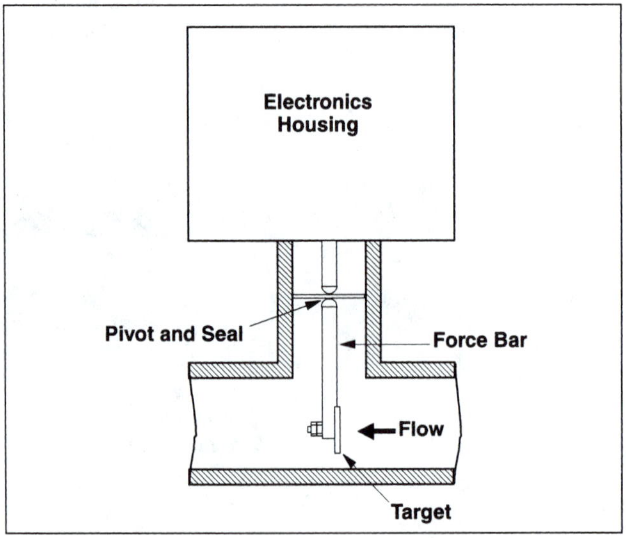

Figure 18–9 Target meter. *(Figure published in the Rosemount Product Catalog, Publication No. 00805-0100-1025 © 1998, Rosemount, Inc: Used with permission.)*

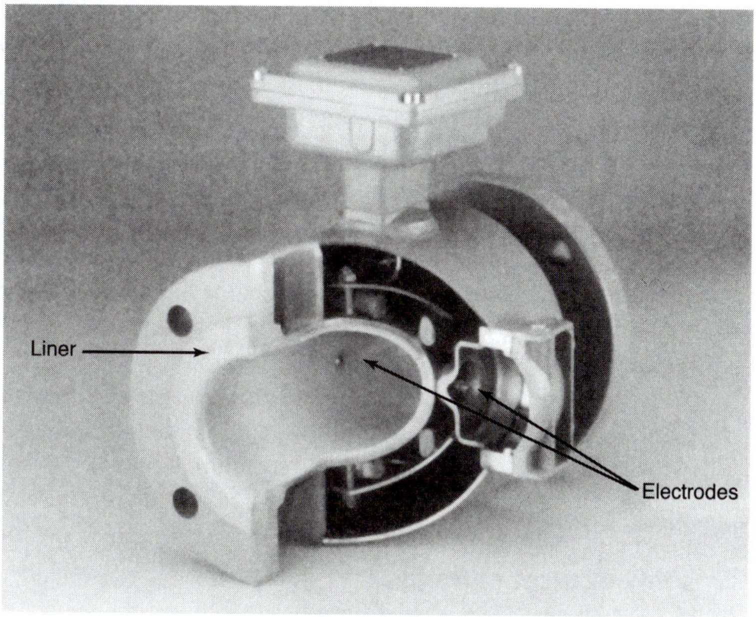

Figure 18–10 Magnetic flowmeters. The top photo shows a cutaway section of a flowmeter. The bottom photo shows a typical magnetic flowmeter with a digital transmitter and readout. *(Figure published in the Rosemount Product Catalog, Publication No. 00805-0100-1025 © 1998, Rosemount, Inc. Used with permission.)*

flow rate. This change in frequency is converted to a voltage or current change as an input to the controller. Magnetic flowmeters do not interfere with the flow, nor do they put any restrictions in the line.

An ultrasonic flowmeter with a transmitter and receiver measures flow (see Figure 18–11). The two main types of ultrasonic flowmeters are transit time (pulsed type) and Doppler shift (frequency shift) as shown in Figure 18–11. The transit type utilizes sound waves through the flow. The transducers are mounted at either 45 or 90 degrees to the flow. The velocity of the first pulse, directed downstream, increases the speed of the flow. The speed of the next pulse directed upstream is decreased by the flow. The average of the pulse time between the two is proportional to the average velocity of the flow.

Doppler devices measure a shift in frequency caused by the flow of a fluid, which is moving, so it causes a shift in the frequency of the returned pulse. The shift is proportional to the velocity of the fluid. Bubbles, particulates, or any discontinuities cause the pulse to be reflected back to the receiver.

Flow Control of Solids

Many industrial processes need to control the flow of solids (particulates), especially when batching for recipe control. Solid flow control usually involves weight measurement. Figure 18–12 is a simple example of a flow control system for solids. Flow in weight systems is expressed as weight per unit of time:

$$\text{Flow rate} = \text{weight} \times \text{time}$$

Assume that the particulates are relatively consistent and that a uniform amount flows from the hopper onto the conveyor. The controller can control the flow by adjusting, the drive speed. A faster rate of speed on the conveyor allows more partic-

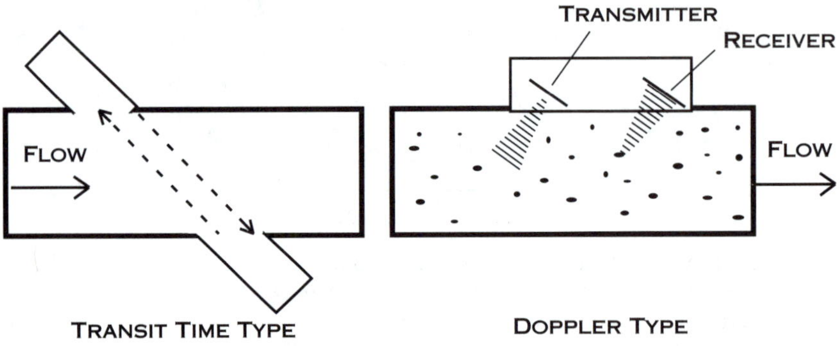

Figure 18–11 Two types of ultrasonic flowmeters.

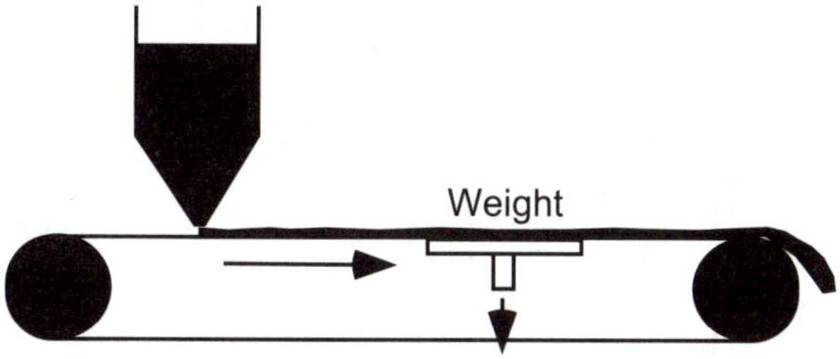

Figure 18–12 Simple flow control system for solids.

ulates to drop from the hopper onto the conveyor. In this case, the weight per unit of time increases. The weight of the particulates is measured on the conveyor belt. The weight measurement must be taken a sufficient distance from the hopper so that the material is not in a turbulent state from dropping out of the hopper and onto the conveyor, but has settled on the conveyor. The weigh station must be positioned so that it will not be influenced by a lifting effect. In other words, conveyors have areas where the belt may actually be lifted partially off the weigh station. This situation causes erratic and inaccurate feedback and thus output.

The weigh stations utilize load cells (strain gage) or linear variable displacement transducers (LVDTs). The load cells are mounted under a weigh station and generate an output base that is proportional to the weight of the particulate on the weigh station. LVDTs can measure the droop in the conveyor. The heavier the load, the more the conveyor droops. The more the conveyor droops, the more the LVDT is displaced. The distance that the LVDT is displaced is proportional to the weight of the particulates on the conveyor.

Level Control

Level control for liquids or solids is common in industry. *Level* is the interface between two material phases. The different phases that need to be measured and controlled can be between two solids, a liquid and a solid, a liquid and a gas, two liquids, and so on. Level is controlled using a reference level (setting). Many level control applications are simple and require only rudimentary control, for example, a switch that senses only whether the level is above or below the reference level. Many other applications, however, require closer control.

Level Control Methods

Bubbler Method The bubbler method is one common method to measure a liquid level (see Figure 18–13). This method uses one or two tubes. The basic principle is

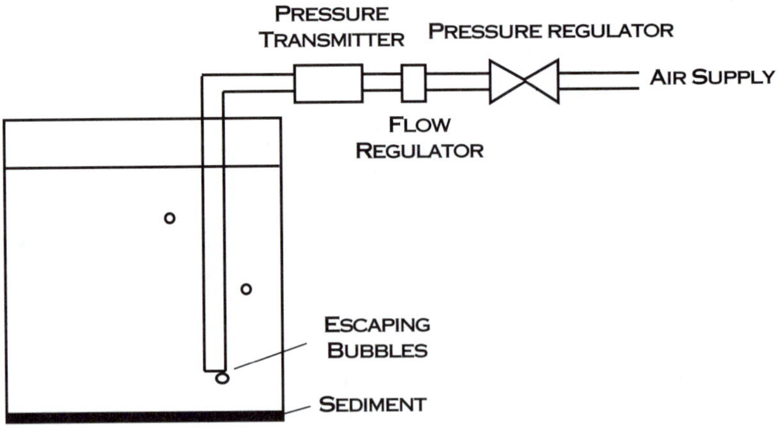

Figure 18–13 Bubbler method of level measurement. It may be necessary to keep the bottom of the bubbler tube above any sediment that is present in the tank.

simple. Imagine blowing air through a tube into a tank of water. The deeper the tube is placed in the water, the more pressure it takes to blow bubbles. The higher the tube is raised, the easier it is to blow bubbles.

The bubble method puts air pressure into the bubbler tube. Just enough pressure is applied to produce one or two bubbles per second when the fluid is at the desired level. A simple needle valve controls the air pressure, or a constant airflow device more closely controls the pressure that is allowed into the tube. The pressure in the bubbler tube can then be measured by a pressure switch or a pressure transmitter. Note that the higher the level of fluid, the higher the pressure of the fluid in the bottom of the tank. This means that fewer bubbles escape and the pressure in the tube is higher. A low level of water decreases the pressure of the fluid at the bottom of the tank, and bubbles escape more readily, decreasing the pressure in the tube. The pressure transmitter can measure these changes in bubbler tube pressure and transmit the values to a controller.

Many applications require two tubes, as in enclosed, pressurized tanks. One tube is located at the bottom of the tank and the other at the top of the tank (see Figure 18–14). A differential pressure transmitter (discussed earlier in the flow control section) is used in this case. The lower tube pressure is attached to the high pressure side, and the tube at the top of the tank is connected to the low pressure side. The difference in these pressures can be used to measure density. It is usually necessary to provide a cleanout in the bubbler tubes. A disadvantage of this pipe is the buildup of solids in the bubbler tubes. Buildup in the tube negatively affects the system's accuracy.

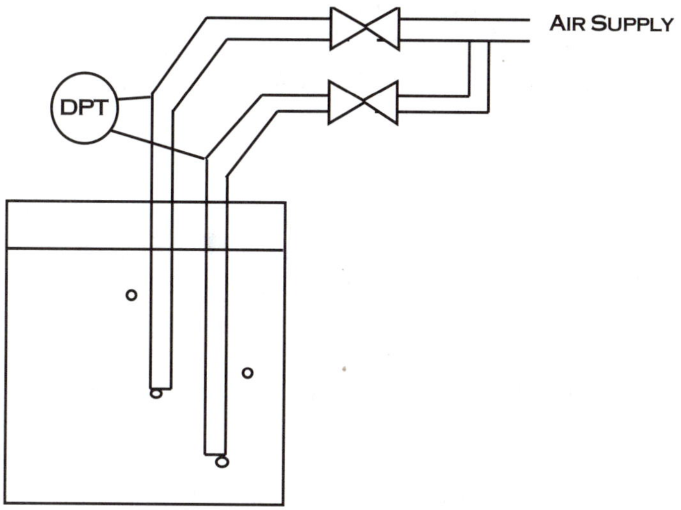

Figure 18–14 Bubbler method of density measurement.

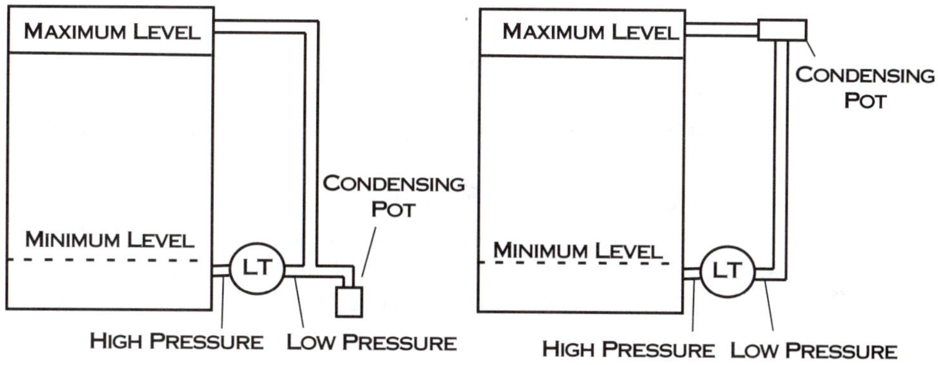

Figure 18–15 A dry-leg system on the left and a wet-leg application on the right.

See Figure 18–15 for two types of level applications. The application on the left is called a *dry leg*. Its top pressure tube is exposed only to air in the system. Condensation and collection in the low pressure side distorts the reading. A condensing pot below the low pressure input side to the differential pressure transducer collects any condensate, which must be drained periodically. The example on the right side of Figure 18–15 shows a wet-leg application. This application has a vapor over the fluid in the tank. The vapor in the low pressure side distorts the reading. In this case a condensing pot is placed at the top of the dry leg to condense and remove the vapor that would distort the reading.

The bubbler method requires the fluid to be of somewhat constant density. The temperature or the mix of the fluids in the tank can affect the density. To illustrate this, imagine a tank of molasses. If the temperature is 150 degrees Fahrenheit, the bubbles flow quite freely. If the temperature is 40 degrees Fahrenheit, there are few bubbles. In this case the temperature should be kept somewhat constant.

Ultrasonic Sensor Method Ultrasonic sensors can also be used to measure the level of liquids or solids. A sound wave is transmitted into the tank and reflected off the solid or liquid back to the sensor. The amount of time it takes for the sound wave to return is converted to an analog signal that represents the level of the tank. There are some problems with the use of ultrasonics. Air temperature affects the performance of ultrasonics, but this can be compensated for. Ultrasonics are not appropriate for foaming or bubbling fluids, as in a boiling application, because they can produce inaccurate readings. They may also be inappropriate for solids whose levels are not consistent (see Figure 18–16).

Radar Method A radar gage can be used to provide level feedback (see Figure 18–17). Notice that this gauge can be used in difficult applications that have high agitation and/or foaming.

Weight Method Load sensors can be used to measure the weight of a tank. Obviously, the more material a tank contains, the heavier it is. A weight measurement level system is illustrated in Figure 18–18. A load cell sensor (Figure 18–19) placed under each column supports the tank and sends a signal to the controller. The heavier the load, the larger the signal. The sensor use is based on the principle of the strain gage. The weight method of level measurement is usually more expensive and more difficult to design and build than other level measuring devices. The input and output piping cannot interfere with the upward and downward movement of the tank.

Figure 18–16 Level of solids used in this tank is not consistent.

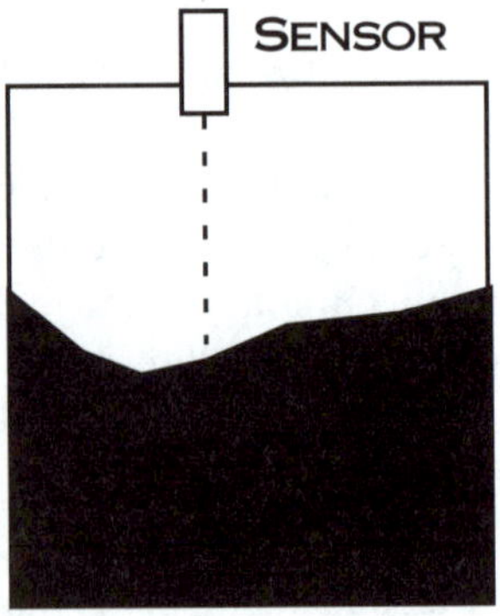

Figure 18–17 Radar gauge measuring the level of the fluid in the tank.
(Figure published in the Rosemount Product. Catalog, Publication No. 00805-0100-1025 © 1998, Rosemount, Inc. Used with permission.)

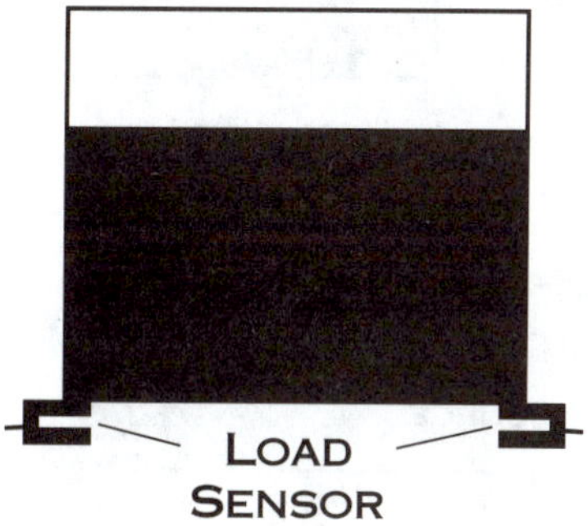

Figure 18–18 Weight method system.

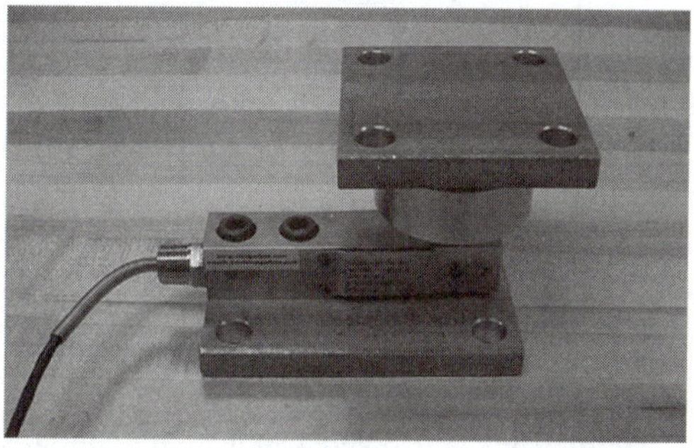

Figure 18–19 Load sensor.

Capacitive Level Sensing Method Capacitive sensors can also measure a liquid or solid level. See Figure 18–20 for examples of capacitive-level measurement for conductive and nonconductive materials. The insulated metal probe is one side of the capacitor and the tank is the second side. The capacitance varies as the level around the probe varies. A capacitance bridge measures the change in capacitance.

Radioactive Method Figure 18–21 shows two examples of radioactive level measurement. Radioactive methods use cobalt 60 or cesium 137 to provide a source

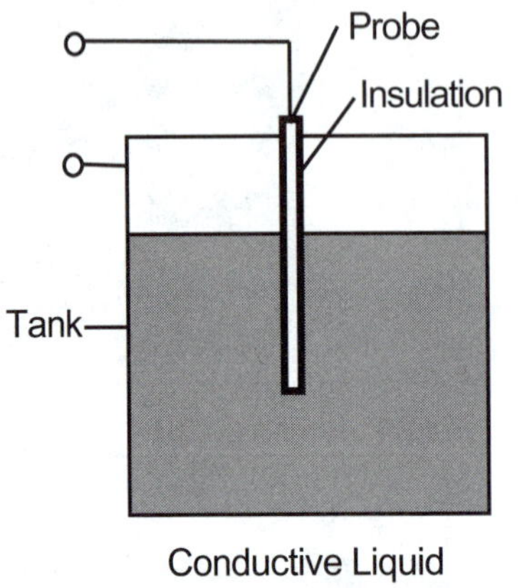

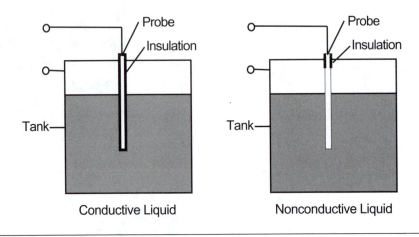

Figure 18–20 Capacitive level sensor.

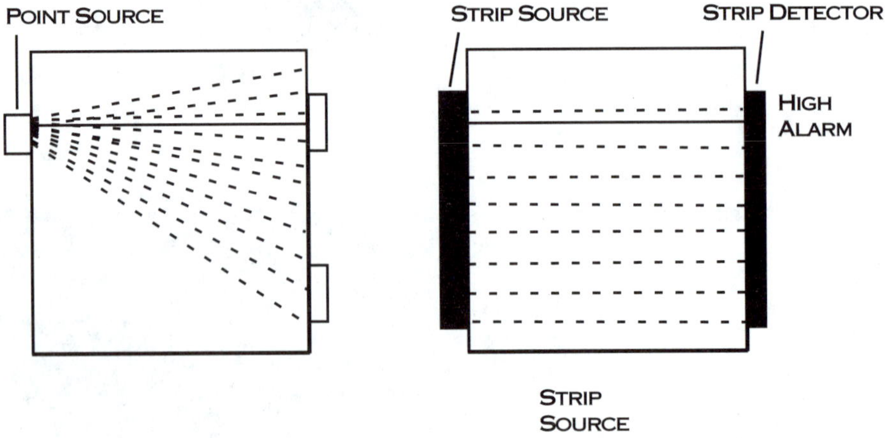

Figure 18–21 Radioactive level measurement (left); point and strip source radiation level measurement (right).

of radiation. The radiation is emitted on one side of the tank and sensed on the other. The higher the level, the more radiation is absorbed before it reaches the sensor. The amount of radiation sensed can be converted to a level reading. The level of radiation is low and the danger is relatively low, but the use of radiation technology requires its operator to be a licensed technician.

Control Valves

Control valves regulate the flow rate of fluid in a system. They control the flow rate by varying the size of the passage through which the fluid flows. Control valves typically consist of either two or three components: a valve and an actuator; many also have a positioner. The control valve on the left of Figure 18–22 consists of a valve body and an actuator. The photo of the valve on the right of the figure also has a positioner to control the position of the valve.

The positioner is in the middle of the assembly and has two pressure meters on the front. The typical positioner must be supplied with approximately 20 psi to operate. A plastic vent cap threaded into the top of the actuator allows the diaphragm to move freely.

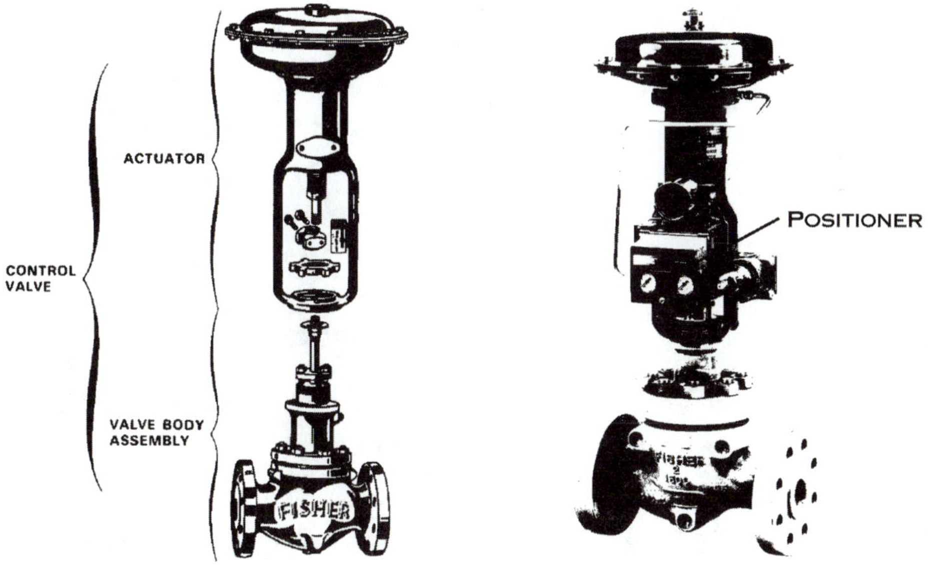

Figure 18–22 A typical control valve assembly on the left. The control valve on the right consists of a valve body, actuator, and positioner (Courtesy Emerson Inc.).

There are two pressure indicators on the front of a typical positioner. The indicator on the right of the positioner in Figure 18–22 shows the supply line pressure. The indicator on the left indicates the current actuator pressure. The input range is 4–20 mA and the output psi to the valve positioner is 3–15 psi. This also corresponds proportionally to a 0–100 percent valve opening.

The valve body houses the valve components and also provides inlet and outlet connections for the fluid (see Figure 18–23). The components inside the valve that regulate the flow are called trim elements. Trim elements include a valve seat. The seat is in a fixed position. A valve plug is used to open and close against the valve seat. The valve plug is moved by the valve stem. Packing

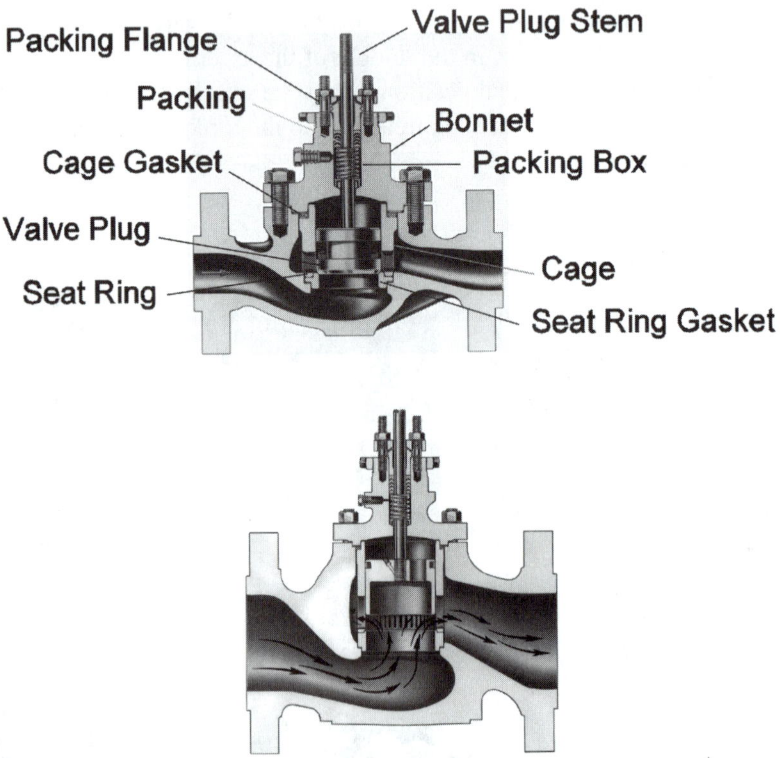

Figure 18–23 Typical industrial valves. Note the main components of the valve on the top. The valve on the bottom has arrows showing how fluid flows through a valve (Courtesy Emerson Inc.).

around the valve stem prevents leakage. Valves can be rebuilt by replacing the seat, valve plug, and packing.

Valve Actuation

Valves are typically actuated by electric or pneumatic actuation. Pneumatic actuation is common. The typical industrial actuator is controlled by air pressure. A diaphragm is used to move a stem in one direction. A spring returns the stem to the other position if pressure is decreased. The shaft opens and closes the valve. Three to fifteen psi is typically used to actuate the valve. Pneumatic valves are also safe because they typically move in one direction with air pressure and return with spring pressure. If there is a loss of signal or air pressure to the valve, the spring acts to move it to a safe position.

There are two types of valve actuators: direct acting and reverse acting (see Figure 18–24). A direct-acting valve uses pressure to close the valve. A reverse-acting valve uses pressure to open the valve. (Corresponding terms are air-to-extend and air-to-retract, respectively.) The choice of which type of valve to use is important. What if control is lost somehow or air pressure is lost? We would like the spring pressure on the diaphragm to move the valve stem to the correct position, either open or closed. For example, if we want the flow to stop in the event of a loss of control or air pressure, we would choose an air-to-open type.

Valves are also available in push-to-close or push-to-open styles. Figure 18–25 shows a push-to-close valve.

Valve Positioners

Many industrial valves have a positioner to control the position of the actuator that controls the valve position. Valve positioners typically have a current-to-pressure (I/P) converter built in. An I/P converter takes an input of 4–20 mA and converts it to 3–15 psi to move the diaphragm in the actuator in one direction. The return is done by spring pressure.

Figure 18–26 is a simplified illustration of how I/P positioners work. Study the figure and note that the diaphragm is attached to the stem of the valve. Air pressure is used to force the diaphragm down, and spring pressure is used to return the diaphragm up. Next notice that 20 psi is supplied to the valve actuator through a pressure regulator. The air splits into two directions below the regulator. The pipe to the left supplies pressure to move the diaphragm. The pipe to the right goes to a nozzle that touches the balance arm. The electric coil is used to pull the magnet and balance arm up. The signal to the I/P converter is between 4 and 20 mA; 20 mA exerts the most upward force on the balance arm. This allows less air to escape form the nozzle and thus the pressure is also higher on the diaphragm. If the input to the I/P is

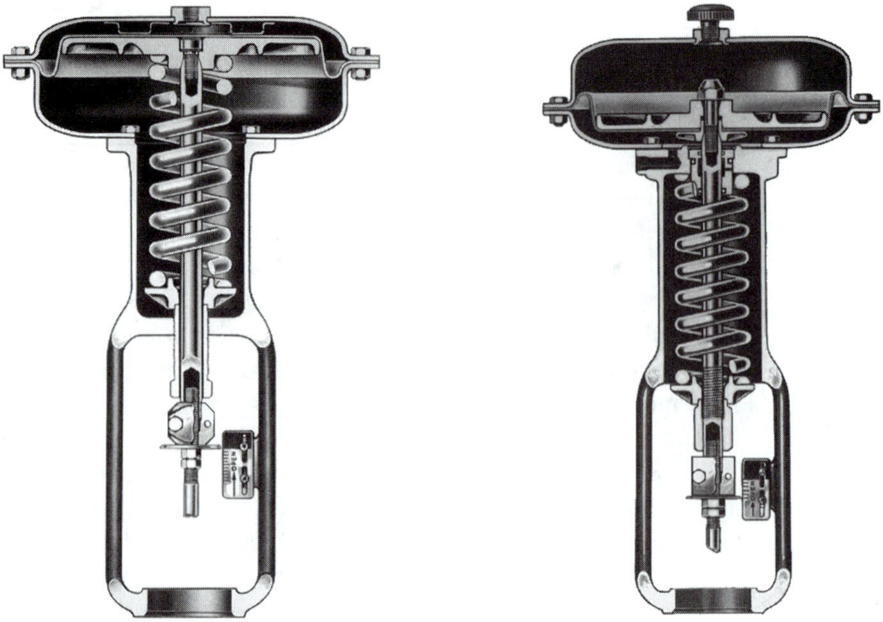

Figure 18–24 A direct-acting actuator is shown on the left; a reverse-acting actuator is shown on the right (Courtesy Emerson Inc.).

Figure 18–25 A push-to-close valve (Courtesy Emerson Inc.).

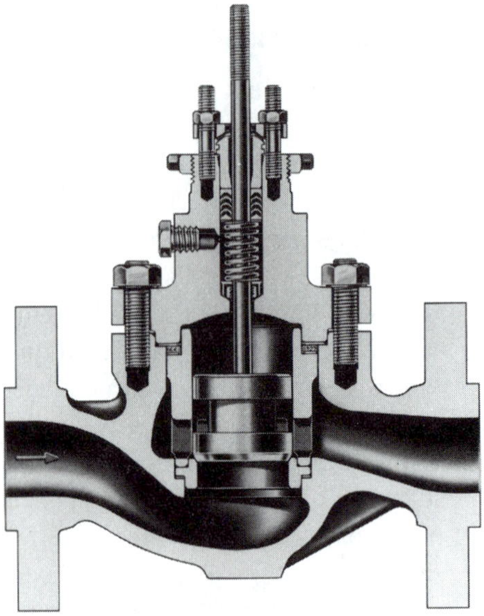

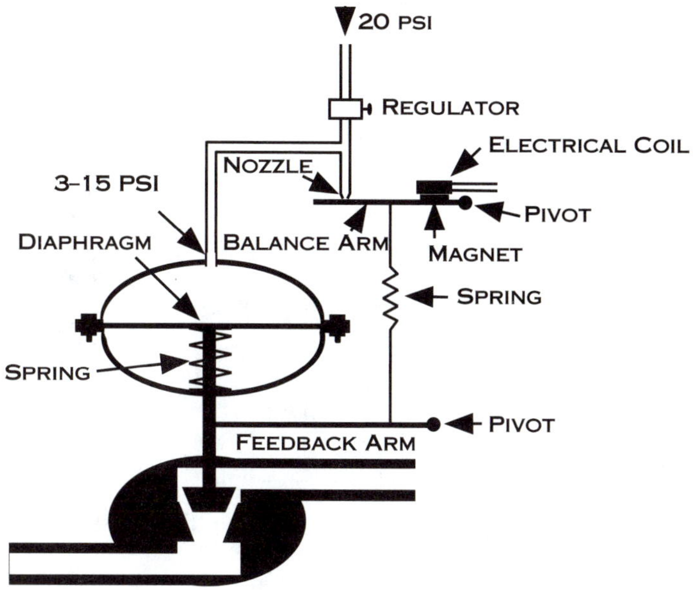

Figure 18–26 How a positioner works.

4 mA, there is not much upward pressure on the balance arm and more air escapes the nozzle. Thus, less pressure is exerted on the diaphragm. In effect the amount of input current to the IP controls how much air is allowed to "leak" out of the nozzle to keep the air pressure at the commanded pressure. Figure 18–27 shows a diagram of an actual positioner. Figure 18–28 shows a positioner connected to an actuator.

Maintenance of Valves

Examine the valve and actuator in Figure 18–29, the valve stem is the shaft that is moved to open and close the valve. The actuator stem is attached to the actuator diaphragm. These two stems are connected in one of two ways. One way is for the valve stem to be threaded into the actuator stem. There are two lock nuts to keep the desired position from changing. Another common method is to have a stem connector on the outside that clamps the two stems together at the correct position. The relationship between these two stems ensures that the valve opens and closes completely within the range of the actuator travel.

General Procedure for Valve Maintenance Note that the valve must be removed from service. Even when removed from service, the valve packing may contain

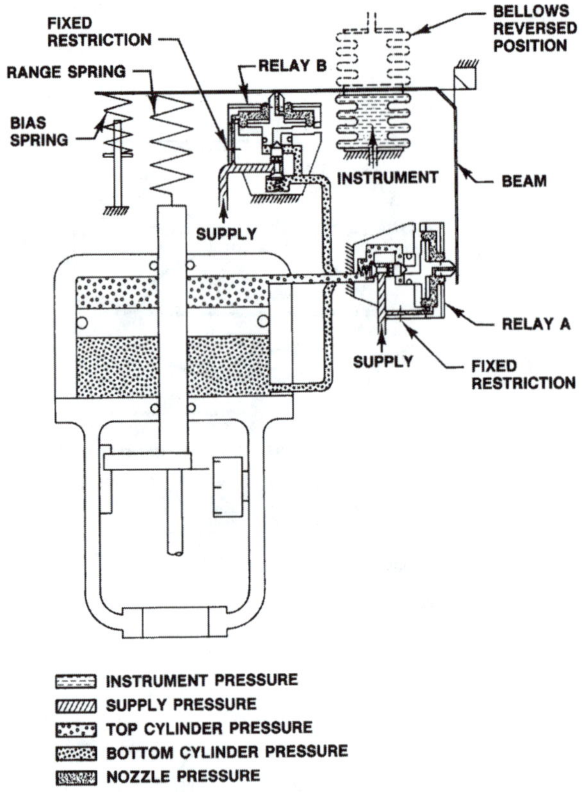

Figure 18–27 Positioner (Courtesy Emerson Inc.).

process fluids that are pressurized. Process fluids may spray out when removing packing hardware or packing rings. Follow all lockout/tagout procedures.

1. You must apply enough pressure to the actuator to move the valve to an intermediate position so there is no load on the stem. Disconnect the valve stem from the actuator stem. Remove the air pressure.
2. Separate the valve from the actuator by removing the yoke nut (see Figure 18–29).
3. Loosen the nuts on the packing flange to ensure that the packing is not tight on the valve plug stem (see Figure 18–27). Remove the travel indicator disk and lock nuts, if there are any. When you remove the

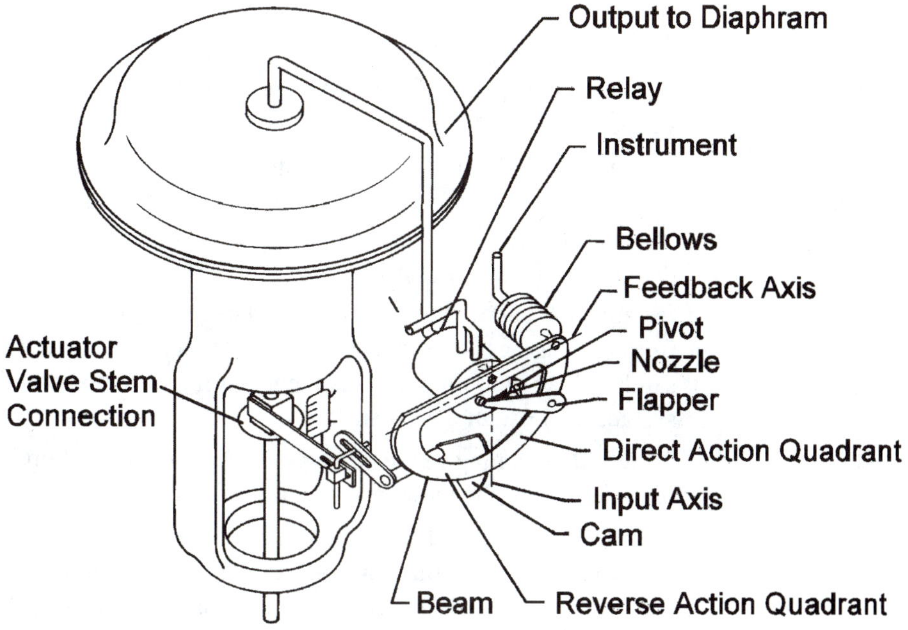

Figure 18–28 A positioner connected to the actuator (Courtesy Emerson Inc.).

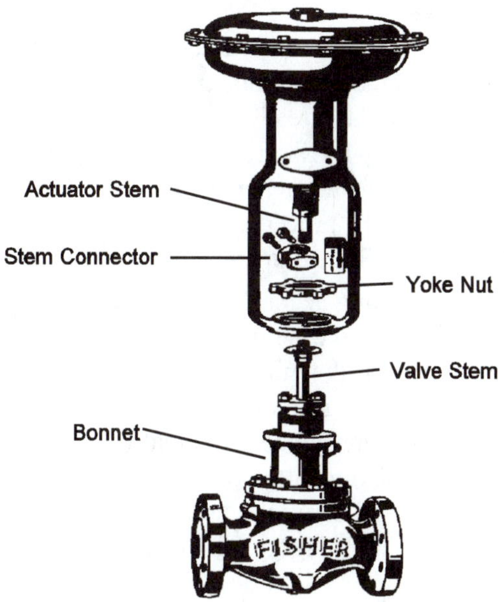

Figure 18–29 Typical valve assembly (Courtesy Emerson Inc.).

bonnet, be sure that the stem and valve plug assembly remains on the seat ring. This precaution helps to prevent damage to the seating surfaces.

4. Remove the bolts that hold the bonnet in place and lift the bonnet off the valve stem. If the valve stem begins to lift with the bonnet, use a lead hammer to tap it down again. Be careful with the bonnet to prevent damage to the gasket surface.
5. Remove the valve plug, seat ring, and cage. Remove all gasket material from the cage gasket surfaces. Do not damage the surfaces. Use 360-grit sandpaper to smooth and polish any scoring or damage to the gasket surfaces. Clean all gasket surfaces.
6. Protect the opening in the valve body. A cloth helps keep foreign particles out of the valve body while you are working.
7. Remove the packing flange nuts, flange, upper wiper, and the packing follower. Next, carefully push all remaining packing from the bonnet using a tool that will not damage the walls of the packing box. Clean the packing box and related parts.
8. Inspect the valve stem threads for any sharp edges that could damage the new packing. Any burrs or sharp edges can be removed with a deburring stone.
9. Remove the cloth from the valve body and install the cage using new gaskets. Install the plug. Slide the bonnet over the stem and onto the studs. Lubricate the threads and surface of the nuts and install them. Torque the nuts in a crisscross pattern to no more than 25 percent of the specified torque value. When all are done, repeat with another 25 percent. Repeat the procedure until all are at the specified torque. Check all nuts. If any still turn, tighten every nut again.
10. Install new packing according to the instruction sheet. Packing parts can be lubricated to ease installation and reduce the chance of damage. A piece of pipe with smooth surfaces should be used to tamp each soft packing part gently into the packing box. Slide the packing follower, wiper, and packing flange into position. Lubricate the packing flange nuts and studs. Tighten the packing flange nuts to the specified torque. If the packing set is spring loaded, adjust the springs as indicated in the instruction manual. Figure 18–30 shows a typical packing arrangement.
11. Reconnect the actuator and valve body. Note that the bench set should be performed before reassembly to be sure that the actuator is adjusted correctly.

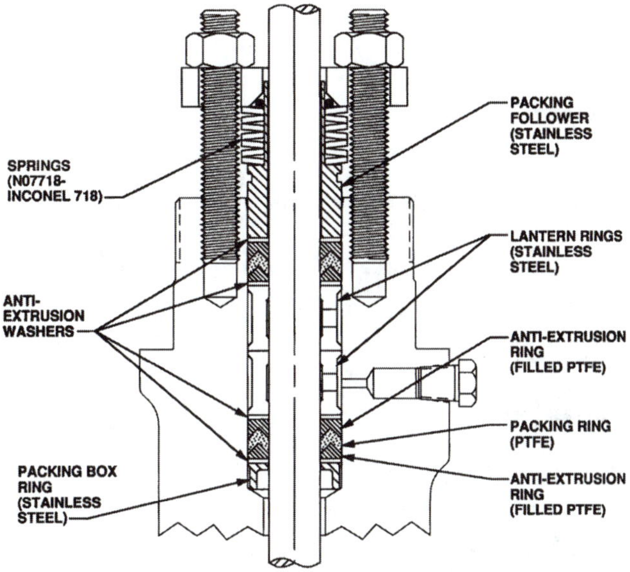

Figure 18–30 Typical packing arrangement (Courtesy Emerson Inc.).

Two types of packing meet most application requirements: PTFE and graphite. PTFE is used in application temperatures of between −40 and 450 degrees Fahrenheit. Graphite is generally used for temperatures above 450 degrees Fahrenheit. PTFE packing is composed of solid rings of PTFE material. There are generally two or more packing rings with a V cross-section. Graphite packing usually consists of graphite rings and sacrificial zinc washers. The zinc washers are used to protect the stem from galvanic corrosion. Figure 18–31 shows a typical graphite packing arrangement.

Bench Set Next, the valve is reassembled to the actuator and needs to be adjusted correctly so that the relationship between the valve and actuator is correct. This is called bench set. This generic bench set example is for an air-to-retract valve. It is a generic example only; you should check the manuals for the valve you are working on to find the correct procedure for your specific valve.

Bench set involves adjusting the initial compression of the actuator spring. The term *bench set* comes from the fact that it is done on the bench. During bench set, the actuator is not connected to the valve or other loads. Bench set ensures that the spring tension is such that the actuator will start to overcome spring pressure at a specific psi. (You must check the instruction manual for your valve to find the correct pressures and procedure.)

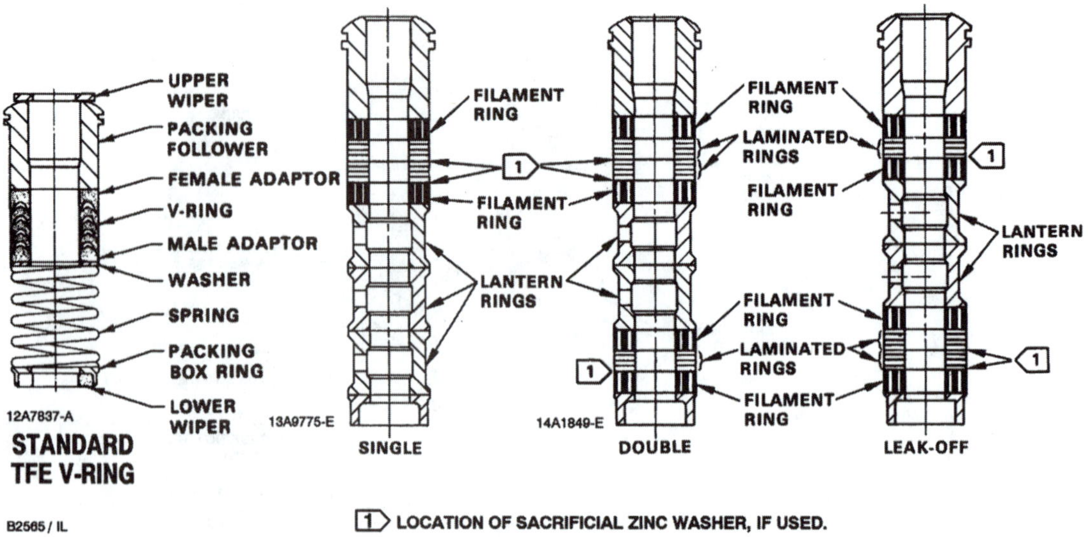

Figure 18–31 Graphite packing arrangements (Courtesy Emerson Inc.).

Remember that the air pressure moves the diaphragm, and that the spring on the opposite side of the diaphragm takes the slack out of the system and returns the diaphragm in the opposite direction as air pressure is decreased. For this example, assume that the bench set pressure is 5 psi. This means that we want the valve stem to begin moving (opening the valve) at approximately 5 psi. We want to make sure that the valve is completely closed from 0 to 5 psi. You should use a certified pressure gage to ensure that you are using the correct pressure when performing the bench set.

1. Connect the air source to the actuator. Beginning at 0 psi, slowly increase the pressure until the valve stem begins to move.
2. If the pressure is less than the bench set pressure you looked up (5 in this case), the actuator and valve stem assembly should be longer. Turn the valve stem out of the actuator stem one half-turn. Make sure you countertighten the two nuts on the valve stem and use a wrench on them to turn the valve stem. If the pressure was too high when the valve stem began to move, the actuator and valve stem assembly should be shorter. Turn the valve stem into the actuator stem one half-turn. Repeat the adjustments and checks until the bench set pressure is correct.

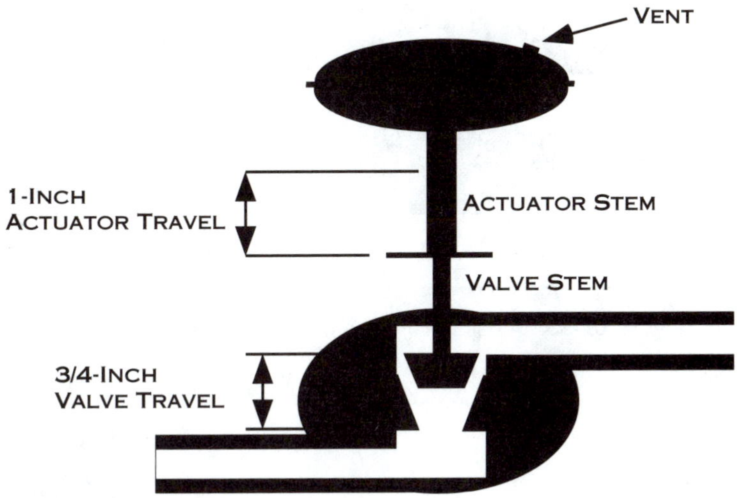

Figure 18–32 Diagram showing travel of actuator and valve.

Study Figure 18–32. Note that the actuator stem and valve stem have different ranges of motion. In this example, the actuator stem has a 1-inch travel and the valve stem has only a 3/4-inch travel. The valve should be completely seated (closed) at about 11 psi.

Types of Valves

A ball valve has a spherically shaped plug that rotates. As the plug rotates, the flow is varied minimum to maximum flow. Ball valves are good for use with corrosive materials. Ball valves also have the greatest flow capacity of all types of valves.

A butterfly valve has a round-plate restrictor that controls flow (Figure 18–33). An advantage of a butterfly valve is that the fluid passes straight through the valve without turns or pockets that can accumulate sludge. They are convenient when space is a concern because the valve body that houses the valve is small compared to other valve styles.

The valve on the left of Figure 18–34 shows a reverse-acting, double-ported, globe-style valve body. The valve on the right of Figure 18–34 shows a single-ported, globe-style valve.

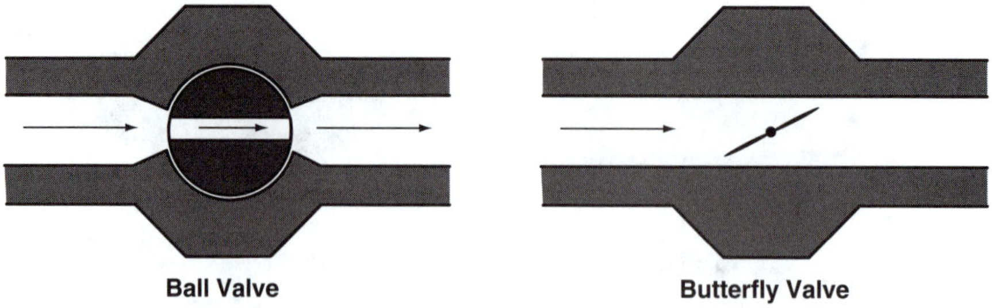

Figure 18–33 A ball valve and a butterfly valve.

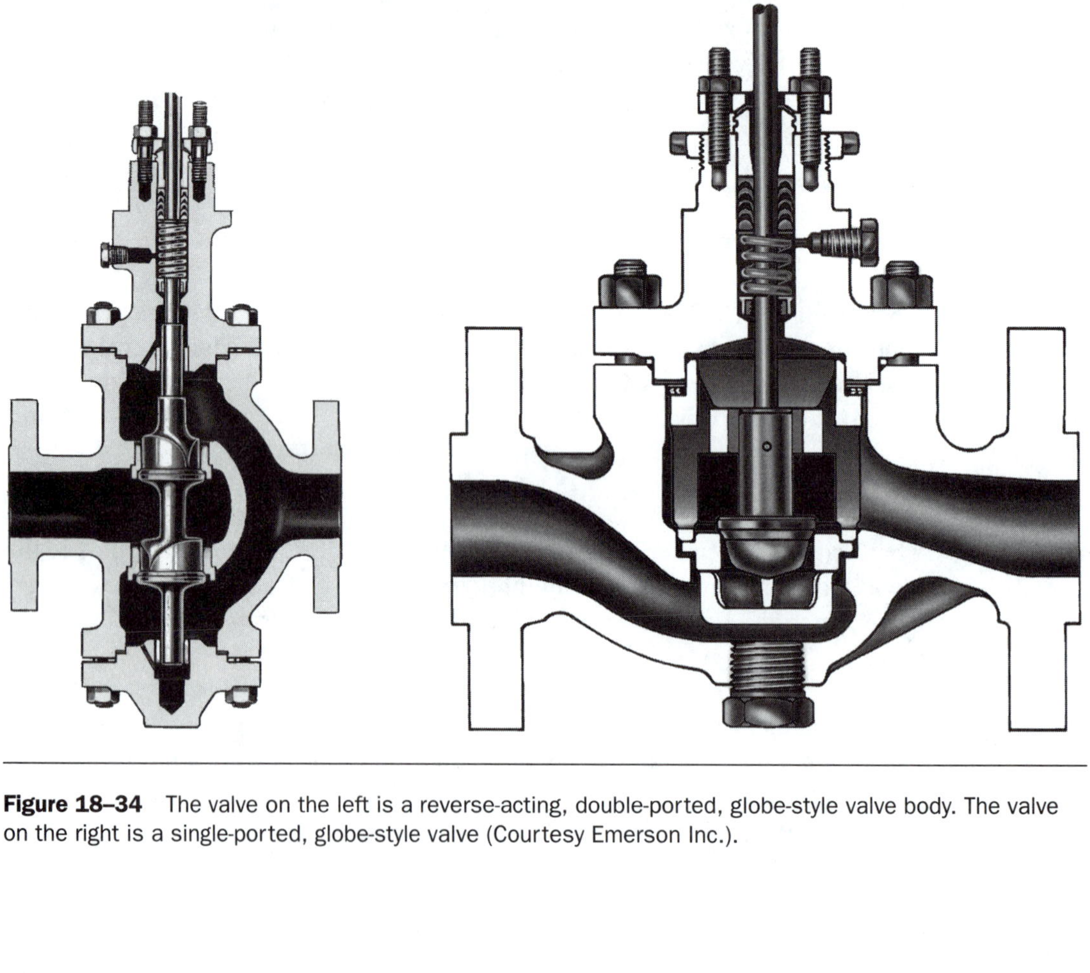

Figure 18–34 The valve on the left is a reverse-acting, double-ported, globe-style valve body. The valve on the right is a single-ported, globe-style valve (Courtesy Emerson Inc.).

Choosing a Valve

You must consider the following when choosing a control valve:

- Operating temperature: Temperature affects valve selection in a couple of ways. Not all valves are available in all temperature ranges. The selection of the valve packing is also temperature dependant.
- Material compatibility: You must make sure that the valve trim and packing are compatible and will hold up with the material that will flow through them.
- Body pressure requirements: A valve must be chosen that can withstand the process pressure in the application. Not all types of valves are compatible for high- and low-pressure applications.
- Flow requirements: Valves must be the correct size for the application flow rates.
- Cost: Cost is also an important factor. If expensive materials are involved, large savings may be possible from the correct type of valve, even if it is more expensive. You must also consider how much maintenance a valve will require when you consider total cost. Another consideration is whether the valve should have communications capability. New technology valves can communicate with calibration tools and computers. The performance of valves can be monitored remotely. Parameters for valve operation can also be uploaded and downloaded automatically. This ability to communicate and monitor devices enables a plant to replace or repair devices before they break down. This can dramatically reduce downtime and unnecessary scheduled maintenance.

Several communications standards are available. Profibus and HART are two of the more common standards. Many valves today are HART-compatible. These valves are "smart valves" and can self-calibrate and communicate with computers and handheld HART communication devices. The HART protocol is a standard for devices. Most industrial process input and output devices operate on a 4–20 mA analog signal. The HART communication digital communications signal "rides" on top of the current signal. This capability allows a computer to upload or download parameters and current operating conditions. It enables the computer to monitor the devices and status. Monitoring these devices makes it possible to apply predictive maintenance techniques. This capability can dramatically reduce maintenance in a plant.

CALIBRATION

Calibration is done to ensure that measurement and control devices are accurate. In a manufacturing plant, all measurement devices that affect quality are calibrated regularly. It might be easiest to think of a micrometer. In the plant, micrometers can be dropped and damaged, or they may become inaccurate over time. Calibration procedures in this case require that micrometers are calibrated on a regular basis. A sticker on the micrometer explains when the next calibration is required. The operator cannot use the micrometer after that date if the micrometer has not been calibrated. The micrometer is taken to the quality department, where its accuracy is checked against an accurate micrometer standard. If it is inaccurate, it is adjusted if possible. If it cannot be made accurate, it is disposed of. The standard is also checked on a regular basis to make sure it is accurate. The standard is checked against an even more accurate, traceable standard.

Process instruments affect quality and must also be calibrated. Sensors and output devices must usually be calibrated in a steady state. This means that a sensor is compared to a standard to make sure the output from the sensor is within acceptable limits. The accuracy and efficiency of any process control system depends on each of the devices in the loop. Calibration is performed to ensure that the output from each device represents the input accurately.

Measurement and control devices usually receive an analog input and output an analog output based on the input value. For example, a typical pressure-to-current converter takes an input pressure of between 3 and 15 lbs of pressure. The span (or range) of input values is 12 lbs (15 – 3). It outputs 4–20 milliamps over the 3–15 lb input pressure. The output span (or range) is 16 milliamps.

Devices must be calibrated when they are first installed or after they have been repaired to make sure the output reflects the input accurately. Devices must also be regularly calibrated during their life. Instruments can wear, drift, or be damaged. They must be calibrated regularly to be sure they are accurate. Inaccurate devices can cause quality problems and can also add unnecessary cost. For example, imagine a process where an expensive chemical is added. If the device is inaccurate, more chemical may be added than is necessary, thus raising the cost of our product. To calibrate an instrument properly, you must give the device a known input and measure the output.

Figure 18–35 shows a typical setup for calibrating a pressure-to-current transducer. A process calibration meter is used to provide the 4.2 psi input signal to the transducer, and the output of the transducer is measured with an accurate current meter. Note the zero and span adjustment screws on the transducer.

In a calibration, several input values must be tested to be sure the device is accurate throughout its range. At least five input values should be tested. One of the five must be at or within 10 percent of the lowest input value and one must be at or within

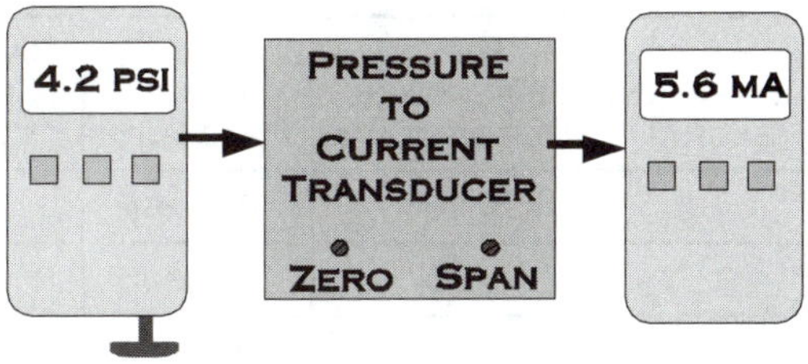

Figure 18–35 Calibration setup.

90 percent of the highest input value. Using at least five test values helps ensure that the instrument's output is linear and that we get a true picture of the performance.

The calibration test is performed from the bottom to the top input value (see Figure 18–36) and then from the top to the bottom value (see Figure 18–37). This helps ensure that the device is accurate in both directions and shows any effects that may be due to hysteresis. Note that values of 0 percent or 100 percent should not be used because they cannot be approached from below 0 percent or above 100 percent.

Note that the actual upscale calibration procedure has the technician begin below the first pressure and carefully adjust pressure up to the first test point. This step is repeated for each test point. Each is from below the required pressure value. This helps eliminate the possible effects of hysteresis. If the technician adjusts the input too high for an upscale measurement, the value must be reduced and approached from below the test value to ensure accurate calibration.

The downscale procedure is done in the opposite manner. The technician starts with the pressure higher than the first test value and carefully adjusts pressure down until the correct value is reached. If the technician adjusts the input too low for a downscale measurement, the value must be increased and approached from above to ensure accurate calibration. Next, a simple graph is made to show the input versus the output values. See Figure 18–38.

In the figure, two errors need to be adjusted for: zero and span. As you can see in the graph, the measured value for the actual output response is not at 0,0 on the graph, it is shifted up. The zero needs to be adjusted. Note also that the actual response at 100 percent input is too low, which means that the span is too small. There are usually two adjustments on devices: the zero and the span adjustment. To calibrate this instrument correctly, the technician first gives the device a 0 input and adjusts the input to the 10 percent value and then adjusts the zero adjustment until the output is correct. Next the technician increases the input value to 90 percent. The technician then adjusts the span adjustment until the output value is correct.

The technician must then recheck the 10 percent value to make sure it has not changed. If it has changed, the technician must go through both adjustments again.

Calibration Reading	Input Pressure Percentage	Input Pressure	Calculated Output Current	Actual Output Current	Percent Difference
1	10	4.2 psi	5.6 mA		
2	30	5.6 psi	8.8 mA		
3	50	9 psi	12 mA		
4	70	11.4 psi	15.2 mA		
5	90	13.8 psi	18.4 mA		

Figure 18–36 Values for upscale procedure.

Calibration Reading	Input Pressure Percentage	Input Pressure	Calculated Output Current	Actual Output Current	Percent Difference
1	90	13.8 psi	18.4 mA		
2	70	11.4 psi	15.2 mA		
3	50	9 psi	12 mA		
4	30	5.6 psi	8.8 mA		
5	10	4.2 psi	5.6 mA		

Figure 18–37 Values for downscale procedure.

Questions

1. List the three main types of flow.
2. Explain the principle of the measurement of flow rate by measuring differences in pressure.
3. What is an orifice plate?
4. What is a flow nozzle?
5. What is a venturi?

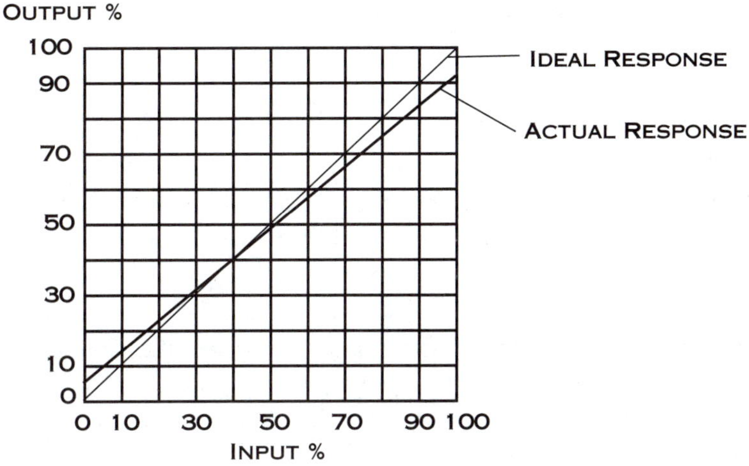

Figure 18–38 Graph of actual output versus ideal output.

6. Name one type of mass flowmeter and explain the principle of its operation.
7. Describe the purpose and use of a differential pressure transducer.
8. Describe the bubbler method of level measurement.
9. Explain the principle of density measurement and control.
10. What are the two main components in a control valve assembly?
11. Describe the difference between direct-acting and reverse-acting valves.
12. What is a positioner used for on a valve?
13. What is bench set and why is it done?
14. List and explain at least four actions that should be taken when choosing a valve.
15. When performing an upscale test, the input test point should be approached from _____.
    a. above
    b. below
    c. either direction is fine

16. If zero shift and a span error is detected, the _____ is usually corrected first.
    a. zero
    b. span
    c. either can be done
    d. neither
17. The graph below shows the results of a calibration procedure. Explain the sequence of adjustments that need to be made.

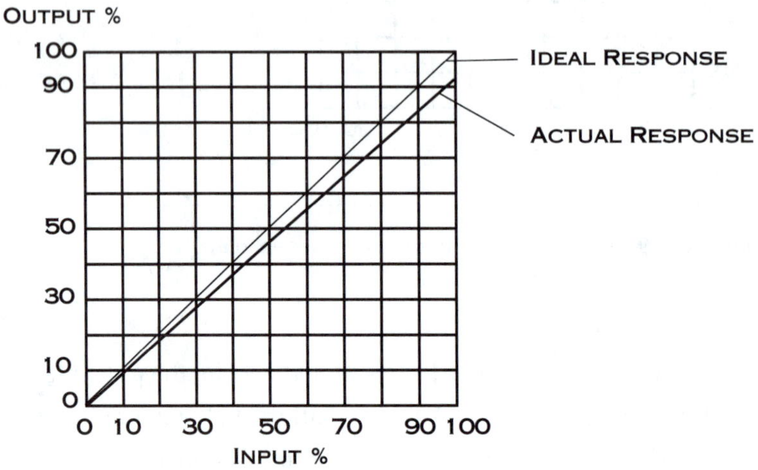

18. The graph below shows the results of a calibration procedure. Explain the sequence of adjustments that need to be made.

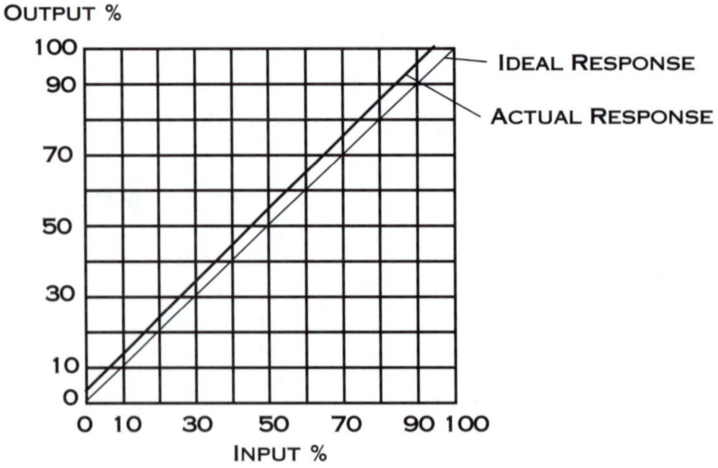

19. The graph below shows the results of a calibration procedure. Explain the sequence of adjustments that need to be made.

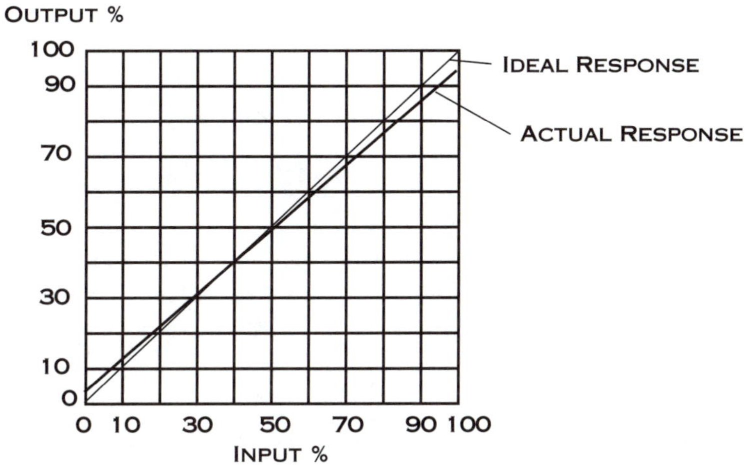

20. Assume you are given a current-to-pressure (I/P) transducer to calibrate. The input current range is 4–20 mA. The output from the transducer should be 3–15 psi. Develop a calibration procedure to calibrate the transducer. Calculate the settings and output for each reading that needs to be taken.

chapter 19

Batch and Continuous Processes

This chapter will examine various types of process control systems and instrumentation.

OBJECTIVES

Upon completion of this chapter, you will be able to:

1. Define terms such as *batch, continuous, separation,* and *distillation.*
2. Describe variables that are typically controlled in processes.
3. Describe a reactor.
4. Explain the process of fractional distillation.
5. Describe a distillation column.

BATCH PROCESSES

Batch processes are common in industries such as food, pharmaceuticals, and chemicals. Many products are made in batch processes. Batch processes are typically used when the product cannot be made any other way or when small quantities of specialized products are required.

A practical batch process we all understand is making hot chocolate. To make hot chocolate, you mix milk and chocolate syrup together and then heat the mixture.

This would be a typical batch process. The quality of the final product depends on the correct quantities of raw materials, the mixing, and the temperature. The critical variables that typically need to be controlled in batch processes include temperature, pressure, and flow. Some may also include some measure or degree of completion. Beer brewing is another example. The beer needs to be fermented to achieve the correct alcohol content. In our chocolate milk example, temperature was important. In fact temperature could have helped our process. Mixing the two raw materials (milk and chocolate syrup) would have been easier had the milk and syrup been heated first.

Making a pot of coffee is another example of a batch process. The first step is accurate control of the quantity of raw materials (water and coffee). Next, the temperature of the water must be correct for proper extraction of the flavor from the coffee. The flow rate of the water also affects the flavor extraction from the coffee. Other variables include the quality of the coffee and the water.

In batch processes the knowledge of what goes into making a good product are key to the control of the process. The raw materials must be chosen carefully and quantities must be carefully controlled. Temperature is key to many batch processes and also needs to be carefully controlled. Rate is another variable that needs to be controlled in batch processes.

Let's examine batch processes more closely by examining a product we are familiar with: Jello. The process of making Jello is as follows: stir one cup of boiling water into the gelatin for two minutes until completely dissolved, add one cup of cold water, then refrigerate for 4 hours until firm. To make Jello we utilize a pot on the stove. We put a cover on the pot to speed the heating process and also to keep materials from escaping into the environment. In industry the kettle is similar and is called a reactor (see Figure 19–1). When we are making Jello it is important to agitate the materials by stirring. An industrial process is very similar. Industrial reactors typically have agitators. Heat must be added to many batch processes. Heat usually speeds the process, enables a process to change, and usually reduces the required agitation.

The kettle on the stove is not very efficient. Only the material on the bottom of the kettle is in direct contact with the heated portion of the kettle. The material in the kettle above the bottom is heated by heat transfer from the sides of the kettle or through the material. Industrial reactors typically have jackets around the kettle to make the heating and/or cooling more efficient (see Figure 19–1).

Industrial reactors are usually pressurized. Increased pressure raises the boiling point of fluids and can speed the heating process. Note in Figure 19–1 that the reactor is sealed at the top by a stuffing box that allows the agitator to turn but also keeps the pressure inside the reactor. This is the same principle as the pressure cooker at home. In addition to the heating jacket, coils can also be added to speed

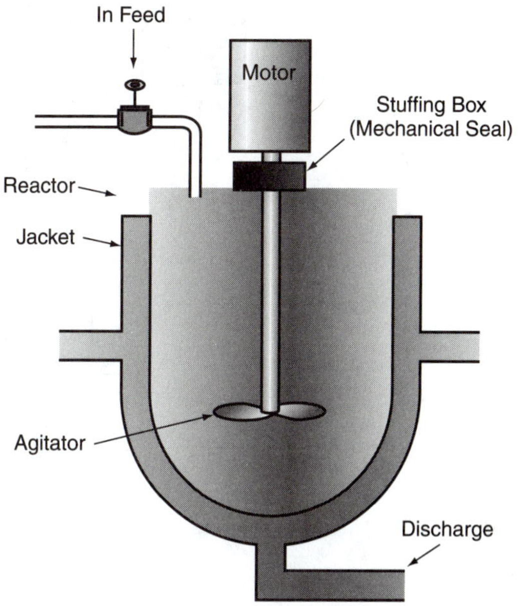

Figure 19–1 An industrial reactor.

the heating or cooling. The process can be exothermic or endothermic. Processes that require an external source of heat are called endothermic. Cooking is an example of an endothermic process. Processes that generate heat are called exothermic. The use of Drano to clear a clog in a pipe is an example of an exothermic process. Figure 19–2 shows an example of a batch process. Note the two tanks that hold the raw materials. A valve is used to control the quantity of each raw material used in each batch. Note also the discharge at the bottom of the tank for discharging the finished product.

In batch processes, the quantities of raw materials used are called a recipe. Industrial reactors are typically used for many different recipes or "mixes" of product. Paint is an example. Different colors of paint require different types and quantities of raw materials. Different qualities of paint may also require different raw materials and quantities. Close control of raw materials is important for product quality but can be just as important for cost control. Some raw materials are expensive. Close control of batch processes can be profitable. Expensive instrumentation and controls can often be justified by the savings in raw materials brought about by closer control.

Types of Typical Batch Processes

Mixing and Blending Mixing and blending are commonly done in batch processes. Making a cake is an example of a mixing and blending process. Heating

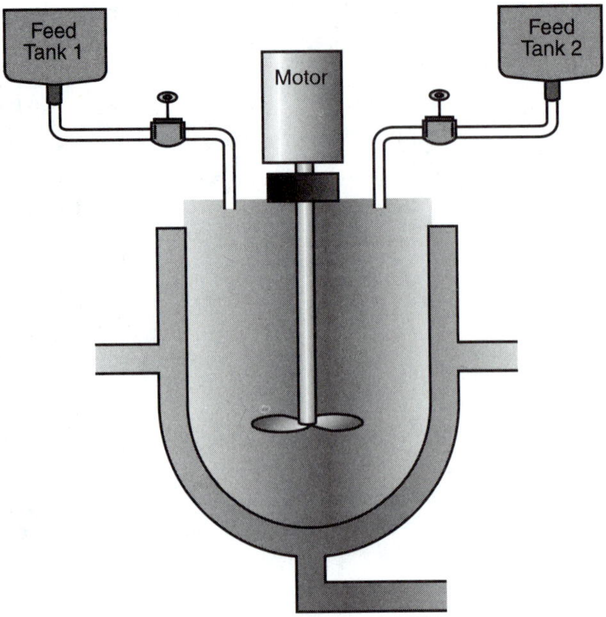

Figure 19–2 Typical batch process.

or cooling is common in mixing and blending processes. Heating can often aid the process. An example would be instant coffee. Instant coffee mixes much easier in hot water than in water at room temperature. The heat decreases the required mixing and shortens the time. Mixing and blending may also include chemical reactions.

Chemical Reaction A chemical reaction involves bringing two or more reactants together under pressure or heat to form one or more products. Chemical reaction processes are common in pharmaceuticals, chemicals, and agricultural products such as fertilizers.

Polymerization Polymerization is the combination of a large number of the same type of molecules to form a product. Plastics and synthetic fibers are examples of polymers. Polymerization may require a catalyst, heat, and pressure to enable the polymerization to occur.

Separation In a separation application, a component is removed from a material or fluid. Removing oil from water is one example (see Figure 19–3). In the case of oil removal from water, oil rises to the top of a reactor. Heating or cooling

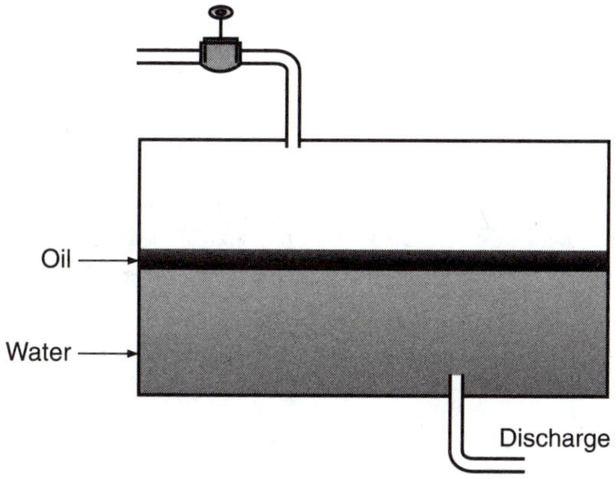

Figure 19–3 Separation process.

may improve some separation processes. Agitation is usually not desirable in separation processes.

Crystallization Crystallization is the formation of a solid from a vapor, a solution, or a liquid. Many pharmaceuticals are made by crystallization. Temperature is usually important in crystallization processes.

Batch Process Example Beer is made in a batch process, and a simplified example of the process is explained here. Germinated barley is dried in a kiln. It is then called malt. Malt gives color and flavor to beer as well as providing the sugar for the yeast. The malted barley is then milled into a course flour that is called grist.

The first batch process converts the starch in the grist to sugar. Grist and hot water are added to a large kettle called a mash tun (see Figure 19–4). The temperature is reduced from approximately 170 degrees Fahrenheit to 130 degrees Fahrenheit. This changes the starch in the grist to sugar. The liquid that results has the sugar, flavor, and color of the grist. The liquid is called wort.

The wort is then transferred to another kettle. The second batch process is a boiling process. Hops are added before the boiling starts. The hops in the boiling wort add bitterness and flavor. The wort is boiled until all the flavor of the hops has been released.

The third batch process is fermentation. After the wort cools, it is transferred to a fermentation vessel and yeast is added. The yeast multiplies rapidly as it interacts with the sweet wort, converting the sugars into alcohol and carbon dioxide. When fermentation is complete, the yeast is drawn off for reuse and the beer is matured in maturation vessels. Beer normally contains about 5 percent alcohol. The beer is chilled and stored to allow the flavor to develop.

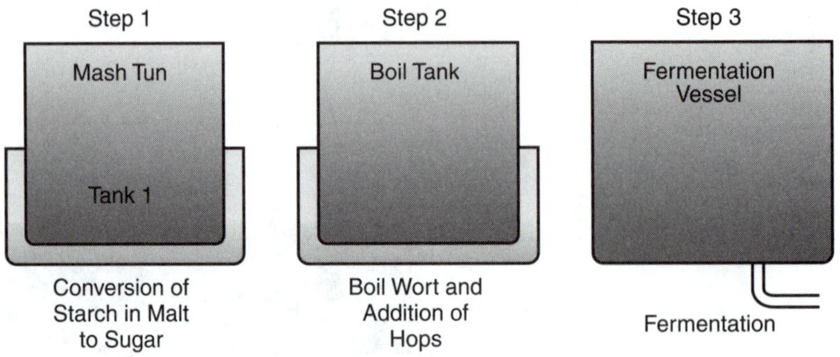

Figure 19–4 Beer making in batch processes.

Continuous Processes

Continuous processes are used to produce large quantities of a product. The main variables in continuous processes are usually flow, pressure, level, and temperature. Many products are made in continuous processes. Chemicals, petroleum, paper, cardboard, packaging materials, and so on, are made in continuous processes. Continuous processes are designed to produce a large quantity of one particular product. In a continuous process, raw materials are continuously run through processing equipment. Several variables are normally monitored to ensure the quality of the product. The setpoint of each variable must be continuously controlled close to setpoint, regardless of disturbances to the process.

Figure 19–5 shows a simple continuous process. Note that this process looks much like a batch process. There is a reactor with a heating element (steam in this case). The difference in this example is that raw materials are fed continuously into the process. The process in the figure represents an evaporation process. The fluid is heated to the boiling point in the reactor, and the vapors rise to the top and exit. There is a continuous outflow of finished product also. This type of process is called an evaporator.

Typical Controlled Variables in Continuous Processes

Generally two or more raw materials are fed into continuous processes. It is crucial that the quantities of each are controlled to ensure product consistency and quality. The most common method of controlling the raw materials is through controlling their flow into the process. *Composition* is the term that is used to describe the right mix of product.

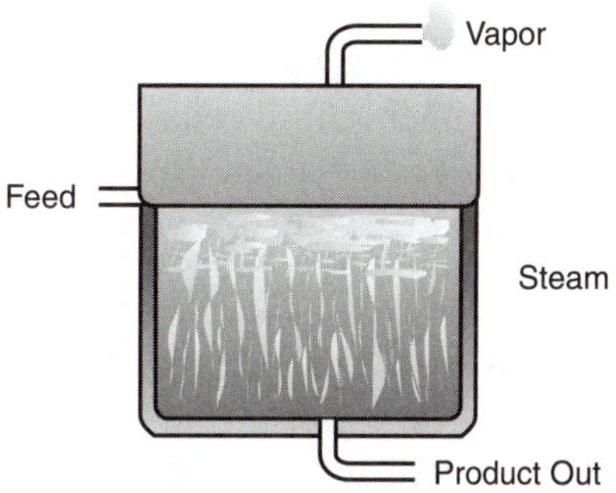

Figure 19–5 Simple continuous process.

As the raw materials pass through the process, it may be important to control several variables. Composition needs to be controlled. Temperature is often a critical variable. Flow rate is also a critical process variable. Pressure and level may also be important.

Composition and/or Concentration Composition may be the proper concentration of two or more materials, the concentration of solids in a liquid, completion of a chemical reaction, or even consistency of a blending in a process. In our simple system in Figure 19–1, it may be important to measure the concentration of the product as it leaves the reactor to be sure it meets specifications. Measurement can be done with an online analyzer that monitors the concentration continuously.

Temperature Temperature is often a critical variable in batch and continuous processes. Some processes require heating and others may require cooling. Many processes have a combination of heating and cooling cycles. Heat exchangers are often used to supply heat or cooling to continuous processes.

Temperature can also be used indirectly to measure concentration. The concentration of finished products can be checked by its temperature. The concentration can be inferred from the temperature. The product temperature is an indication of the boiling point of the fluid in the reactor. The boiling point is one way to measure the concentration. As concentration changes, the boiling point changes; so if we measure the temperature, we can infer the concentration of the fluid.

Flow Flow needs to be controlled in almost all continuous processes. The flow rate of the raw materials must usually be controlled. The flow rate of the raw material into the process directly affects how much heat must be added to the process. If heating or cooling is involved, the flow rate of steam through a heat exchanger must be controlled. Examine the heat exchanger in Figure 19–1. The flow rate of the finished product may also need to be controlled because it also affects the temperature of the product.

Pressure In many cases, reactions occur more quickly at elevated temperatures. Pressure elevates the boiling temperature and enables faster processing and shorter reaction times. Close control of the pressure can help ensure product quality and the safety of personnel. Pressure must be monitored to ensure that operating pressures remain at safe levels.

Level If a continuous process takes place in a reactor, the quantity may be measured by measuring the level in the reactor. Level is often easier to measure than quantity. The correct quantities of raw materials must be present to ensure product quality. There must also be enough material in the reactor for safety reasons. There must be enough material present to cover any heat or cooling exchangers present in the reactor. There must not be too much material because it could cause an overflow or it could result in pressures being too high in the reactor. Many types of sensors measure levels in a system.

Continuous Process Example Distillation can be a batch or a continuous process. Let's consider the process of whiskey making. The process is actually quite simple. Whiskey is made in a distillation process. Whiskey was made only in batch processes until the 1820s. In the late 1820s, a new type of still was invented by Robert Stien, which produced liquor in a continuous stream as long as wine, beer, or some mildly alcoholic wash was fed into it. Aeneas Coffey studied the process and improved it. Coffey's version of the continuous still, the continuous or Coffey still, caught on worldwide.

The process is quite simple. An alcoholic *wash* is fed continuously into a distillation column (see Figure 19–6). The temperature is raised. Alcohol boils at a lower temperature than water. The alcohol vapors rise in the column and condense to form a liquid with higher alcohol content. One-hundred-proof liquor is 50 percent alcohol. The product is then filtered and aged, for years, in charred oak barrels to give it a unique flavor.

Fractional Distillation Refining crude oil is a good example of fractional distillation. Distillation is a separation process. Crude oil consists of many components that can be separated out. The various components that make up crude oil have different weights and boiling temperatures. Distillation is a separation process that takes advantage of different boiling points. Boiling water is actually a distillation process. At 212 degrees Fahrenheit water begins to boil and water vapor forms. This temperature is the separation point. The water vapor rises and as it cools, it condenses.

If we have a mixture of liquids with different boiling points, we can use the distillation process to separate them. Figure 19–7 shows a fractional distillation process. In this sample system, crude oil is pumped into the process and heated in

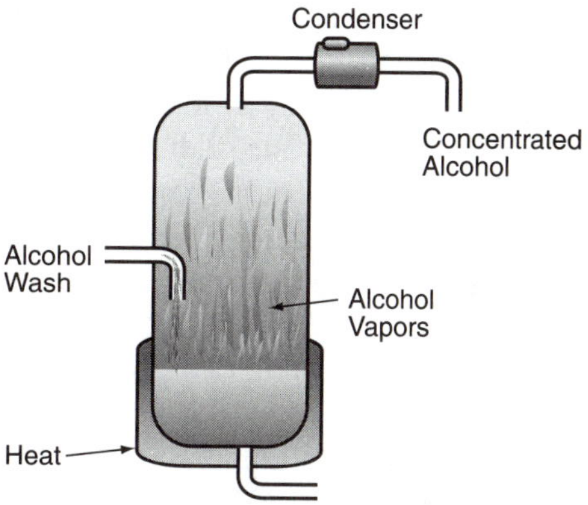

Figure 19–6 Alcohol distillation.

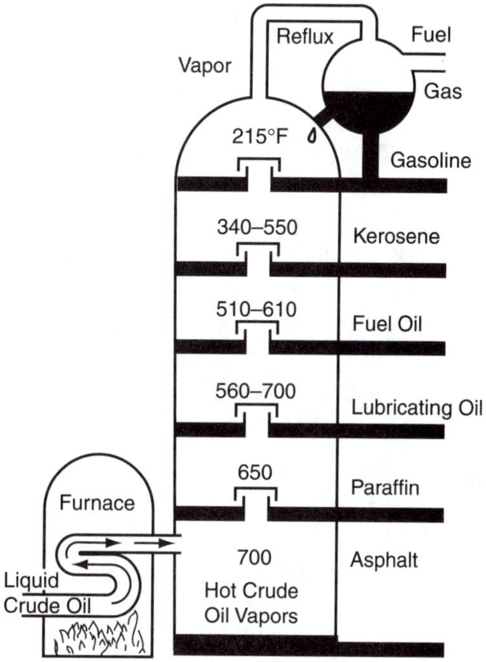

Figure 19–7 Fractional distillation process.

a boiler before it enters the distillation column. The boiler has an internal heat exchanger through which superheated steam is passed to heat the crude oil. The crude oil is heated to approximately 1112 degrees Fahrenheit.

As the crude oil boils, most of the substances that make up crude oil go from the liquid to the vapor phase. The vapors enter the bottom of the distillation column. The column is hotter at the bottom and cooler at the top. The vapors rise up the column. "Trays" at various levels in the column have holes in them through which the vapors can rise through the column. As the vapors rise through the column, they cool. When they reach the level at which they condense to a liquid again, they gather on the tray at that level and flow out of the process. Each substance that makes up the crude oil has a different boiling point, so each condenses at a different level; thus, each is separated in the distillation column. The substance with the lowest boiling point condenses at the highest point in the distillation column. The substance with the highest boiling point condenses at the lowest point in the distillation column. These liquids that are separated are sometimes called fractions, thus, the term *fractional distillation.* Fractional distillation is the most important step in the oil-refining process. It is used to separate a mixture of substances with narrow differences in boiling points.

Many of the liquids are chemically processed to make other fractions. Gasoline is approximately 40 percent crude oil. Some of the liquids from the distillation process are further processed to increase the yield of gasoline from the process.

Piping and Instrumentation Diagrams

Piping and instrumentation diagrams (PI&Ds) are used in process control to represent process control systems graphically. They use standard symbols common to all process industries. Instruments in a system are represented by circles (also called balloons). Balloons can contain numbers, lines, and letters that identify function and location in a process control system. Figure 19–8 shows some of the symbols for various types of devices. A circle represents a stand-alone instrument and is used to represent a transmitter, alarm, or sensor. If the circle is in a square, the device performs multiple functions such as measurement and display and/or control. A diamond in a square represents a PLC. A hexagon represents a computer function.

Lines within the symbols indicate how the devices are mounted (see Figure 19–8). This provides a technician with valuable information about the location and accessibility of the instrumentation. No line in the symbol means that the device is installed in the field, close to the final control element or near where the variable is being measured. A solid line indicates that the instrument is board-mounted in the control room and should be easily accessible to the operator. Double lines indicate that the device is not located by the process. A dashed line indicates that the device is behind a panel (see Figure 19–9). In this case it may be difficult to access.

Tag Numbers

Alphanumeric codes are used to provide information about the instrument for the technician. Letters appear at the top of the symbols. The first letter is used to indicate what the instrument measures, for example, P is pressure. Figure 19–10 shows

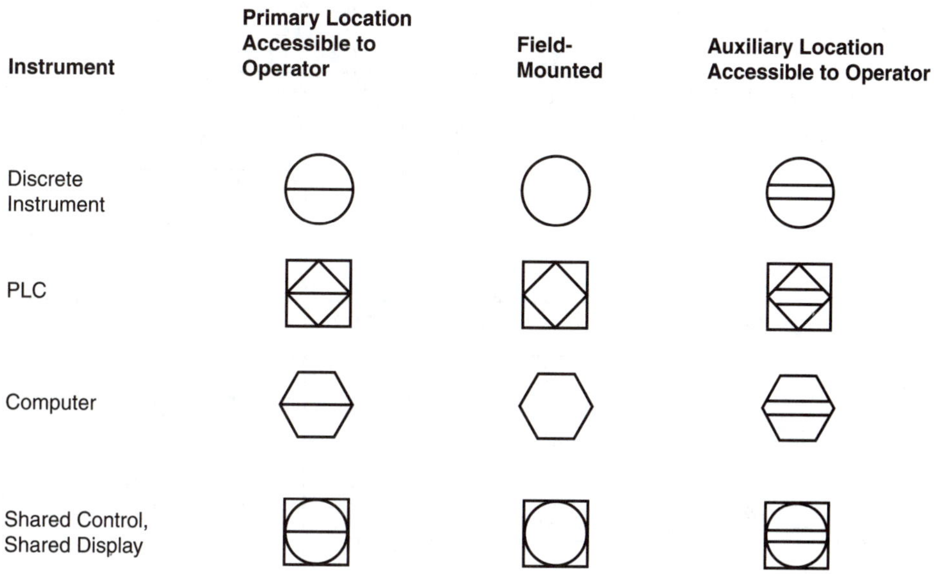
Instrument
Primary Location Accessible to Operator
Field-Mounted
Auxiliary Location Accessible to Operator
Discrete Instrument
PLC
Computer
Shared Control, Shared Display

Figure 19–8 Instrument or functional symbols.

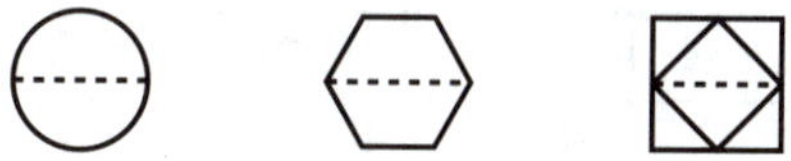

Figure 19–9 A dashed line indicates that the device may be inaccessible or behind a panel.

a list of the function identifiers. The second letter is used to indicate the actual function of the instrument; for example, a C indicates that the device is used for control. If there are three or four letters, the second letter is used to provide additional information about the first identifier; for example, PDT indicates a pressure differential (differential pressure) transmitter. A four-letter example would be PDAL, which indicates a pressure differential alarm that is activated if pressure falls too low.

	First Letter		Succeeding Letters		
	Measured or Initiating Variable	Modifier	Readout or Pasive Function	Output Function	Modifier
A	Analysis		Alarm		
B	Burner or Combustion		User's CHoice	User's CHoice	User's CHoice
C	User's Choice			Control	
D	User's Choice	Differential			
E	Voltage		Sensor		
F	Flow Rate	Ratio			
G	User		Glass Viewing DEvice		
H	Hand				High
I	Current		Indicate		
J	Power	Scan			
K	Time	Time Rate of Change		Control Station	
L	Level		Light		Low
M	User's CHoice	Momentary			
N	User's CHoice		User's CHoice	User's CHoice	User's CHoice
O	User's CHoice		Orifice Restriction		
P	Pressure or Vaccum		Point Connection		
Q	Quantity	Integrate			
R	Radiation		Record		
S	Speed or Frequency	Safety		Switch	
T	Temperature			Transmit	
U	Multivariable		Multifunction	Multifunction	Multifunction
V	Vibration or Mechanical Analysis			Valve or Damper	
W	Weight or Force		Well		
X	Unclassified	X Axis	Unclassified	Unclassified	Unclassified
Y	Event or State	Y Axis		Relay or Compute	
Z	Position or Dimension	Z Axis		Final Element, Actuator, or Driver	

Figure 19–10 Function identifiers.

Loop Identifiers

Numbers under the letters in symbols identify which process loop the device is in. A process loop contains one or more instruments used to control the process. All instruments in the loop are given the same identifier. The loop number can be used to indicate the location and specific information about a device. These numbers are often listed on the actual device to make identification easily available.

Line and Actuator Symbols

Figure 19–11 shows some of the more common symbols for different types of connection lines in a process system. Figure 19–12 shows some of the common actuator symbols.

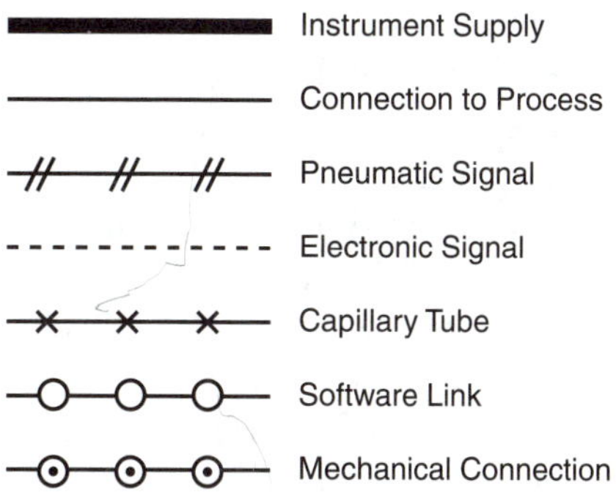

Figure 19–11 Connection line symbols.

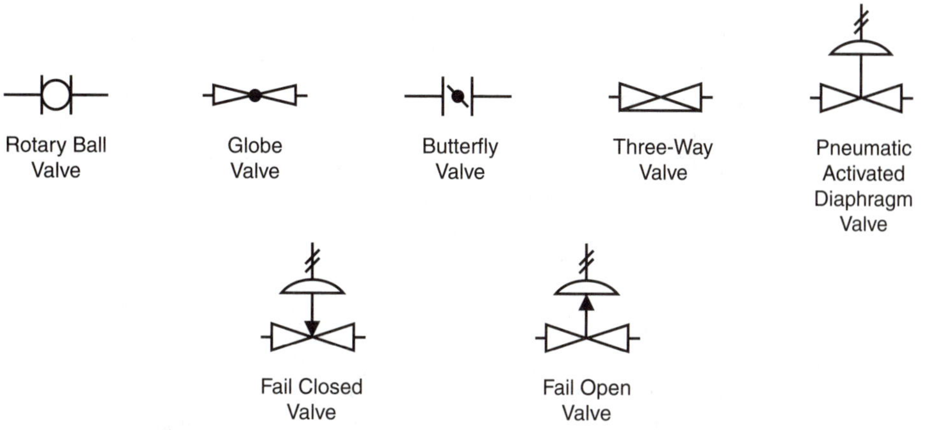

Figure 19–12 Common actuator symbols.

Flow Control System Example

Consider a system that measures differences in pressure to approximate the flow rate. If a restriction is placed in a pipe with flow, the pressure on the downstream side of the restriction will be lower than the pressure on the upstream side of the restriction. An orifice plate used in the system shown in Figure 19–13, can be used to create the pressure differential. If we measure that difference in pressure, we can approximate the flow. The higher the flow, the greater the pressure difference between the two.

The orifice plate (see Figure 19–14) is placed in the pipe. The pressure is then measured on each side of the restrictor plate; it is lower on the downstream side of the plate. One pipe is connected to the upstream side and another is connected to the downstream side. These two pipes are then connected to a device called a *differential pressure transducer (DPT).* The DPT is shown as FT-101 (flow transmitter 101) in Figure 19–13. A DPT has two inputs: a high pressure side and a low pressure side. Imagine a membrane in the center of these two pressure ports. If there is a pressure difference between the two, the membrane yields to the low pressure side. If the membrane has a strain gage, it can measure the difference in pressure. The DP then converts the strain gage signal into a 4–20 mA signal. The dotted line in the system drawing shows that the signal from FT-101 is connected to FC101 (flow controller 101). The 4–20 mA becomes the feedback signal to a controller. This measures pressure with the DPT but we really want to measure flow. This would not be a problem if a linear relationship exists between pressure and flow, but it does not. The pressure increase is the square of the flow increase. Fortunately, we can buy DPTs or signal conditioners *square root extractors* that take the pressure reading and the square root of the difference and convert the result to a 4–20 mA signal.

Next, the flow controller (FC-101) must control the flow rate. We can use a proportional valve to control the flow rate (see Figure 19–15). We have now added a proportional valve to the flow control system. The proportional valve opens between 0 and 100 percent, depending on the input it receives. The valve in this system requires a 4–20 mA input signal. At 4 mA, the valve is closed; at 20 mA, it is 100 percent open. The 4–20 mA signal is converted to air pressure to move a shaft and control the position of the valve.

Density Control Example

To explain density control, we use chocolate milk as our product. The process is quite simple: mix milk with chocolate syrup. The densities of milk and chocolate syrup are quite different. This is a good example of a density control system. Accomplishing this task requires measuring and controlling the density of the product. First, we mix batches until we find the perfect taste. Then we measure the density of the perfect batch and use that as our setpoint.

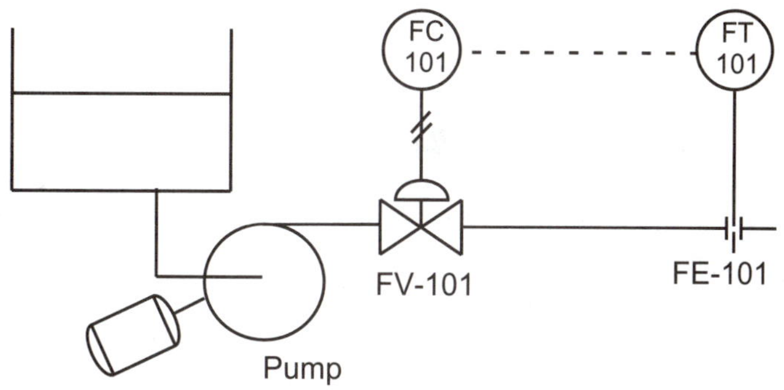

Figure 19–13 Measurement of flow rate by differential pressure measurement.

Figure 19–14 Orifice plate.

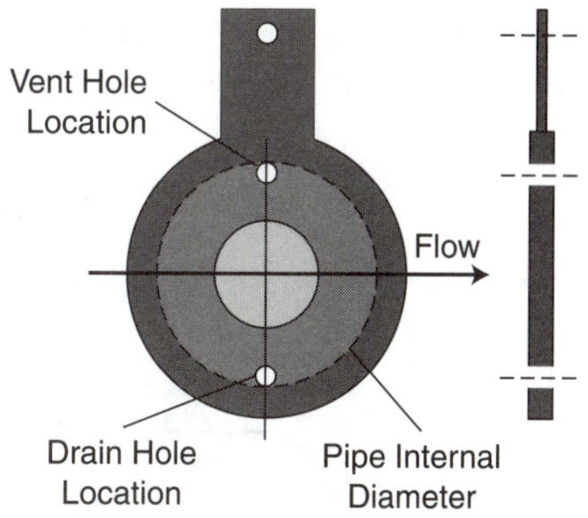

Figure 19–15 Proportional valve.

This system uses a Coriolis meter to measure the product's density because it can measure extremely small changes in density. We control the flow of milk and syrup into the process (see Figure 19–16) using a static mixer to mix our milk and syrup thoroughly (see Figure 19–17). The Coriolis meter measures the density after using the static mixer.

The first step is choosing a desired flow rate for the milk, assumed here to be 1 gallon per minute. The process controller controls a valve to set a flow of 1 gallon per minute for the milk.

The next step is to control the flow rate of the chocolate syrup accurately. The process loop (density) controls the flow of syrup into the process to keep the density of the product at setpoint. The Coriolis meter provides feedback about the density to the controller. The controller controls the valve position for the chocolate syrup to ensure the correct density of chocolate milk.

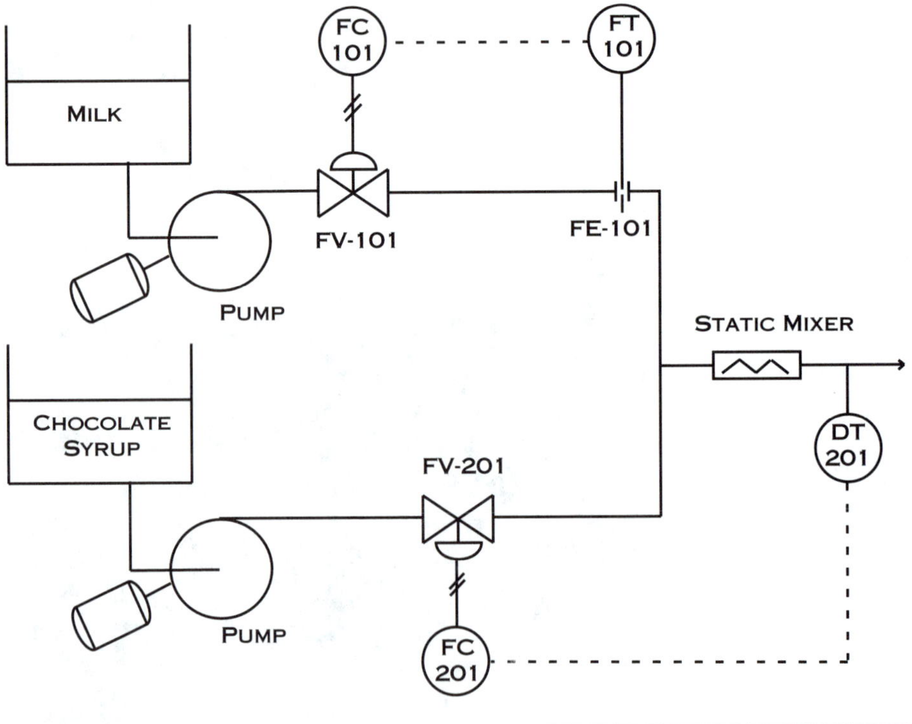

Figure 19–16 Density control system.

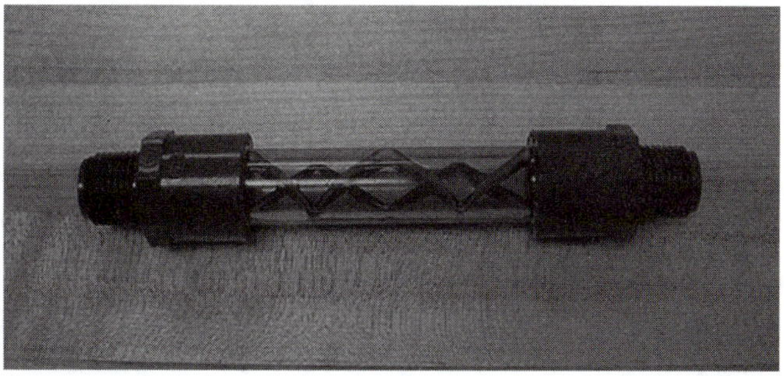

Figure 19–17 Static mixer.

Temperature Control Example

This process controls the temperature of a fluid in a tank. Heated fluid comes into the tank and then must be cooled to 95 degrees Farenheit for the specific process. Actually, this process has two loops: a level control loop and a cooling loop (see Figure 19–18). Note that each loop has a specific range of numbers. The chiller loop, for example, is 201. The level loop is 301.

The *level control loop* consists of an inductive transducer and transmitter to measure the height of fluid. The level sensor's signal is sent to the controller, which controls a ball valve that regulates the flow of fluid into the tank.

The level control loop also consists of a level transmitter (LT-301) and a flow controller (FC-301) that regulates a valve. The dashed lines connecting the flow controller to the valve indicate an air line. In this system heated fluid is allowed into the tank until it reaches the desired level. Then the cooling process begins.

The *cooling loop* consists of a pump, a ball valve, an RTD, and a heat exchanger (Figure 19–19). The pump circulates the water from the tank through the heat exchanger loop, and a ball valve controls the flow through it. The RTD temperature element 201 (TE-101) and the temperature transmitter (TT-201) provide temperature feedback. The output from the temperature transmitter is connected to the flow controller (FC-201) to control the ball valve (with a pneumatic signal). Chilled water and heated tank water are circulated through the heat exchanger separately. The cooled water returns to the tank. This process continues until the temperature reaches the setpoint in the tank, when the tank is emptied so a new batch can be processed.

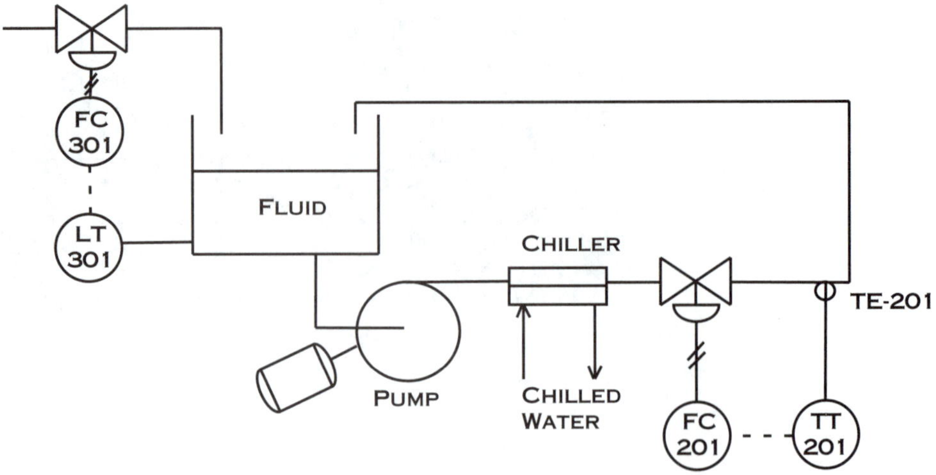

Figure 19–18 Temperature control system.

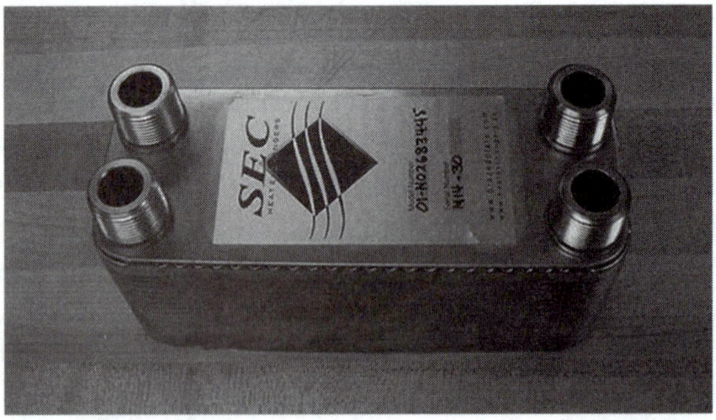

Figure 19–19 Heat exchanger. Two ports are used for hot water from the tank and two are used for the chilled water.

Questions

1. What are the main differences between a batch process and a continuous process?
2. Why would a continuous process be chosen over a batch process for a product?
3. Draw a picture of a reactor and label the main parts.
4. Coffee brewing is an example of a(n) _____ process.

5. Draw and explain the operation of a distillation column.
6. Name at least four types of typical batch processes.
7. What does P&D stand for?
8. What does a dashed line in a symbol mean?
9. Draw a picture of an air-operated valve symbol.
10. Draw a picture of an air-operated fail closed valve.

chapter

Embedded Controllers 20

Embedded controllers are becoming more widely used for industrial control. This chapter will examine the programming and use of embedded controllers.

OBJECTIVES

Upon completion of this chapter, you will be able to:

1. Describe the use of an embedded controller.
2. Compare and contrast the advantages and disadvantages of embedded controllers versus programmable logic controllers.
3. Describe the programming of an embedded controller.
4. Define *multitasking*.

OVERVIEW

Relay ladder logic (RLL) was intended to replace the relays, timers, and counters common to control systems of the 1960s and 1970s with a more flexible software implementation. RLL was successful in achieving and surpassing those modest goals. RLL is still a powerful tool when applied to applications for which it is appropriate.

The industrial world has become more complex. It is no longer sufficient for a machine to perform the same series of actions endlessly. Systems must now be optimizing to maximize production. Many systems must keep basic statistics, which are a requirement for quality assurance. Many systems need to interact with one or more third-party devices whose interfaces are alien to the controlling PLC. More and more systems need to interface with enterprisewide networks, which provide management with up-to-the-minute information.

RLL was not designed for these complications. Some RLLs have been retrofitted with features to handle some of these situations. At best many of these additions have been awkward and slow. They have also been difficult or nearly impossible to use. RLL was simply not designed to solve all of today's industrial problems. These requirements have made other types of control devices more useful. Single-board controllers are one of these technologies.

Single-board controllers are gaining acceptance in industrial applications. They are also called embedded controllers. They are widely used for machine control. Original equipment manufacturers (OEMs) use this technology to control machines that they build. As more and more people are exposed to this technology, it will become more and more accepted and prevalent in industrial control.

Many computers do not look like desktop PCs simply because you cannot see them at all. These invisible computers are hidden or embedded within a larger system or product. Embedded controllers provide better control for a larger system. The components of many embedded computers fit on a single printed circuit board, making an embedded computer compact and economical. The stringent real-world, real-time requirements of control applications mean that the hardware and software of embedded computers are distinctly different from ordinary desktop PCs. Many embedded controllers are available. Many choices of programming languages are also available. Programming languages include C, visual basic, BASIC, assembly, and others.

RLL was designed to perform specific tasks, which it does well. C, on the other hand, was designed to be a general-purpose programming language. Not only does C dominate the development of applications for personal and corporate computing, it also dominates the development of embedded controllers. Thus, C is used to develop programs for everything from word processors and databases to microwaves and TVs.

C's general-purpose nature is the reason for its popularity. Because it makes no assumptions about its environment or what tasks it will perform, programs written in C use resources more efficiently than equivalent RLL systems. For example, most PLCs provide the user with a fixed number of auxiliary relays, a fixed number of registers, a fixed number of timers, and so on. Eventually every RLL programmer encounters a situation where this fixed allocation of resources is unacceptable for the application. A particular program may require only a few auxiliary relays and timers but more than the available number of registers. The most aggravating aspect of this situation is that the PLC often has the needed memory but unfortunately has allocated it to unneeded auxiliary relays and timers.

In C this situation is not a problem. C allocates resources only as requested. As long as the programmer doesn't attempt to allocate more resources than are available, C lets him or her decide how best to use them. Another important advantage of C is a definite focus of execution. Remember, RLL was written to simulate hardware. Because hardware inherently operates in parallel, RLL had to simulate this parallel operation. This is accomplished by executing every rung of the RLL program on each scan. While this approach achieves the desired effect, it has one major drawback: scan times increase as the size and complexity of the program increases. Those attempting to use RLL programs to perform complex tasks can find scan times increasing well beyond 10 milliseconds, which can be too slow for some applications.

C, on the other hand, has the notion of a program counter, which starts at the main (the first function executed in a C program). The program counter works its way through the code as directed by the flow control statements written by the programmer. At any given time, a processor running a C program executes only one sequence of instructions. While doing this, the rest of the program sits idly by, waiting for its turn to execute. This single focus of execution gives the programmer incredible power to decide what aspects of the application need the most intense attention and which need to be checked periodically. Without such control, handling of high-speed events (such as the deployment of airbags) would be impossible.

C is powerful. But like all powerful tools, it also has dangers. Like most tools, its strengths are often its dangers:

- C's flexibility allows so many solutions to a problem that beginners often get lost in its possibilities. RLL is so restrictive that the simplest problems often seem to solve themselves.
- C's ability to focus execution on a small portion of code also means that it is easy to write programs where other tasks are not done often enough or are skipped entirely. RLL executes everything on every scan with no chance of code not executing.
- C is less intuitive for beginners and a little harder to learn than RLL. All of RLL can be learned relatively quickly. With C, a useful set of basics can be learned quickly, but the more complicated aspects of the language can be quite confusing even to programmers with moderate experience.

Even with these problems, C is well worth the investment. C can be used to solve most problems and is supported on all major programming platforms. Ultimately, a good C programmer always outperforms an equally competent counterpart who uses RLL.

Software programs that an embedded controller executes are generally developed on a PC using a software development system. For many embedded controllers, the development system consists of a software editor, a compiler, and a debugger, all of which are separate items that are not always guaranteed to cooperate. An alternative to utilizing separate editors, compilers, and debuggers is to use a software development system that integrates all the components into one software development tool. This approach alleviates potential problems, guaranteeing all three development tools will work properly. An integrated software development system, such as Z-World's Dynamic C, provides a direct and fast means of developing and debugging programs for an embedded controller. And because the target controller can be used during program development, the software and the hardware can be considered as one unit.

Embedded controllers are available with a wide variety of capabilities, from very simple digital I/O to very complex motion control capability. Embedded controllers are flexible. Their capabilities are really only limited by the ability of the programmer. Programming has become easier as companies have added special functions to accomplish many standard tasks. Remember that a PLC or a computer is just a microprocessor with memory and associated peripheral hardware. An embedded controller is no different. It is a microprocessor with memory and associated peripheral hardware for I/O. The PLC is programmed with ladder logic, and the embedded controller is programmed with some other language.

Embedded controllers usually offer a price advantage over PLCs, especially when multiple machines are to be built. PLCs still offer the easiest programming language for those familiar with electrical diagrams. In this chapter we will consider one brand of controller made by Z-World, which manufactures a wide range of C-programmable controllers. C-programmable controllers enable relatively easy design of control systems requiring multitasking or floating-point math calculation. This chapter will examine the use of embedded controllers by examining the use of the Z-World BL2100 SMARTCAT series controller.

I/O Capabilities of the BL2100 SMARTCAT

Figure 20–1 shows the BL2100 controller. Embedded controllers have extensive I/O capabilities. They offer analog, digital, communications, operator I/O, and many other capabilities. The BL2100 SMARTCAT controller has the I/O capabilities described in this section. Figure 20–2 shows the pinout for an embedded controller.

Digital Inputs

Twenty-four digital inputs are protected in the range from −36 V to +36 V. These inputs can be used to configure the inputs as either sinking or sourcing. Additional

Figure 20–1 A BL2100 SMARTCAT single-board computer. *(Courtesy Z-World.)*

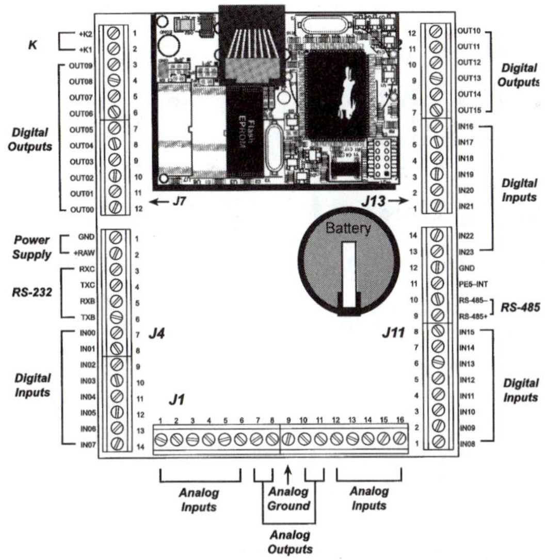

Figure 20–2 The pinout for an embedded controller. *(Courtesy Z-World.)*

boards can be added to increase the number of inputs. Here is an example of the use of an input in C code:

```
waitfor ( zDigIn (5) == 1);
```

This statement causes the program to wait until input 5 becomes a 1 (ON). The waitfor comand causes the wait. zDigIn is a function that gets the value of the input it references, in this case 5. The == is used in C code to check for equality. The semicolon is normally used at the end of every line of code in C. So if digital input 5 is equal to 1, the program will continue. If it is not, the program waits until digital input 5 becomes true (1).

Digital Outputs

Sixteen digital output lines are capable of controlling inductive loads such as relays, stepping motors, and solenoids. Each line can normally source and/or sink up to 200 mA at 36 V. Additional boards can be added to increase the number of outputs. Various types of digital outputs are available, including relay and high voltage. Here is an example of the use of an output function in C code:

```
zDigOut (3,1);
```

This statement uses the zDigOut function. This statement turns on output 3 (1). If the statement is zDigOut (3,0);, the output would be turned off.

Analog I/O

The BL2100 has eleven 12-bit analog inputs and four channels of 12-bit analog output.

Watchdog Timer

When enabled by a jumper, the watchdog timer resets the controller after a short period (about 1.6 seconds), if running software has failed to reset the watchdog. This feature allows your system to recover should a program "hang" or stop operating correctly.

Handling Power Failures

Even if the code is perfect, a controller's program may crash because of conditions that are beyond the control of the software engineer: complete power failures, brownouts and spikes on the incoming power lines, or even human error. Software cannot identify or handle all hardware-related failures reliably because, after all, the software runs on the failing hardware. Detecting hardware failures requires special-purpose hardware. For example, a watchdog timer is an independent hardware device that monitors the health of the software. The watchdog timer resets the controller unless it receives a signal from the software within a specified time, usually about 1 second. On most boards, when input voltage falls below an acceptable threshold, a nonmaskable power-failure interrupt takes place. The program can have a function to perform a short shutdown procedure prior to reset.

Memory

Random access memory (RAM), or static random access memory (SRAM), is a common form of memory. Although this type of data memory is inexpensive, it is volatile. Data residing in RAM disappears when power is removed. This behavior is not acceptable in control applications. Nonvolatile memory in the form of electrically eraseable programmable read-only memory (EEPROM), battery-backed RAM, or flash EPROM provides varying degrees of security for mission-critical data. The type of low-power, read-write SRAM used in embedded controllers retains its data in a sleep mode as long as a modest power level is applied. Consequently, a single lithium battery can possibly keep data alive many years, if necessary. Once written to EEPROM, data is not lost no matter how long power is removed. Given the EEPROM's combination of slow write speeds, high cost, and small capacity, however, the best use for an EEPROM is storing small amounts of mission-critical data at infrequent intervals.

Nonvolatile Memory

Flash EPROM is a variation of EEPROM that is becoming more and more popular. Clever designers have reduced the amount of on-chip circuitry needed for writing to flash EPROM, lowering cost substantially while retaining the inherent advantages of EEPROM. Flash EPROM is much less expensive than an equivalent EEPROM and not much more expensive than an equivalent EPROM. Flash EPROM can be reprogrammed without removing it from the controller, eliminating a separate programming step and making software development much easier. And field upgrades become simple and efficient. The BL2100 has 128K SRAM/256 flash (standard).

Power Failures Personal computers react to a power failure by simply shutting down and losing any data that may reside in the RAM. This situation is not acceptable for an embedded controller supervising a critical process. A special-purpose circuit in an embedded controller constantly monitors the input voltage as well as the regulated supply for the controller's internal components. A circuit called a supervisor gives the controller advanced warning of an impending power failure, allowing the controller to shut down properly. If the controller has a backup battery, the supervisor can switch critical components, such as RAM, over to battery power. If the controller has some form of nonvolatile memory, the data can then be saved.

Mission-Critical Data A common problem in industrial applications is retaining mission-critical data during power outages or other emergencies. To understand how an embedded controller handles outages, one needs to know the nature of interrupts. When a controller's microprocessor recognizes an interrupt, the microprocessor almost immediately shelves what it is doing and jumps to the appropriate interrupt service routine. Software engineers use this built-in, fast hardware and software facility in three ways:

- To respond quickly to unscheduled or anomalous events.
- To acknowledge some input but defer its processing until later.
- To respond to interrupts generated by some regular timing mechanism to achieve the regulated, orderly operation of some sequence.

Real-Time Clock

In many applications, software engineers use simple hardware counters to keep track of time intervals. But if the application needs to log data or schedule operations by time and date, then a special-purpose real-time clock may be beneficial because it relieves the controller's microprocessor of much of the timekeeping overhead.

Communications

The Electronic Industry Association's RS-232 is the oldest and still most widely used standard for exchanging data between two devices. A vast array of equipment has RS-232 ports available. RS-232 serial communication transmits data one bit at a time over a single data line. Single data bits are grouped together into a byte and transmitted at a set interval, or baud rate. Serial transmission can be over a simple connection consisting of three wires for transmit, receive, and ground signals. There are four serial ports on the BL2100: two ports are RS-232, one is RS-485, and one is 5 V CMOS compatible for programming.

The controller also has Ethernet capability. The Ethernet port can enable the card to monitor and supervise another system remotely or to enable an application to communicate.

Figure 20–3 Keypad, display option. *(Courtesy Z-World.)*

Operator Interfaces

Embedded applications often do not need and cannot use a typical PC monitor, keyboard, and mouse. Instead, a simple keypad and LCD (see Figure 20–3) is often more appropriate and less expensive.

A 122 × 32 graphic display with seven relegendable keys is available as a cost-effective user interface. This means that the user can label the keypad to match the application's needs. User-programmable LEDs provide quick-status feedback, and the entire board/display/keypad assembly mounts in an integrated enclosure.

Many functions are available in Dynamic C to make programming the display easier. The operator can change system parameters, scanning rapidly through multiple menus and submenus. The program can monitor and make sure that the operator enters only values that are acceptable for the application. Besides allowing an operator to change system parameters, it periodically displays messages or parameters of your choice.

Simple Tank Application

Consider the tank system shown in Figure 20–4. There are two level switches; one for when the tank level gets too full and one for when the tank level is too low. There is one output in this simple system. The output controls the pump. For this example we will connect the low-level switch to input 1 (PIN01) and the high-level switch to input 2 (PIN02). These are the first two inputs on the controller. The pump is connected to the first output (HV01) on the controller.

Imagine that a worker operated this system. Our instructions to an operator would be:

- If the tank is almost empty, open the valve.
- If the tank is almost full, close the valve.

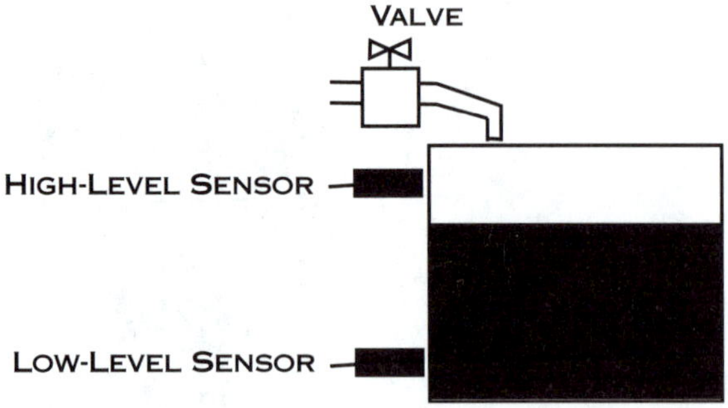

Figure 20–4 Simple tank-level application.

This system is obviously simple. An operator can perform this operation easily. There are some problems, however. What does "almost empty" or "almost full" mean? Will these terms mean the same to every operator? Obviously not. There will also be times when the tank overflows and other times when the tank runs empty. These conditions may be caused by boredom and lack of attention on the operator's part. It is also an expensive proposition to have an operator watch a tank level. Figure 20–5 shows the I/O data used in this application. There are two inputs and one output.

Figure 20–6 shows what the program would look like if we wrote it in English IF statements. The I/O states are shown in parentheses after each statement. Figure 20–7 shows what the program would be in C language. Let's examine a line of code. The *if* statement evaluates zDigIn (PIN01) to see if it is equal to (==) 1. In the C language == is used to check for equality. PIN01 is input 1. Input 1 is our low-level switch. If it is a high (1), the if statement is true. If the first part of the statement is true, the rest of the statement will be executed. In this case the second portion of the statement is zDigOut (HV01,1). This statement turns output HV01 on. The second statement in Figure 20–7 examines input 2 (the high-level switch) to see if the tank level is too high. If it is on, the second portion of the statement turns output 1 (pump) off.

We simplified the program to gain an understanding of what a program looks like. The previous code would have had to be in a loop so that the controller monitored the tank level continuously. A more complete program is shown in Figure 20–8. Everything between the BEGINTASKS and ENDTASKS executes endlessly. This loop ensures that the tank level is monitored continuously. The MAIN statement is used to begin all C programs. The main_init(); and MasterInit(); statements are functions that initialize the system.

I/O Type	Use
Input 1 (PIN01)	Low Level Switch
Input 2 (PIN02)	High Level Switch
Output 1 (HV01)	Pump

Figure 20–5 The I/O data used in the tank application.

If the tank is nearly empty (PIN01 is HIGH),
turn on the pump (enable HV01).
If the tank is almost full (PIN02 is HIGH),
turn off the pump (HV01).

Figure 20–6 The program written with IF statements.

```
if (zDigIn (PIN01) == 1 zDigOut (HV01,1);
if (zDigIn (PIN02) == 1 zDigOut (HV01,0);
```

Figure 20–7 Examples of input and output functions.

```
MAIN
main_init();
MasterInit();
BEGINTASKS
   if (zDigIn (PIN01) == 1 zDigOut (HV01,1);
   if (zDigIn (PIN01) == 1 zDigOut (HV01,0);
ENDTASKS
```

Figure 20–8 Example of an input/output C program.

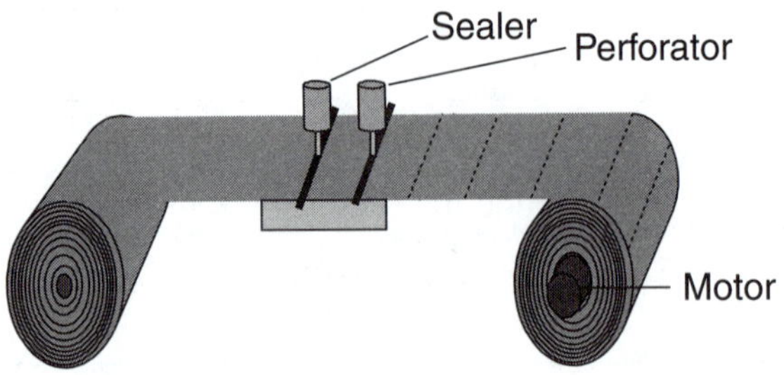

Figure 20–9 Indexing and perforating application.

Indexing, Sealing, and Perforating Application

This application involves a machine that indexes and perforates material. The machine produces plastic bags. The bags are wound on a continuous roll (see Figure 20–9). They are sealed on one end and perforated for easy removal by the user. A roll of garbage bags is an example of this type of product.

The Process

The motor is used to advance the material the correct length for cutting (perforating) and sealing. The machine consists of pneumatic cylinders to extend and retract the perforating knife and sealer. The sealer is a hot wire used to seal one end of each bag. The cutter knife is then advanced to perforate the bag. The controller ensures the safety of the machine and operator by monitoring inputs for knife position, operator location, material shortage, and preventive maintenance schedules. The machine also records and displays dates and times of operation. Figure 20–10 shows a diagram of the em-

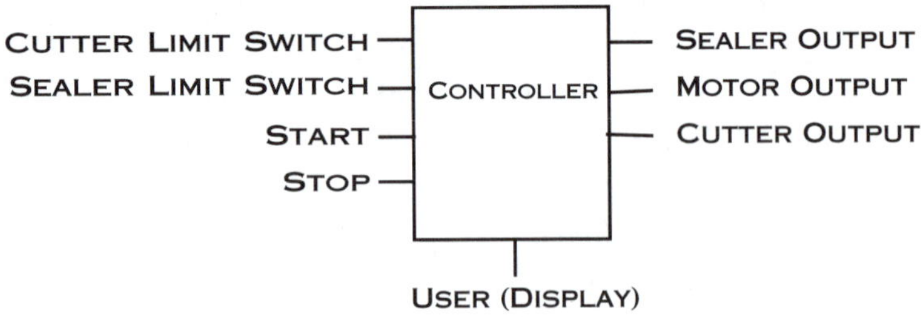

Figure 20–10 A block diagram of the I/O and controller for the indexing, sealing, and cutting application.

bedded controller and I/O. The inputs are shown on the left and the outputs are shown on the right of the controller. Figure 20–11 is the program to control this application.

The first part of the program creates three integer variables and assigns them a number. Motor is set to equal 1, Sealer is set to equal 2, and Cutter is set to equal 3. MOTOR, SEALER, and CUTTER are all digital outputs. MOTOR is digital output 1, SEALER is digital output 2, and CUTTER is digital output 3 (CUTTER is the perforating blade). This part of the program is shown below:

```
MAIN
int MOTOR = 1;
int SEALER = 2;
int CUTTER = 3;
```

The first task is used for safety. If a dangerous condition occurs, the task will display Safety Mode and the display of the embedded controller; sound the beeper; and then turn the motor, sealer, and cutter off. Note that the names MOTOR, SEALER, and CUTTER were used in the output statements. The actual number of the output could have been used also, but the use of a name makes the program much more understandable. The lc_printf is a function used to display information to the controller display. The up_beep function is used to sound the beeper, as you can see here:

```
BEGINTASKS
     task
         {
          while (Safety_Mode)
              {
               lc_printf ("Safety Mode");
                up_beep (200);
               zDigOut (MOTOR, OFF);
               zDigOut (SEALER, OFF);
               zDigOut (CUTTER, OFF);
              }
         }
```

The first part of the next task (shown below) checks to be sure there are no safety problems (knife safety switch or sealer safety switch). The if statement looks at the two safety switches and if either of them are ON, Safety_Mode is set to 1 and the yield causes the program to leave this task and return to the first task, which then

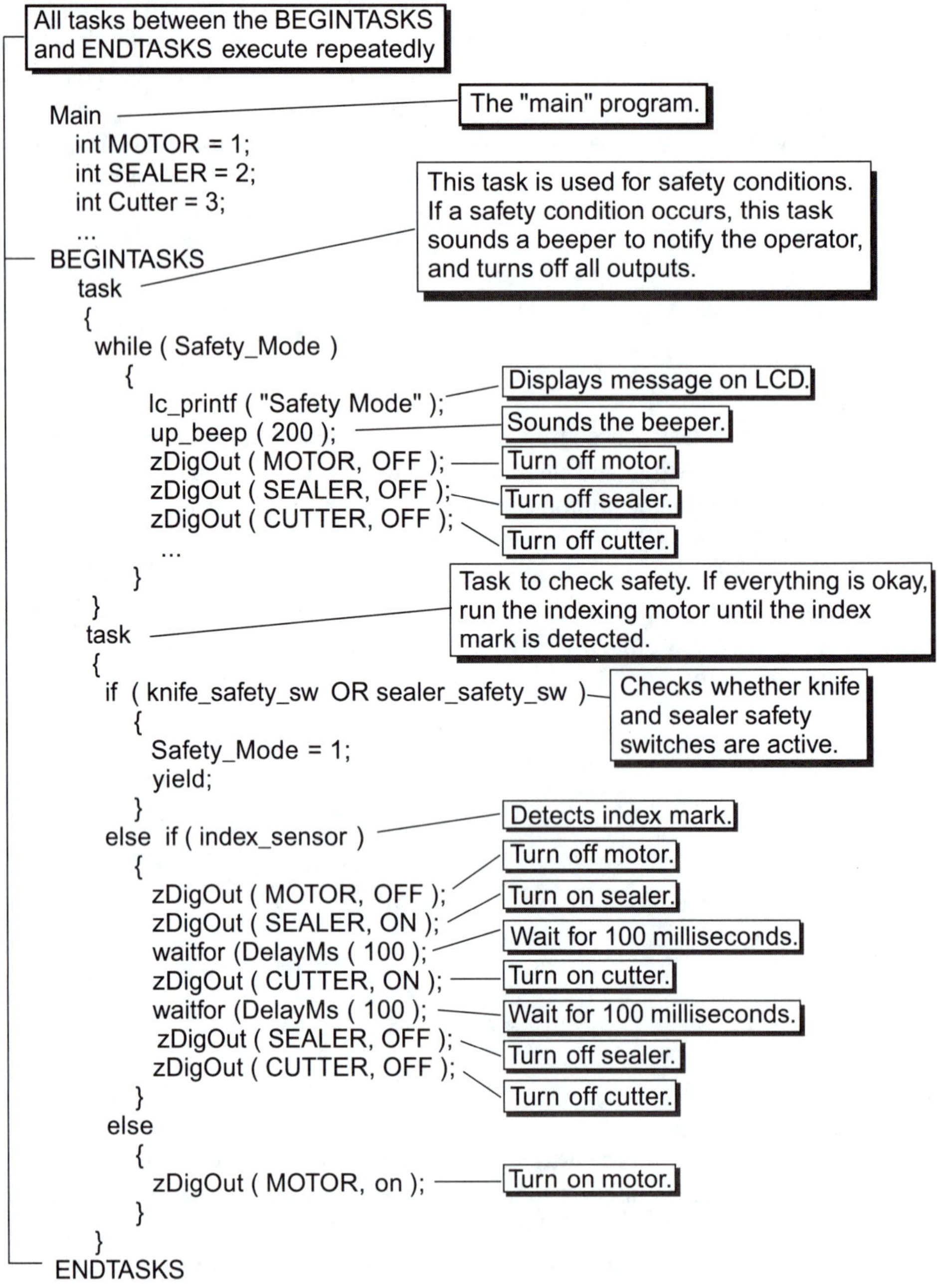

All tasks between the BEGINTASKS and ENDTASKS execute repeatedly
Main
int MOTOR = 1;
int SEALER = 2;
int Cutter = 3;
...
BEGINTASKS
task
{
while (Safety_Mode)
{
lc_printf ("Safety Mode");
up_beep (200);
zDigOut (MOTOR, OFF);
zDigOut (SEALER, OFF);
zDigOut (CUTTER, OFF);
...
}
}
task
{
if (knife_safety_sw OR sealer_safety_sw)
{
Safety_Mode = 1;
yield;
}
else if (index_sensor)
{
zDigOut (MOTOR, OFF);
zDigOut (SEALER, ON);
waitfor (DelayMs (100);
zDigOut (CUTTER, ON);
waitfor (DelayMs (100);
zDigOut (SEALER, OFF);
zDigOut (CUTTER, OFF);
}
else
{
zDigOut (MOTOR, on);
}
}
ENDTASKS
The "main" program.
This task is used for safety conditions. If a safety condition occurs, this task sounds a beeper to notify the operator, and turns off all outputs.
Displays message on LCD.
Sounds the beeper.
Turn off motor.
Turn off sealer.
Turn off cutter.
Task to check safety. If everything is okay, run the indexing motor until the index mark is detected.
Checks whether knife and sealer safety switches are active.
Detects index mark.
Turn off motor.
Turn on sealer.
Wait for 100 milliseconds.
Turn on cutter.
Wait for 100 milliseconds.
Turn off sealer.
Turn off cutter.
Turn on motor.

Figure 20–11 Program for the indexing, sealing, and perforating application.

notifies the operator that there is an unsafe condition. If neither switch is ON, the program continues to the next lines of code.

```
task
    {
       if (knife_safety_sw OR sealer_safety_sw)
            {
              Safety_Mode = 1; yield;
            }
    }
```

The next portion of the task actually runs the process (refer to the code below). This portion begins with an else–if statement, which is used to detect when the plastic material has moved to the correct position. A sensor is used to look for the index mark on the material. When the index sensor is ON, the material is in position, so the motor is turned off and the sealer is turned on to seal the bag. Then the waitfor function causes the program to delay for 100 milliseconds. Next the cutter is turned on and the bag is cut. The *waitfor* causes another delay of 100 milliseconds, and then the sealer and the cutter are turned off.

```
else if (index_sensor)
     {
      zDigOut (MOTOR, OFF);
       zDigOut (SEALER, ON);
       waitfor (DelayMs(100));
       zDigOut (CUTTER, ON);
       waitfor (DelayMs(100));
       zDigOut (SEALER, OFF);
       zDigOut (CUTTER, OFF);
     }
```

If the index sensor was not on, the next portion of the code executes (see the code below). The *else* statement executes if the previous *if* statements have not been true. This else is used to turn the motor on. This code turns the motor on, and then program execution goes to the top of the program again to check for safety conditions and to watch for the index mark on the material. The motor then turns and advances the material until the index sensor senses the index mark on the plastic.

```
          else
                {
               zDigOut (MOTOR, ON);
                }
ENDTASKS
```

Analog I/O

The SMARTCAT BL2100 11 analog inputs at 12-bit resolution, which means that an analog signal can be converted to an integer value between 0 and 4095. Four 12-bit analog outputs are available on the card. Below is an example of an analog statement:

```
zAnOut (DAC_0 + 1, 725);
```

The zAnOut function is used to send the number 725 to analog output 1. In the argument (DAC_0+1,725), *DAC_0* is the name of the analog card that is attached to the embedded controller, and the *1* is the analog output that the number *725* is being sent to.

Below is an example of the use of an analog input statement.

```
var1 = zAnIn (ADC_1_RAW + 2);
```

This function would read the "raw" 12-bit value of input 2 from analog card number 1 (ADC_1_RAW) and assign it to var1 (a variable). Var1 can then be used in the program.

Multitasking

Simple computers usually execute one program at a time. Embedded computers often execute more than one task at a time through the use of multitasking software. Multitasking simply means that a computer can perform multiple tasks simultaneously. In reality, a single processor can execute only one instruction at a time. Multiple tasks interleave their execution, only appearing to execute together. The tasks are actually sharing the processor's time. Two common types of multitasking are preemptive and cooperative.

Preemptive Multitasking Preemptive multitasking means that some top-priority event, usually an alarm, a timer interrupt, or a supervisory task (often referred to as a kernel), takes control of the processor from the task currently running and turns it over to another task. In this scenario, program tasks compete for the processor's time. Tasks have no control over when they may be preempted and usually have no information about other tasks. Cooperation, coordination, and communication among tasks that get asynchronously preempted is a major application problem. Preemptive multitasking requires special attention during software development.

Cooperative Multitasking Cooperative multitasking solves some of the development headaches associated with preemptive multitasking. Under cooperative multitasking, each task voluntarily yields control so other tasks can execute. The advantages of cooperative multitasking include:

- Explicit control of the points where a task begins and ends logical subsections of its overall job.
- Ease of task creation.
- Lower and less indeterminate "interrupt latency."
- Simplified programming.

Networking/Multitasking Multiprocessing refers to hardware; multitasking refers to software. The two often go together, however, and have much in common. Multiprocessing means, literally, having multiple processors. Some applications may require more input and output ports than a single controller can provide, but the application may not support two separate programs running on two separate controllers. In other cases, running numerous input and output lines throughout a machine to and from a single controller might be physically unsafe or impractical. Although software can execute across multiple processors in many different fashions, a simple and straightforward approach to multiprocessing is to create a multiprocessing system by establishing a network of controllers communicating over industry-standard RS-485 serial lines or over Ethernet.

Z-World's programming language offers extensions to the C language. One of the most powerful is the costatement. Costatements are extensions to C that facilitate cooperative multitasking. Let's consider an example. Industrial applications often involve a simple, linear sequence of events such as the one shown in Figure 20–12. One way to do this task is to create the wait states by making the controller wait for

Start:

1. ***Wait for cycle start push button to be pressed.***
2. ***Turn on pump 1.***
3. ***Wait 30 seconds.***
4. ***Turn on pump 2.***
5. ***Wait 45 seconds.***
6. ***Turn off pump 1 and pump 2.***
7. ***Go back to start.***

Figure 20–12 Linear sequence.

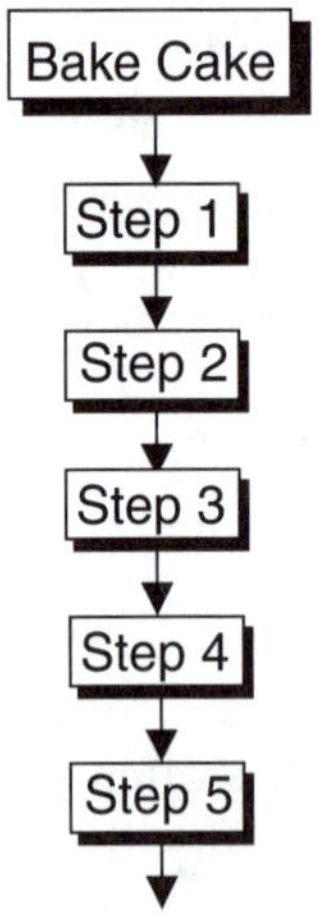

Figure 20–13 An example of a single-task sequence of operations. The task in this case is baking a cake. Note that the process has several steps.

30 seconds in step 3 and then wait for 45 seconds in step 5. With this method, the controller cannot perform any other tasks while it is waiting. All of this wait time is essentially wasted.

Another method is to organize the task as a single task, like baking a cake. The cake-baking task has several steps (see Figure 20–13). Each step takes time. The baker is initially busy as he finds the ingredients, measures them, mixes them, and so on. But when the baker puts the cake in the oven, he has some time with nothing to do. If there is nothing else to bake, he waits and occasionally looks at the temperature and time to be sure he takes the cake out in time. In this example the baker has a lot of wasted time as he waits for the cake to bake.

The second type of situation involves several tasks, all of which must operate concurrently. Imagine a cook in a restaurant. A waiter brings the orders (four of them) from table 12 to the kitchen (see Figure 20–14). All are received at the same time in this case and all should be finished at about the same time so that everyone receives hot food. The cook must process these orders simultaneously. In the previous example the baker could relax during wait times (while the cake is in the oven). The cook, on the other hand, must use wait times in one task to work on the others.

In industrial applications concurrent tasks are often required. If other tasks need to be run, we need to write a program that can process several tasks concurrently (see Figure 20–14). When there is idle time in one task, the processor can do other tasks. Each task can relinquish its control when it is waiting. This allows other tasks to be processed. Every task's work is done during the idle time of other tasks.

A more industrial example of concurrent processing is a worker in a machining cell. The cell has several machines. The worker loads a part into a machine and the machine starts its cycle. The worker has free time now, so she moves to another machine and loads or unloads a part. The worker then has time to inspect a part from

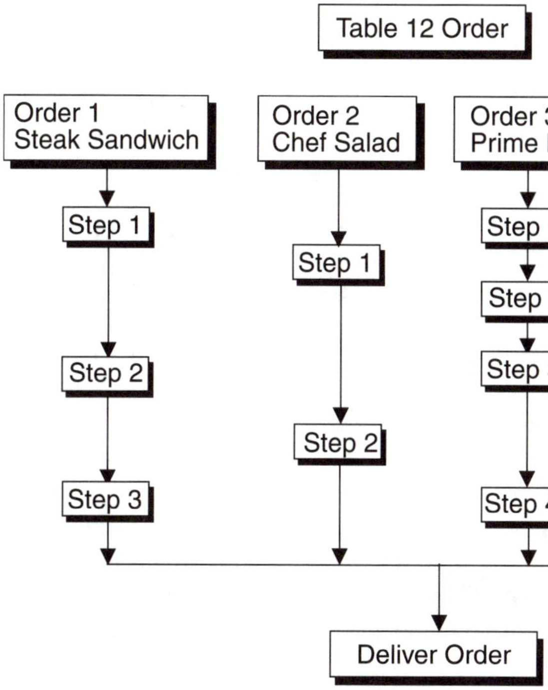

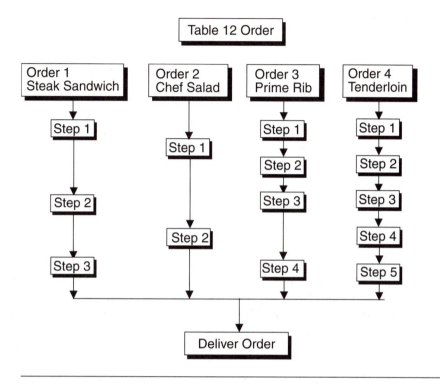

Figure 20–14 Several tasks that need to be executed simultaneously.

another machine. The worker can thus stay productive by moving between the machines when needed. The worker is like the processor. The machines are like the tasks. The worker is free to work among the machines as needed and so is much more productive than if she had to start one machine and stay until the sequence was complete before moving to the next machine.

A ladder logic program is usually hard to modularize. It is essentially one long program. Every rung of logic is evaluated every time unless master control relays or zone control logic is used. Even portions of the ladder diagram that are not active during a portion of a sequence are examined every scan. This is usually not a problem in most applications, but it does make the logic quite confusing.

It is normally quite difficult to write C code for concurrent tasks. Z-World has created C language extensions to make writing programs with concurrent tasks easy. They are called costates (see Figure 20–15). They also help to modularize a program, which helps break a program into manageable and understandable blocks.

Figure 20–15 Simplified example of the use of a costatement.

```
costate {task1}
   {
     ( C code for task 1 )
   }
costate {task2}
   {
     ( C code for task 2 )
   }
costate {task3}
   {
     ( C code for task 3 )
   }
}
```

Costatements are powerful because they can wait for events, conditions, or the passage of time. They can yield temporarily to other costatements. They can also abort their own operation. They can then resume operation from the point at which they left the costate. Costatements are also powerful for the programmer because they take the complexity out of programming concurrent activities and they help modularize programs.

You can see that embedded controllers offer many advantages for industrial control. Today's embedded controllers are fast and powerful, and they offer almost unlimited I/O capability. Networking, communications, and operator I/O have been made simple for the programmer through the use of functions.

QUESTIONS

1. Define *embedded controller* and explain its use(s).
2. What are the most common languages used to program embedded controllers?
3. List at least three advantages of embedded controllers.
4. Explain how embedded controllers can be expanded.
5. Explain how embedded controllers can be networked.
6. What are costatements and what are they used for?

Chapter 21
Introduction to Robotics

You will gain a practical understanding of robotics. This chapter will examine the types of robots and the suitable application for each. We will also consider the programming and interfacing of robots with other devices.

OBJECTIVES

Upon completion of this chapter, you will be able to:

1. Define the terms *homing, axis, interpolation,* and *harmonic drives.*
2. Describe types of robot work envelopes and robot motion.
3. Describe types of robot drive systems.
4. Explain typical robot applications.

OVERVIEW

Robots are used in manufacturing for many reasons. With the costs of labor increasing and world competition forcing faster delivery, higher quality, and lower prices, robots have taken on new importance. Their use to relieve people of dangerous tasks also has increased. The environment can be dangerous because of temperature or fumes, but robots can function very well in such environments. Robots also perform repetitive operations that tend to be tedious. For example, spot welding auto bodies all day is a very tedious job, one that can make people careless. Carelessness can reduce the quality and increase the variability of the product.

Some jobs are too difficult for humans to keep up the speed required to perform certain applications. Assembling printed circuit boards is an example. Many welding applications are inappropriate for people to do because it is difficult to reach certain areas of a part comfortably for long periods of time. Operators must wear weld masks to protect their eyes. Raising and lowering the mask every time the operator begins a weld to position the torch slows the operator and adds variability to the parts.

Robots and other automation are also appropriate when qualified labor is not available for various reasons. For example, a company's pay constraints affect the quality and availability of labor. Robots and automation provide a labor alternative. The use of automation provides an opportunity to compete in the world economy.

Automation in general affects the type of personnel required by manufacturing. One result is a gradual drop in the need for low-skill manual labor with increased demand for manufacturing engineers and skilled technicians able to implement, use, and troubleshoot new technology rapidly.

Robots are typically used for repetitive functions such as moving and positioning parts between devices and production tasks such as welding parts together. They are fast and accurate. Robots are tools, however, and comprise a small part of a system. In general, skilled operators program them. A company can train its best welder to program the robot.

Robots are flexible tools in automated systems. They can change tasks immediately, which is becoming a more important characteristic as demands for flexibility increase.

Robot Axes of Motion

Robot axes are also referred to as *degrees of freedom.* Each axis represents a degree of freedom; for example, a five-axes robot has five degrees of freedom. Tool orientation normally requires up to three degrees of freedom, and three to four are needed to position the robot in the work space. A five-axes robot is not adequate for making complex and intricate moves such as moving the welding rod in many directions, getting into tight spots, and avoiding hitting the fixture. Six-axes robots are usually required for welding (see Figure 21–1).

Wrist Motion

A robot's wrist is used to aim the tool, or end effector, at the work piece. The wrist has three main potential motions: yaw, pitch, and roll (see Figure 21–2).

Figure 21–1 The axis of a typical six-axes robot. *(Courtesy Fanuc Robotics Inc.)*

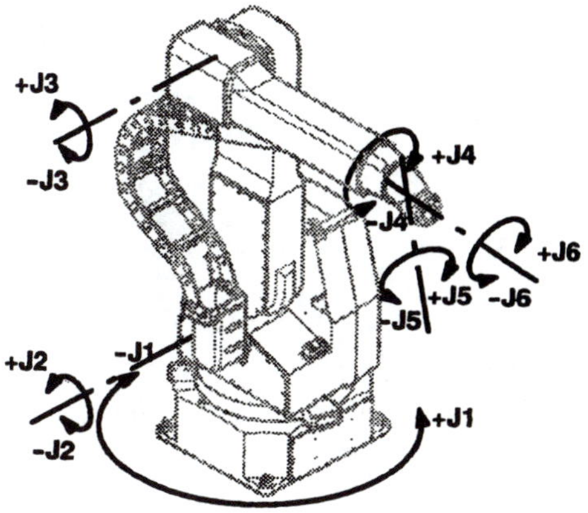

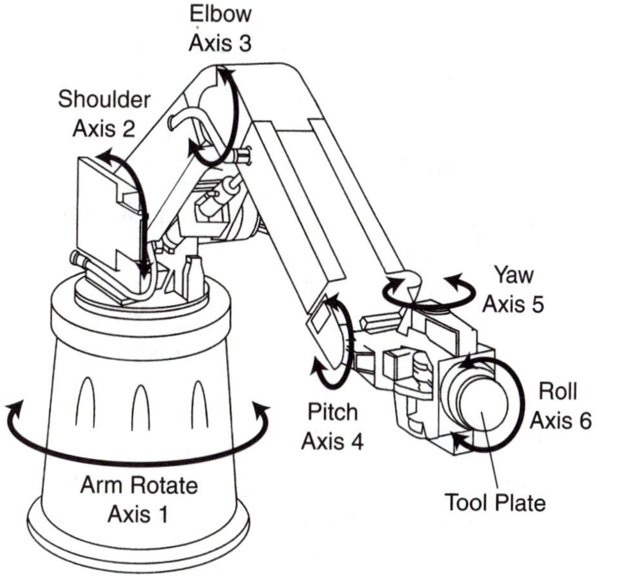

Figure 21–2 Three main motions of a robot wrist.

Robot Geometry Types

The types of physical geometries of robots vary, as the following discussion indicates.

Cartesian Coordinate Robot

The Cartesian system is probably the simplest physical geometry to understand. It is usually used for assembly applications and is particularly well suited for assembly of printed circuit boards (see Figure 21–3). The main motions are on the

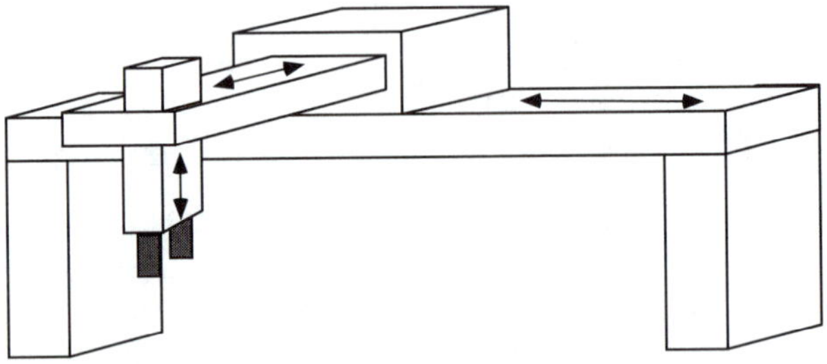

Figure 21–3 Cartesian coordinate–style of robot.

x, *y*, and *z* axes. It is capable of picking up components and inserting them into boards. Its simple Cartesian coordinates make the Cartesian coordinate robot easy to program offline.

Cylindrical Coordinate Robot

The cylindrical-style robot's first axis of motion is rotate, the second is an extend axis, and the third axis is the *z* axis. Other cylindrical styles are available with more axes of motion, but they all have a rotate as the first axis and an extend axis as the second axis (see Figure 21–4). Figure 21–5 shows the work envelope for a cylindrical robot. The work envelope is the area in which the robot can move.

Spherical Robot

The spherical robot has a rotate for the first and second axis of motion and an extend for the third (see Figure 21–6). This configuration was the most popular for the first hydraulic robots. See Figure 21–7 for the work envelope.

Figure 21–4 A cylindrical-style robot whose first, second, and third axes are rotate, extend, and *z*.

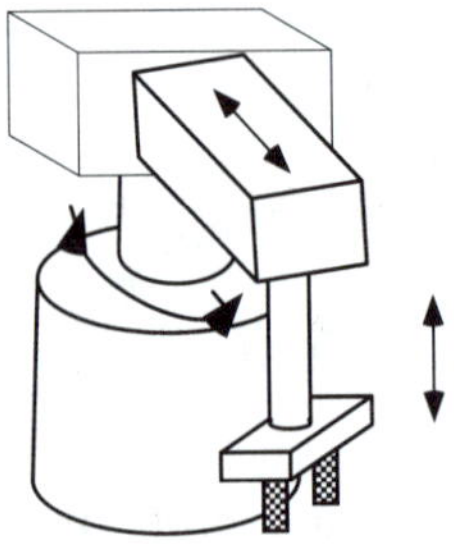

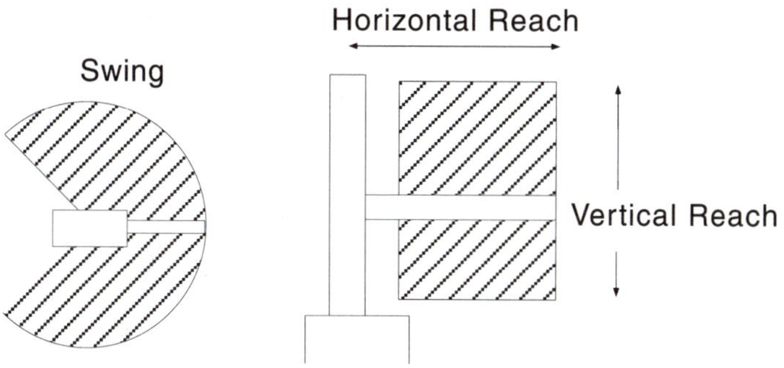

Figure 21–5 Work envelope for a cylindrical-style robot.

Figure 21–6 A spherical robot.

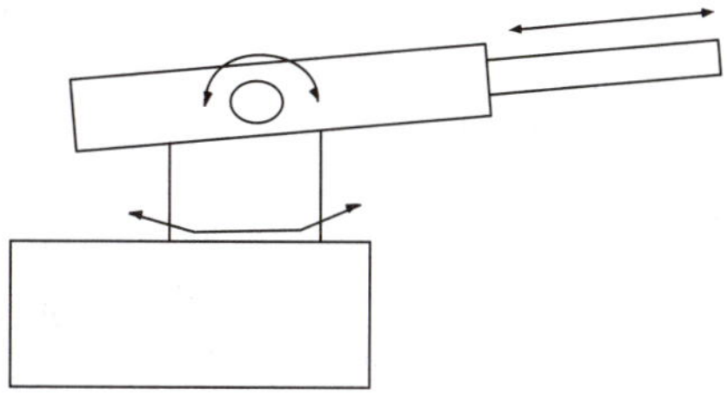

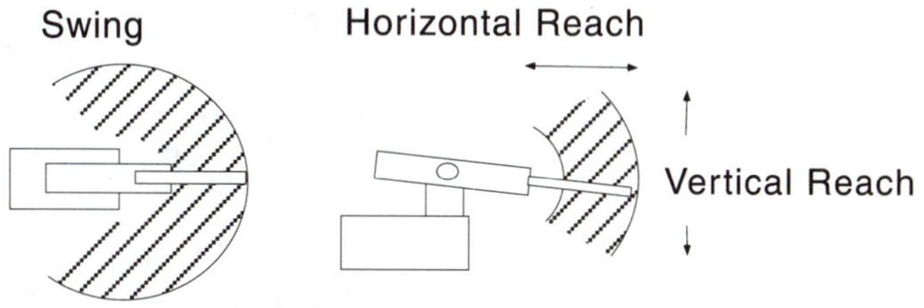

Figure 21–7 Work envelope for a spherical robot.

Selective Compliance Articulated Robot Actuator

The selective compliance articulated robot actuator (SCARA) typically has three vertical rotate axes and a *z* axis. See Figure 21–8 for a SCARA-type industrial robot. SCARA robots are very good at working in the *x*, *y*, and *z* orientation. They are extremely fast and particularly well suited to picking and placing electronic components.

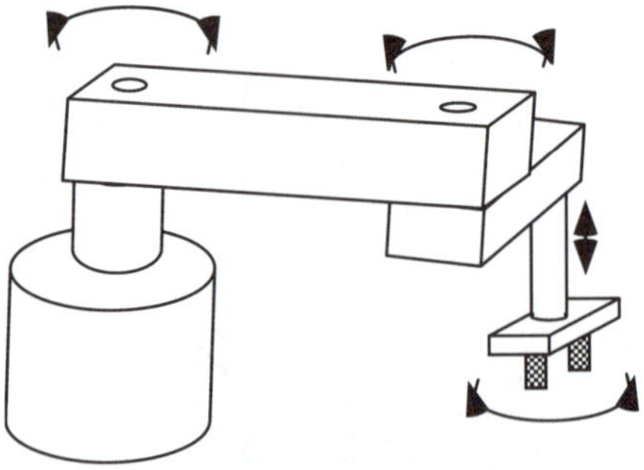

Figure 21–8 SCARA-type robot.

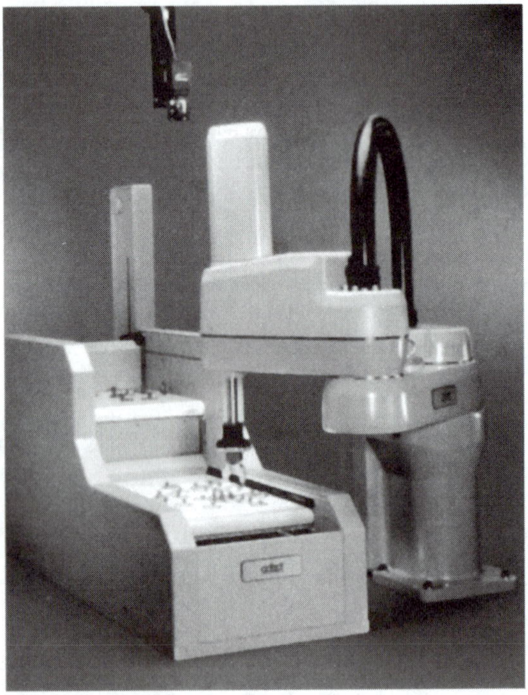

Figure 21–9 Adept SCARA-style robot with a FlexFeeder, a flexible parts presentation system that can find the parts and their orientation for the robot.

The SCARA style is well suited to electronic assembly because the three vertical rotational axes allow for minor location differences in the application. See Figure 21–9 for an Adept SCARA-style robot with a FlexFeeder. Figure 21–10 shows the work envelope for a SCARA-style robot. In Figure 21–11 a pin is to be pushed into the hole, but it is misaligned slightly. It can still start into the hole because of the chamfer (angle) on the pin. A normal style of robot would have difficulty pushing the pin into the hole because the robot axis would give in many

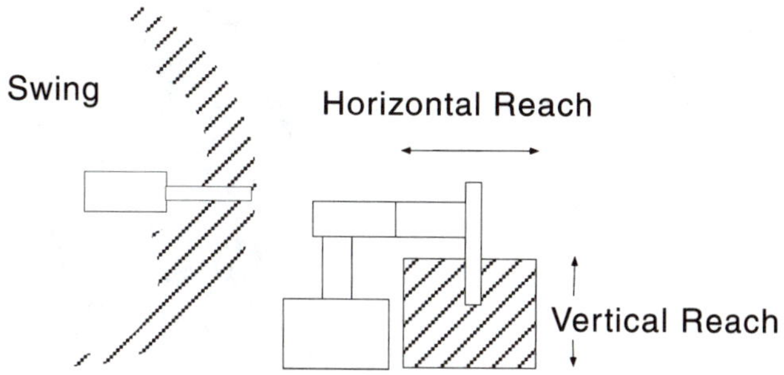

Figure 21–10 Work envelope for a SCARA-style robot.

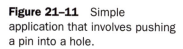

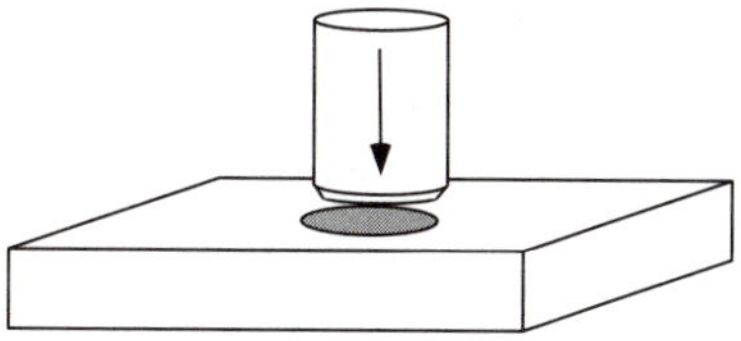

Figure 21–11 Simple application that involves pushing a pin into a hole.

different planes. A SCARA robot would allow movement only in the *x*-*y* plane. It would keep the *z* axis in a vertical orientation. This means that the three-rotational axis would allow small movement as the pin was pushed into the hole.

Articulate Robot

The articulate robot is the most common style of robot configuration and is most like the upper body of humans. The rotational movement of the human waist is similar to the robot's base rotate and the waist bend motion is similar to the robot's next axis. The human arm is similar to the rest of the axes. Note that humans have many more axes of motion than a robot does, which limits the robots. At least six axes are required for most applications (see Figure 21–12).

Although they are limited, articulate robots are versatile. They can reach into tight and confined areas such as auto interiors. They can move at odd angles to weld parts together or to apply a bead of sealant in any plane. They can assemble parts and place assemblies at any angle. The articulate work envelope is called a *revolute work envelope* (see Figure 21–13).

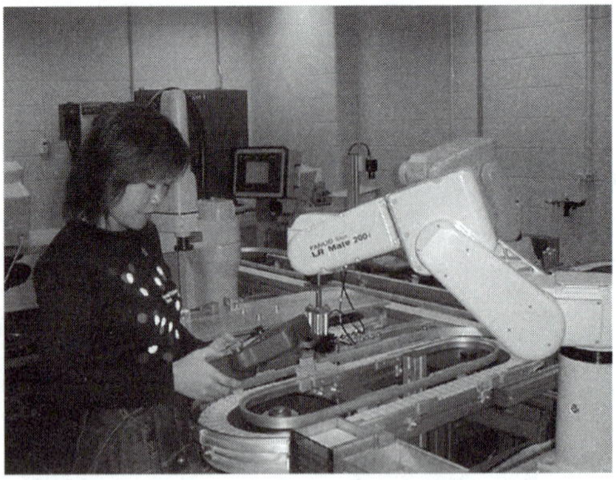

Figure 21–12 Fanuc articulate-style, six-axes robot. *(Courtesy Hiromi Kugimiya.)*

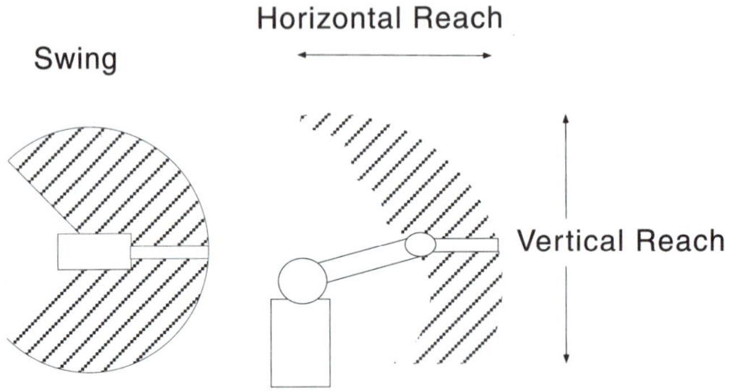

Figure 21–13 An articulated-style robot.

Reference Frames

This section examines how the robot relates to its physical world. To do so, first imagine yourself sitting in a chair in a small room. The room represents your world coordinates. The fixed chair location establishes your right, left, front, back, up, and down. When a robot is fastened into its work cell, its *world coordinates* are established (see Figure 21–14). This world coordinate system can also be called a *world reference frame.* A robot programmer can utilize the world reference plane

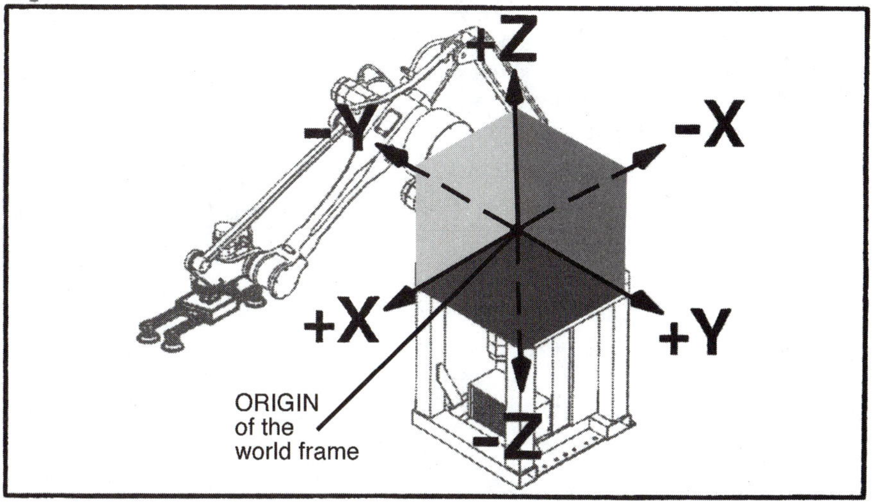

Figure 21–14 Robot's typical world coordinate system. *(Courtesy Fanuc Robotics Inc.)*

in a program. If we think in terms of a Cartesian coordinate system, the *x*, *y*, and *z* axes are fixed in relation to the robot.

User Reference Frame

In many cases, however, it would be easier for the user to establish a different coordinate frame (see Figure 21–13). Many robots allow the user to set its own reference frames, which Fanuc Robotics Inc. calls *user frames,* to correspond with the work-cell space. The user would prefer to make the robot's *x, y,* and *z* axes correspond to the fixtures. A user who can make the *x, y,* and *z* axes of the robot correspond to the fixtures can utilize CAD data to program the robot. This also makes it easier to move the robot because a move in an *x, y,* or *z* plane now follows the fixtures axis instead of the robot's world frame. This transformation to a user reference frame is easy to accomplish on many robots. With a Fanuc robot, the user simply teaches three points, as shown in Figure 21–15. The user can create and name several user frames and then use the one that is most appropriate for the application.

User reference frames are sets of Cartesian coordinates that describe the relationship between the robot tool center point and the work space. The positional relationships between the robot and the work space are described by mathematical equations called *transformations.* The robot controller calculates the necessary joint angles that will align the robot axis to orient the tool to the work space.

Figure 21–15 User teaches three points to a Fanuc robot to create a user reference plane. *(Courtesy Fanuc Robotics Inc.)*

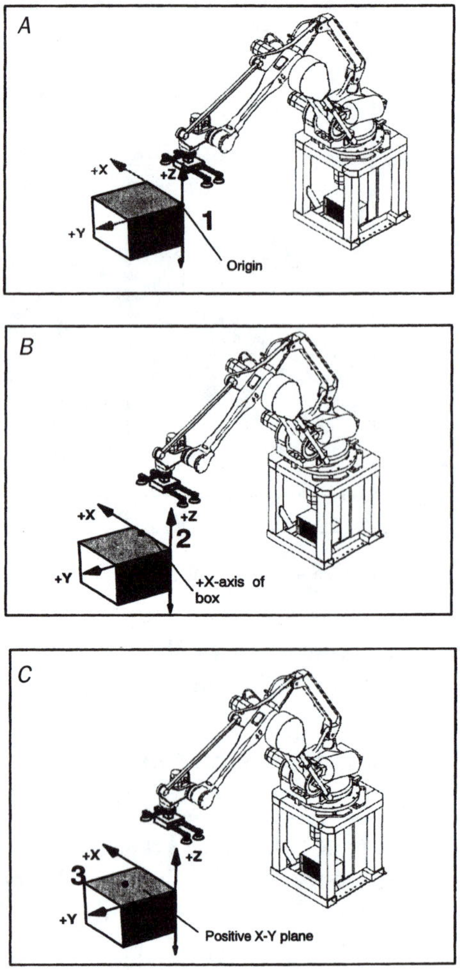

In other words, the robot controller aligns the robot reference frame with the work-space reference plane. The transformation equations depend on the type and number of the robot's axes of motion. These equations are transparent to the programmer.

User reference frames can also be very helpful to move equipment in a cell. A new user frame can be created, and the program can be transformed to the new user frame and run. Fanuc also has a program shift function that teaches the robot the three points to locate the new position of the equipment, and the robot then transforms the program to the new position.

Tool Frames

Many robots can establish a tool frame. For example, in welding a circular shape, it is important to keep the torch at the proper orientation to the weld as the robot moves the torch around the weld. A *tool frame* tells the robot what the tool's center point is. If this is not done, the robot control makes its calculations based on the robot's gripper face plate. See Figure 21–16 for an example of a tool frame.

Three methods—the three-point, six-point, and direct entry—can teach a tool frame to a Fanuc robot. The three-point and six-point methods involve bringing the tool center to a known position and recording either three or six points, respectively, as the point is approached from different orientations. The robot combines these different orientations to the same point and the robot's known positions, and the robot calculates the tool orientation data to create the tool frame. Values can also be entered directly into the controller to create a tool frame.

Methods for Moving the Robot During Teaching

The way the robot moves during jogging and programming can be chosen. Fanuc robots offer three modes: joint, XYZ, and tool. Each can be used to move and position the robot. The user chooses the method that is easiest to move to given positions during programming.

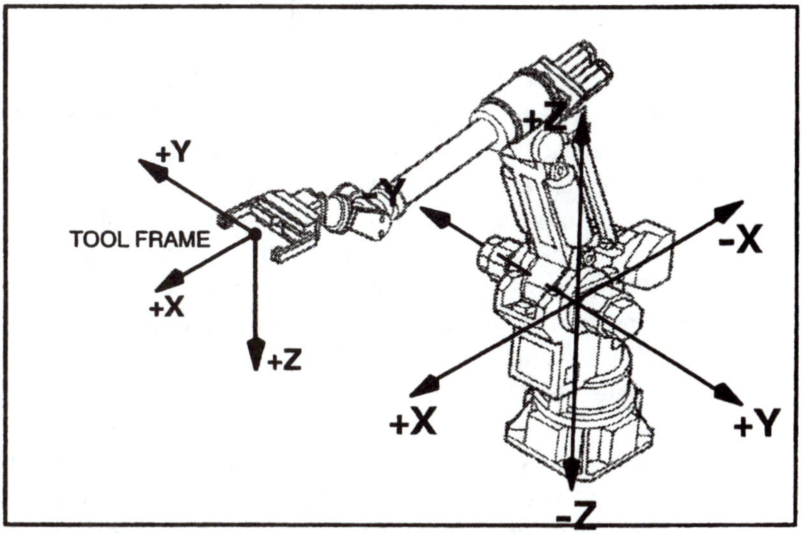

Figure 21–16 Tool frame. *(Courtesy Fanuc Robotics Inc.)*

Joint Mode In the joint mode, each axis moves independently. A typical robot might have six axes, each of which can be moved independently. Each axis usually has two keys: a forward and a backward key. The user holds one of them down and the robot moves one axis.

XYZ Mode The XYZ mode is often used to program robots. In the XYZ mode, two keys are assigned to the x axis, two to the y axis, and two to the z axis. If the user chooses the $+x$ axis, the robot coordinates the motion of several axes to move the robot in the $+x$ direction. The robot maintains its y and z positions while the move is made.

Tool Mode In the tool mode, the tool center point is moved in the x, y, or z direction similar to the X, Y, Z except that the tool center point frame is used for orientation.

Types of Robot Motion Just as there are three methods to move the robot when programming, there are also different move choices for the robot while actually running the programs. Fanuc Robotics Inc. offers three: joint, linear, and circular.

The joint mode is the fastest method of motion and the shortest cycle time. It is appropriate for movements around obstructions; the actual path is unimportant as long as the robot misses the obstruction. This mode is the quickest because it is not necessary for the robot to coordinate the motions of all axes together. Linear motion is used when it is important to control the path between two points, as in a weld application. Circular motion is used when the robot's motion must follow an arc or circle.

We can also control how the robot moves when it reaches its programmed points. Fanuc Robotics Inc. has two choices, fine and continuous. In the *fine mode,* the robot stops at each point before moving to the next. In the *continuous mode,* the robot decelerates as it approaches the next point but it does not stop at it before it accelerates to the next point. The user can enter a value between 0 and 100 to determine how closely the robot comes to the point. A value of 0 with a continuous type is a fine move. The larger the value with a continuous move, the farther the robot stops from the position with the least deceleration. Continuous mode provides smoother and quicker moves than the fine type, but it does not go to the exact intermediate points.

Compliance Adaptor devices can also be used in automation to allow some controlled movement in the end effector and then to return the effector to the normal position. On a robot, these devices are placed between the robot and the end effector and usually allow for lateral and rotational compliance (see Figures 21–17 and 21–18).

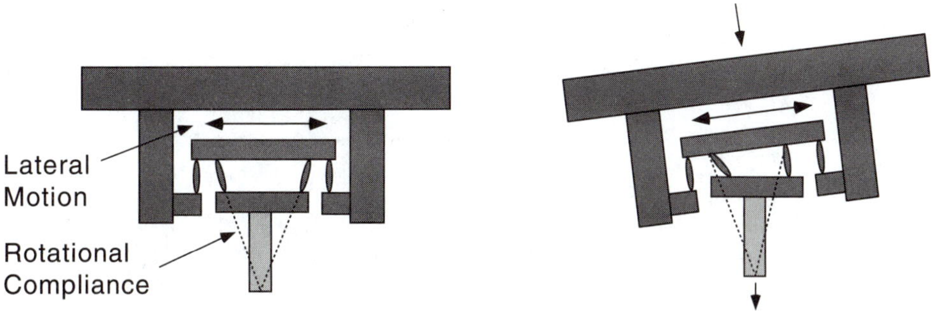

Figure 21–17 A device capable of lateral and rotational compliance.

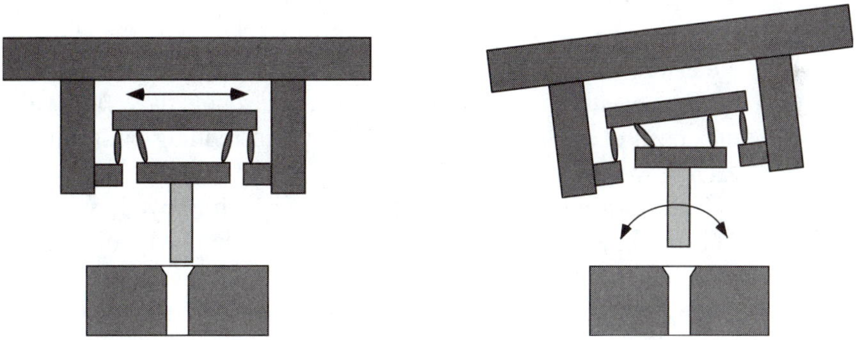

Figure 21–18 Lateral and rotational examples of compliance. Some compliance devices are capable of both.

ROBOT DRIVE SYSTEMS

Robots have several types of drive systems: pneumatic, hydraulic, and electric. Each has its own advantages and is best for certain applications. "Best" means that the robot can do the job and is the lowest total cost solution.

Pneumatic Robots

The simplest and least costly robot drive system is pneumatic. It is sometimes called a *bang-bang robot* because of the way the axes operate (see Figure 21–19). It is appropriate for simple positioning tasks such as moving parts between devices. This type is inexpensive, fast, and accurate. Pneumatic robots are limited in some ways, however, such as in the number of positions to which they can move. They

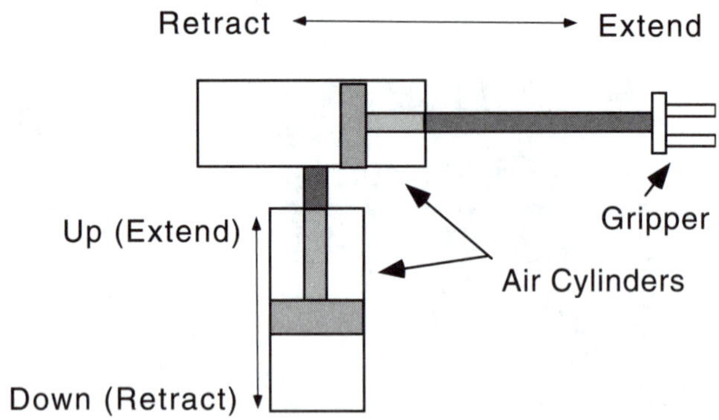

Figure 21–19 Simple pneumatic robot that can move to only four different positions with two axes. One axis is for up and down and one axis is for extend and retract.

are also generally limited to two positions per axis, nor can they control the path taken between positions. Some pneumatic robots have flow valves to help control the speed of each axis. Some also have cushioning built into the cylinder so that each move can stop more gently.

The versatility of pneumatic robots can be improved by adding additional stops. This might permit an inexpensive, fast, simple pneumatic robot to meet the needs of a specific application. For example, an application involves basically simple moves except that it requires three positions for the *z* axis (up-and-down axis): one "up" position and two "down" positions (see Figure 21–20). These positions are accomplished by adding another programmable stop and a cylinder that can extend and retract. If the cylinder is extended, it provides another hard stop for the robot. When the robot moves down, it hits this extended cylinder and stops. If the cylinder is retracted, the robot moves all the way to the bottom of its travel. A clever engineer or technician can make a pneumatic robot even more versatile.

Speed and Accuracy Pneumatic robots are fast and can be extremely accurate. An air cylinder can extend and retract quickly and is also very accurate, but it can only extend and retract (see Figure 21–19). They move to hard stops and no farther. Pneumatic robots have excellent repeatability, which is the ability of a device to return to the same position every time, because they move to a hard stop every time. The actual stop position can usually be adjusted mechanically.

Programming Pneumatic Robots Because of their simplicity and control ease, pneumatic robots can be purchased with or without a controller. Control is performed by a PLC or a device that controls and integrates the other equipment in the cell.

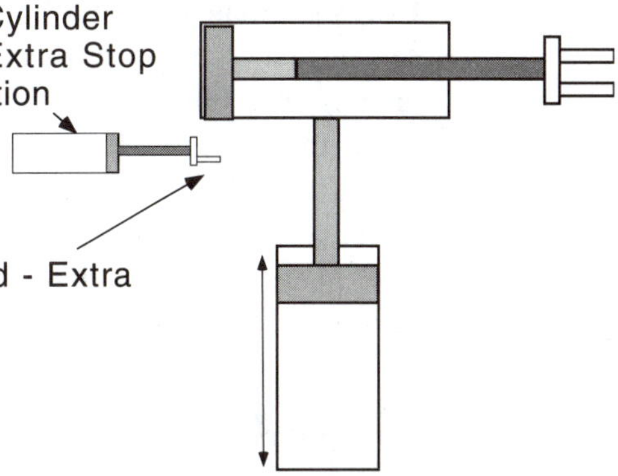

Figure 21–20 An extra stop added to make a pneumatic robot more versatile.

Pneumatic robots with controllers are normally sequential types that can be used to pick up a piece in one position and move it to another. The sequence typically is the same every time. This type of controller is programmed by using the robot to make each move and using a teach key to make the controller remember the move.

Programming can be more complex if a different device such as a PLC is used for control. Sensors and other devices provide feedback to the PLC, which can then make a decision and turn on outputs to cause the robot to make appropriate moves. A PLC or other complex controller is not limited to one sequence of moves.

Limit Switches Limit switches provide robot axes with feedback to ensure that the robot actually makes the move it is commanded to make. Proximity sensors are usually used on pneumatic robots to do this. A *proximity sensor* is positioned at each end of the cylinder's travel. When the cylinder extends, it closes the extend switch, which the PLC can see. When it retracts, it closes the retract switch, which the PLC can then see. If the PLC commands the cylinder to extend, it should easily see the extend limit switch.

Pneumatic Robot Maintenance and Troubleshooting Because pneumatic robots are really just air cylinders, maintenance is similar to that for any pneumatic system. Air lines can become loose or worn and leak air.

Solenoid valves, which consist of moving parts and a coil, can malfunction or fail. Solenoid valves can be taken apart and cleaned very easily. The coils can be tested easily by continuity and then applying power. Spare parts for solenoid valves should be kept in stock so that they can be repaired quickly and the robot can be put back into service.

If a limit switch fails, it does not provide feedback to the controller, so the sequence stops. Limit switches can be tested by looking at the input LEDs on the PLC's input module. If a robot axis is in the extend position, the extend input to the

PLC should be on. The robot can then be moved to a different position to check that sensor. Such checks must be performed very carefully and not in the robot work area. If the robot is waiting for an input before it makes a move and the input is supplied, the robot immediately makes the next move, which can be dangerous for the operator. For this reason, robot maintenance should be performed with electric, pneumatic, and other equipment locked out.

A pneumatic robot that is losing accuracy is experiencing wear or its hard stops need to be adjusted.

Hydraulic Robots

Hydraulic drive systems for robots, among the first robots developed, were capable of moving heavy loads and working at fast speeds. Some of these robots work in very hot environments such as foundries and diecasting plants to place and remove hot parts from machines. Hydraulic robots are good for painting and moving heavy objects, and they can make very smooth moves. They are also good for dangerous applications where an electrical spark could cause an explosion.

Electric Robots

Robots with electric drive systems are versatile, fast, and accurate. Electrical robots can use either AC or DC motors to provide their motion and typically are available with four or more axes of motion. Most robots have four or more axes. The robot controller typically contains the drives and the robot program; there is one drive for each axis of motion (see Figure 21–21). Each step of the program contains position and speed information for each axis of motion. The controller calculates the speed and distance of each move and then outputs proper voltage to each axis motor drive. The magnitude of the voltage corresponds to the desired velocity, and its polarity corresponds to the direction of travel.

The drive compares (sums) the velocity reference command (voltage) from the control device with the feedback from the tachometer. The drive then outputs an error signal, which is the difference between the command and the feedback. The drive also converts the small voltage and current to larger voltage and current to drive the motor.

Meanwhile, the encoder feeds position information back to the control device, which monitors the actual and commanded positions. If a wide variation occurs during a move, the controller assumes that the arm has quit moving and inhibits the drive. This is a safety feature. The control device also controls the drive's speed of acceleration and deceleration, called *ramping* (see Figure 21–22). The controller ramps the velocity command up to the required speed, monitors the actual position of the drive, and then ramps the velocity down as the axis approaches the commanded position.

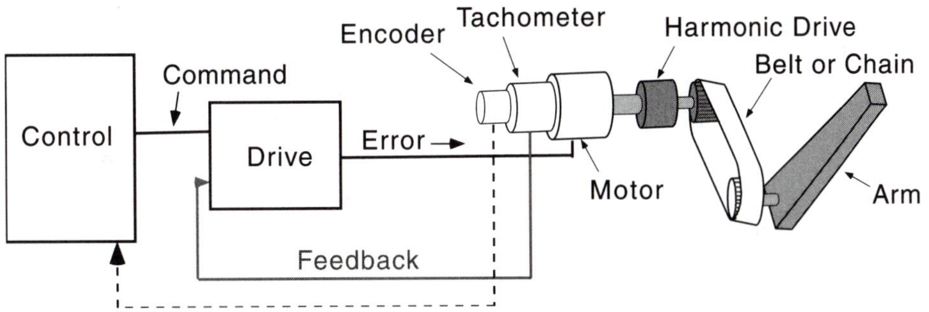

Figure 21–21 Typical axis of motion for a robot.

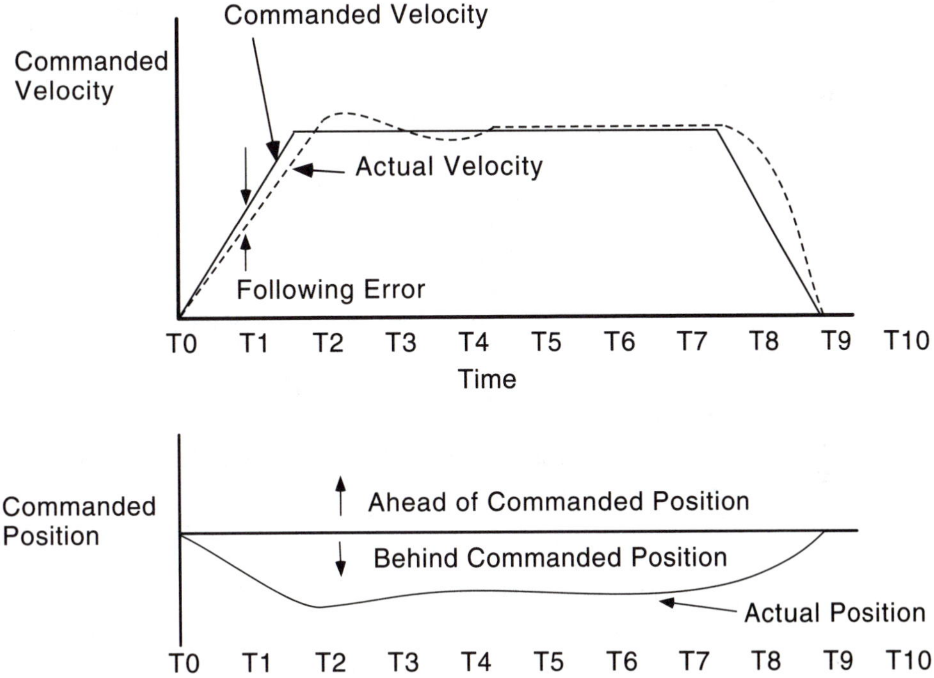

Figure 21–22 Example of ramping.

The actual drive position during a move is never exactly the same as the commanded position. The reasons include backlash in the drive linkage, the mass of the moving object, friction, and delays, all causing the actual position to lag the commanded position. This error is a *following error.* The control device constantly monitors the amount of following error during a move and inhibits the drive if the following error goes beyond a certain value. In Figure 21–22, the top of the graphic shows the commanded velocity and the actual desired velocity. The actual velocity lags the desired velocity for some time. The bottom part of the graph shows the actual position over time. If the drive were perfectly in position, it would be on the horizontal line. If it is behind position, it is below the line. Note that because of the following error, the actual position is behind the commanded position until the end of the move, when it catches up.

HOMING PROCESS

When a robot is initially powered up, it does not know what position it is in. It must be *homed* to establish its own position. Many robots, such as CNC machines, must be homed before operation. To do this, the operator must move the robot to a position in each axis that will be safe for homing because if a fixture, wall, object, or person is in the way, the robot will hit whatever is in its path when the axis begins to move toward its home position.

Each axis of a robot consists of a motor and an encoder for position feedback. Robots that have absolute encoders do not need to be homed, but most equipment has incremental style encoders and must be homed. When the robot is powered, it does not know its position but begins to move toward home, moving each axis until the robot encounters the home switch of that axis (see Figure 21–23). The purpose of the switch is to let the robot know that it is close to home. The home switch will move the robot to within one revolution of the encoder. When each robot axis finds the home switch, the robot starts slowly moving in the other direction until it finds the home pulse on the encoder. Then it knows exactly where it is. Many newer robots utilize absolute encoders and do not need to be homed when they are started.

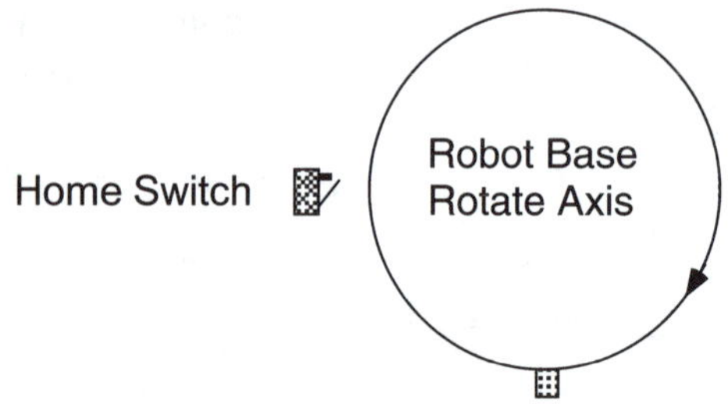

Figure 21–23 One axis of a robot. When a robot is homed, each axis moves slowly in one direction until it finds its home switch. This provides a rough position for a robot. The robot then moves each axis in the opposite direction to find the home pulse on the encoder.

Harmonic Drives

A harmonic drive is a unique and interesting device that can create a large gear reduction in a small space. It consists of three main parts: a flexible can, a solid ring, and a wave generator. The flexible can has teeth on the outside called the *flexible spline* (see Figure 21–24). A solid ring with teeth on the inside is usually held in a fixed position on the outside of the can so it cannot move. The wave generator can be thought of as an ellipse that is attached to the motor shaft inside the flexible can. As it turns, the wave generator forces the flexible can (spline) to distort. The can's distortion forces a few teeth on two opposite sides of the flexible spline to mesh with teeth in the fixed ring.

The secret to the harmonic drive's gear reduction is that the fixed ring and the flexible spline each have a different number of teeth. The ratio of the reduction is based on the number of teeth on the fixed ring and the difference in the number of the teeth between the fixed ring and flexible spline. For example, one may have 200 teeth and the other 201 teeth. This is a ratio of 200 to 1. If the fixed ring had 300 and the flexible spline had 298, the ratio would be 150 to 1. The wave generator turns inside the flexible spine and distorts it, causing the mesh point for the teeth to move around the fixed ring (see Figure 21–25). In this case, for every revolution of wave generator, the flexible spline moves one tooth (the difference in the number of teeth between the fixed ring and the flexible can) for a 200-to-1 reduction. The motor must move 200 revolutions to make the flexible spline move one rotation. The movement is very smooth, which allows the axis to be more stable. The device is very accurate and experiences little wear and almost no backlash. This setup allows small high-speed motors to move heavy loads.

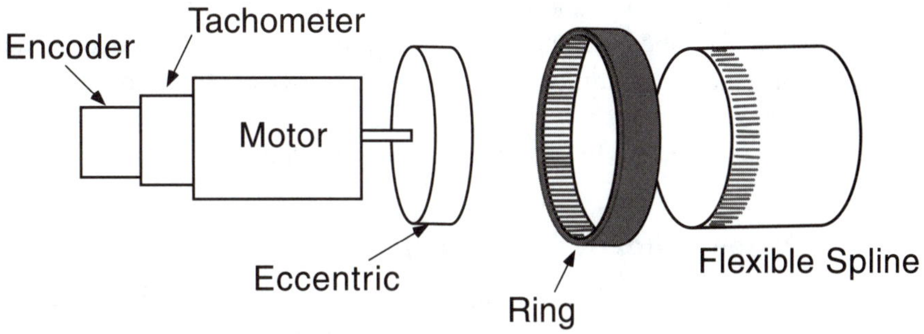

Figure 21–24 Main components of a harmonic drive.

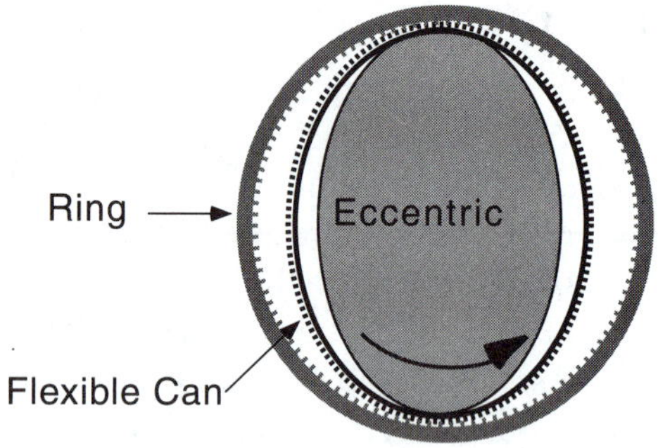

Figure 21–25 How a harmonic drive works.

Mechanized Linkages

Robots have three different styles of mechanical linkages: chain, timing belt, and gears. Chain drives utilize a sprocket on the motor shaft and a sprocket attached to the axis with a chain around both. This arrangement is very common and works well but provides a certain amount of backlash. Chains must be retightened occasionally or the robot will begin to lose repeatability. Timing belts are special belts and are also called *synchronous belts.* They are belts with teeth that mesh with sprockets. An idler pulley normally is used to adjust tension on the belt. Geared linkages for some robot axes are adjustable to minimize backlash for accuracy.

Robotic Accuracy and Repeatability

Accuracy refers to a robot's ability to move to a programmed position, which is often determined by offline programming. If a robot is programmed to move to one point in its work space, its accuracy is determined by how close to the programmed point it can move. In shooting at a target (see Figure 21–26), accuracy is a measure of how close the shots come to the center of the target.

Repeatability is a measure of how precisely a robot can return to a given position. Robotic repeatability is always much better than robotic accuracy. For example, the repeatability of a pneumatic robot is almost perfect because it hits hard stops at the end of each axis of travel. In the case of target shooting, repeatability measures how closely grouped the shots are (see Figure 21–27).

Types of Robot Programming

Robots are easy to program. Two types of programming, teach type and language type, are used. Some robots have a combination of the two. A robot program is sequential. It is a series of steps, each of which contains a move and speed and other information, such as required inputs or output signals or time delays.

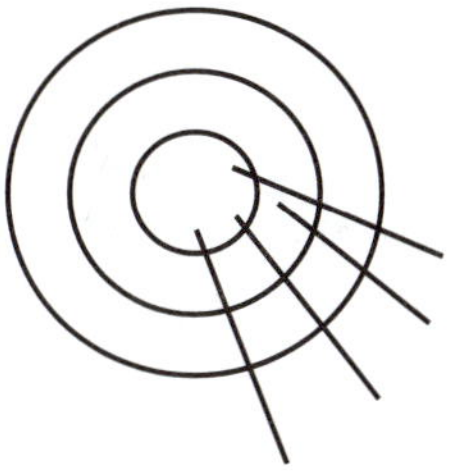

Figure 21–26 Example of accuracy.

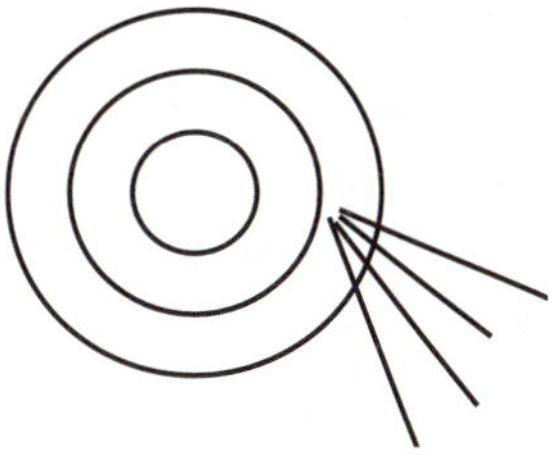

Figure 21–27 Example of repeatability.

Teach-Type Programming The teach-type robot is programmed by moving the robot to the desired position. The programmer then inputs the information for that particular step and the desired speed. If an input is required before the next move is made, it is entered in the step. The programmer also enters any outputs required, such as the gripper or welding output or other outputs that send a signal to other equipment.

A derivation of the teach-type robot is the *lead-through type*, which is particularly suited to painting applications. The best painter in the shop would put the robot into teach mode and then grab the end of the arm and paint a part. The robot remembers the speeds and motions that have been taught and can repeat the motions and speeds.

Language Programming The other type of robot programming uses a language. It might be similar to a Pascal or C language program with some special functions for moves, loops, and palletizing routines. A programming language makes the robot more versatile but is somewhat more difficult to learn to program. The use of a programming language allows the program to be written offline on a computer and downloaded to the robot. Typical robot languages have if/then/else statements for loops, while and do loops, do until loops, and case statements.

Following is a partial listing of a program for a Fanuc robot. The commands are quite easy to understand and use. Note that each line has been commented to explain its function.

```
    ! Comment—Line 1 turns digital output 7 on.
1.  DO{7} = ON
    ! Comment—Line 2 is a move command. The command is to
    move in joint mode to position 1 at 100% of the
    programmed speed.
2.  J P[1] 100% Fine
    ! Comment—Wait for digital input 1 or 3 or 4 before
    continuing the program.
3.  WAIT DI[1] = ON OR DI[3] = ON OR DI[4] = ON
    ! Comment—Turn robot output 1 on.
4.  RO[1] = ON
    ! Comment—The command is to move in joint mode to
    position 1 at 100% of the programmed speed.
5.  J P[1] 100% Fine
    ! Comment—The next line is a conditional statement. If
    digital input 5 is on, the program will jump to the
    section of the program called LBL 2. It will skip all of
    the program lines up to LBL2.
6.  IF DI[5] = ON JMP LBL[2]
    ! Comment—The command is to move in joint mode to
    position 3 at 100% of the programmed speed.
7.  J P[3] 100% Fine
    ! Comment—The command is to move in joint mode to
    position 4 at 100% of the programmed speed.
8.  JP[4] 100% Fine
    ! Comment—This is label 2. Labels are a way to identify
    sections of a program and can be used to identify where
    a program should jump to.
9.  LBL[2]
    ! Comment—Turn robot output 1 on.
10. RO[1] = ON
    ! Comment—Wait until digital input 2 is on.
11. WAIT DI[2] = ON
    ! Comment—The command is to move in joint mode to
    position 9 at 100% of the programmed speed.
12. JP[9] 100% Fine Offset
    ! Comment—Wait 2 seconds before continuing the program.
13. WAIT 2.00 (sec)
    ! Comment—The command is to move in joint mode to
    position 10 at 100% of the programmed speed.
14. JP[10] 100% Fine Offset
    ! Comment—The programmer would then teach each of the
    points in the program by moving the robot to the desired
    locations and recording each. Note that in this program
    points were given numbers as names. They could have been
    named alphabetically instead.
```

Many of the languages also provide special functions for operations such as palletizing. The operator simply supplies information about the number of rows and columns and the location of corners. Welding robots have special functions available to make weaving easy, and others have functions that help integrate a vision system. The vision system can help find the parts and or inspect the parts and then share this information with the robot.

Communications Many types of communications are available for robots today. The most limited is RS-232, which provides basic communications capability. RS-232 has historically been used to connect a robot to a microcomputer for programming and uploading/downloading programs. Its speed and noise immunity are quite limited, which restricts the distance over which communication can take place.

Robots can now be networked with Ethernet, DeviceNet, ControlNet, and several other systems. These communication networks are fast and powerful and are readily available for computers, PLCs, and other devices. They make integration an easy task.

Robotic Applications

Disaster stories abound about companies that purchased robots that failed miserably in their application. Many of these robots ended up sitting in warehouses gathering dust, often because the application was not appropriate for the robot or it was too complex. One of the first things companies learn is that robots are very unforgiving. They also quickly learn that they need to improve the quality and consistency of their part production. They realize that parts have to be positioned accurately and consistently for a robot to be able to pick them up consistently.

Welding

Arc Welding Arc welding is appropriate for a robot. The robot uses an electric arc to generate the heat to fuse metals together, with wire used as a filler material. Robotic arc welds are much more consistent than human welds. Robots also use a flame-cutting torch or laser as a tool to cut out shapes in metal. For example, motorcycle manufacturers use a laser tool to cut shapes into exhaust heat shields.

Manufacturers make parts that vary very little but must provide good fixtures to hold the parts for the weld. The fixtures must accurately locate the parts and position them for the weld. Parts must generally be held accurate to one-half of the material thickness to be a candidate for robotic arc welding.

Some unique robotic methods are used to compensate for part inaccuracies. Vision has been tried but has not been very successful because of the dirty environment that welding creates. Another method has been to sense the current during the weld. A weld generally uses a weave pattern to weld pieces together. By sensing the current, the robot path can be adjusted automatically to compensate for inaccuracies and gaps.

Spot Welding Spot welding fuses two pieces of metal together, generally along a seam. It is an excellent application for robots. An electrical current passing through the parts creates the heat. A spot welder is a tool that resembles the human thumb and a finger. The parts are positioned in a fixture, and the robot closes the spot welder (thumb and finger) over the desired area and clamps it. Current is passed through the metal, creating heat that fuses the two pieces together. The spot welding attachment is tolerant of positional inaccuracies, and the robot can do the welding consistently and accurately.

Investment Casting

Investment casting is an application that primarily involves part handling. The robot picks up a wax impression of the parts to be made, such as steel golf club heads. The parts are often in a treelike configuration. The robot then moves the wax tree part into a slurry of fine refractory material, which dries on the wax parts. More coats of the refractory material are added. The robot spins the molds to ensure consistent thickness of refractory material. The robot is finished at this point, and the lost wax process is used to mold the parts.

Forging

Forging is the process of hammering metal into a required shape, generally by presses. Forging is done at high temperatures so that the metal is in a malleable state. This heavy, dangerous, and noisy work is well suited to a robot. The robot moves the part from furnace to press and then to an outgoing bin.

Die Casting

Robots act as part movers in a die cast application, which typically is performed in a hot, dangerous environment. The robot removes very hot finished parts from a die cast machine. The robot removes the part and puts it in a press that trims the part. This removes excess material and flash from the part.

Painting/Sealing/Gluing

Spray painting is a good application for robots. As noted earlier, hydraulic robots are generally used for spray painting applications because they do not generate sparks that could ignite the fumes in a paint booth. Hydraulic robots also apply paint very smoothly. Robots also apply glue and/or sealants to surfaces such as an automobile windshield. One robot applies sealant to the rim of the window and another robot puts the windshield into the car.

Loading/Unloading

Diecast, forging, and loading/unloading applications are load/unload applications. CNC machines can be fully automated with the addition of a robot that loads parts into the machine and unloads parts from it.

Assembly

Robots often perform assembly operations. Robotic machinery often assembles printed circuit boards. They also assemble other parts, such as motors and disk drives.

Finishing/Deburring

Robots' ability to make complex three-axis moves is very useful for finishing and deburring applications. A robot is a natural choice to move a complex part, such as a rotor with many sharp edges, and deburr the part. Robots hold parts against belts to shape them or remove burrs.

Inspection

Robots are adept at inspection tasks. One good example is in automobile inspection. They can move cameras into position to move and triangulate sizes. They can do this quickly right on the production line. Robotic inspection enables every car to be inspected instead of a random inspection. It also eliminates the need for an inspection fixture to hold and locate the part. The vision system finds the car and adjusts to its position for the inspection. This avoids disrupting the assembly line for hours of inspection.

Automated Guided Vehicles (AGVs)

Automated guided vehicles (AGVs) automatically move parts and materials through a plant. They are controlled in two ways. One is for the robot to sense a wire in the floor. The other method is by radio frequency, which can transmit commands to the AGV. These methods are quite often used in automated storage facilities. They are guided and controlled by a central computer.

Automated Storage/Automated Retrieval (AS/AR)

An automated storage/automated retrieval (AS/AR) system is really a rectangular style robot. It is used to load or unload materials into bays in storage facilities. It can be extremely large. A computer optimizes the storage and retrieval to ensure the security of inventory and the use of the oldest first. It also helps optimize the use of storage space.

Application Considerations

The application usually dictates the type of robot to be utilized. The best choice is usually the most inexpensive one that will accomplish the task. Some factors to be considered are type of application, speed required, complexity of task, type of motion required, need for flexibility, weight involved, and hazards. Some applications may not require a robot. A linear drive (axis) or two may be able to automate a simple part of an application cost effectively. The main types of motion are SCARA or articulate.

Electronic assembly often dictates the use of a SCARA-style robot because it is good working in x, y, and z axes. These robots are very fast and very good at picking and placing components in an x, y, and z orientation.

Painting, applying a sealant, deburring or polishing, machine loading/unloading, or welding applications usually dictate an articulate-style robot. This robot is capable of working in any orientation. Electric robots are the most common choice, but hydraulic robots are also good for painting applications because they are very smooth.

Pneumatic robots are a great choice for simple repetitive tasks. They are generally more suited to applications that require little flexibility.

Hazardous and very heavy or hot applications may require a hydraulic robot. If the robot is working in an area with flammable materials, the hydraulic type is a good choice. The robot control can be located outside the area and only the actual robot is in the area. There is no danger of sparks igniting the flammable materials. It should be noted, however, that electric robots have also been greatly improved and spark proofed for hazardous environments.

Robot Integration

Some robots are programmed using a specialized robot language. These robots, which have digital (on/off) inputs and outputs (I/O) available, are very easy to connect to control devices such as PLCs, computers, or other devices. Most robots wait

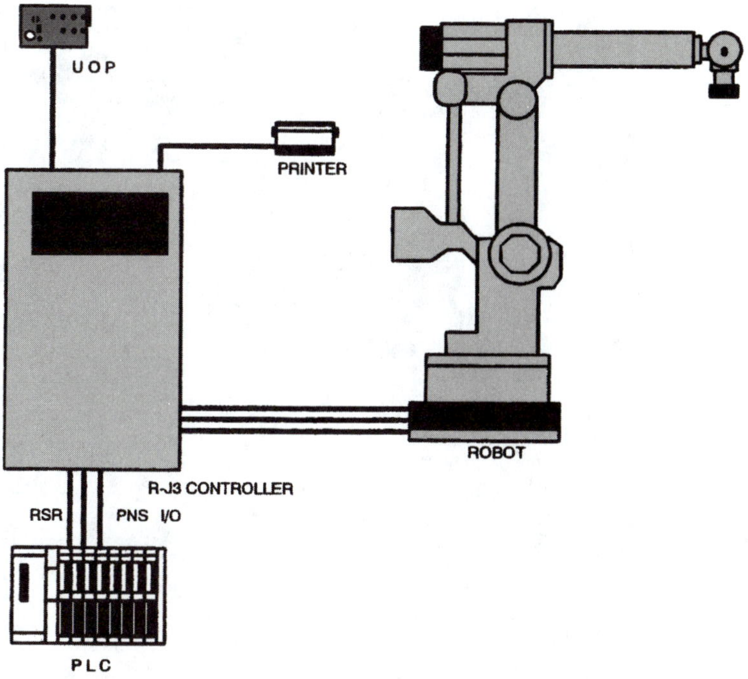

Figure 21–28 A robot communicates with a PLC. A digital output from the robot becomes an input to the PLC. A digital output from the PLC becomes an input to the robot. *(Courtesy Fanuc Robotics Inc.)*

for an input to tell it what task to do, and then perform the task and send an output to a cell controller to tell the control device that it has completed the task and is waiting for another one (see Figure 21–28).

End-of-Arm Tooling

Much design goes into the tooling on the end of a robot. Grippers activated by a pneumatic cylinder are used to pick up and place one size of a part. Often, however, many parts of different shapes and sizes must be picked up and moved. Different areas of the gripper can be configured to pick up different parts. Grippers can also be designed to work in a vacuum. Many types of suction cups are available. Figure 21–29 shows an example.

Weld torches are also a type of end-of-arm tooling. Figure 21–30 shows an example of an articulate robot with a weld torch installed for wire welding. Other

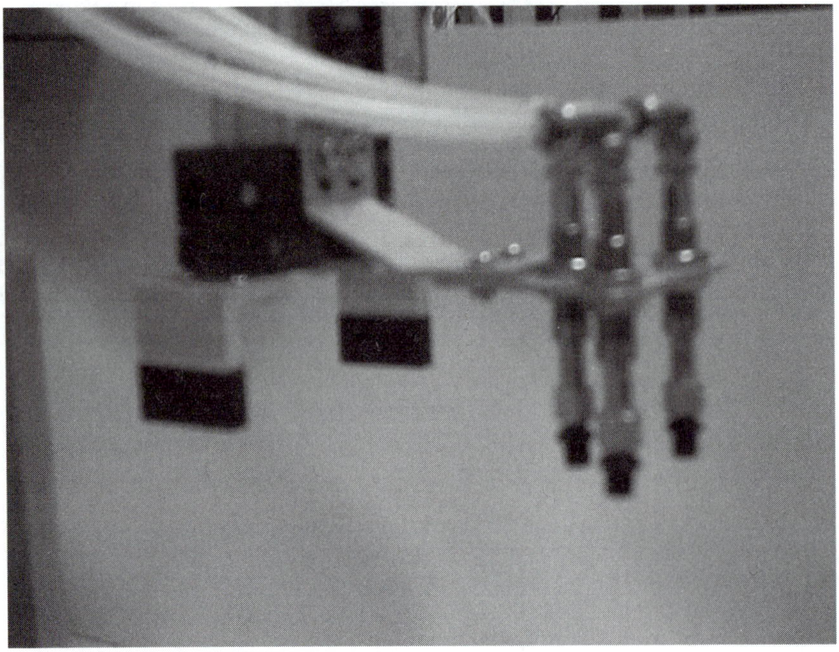

Figure 21–29 This gripper has three suction cups. The arm also has a simple open/close gripper.

types of tooling include automatic screw guns widely used in assembly operations to dispense a fastener and then tighten it.

In complex applications, robots may use several tools. The robots in these applications change their tooling as needed to accomplish their tasks. Air lines and electrical connections are standardized and reconnect to each new tool as it is loaded.

Robot Maintenance

Robot maintenance should first be preventive maintenance. Preventive maintenance should be planned to avoid interfering with production. The manual for the robot or other automated equipment discusses procedures for preventive maintenance. Proper preventive maintenance can reduce the number of breakdowns. One of the most important tasks is to replace the backup battery on the recommended schedule. A loss of battery power causes the robot to lose valuable information. Robots with absolute encoders will lose positional information and will need to be retaught their home position.

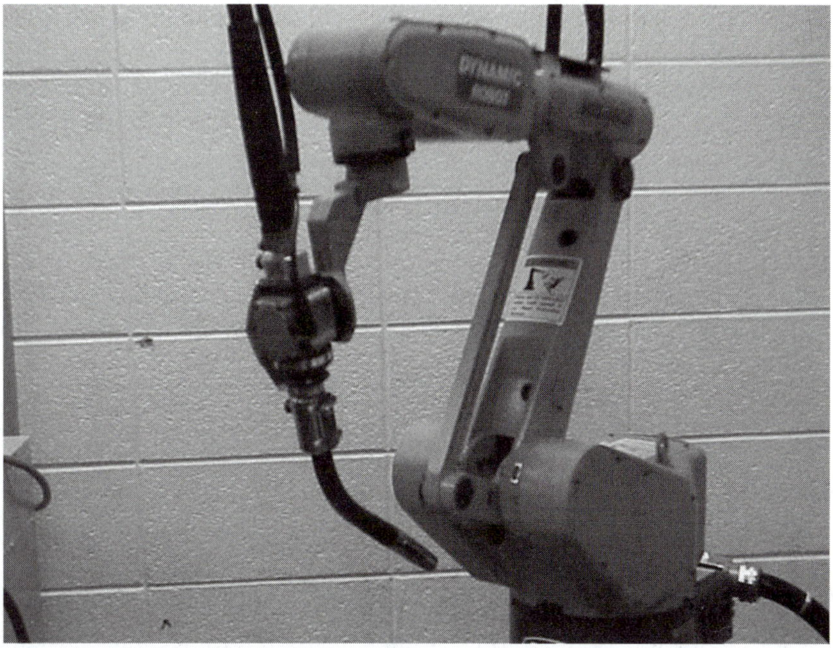

Figure 21–30 Weld torch for wire-feed welding application.

Robot Troubleshooting

Newer computerized equipment, such as robots, and CNC equipment have troubleshooting functions built into the control. The technician can have the control unit run diagnostics to help find a problem. The manual for the equipment normally has a troubleshooting sequence to lead the technician to the problem. In some cases the technician must call the equipment manufacturer for advice. The manufacturer's expert tells the technician what to check, which usually identifies the problem quickly and usually at no cost.

Crashes

Crashes can cause physical and electrical problems. The robot control function can usually help identify the problem. Modern controls provide error messages with specific suggestions for correcting the errors. Serious robot crashes can cause the axes to be misaligned. The operator must test the program of a robot restarted after a major crash very carefully for the first use after the crash. It may be best to single-step the program at very low speed. If the axes have become misaligned or have shifted, slow movement will prevent more damage.

A severe crash can cause physical damage to the robot. Chains, belts, keys, or other linkages can be broken. In some cases part of the robot housing may even break. The electrical protection that is built in can sometimes prevent serious damage. Current limits built into the drive of each axis should cause the drive to inhibit motion if the current limit is exceeded. When the drive is reenabled, the robot should run again. In the case of a rapid, severe crash, the current limit may not protect the robot from further damage.

One of the more common electrical problems after a crash occurs with the drive output transistors. A large, quick overcurrent condition caused by a crash can ruin one or more of the output transistors in the drive (each axis of motion has one drive). Many drives have some built-in troubleshooting device such as an LED on the front of the drive that should indicate that the drive was not harmed.

When a robot or control device is down, production normally stops. Downtime is expensive. To avoid long downtime, spare parts for important equipment and parts that are most prone to fail must be kept in stock. A maintenance kit that contains the parts most likely to fail is generally available when the robot is purchased. A spare motor, tachometer, encoder, electrical boards, and drive should usually be kept. A robot may have four or more axes, all using the same drive.

Lost time in industry is too expensive for a technician to take time to repair a board or drive when the equipment is down. The part should be removed and replaced with a new one. The operator can sometimes try to repair the item later, but it is usually best to send the old one to the manufacturer, who will send a new one and repair the old one, usually at a reasonable cost.

Robots and other computerized equipment have multiple axes that often contain the same devices. Sometimes it is possible to swap components to isolate the problem. For example, the drives may be the same. If one drive has the axis fault LED on, that drive might be exchanged with another to see whether the light stays on in the new position (axis) or if it comes on for the drive that was moved to the problem axis. A light that stays illuminated on the same drive indicates a drive problem. If the axis fault light on the good drive that was moved to the problem axis comes on, the drive is not the problem. The technician should check the manual first to ensure that the components are compatible before they are swapped.

QUESTIONS

1. List at least four reasons why robots are used in industry.
2. Describe the terms *yaw, pitch,* and *roll.*
3. What is a reference frame?
4. Describe a Cartesian coordinate–style robot.

5. What is a SCARA-style robot?
6. What is a compliance device used for?
7. Name three types of robot drive systems.
8. List at least two advantages of pneumatic robots.
9. List at least two advantages of electrical robots.
10. If a robot has incremental encoders on each axis, they do not need to be homed. True or false?
11. What is the purpose of a harmonic drive?
12. What is the difference between accuracy and repeatability?

chapter 22

Industrial Automation Controllers

A new type of controller is becoming more popular in industry. It is not a PLC and it is not easily classified as an industrial computer either. This chapter will focus on one brand of industrial automation controller.

OBJECTIVES

Upon completion of this chapter, you will be able to:

1. Describe an industrial automation controller.
2. Describe applications that are appropriate for industrial automation controllers.
3. Explain how a typical industrial automation controller is programmed.
4. Develop a flowchart for a typical program.

OVERVIEW OF INDUSTRIAL AUTOMATION CONTROLLERS

A new breed of industrial controllers does not fit neatly into the PLC or personal computer classifications. They are often used for special application controls such as motion and process control. A controller from Control Technology Corporation (CTC) will be used to explain industrial automation controllers. Other companies such as Giddings & Lewis also offer industrial automation controllers.

Industrial automation controllers have been designed to fill a need. Traditional PLCs and industrial computers were adequate for the vast majority of control applications. Applications have become increasingly more complex and have required more speed, which has led to more and more servo control. Closed-loop servo control has become prevalent in industry. Traditional controllers have not addressed this need adequately.

Industrial automation controllers offer the traditional discrete and analog control capabilities and also extended capabilities. The CTC controller uses a state logic–type programming language called Quickstep State Language. Quickstep State Language is close to English, which makes it user friendly. The language is also designed to help break complex tasks into logical stages or steps. Next we will take a look at a simple application that will explain Quickstep Language through an application that involves discrete I/O and a motor.

Automated Bottle-Capping Application

The bottle capping operation involves moving each bottle into position and then putting a cap on it. Figure 22–1 shows the I/O used in this application. There are two digital outputs and three digital inputs. There is also one servo motor to advance the conveyor and move the bottles into position. Study the I/O chart and Figure 22–2 to understand the application.

Figure 22–3 shows a flowchart of the application. The flowchart breaks the application into logical operation steps. The first step is called INITIALIZE. This step is used to shut the alarm off and profile the servo motor to the correct parameters. The next step is the RETRACT CAPPER step, which ensure that the capper is retracted before the conveyor is moved (indexed).

Tagname (Register)	Controller Resource	I/O Type
Capper_Up/_Down	Output 1	Digital Output
Alarm_Horn_On/_Down	Output 2	Digital Output
Bottle_Ready/_Not ready	Input 1	Digital Input
Up_Confirm	Input 2	Digital Input
Down_Confirm	Input 3	Digital Input
Conveyor	Servo 1	Servo Motor

Figure 22–1 I/O chart for the bottle-capping application.

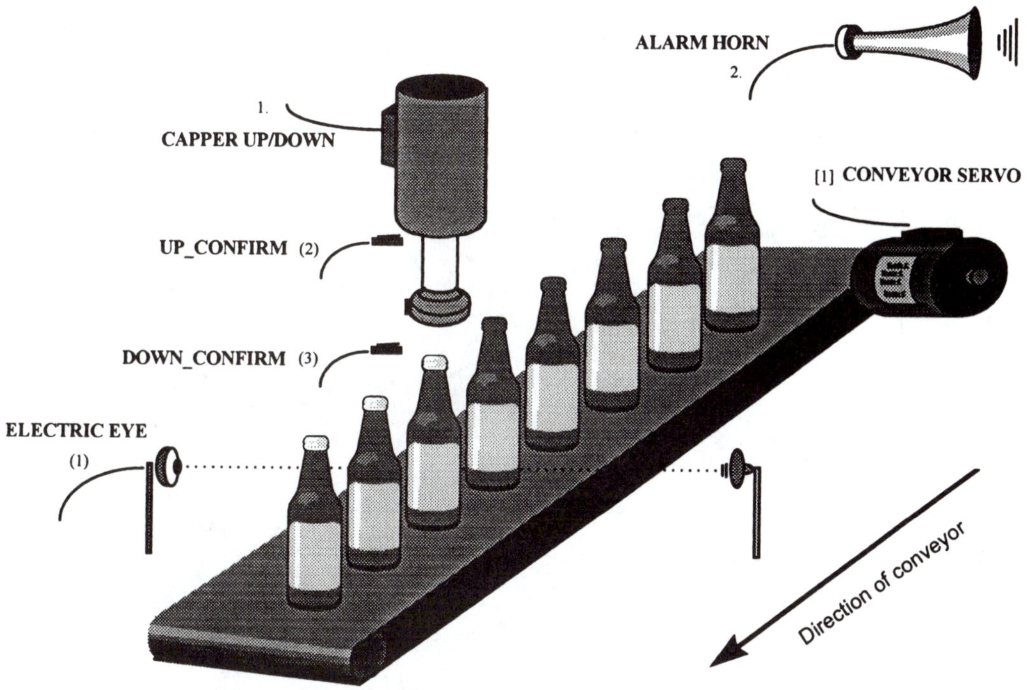

Figure 22–2 The actual bottle-capping application.

The next step is a decision. Decisions can be used to control program flow (see Figure 22–4). The controller must verify that the capper has retracted. It uses input 2 to confirm that the capper is in the up position. If the capper is up, the program moves to the next step. If it is not, the program moves to the next decision step to see if 5 seconds has elapsed. If not, the program goes back to the UP CONFIRM step. If 5 seconds has elapsed, the program goes to the ERROR step. The ERROR step sounds the alarm and halts program execution. If the capper is confirmed as being in the up position, the program moves to the NEXT BOTTLE step.

The NEXT BOTTLE step is used to index the conveyor. The conveyor is driven by a servo motor. This step moves the conveyor the proper amount. The SERVO STOPPED decision step is performed next. This step ensures that the move is complete before the program moves to the next step. When the move is complete, the program goes to the BOTTLE IN PLACE decision step. This step waits up to 100 milliseconds for a bottle. If no bottle is sensed in 100 milliseconds, the program goes to the ERROR step.

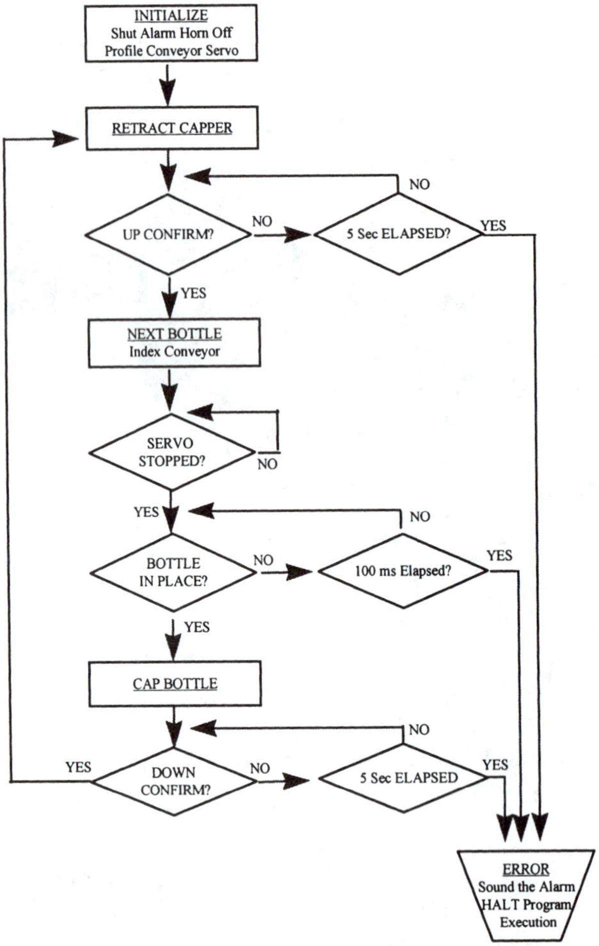

Figure 22–3 Flowchart for the bottle-capping application.

If a bottle is in place within 100 milliseconds, the program moves to the CAP_BOTTLE step. This is the step in which the bottle is actually capped. The next step is the DOWN CONFIRM decision. This step ensures that the capper completed the down motion. If it did, the program goes back to the RETRACT CAPPER step. If the DOWN CONFIRM fails to see the input within 5 seconds, the program goes to the ERROR step.

The actual Quickstep program is similar to the flow diagram. The Quickstep programming language divides a program into logical steps. The steps correspond to

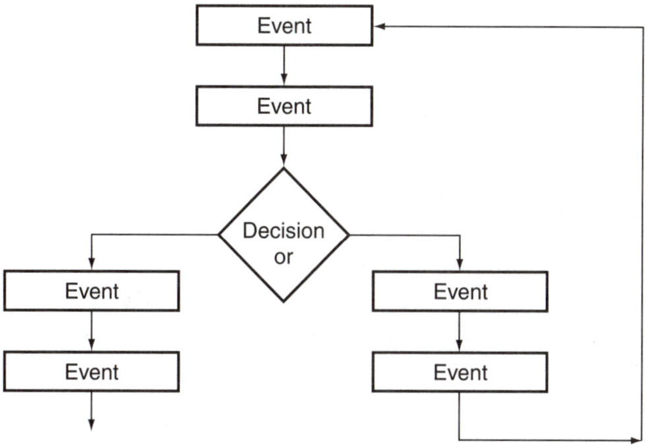

Figure 22–4 Flowchart showing how decisions can be used to control program flow. *(Courtesy Control Technology Corporation.)*

Step number and name	[4] CHK_BOTTLE
Comments	;;; If the photo-sensor does not detect ;;; a bottle in place, branch to the ERROR ;;; step. Operator assistance is needed.
Changes to digital outputs	<NO CHANGE IN DIGITAL OUTPUTS>
One or more statements	monitor Bottle_Ready goto CAP_IT goto Error

Figure 22–5 A typical Quickstep language step.

the process blocks in the flowchart. The decisions that were shown separately in the flowchart can be done right in the steps of the program. Let's look at the actual program, but first let's see what a step looks like (see Figure 22–5).

A step always has the same general format. First is the step number (4 in this case), followed by the step's name. The name of this step is CHK_BOTTLE. The name identifies the step and can be used to tell a program where to branch or move to. In this application, all step names are written in capital letters so that they are easily identifiable.

The next part of a step is optional. A user can add comments to clarify the program or step function. Anything after three semicolons is a comment. It is great programming practice to comment a program adequately. A completed Quickstep program with comments can become a valuable part of the system documentation.

The third part of a step is the section that makes any desired changes in the digital outputs. In this program, all I/O names begin with a capital letter. The fourth part of a step is one or more statements. Statements can be used to monitor inputs, perform branches, and so on.

The bottle capper program is divided into six steps; INIT, RETRACT_CAPPER, NEXT_BOTTLE, CHK_BOTTLE, CAP_IT, and ERROR. The actual program is shown below.

```
[1] INIT
;;; These are remarks that help to document the program. The
INIT step  ;;; turns all outputs off, retracts the capper
and ensures that the alarm horn  ;;; is off.
-------------------------------
<TURN OFF ALL DIGITAL OUTPUTS>
-------------------------------
profile Conveyor servo at position maxspeed=IndexSpeed
accel=RampRate
P=PropVal I=IntegralVal D=DerivVal
goto RETRACT_CAPPER
[2] RETRACT_CAPPER
;;; Raise the capper, then monitor for the Up_Confirm limit
switch. IF we  ;;; don't receive confirmation within five
seconds, we will branch to ERROR ;;; and sound the alarm.
-------------------------------
Capper_Up
-------------------------------
monitor Up_Confirm goto NEXT_BOTTLE
delay 5 sec goto ERROR
[3] NEXT_BOTTLE
;;; Turn the servo 500 increments (steps) clockwise (cw) to
advance the      ;;; conveyor to position the next bottle.
After the conveyor stops, branch to  ;;; the CHK_BOTTLE step.
-------------------------------
<NO CHANGE IN DIGITAL OUTPUTS>
-------------------------------
Turn Conveyor cw 500 steps
monitor Conveyor:stopped goto CHK_BOTTLE
[4] CHK_BOTTLE
;;; If the photo-sensor does not see a bottle in place,
branch to the ERROR ;;; step, need operator intervention.
-------------------------------
```

```
<NO CHANGE IN DIGITAL OUTPUTS>
--------------------------------
monitor Bottle_Ready goto CAP_IT
delay 100 ms goto ERROR
[5] CAP_IT
;;; This step caps the bottle, then branches to the
RETRACT_CAPPER step  ;;; when it detects the Down_Confirm
limit switch.
--------------------------------
Capper_Down
--------------------------------
monitor Down_Confirm goto RETRACT_CAPPER
delay 5 sec goto ERROR
[6] ERROR
;;; This step sounds the alarm and halts program execution.
--------------------------------
Alarm_On
--------------------------------
Done
```

Now let's analyze each step of our program. Study the first step, which is shown below:

```
[1] INIT
;;; These are remarks that help to document the program. The
INIT step  ;;; turns all outputs off, retracts the capper
and ensures that the alarm horn is off.
--------------------------------
<TURN OFF ALL DIGITAL OUTPUTS>
--------------------------------
profile Conveyor servo at position maxspeed=IndexSpeed
accel=RampRate
P=PropVal I=IntegralVal D=DerivVal
goto RETRACT_CAPPER
```

This is step 1 and its name is INIT. The user could have chosen any name for the step. Next the user wrote a few comments to explain what the step does. The next section of the program turned off all outputs to be sure that everything is off when we begin program execution. The next section of the step is the statement section. The first statement was used to define the servo parameters. The statement is a *profile* statement. The profile statement is used to define the maximum speed; the acceleration value; and the proportional, integral, and derivative values for the servo. Maxspeed was set to equal a user-defined variable (IndexSpeed). The other parameters were also assigned variables. Variables were used so that the user can change the value of the variables while in operation without rewriting the program.

Study the second step, which is repeated below:

```
[2] RETRACT_CAPPER
;;; Raise the capper, then monitor for the Up_Confirm limit
switch. IF we  ;;; don't receive confirmation within five
seconds, we will branch to ERROR   ;;; and sound the alarm.
-------------------------------
Capper_Up
-------------------------------
monitor Up_Confirm goto NEXT_BOTTLE
delay 5 sec goto ERROR
```

The second step is named RETRACT_CAPPER. Note that the programmer chose effective step names. An effective step name is one that is descriptive about the purpose of the step. The programmer also added some comments to explain what occurs in this step. In the output section, the programmer typed Capper_Up. This is the name of an output. If you look at Figure 22–1 you will see that Capper_Up/_Down is the name of output 1. The programmer defined Capper_Up as output 1 being OFF. The programmer defined Capper_Down as output 1 being ON. I/O variables can be defined while programming by simply answering a few questions.

Figure 22–6 shows an example of a symbol definition screen. The symbol browser can be used to work easily with any of the symbolic names. Symbolic names are used to represent real-world I/O and variables. Symbolic names make programs user-friendly and understandable. Figure 22–7 shows an example of the symbol browser. This is part of the Quickstep software. The list on the left of the figure shows the available types of symbols. Note that the programmer has chosen to work with the output type, so the output symbols that exist are shown on the right of the figure. Notice the first two output symbols, Punch_On and Punch_Off. Punch_On was assigned an output state of On (1). Punch_Off was assigned a state of Off (0). This allows user-friendly symbol names to be used in the program.

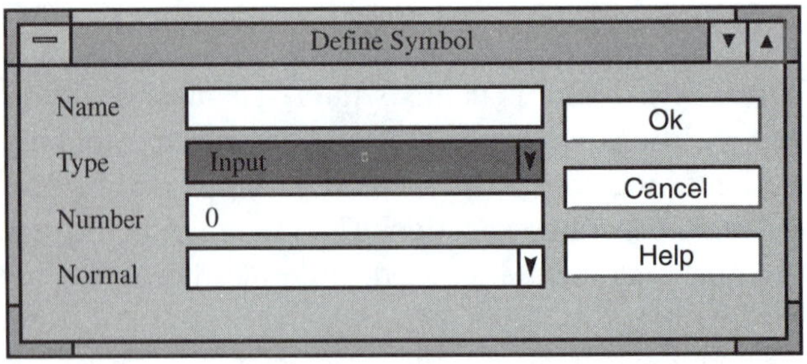

Figure 22–6 Symbol definition screen. *(Courtesy Control Technology Corporation.)*

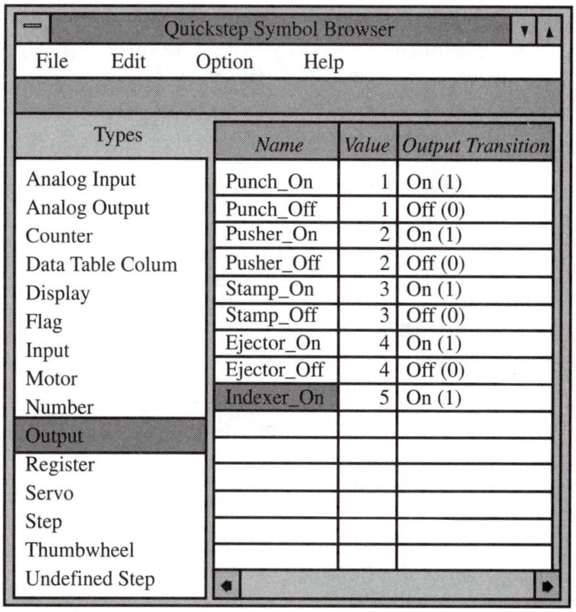

Figure 22–7 Symbol browser. *(Courtesy Control Technology Corporation.)*

For this application, the programmer made sure the capper was up by putting Capper_Up in the output line. Capper_Up is equal to output 1 OFF.

In the statement portion of the step, the programmer examined the state of a sensor. *Monitor* is a statement that causes the controller to look at an input condition. In this case, the program looks at the state of input 2 (Up_Confirm). If it is on, the program goes to the step named NEXT_BOTTLE. If input 1 is not on, the program continues to monitor the input for 5 seconds. If it does not detect the sensor to be on within 5 seconds, the program goes to the step named ERROR.

Study the third step of bottle-capping application:

```
[3] NEXT_BOTTLE
;;; Turn the servo 500 increments (steps) clockwise (cw)
to advance the       ;;; conveyor to position the next
bottle. After the conveyor stops, branch to   ;;; the
CHK_BOTTLE step.
-------------------------------
<NO CHANGE IN DIGITAL OUTPUTS>
-------------------------------
Turn Conveyor cw 500 steps
monitor Conveyor:stopped goto CHK_BOTTLE
```

The third step is named NEXT_BOTTLE. This step moves the servo motor, which moves the next bottle into position. The programmer again typed some comments to explain the purpose of the step. The programmer stated that there were no changes to be made to the digital outputs in the output step. The statement portion of the step was used to move the conveyor motor. The servo motor for the conveyor was given the name Conveyor. The statement Turn Conveyor turns the conveyor motor 500 steps (increments). Steps are equivalent to whatever encoder counts are used in the particular servo system. This is a closed-loop example. The velocity, acceleration, and following error will be monitored with this simple command. Remember that the speed was set by the profile servo command in the first step (INIT). Stepper motors are programmed in essentially the same manner. The second statement (monitor Conveyor:stopped goto CHK_BOTTLE) monitors the move to see when it has finished.

When the move is finished, the program moves to the CHK_BOTTLE step, which is shown below:

```
[4] CHK_BOTTLE
;;; If the photo-sensor does not see a bottle in place,
branch to the ERROR
;;; step, need operator intervention.
-------------------------------
<NO CHANGE IN DIGITAL OUTPUTS>
-------------------------------
monitor Bottle_Ready goto CAP_IT
delay 100 ms goto ERROR
```

The fourth step is named CHK_BOTTLE. The programmer typed a few remarks to explain the step. There were no changes made in the digital outputs. Note that this remark is quite important. We want to maintain the states of the outputs as they are now. The programmer then uses statements to monitor input 1 (Bottle_Ready). Remember that Bottle_Ready/_Notready is input 1. The user defined Bottle_Ready as input 1 being ON. Bottle_Notready was defined as input 1 being OFF. If the input is ON, the program goes to the CAP_IT step. The next statement (delay) causes the program to monitor the input for 100 ms or until it comes on. If the input does not come on within 100 ms, the program goes to the ERROR step.

The fifth step is named CAP_IT:

```
[5] CAP_IT
;;; This step caps the bottle, then branches to the
RETRACT_CAPPER step  ;;; when it detects the Down_Confirm
limit switch.
-------------------------------
Capper_Down
-------------------------------
monitor Down_Confirm goto RETRACT_CAPPER
delay 5 sec goto ERROR
```

The programmer typed a few remarks to explain the purpose of the step. The programmer turned output 1 on in the output section. Remember that output one was named Capper_Down and that Capper_Down was defined as being output 1 ON.

The program then monitors input 3 (Down_Confirm) to see if it is ON. If it is ON, the program goes to the RETRACT_CAPPER step. If it is not on, the delay statement causes the program to continue to monitor the input for five seconds or until it does turn on. If in five seconds the input has not become true, the program will go to the ERROR step.

The final step (6) is the ERROR step, which is shown below:

```
[6] ERROR
;;; This step sounds the alarm and halts program execution.
--------------------------------
Alarm_On
--------------------------------
Done
```

This step simply turns output 2 ON. Output 2 was defined by the programmer to be Alarm_Horn_On. This step also halts program execution.

Industrial Automation Controller Hardware

Control Technology Corporation has several industrial automation controllers available. The simplest is a small controller that controls digital I/O. All of the controllers utilize the Quickstep programming language, so they are easy to utilize. Figure 22–8 shows a starter kit that includes a CTC 2601 Automation Controller. The 2601 Automation Controller is a high-performance control system whose small size conceals a range of resources and capabilities usually found only in much larger and more costly systems. A few of this controller's features include:

- Thirty-two digital I/O.
- Multitasking, with up to twenty-eight parallel tasks running simultaneously.
- 988 storage registers, including 500 nonvolatile registers.
- A data table capable of storing over 8000 numbers in a two-dimensional array.
- Short-circuit and overcurrent-protected outputs.
- Integrated RS-232 communications port.
- Internal counters, which may be software-linked to any input.

The 2601 offers strong communications capabilities, including computer-based data collection, programming, and configuration, as well as the ability to transmit messages to displays and other remote subsystems. Because these messages can

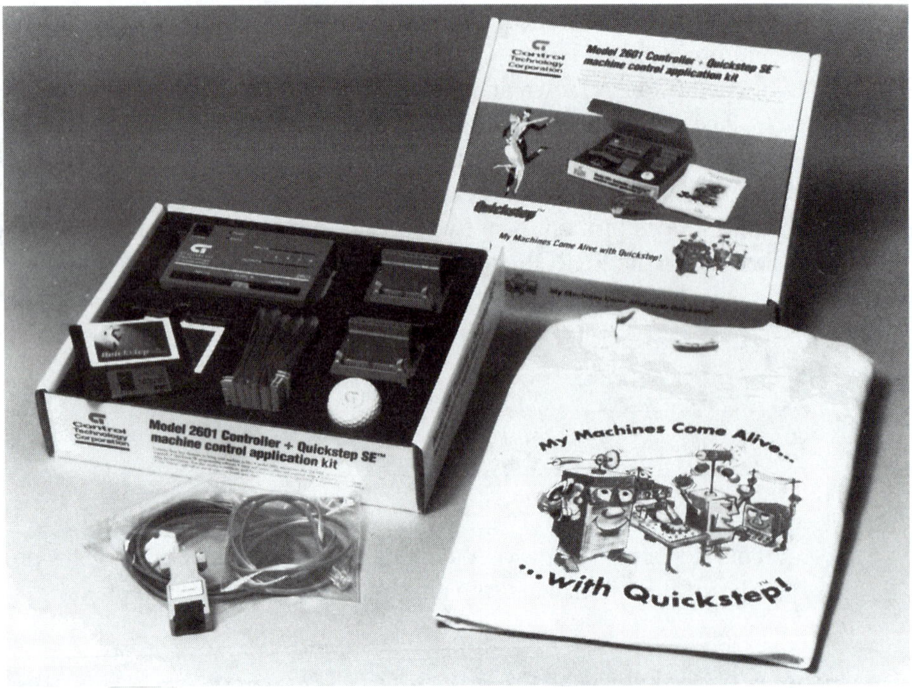

Figure 22–8 The Quickstep starter kit. It includes a 2601 32 I/O controller, cables, input/output simulator, Quickstep software, and manuals. *(Courtesy Control Technology Corporation.)*

contain live data from the controller's registers, the user can create fully interactive operator controls using inexpensive LCD displays or remote terminals. The user can connect an operator interface to the serial port and have full interactive control of machine parameters. The user can also collect production data and communicate it through the communications port to a computer for logging or display. The next level is the 2600XM controller.

Software capabilities include the ability to run up to twenty-eight independent tasks simultaneously, and the instruction set includes time delay, input monitoring, and math and data manipulation commands. These commands make full use of internal resources such as the controller's nonvolatile and volatile registers, user-definable data table, input-linkable counters, and thirty-two flags.

Figure 22–9 shows a modular-style controller. The user can choose modules to meet the application requirements. A wide variety of modules are available. Several types of digital and analog I/O modules are also available. The applications we

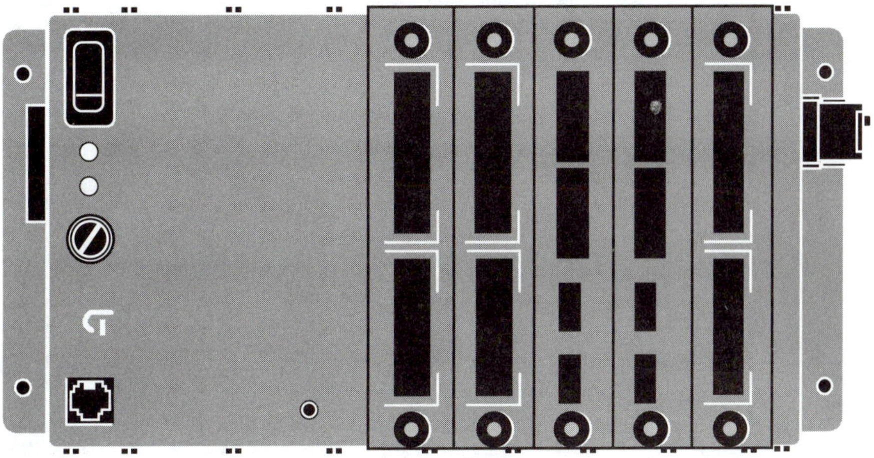

Figure 22–9 A five-rack controller. In this configuration, the controller has three digital I/O cards and two 2-axis servo cards. There is also a serial port on the lower left of the controller. Note that the user chooses the modules that are needed and plugs them into the slots. *(Courtesy Control Technology Corporation.)*

have looked at utilize simple digital I/O and one servo motor. These controllers have almost unlimited application analog capability as well. High-speed counter modules, thumbwheel input, and display modules can be used. There are single- or dual-axis stepper motor modules as well as single- and dual-axis servo control modules. Communications modules are available to add additional RS-232 or Ethernet modules. Figure 22–10 shows a 2600XM controller. This controller has almost unlimited application capability. Ethernet or serial communications can be used to create networks of controllers. The Ethernet network allows the controllers to communicate easily with each other or to communicate with other computers on the network. This allows data gathering and control with SCADA software such as Wonderware.

The 2600XM controller is programmed using the same Quickstep programming language. A variety of user displays, from simple to versatile graphic touch screens, are also available. The servo modules are powerful and easy to use. The dual-axis servo modules can be used to control two-axis or master-slave and/or electronic gearing–type applications independently.

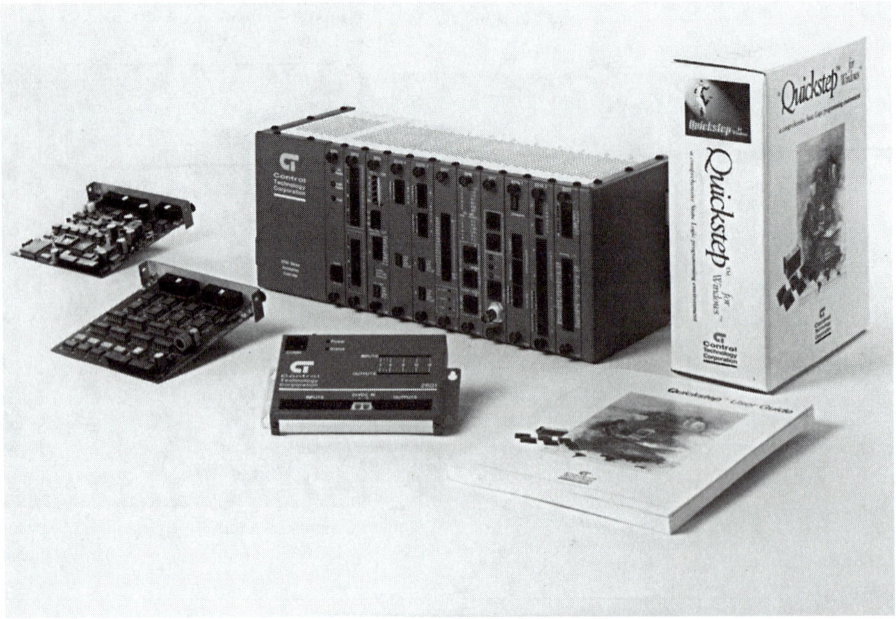

Figure 22–10 The CTC 2600XM controller is shown in the middle of the picture. The small 2601 32 I/O controller is shown in the front for comparison. The 2600XM is available in several different rack sizes. This one has a ten-slot rack. The user can choose any combination of I/Os to fill the rack. Several servo modules can be used to control a complex multi-axis application. Two of the plug-in modules are shown to the lower left of the controller. *(Courtesy Control Technology Corporation.)*

CTC also has a series of controllers called MultiPro automation controllers. Figure 22–11 shows a few of them. The MultiPro controller is aimed at high-speed machine control applications. The MultiPro controller uses multiple processors. Individual processors are dedicated to analog I/O, motion control, communications, and to the main controller. The overall control is exercised by the user's Quickstep program. The English-like Quickstep language makes it easy to program even complex applications. Models are available for stepper or servo control. All have serial communications built in and most have Ethernet communications built in. Figure 22–12 shows an example of how CTC controllers can be networked on an Ethernet network.

Multitasking

Thus far we have looked only at linear processes with Quickstep programming. Multitasking is also possible. The controller can be programmed to control multiple

Figure 22–11 The Control Technology Corporation MultiPro family of Automation controllers. *(Courtesy of Control Technology Corporation.)*

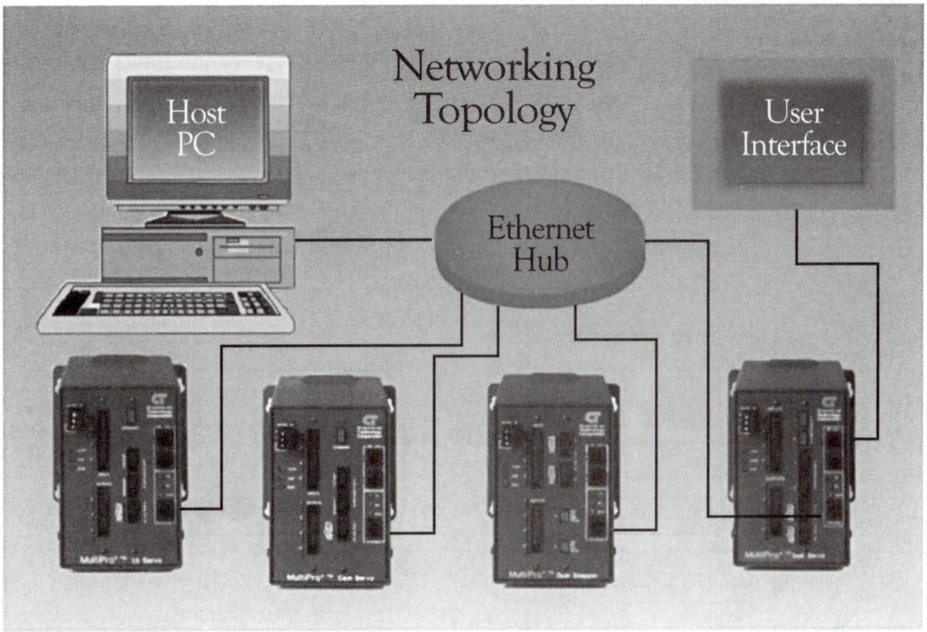

Figure 22–12 Networking topology for CTC controllers on an Ethernet network. *(Courtesy Control Technology Corporation.)*

processes that occur simultaneously. Figure 22–13 shows an example of multitasking. This figure is a block diagram of the steps. Each step has to be defined. Each of the rectangles represent steps in the program. In this example, the START_CYCLE step has a do statement that causes the program to move to step ONload and step OFFload. When the do statement is used, the program should execute all steps that are listed in the parentheses after the do simultaneously.

The second example of concurrent multitasking is shown in Figure 22–14. The top rectangle labeled [42] START_CYCLE has a statement that says do (MOVE_PART

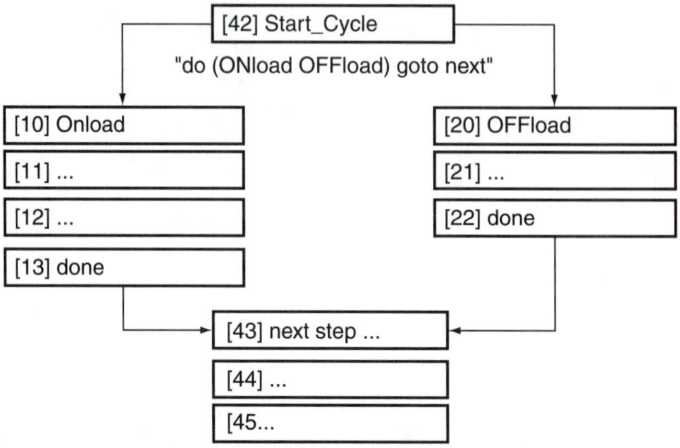

Figure 22–13 Example of multitasking. *(Courtesy Control Technology Corporation.)*

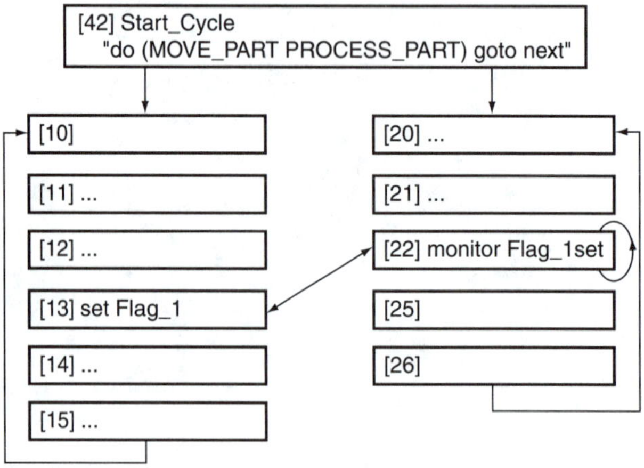

Figure 22–14 Use of a flag to communicate between steps. *(Courtesy Control Technology Corporation.)*

PROCESS PART) goto next. Both sequences then process at the same time. This example shows that data from one step can be shared with other steps. Step 13 sets Flag_1 to a 1. Step 22 contains a statement to monitor Flag_1. This allows concurrent tasks to share information that may be crucial for timing and/or correct sequencing. Flags may be set to a 1 or a 0. A set statement can be used to set flags to the desired state. Monitor statements can then be used to check the state of a flag and make decisions based on the flag's state.

Other memory resources are available for the user. Numeric registers can be used to hold numbers that range in size from 0 to 65,535. Registers 1 to 500 are volatile. Their contents are lost when power is removed from the controller. Registers 501 to 1000 are nonvolatile. These registers maintain their data even if power is removed for up to ten years. Registers 1001 to 24999 are special-purpose registers. These registers are used for advanced programming techniques and give the user access to controller resources such as inputs, outputs, servo parameters, and communications capabilities. The *store* command is used to store data to registers. *If* instructions can be used to test registers.

Data tables can be used to store numeric data or an ASCII message table. A numeric data table can be used to store data from a process. A numeric table can also be used to store recipe information for processes. For example, several different products may be produced in a cell. A data table can be used to store the parameters for each product. The program then reads the needed values from the table, depending on which product is to be produced.

Tables can also be used to store messages in the form of their ASCII equivalents (see Figure 22–15). Messages can then be sent out the communications port to a computer or display or other device by simply storing the row number to a

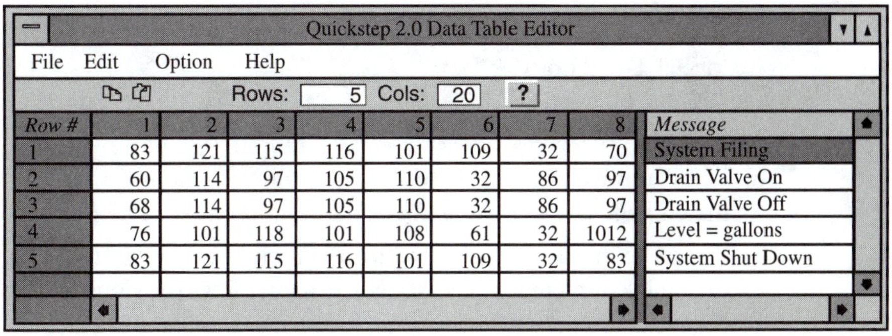

Quickstep 2.0 Data Table Editor
File Edit Option Help
Rows: 5 Cols: 20 ?
Row # 1 2 3 4 5 6 7 8 Message
1 83 121 115 116 101 109 32 70 System Filing
2 60 114 97 105 110 32 86 97 Drain Valve On
3 68 114 97 105 110 32 86 97 Drain Valve Off
4 76 101 118 101 108 61 32 1012 Level = gallons
5 83 121 115 116 101 109 32 83 System Shut Down

Figure 22–15 Data table used as an ASCII message table. *(Courtesy Control Technology Corporation.)*

special-purpose register (reg_12001). For example, if row number 5 was stored to reg_12001, the ASCII characters that represent the message SYSTEM SHUT DOWN would be sent out the serial port. The table view in Figure 22–15 shows only the first eight ASCII equivalents of the message. The user can scroll the screen to see the rest.

This chapter is a brief introduction to the topic of industrial automation controllers. Industrial automation controllers have almost unlimited potential. Remember that PLCs, industrial computers, and industrial automation controllers are all just microprocessor-based computers. Industrial automation controllers provide an easily implemented solution for integrating complex servo, process control, and communications into a single environment. Their strengths will continue to be their ease of use and ease of programming.

QUESTIONS

1. What is an industrial automation controller? Compare and contrast it with PLCs and industrial computers.
2. What are some of the types of applications that industrial automation controllers are particularly well suited for?
3. List and explain the four parts that a step can contain.
4. Explain the following step.

```
[12] FILL_TANK
;;; If the photo-sensor does not see a bottle in place,
;;; branch to the ERROR step,
;;; operator intervention required
------------------------------
Valve_On
------------------------------
monitor Fill_Level goto MIX
delay 20 sec goto ALARM
```

5. Draw a flowchart for the step shown in Question 4.
6. Draw a flowchart that represents the following application: fill a tank, mix the contents, heat the contents, and drain the tank. The I/O is shown below:
7. Write the program to accomplish the tank application from Question 6 using the Quickstep language. Use the correct format for steps and comment your program adequately.
8. Explain how multitasking can be done in the Quickstep language.

Chapter 23

Installation and Troubleshooting

Proper installation is crucial in automated systems. The safety of people and machines is at stake. Troubleshooting and maintenance of systems becomes more crucial as systems become more automatic, more complex, and more expensive. An enterprise cannot afford to have a system down for any length of time. The technician must be able to find and correct problems quickly.

OBJECTIVES

Upon completion of this chapter, you will be able to:

1. Describe safety considerations that are crucial in troubleshooting and maintaining systems.
2. Explain terms such as *noise, snubbing, suppression,* and *single-point ground.*
3. Explain correct installation techniques and considerations.
4. Describe proper grounding techniques.
5. Describe noise reduction techniques.
6. List the steps in a typical troubleshooting process.

Overview

Installation, troubleshooting, and maintenance of an automated system is a crucial phase of any project. It is the point at which the hopes and fears of the engineers and technicians are realized. It can be a frustrating, exciting, and rewarding time. The installation, troubleshooting, maintenance, and operation of any automated system are highly dependent on the quality of the associated documentation. Documentation, unfortunately, is often an afterthought, although it should be developing as the system is developed. The system must be accurately and completely documented to reduce downtime significantly. Fortunately, programming software for PLCs and other control devices has extensive documentation capabilities. Documentation should include the following:

Description of the overall system.

Block diagram of the entire system.

Program listing including cross referencing and clearly labeled I/O.

Printout of the PLC or control device's memory showing I/O and variable usage.

Complete wiring diagram.

Description of peripheral devices and their manuals.

Operator's manual including start-up and shutdown procedures.

Notes concerning past maintenance.

PLC Installation

Proper installation of a PLC is crucial. The PLC must be wired so that the system is safe for the workers. The system must also be wired so that the devices are protected from overcurrent situations. Proper fusing within the system is important. The PLC must also be protected from the application environment. Dust, coolant, chips, and other contaminants are usually found in the air. The proper choice of enclosures can protect the PLC. Figure 23–1 shows a block diagram of a typical installation.

Enclosures

PLCs are typically mounted in protective cabinets. The National Electrical Manufacturers Association (NEMA) has developed standards for enclosures (Figure 23–2). Enclosures are used to protect the control devices from the environment of the application. A cabinet type is chosen based on the severity of the environment in the application. Cabinets typically protect the PLC from airborne contamination. Metal cabinets can also help protect the PLC from electrical noise.

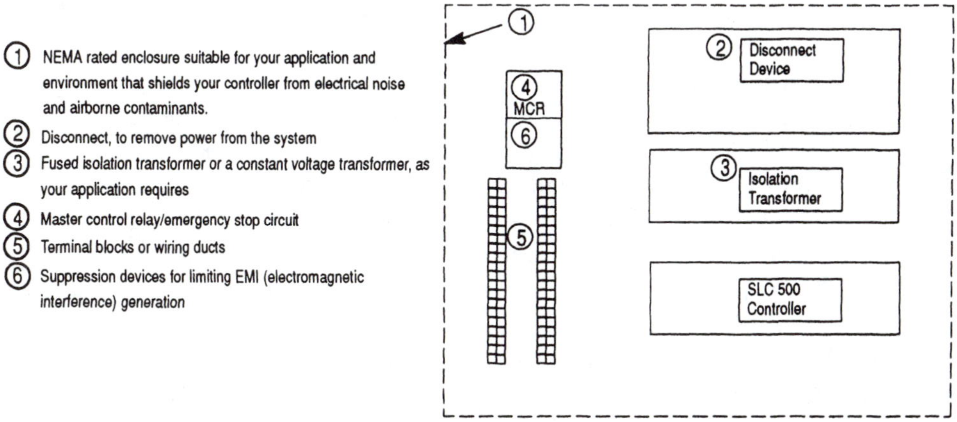

① NEMA rated enclosure suitable for your application and environment that shields your controller from electrical noise and airborne contaminants.
② Disconnect, to remove power from the system
③ Fused isolation transformer or a constant voltage transformer, as your application requires
④ Master control relay/emergency stop circuit
⑤ Terminal blocks or wiring ducts
⑥ Suppression devices for limiting EMI (electromagnetic interference) generation
①
② Disconnect Device
④ MCR
⑥
⑤
③ Isolation Transformer
SLC 500 Controller

Figure 23–1 Block diagram of a typical PLC control cabinet. *(Courtesy Rockwell Automation Inc.)*

Protection Against	Enclosure Type												
	1	2	3	3R	3S	4	4X	5	6	6P	11	12	13
Accidental contact with enclosed equipment	*	*	*	*	*	*	*	*	*	*	*	*	*
Falling dirt	*	*				*	*	*	*	*	*	*	*
Falling liquids, light splashing		*				*	*		*	*	*	*	*
Dust, lint, fibers, (non-combustible, non-ignitable)						*	*	*	*	*		*	*
Windblown dust			*		*	*	*		*	*			
Hose down and splashing water						*	*		*	*			
Oil and coolant seepage												*	*
Oil or coolant spraying or splashing													*
Corrosive agents							*			*	*		
Occasional temporary submersion									*	*			
Occasional prolonged submersion										*			

Figure 23–2 Comparison of the features of NEMA cabinets versus enclosure type.

Heat is generated by devices. One must be sure that the PLC and other devices to be mounted in the cabinet can perform at the temperatures required. Remember that it will be even hotter in the cabinet than in the application because of the heat generated within the cabinet. In some applications, substantial heat may be generated by other devices in the system or cabinet. In this case, blower fans should be placed inside the enclosure to increase air circulation and reduce hot spots within the enclosure. These fans should filter the incoming air so that contaminants are not introduced to the cabinet and components.

The main consideration is adequate space around the PLC in the enclosure, which will allow air to flow around the PLC. The manufacturer will provide installation requirements in the hardware manual for the PLC. Figure 23–3 shows an example for an AB SLC-500. Note that the dimensions shown from each side of the PLC to the cabinet are minimum distances. The actual distance between the PLC and the cabinet should be greater than those shown. Check the specifications for the PLC to determine temperature ranges. In most cases fans will be unnecessary. Proper clearances around devices will normally be sufficient for heat dissipation.

When drilling holes in the cabinet for mounting components, keep chips from falling into the PLC or other components. Metal chips or wire clippings could cause equipment to short-circuit or could cause intermittent or permanent problems.

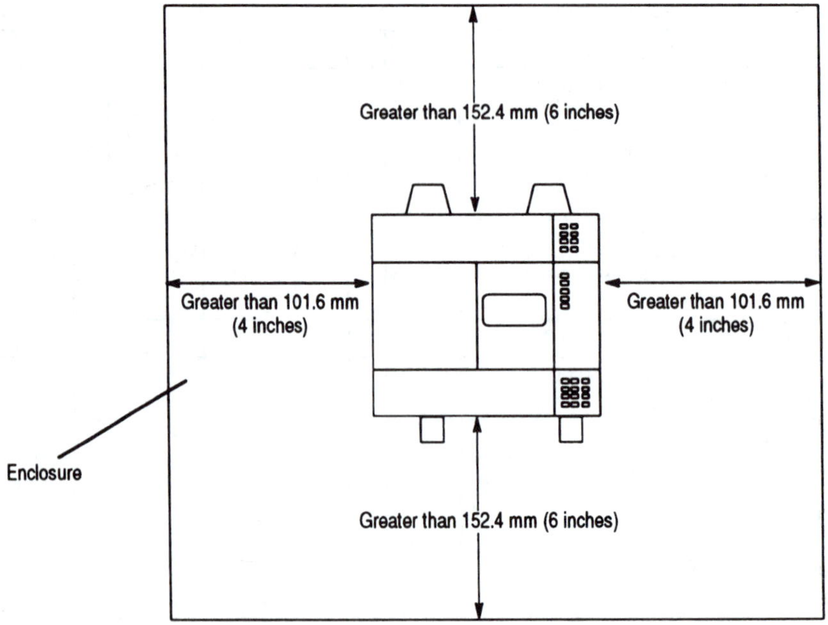

Figure 23–3 Proper mounting for an AB SLC-500. *(Courtesy Rockwell Automation Inc.)*

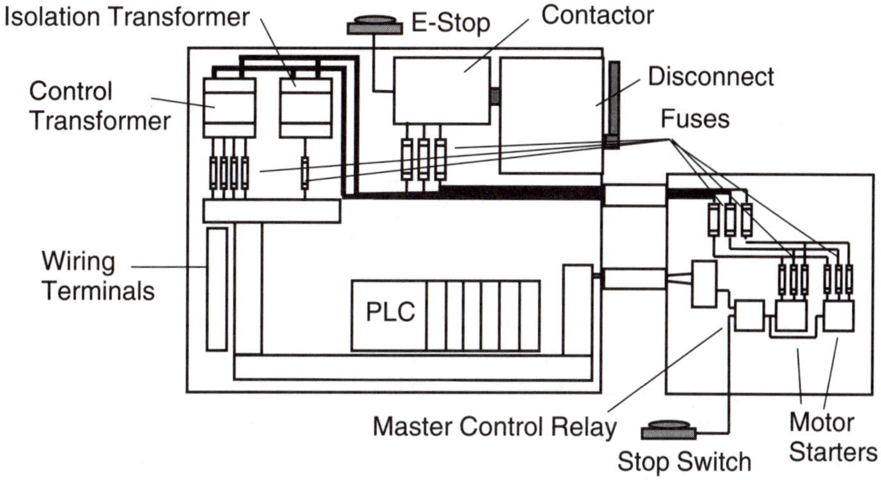

Figure 23–4 Block diagram of a typical control cabinet wiring scheme.

Wiring

Proper wiring of a system involves choosing the appropriate devices and fuses (Figure 23–4). Normally, three-phase power (typically 480 V) will be used in manufacturing. This will be the electrical supply for the control cabinet.

The three-phase power is connected to the cabinet via a mechanical disconnect. This disconnect is mechanically turned on or off by the use of a lever on the outside of the cabinet, and should be equipped with a lockout. This means that the technician should be able to put a lock on the lever to prevent anyone from accidentally applying power while it is being restored. This three-phase power must be fused. Normally, a fusible disconnect is used, which indicates that the mechanical disconnect has built-in fusing. The fusing is to ensure that too much current cannot be drawn. Figure 23–4 shows the fuses installed after the disconnect.

The three-phase power is then connected to a contactor. The contactor is used to turn all power off to the control logic in case of an emergency. The contactor is attached to hardwired emergency circuits in the system. If someone hits an emergency stop switch, then the contactor drops out and will not supply power. The hardwired safety should always be used in systems. This is called a master control relay.

Master Control Relay

A hardwired master control relay (MCR) is used for emergency controller shutdown (Figure 23–5). The MCR must be able to inhibit all machine motion by removing power to the machine I/O devices when the MCR relay is de-energized.

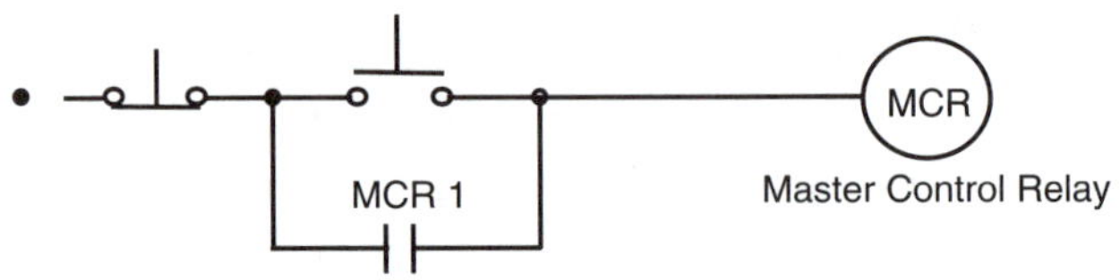

Figure 23–5 Example of a master control relay circuit.

If a DC power supply is used, the power should be interrupted on the load side rather than on the AC supply side to provide quicker shutdown of power. The DC power supply should be powered directly from the fused secondary of the transformer. DC power from the power supply is then connected to the devices through a set of master control relay contacts.

Emergency stop switches can be placed at multiple locations. These should be chosen to provide full safety for anyone in the area. Switches may include limit-type switches for overtravel conditions, mushroom-type push-button switches, and other types. They are wired in series so that if any one of the switches is activated, the master control relay is de-energized. This removes power from all input and output circuits.

Never alter or bypass these circuits to defeat their function. Severe injury and/or damage can occur. These switches are designed to "fail safe": if they fail they should open the master control circuit and disconnect power. It is possible that a switch could short out and not offer further protection. Switches should be tested periodically to be sure they will still stop all machine motion if used.

The main power disconnect switch should be positioned in a convenient and easily accessible place for operators and maintenance personnel. The disconnect should be placed so that power can be turned off before the cabinet is opened.

The master control relay is not a substitute for a disconnect to the PLC. Its purpose is to de-energize I/O devices quickly. Figure 23–6 shows a cutoff system with a hardwired emergency switch connected to the master control relay. The figure also shows the wiring diagram. Note that the system provides for a mechanical disconnect for output module power and a hardwired stop switch for disconnecting the system power. The programmer should also provide for an orderly shutdown in the PLC control program.

The three-phase power must then be converted to single phase for the control logic. Power lines from the fusing are connected to transformers. In the case of Figure 23–7, there are two transformers: an isolation transformer and a control transformer. The isolation transformer is used to clean up the power supply for the

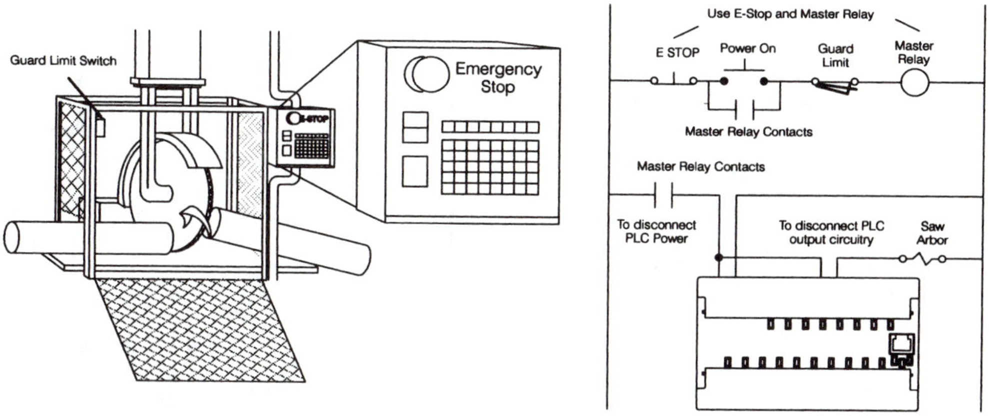

Figure 23–6 The use of a hardwired E-Stop and master control relay. *(Courtesy AutomationDirect Inc.)*

PLC. Isolation transformers are normally used when there is high-frequency conducted noise in or around power distribution equipment. Isolation transformers are also used to step down the line voltage.

The control transformer is used to supply other control devices in the cabinet. The lines from the power supplies are then fused to protect the devices they will supply. These individual device circuits should be provided with their own fuses to match the current draw. DC power is usually required also. A small power supply is typically used to convert the AC to DC.

Motor starters are typically mounted in separate enclosures to protect the control logic from the noise these devices generate.

Within the cabinet, certain wiring conventions are typically used. Red wire is normally used for control wiring, black wire is used for three-phase power, blue wire is used for DC, and yellow wire is used to show that the voltage source is separately derived power (outside the cabinet).

Signal wiring, which is typically low voltage or low current, should be run separately from 120 V wiring. It could be affected by being too close to high-voltage wiring, so when possible, run the signal wires in separate conduit. Some conduit is internally divided by barriers to isolate signal wiring from higher voltage wiring.

Voltage is supplied to the PLC through wiring terminals. The user can often configure the PLC to accept different voltages (Figure 23–8). Here the PLC can be configured to accept either 220 or 110 V. In this case, the shorting bar must be installed if 110 V will be used.

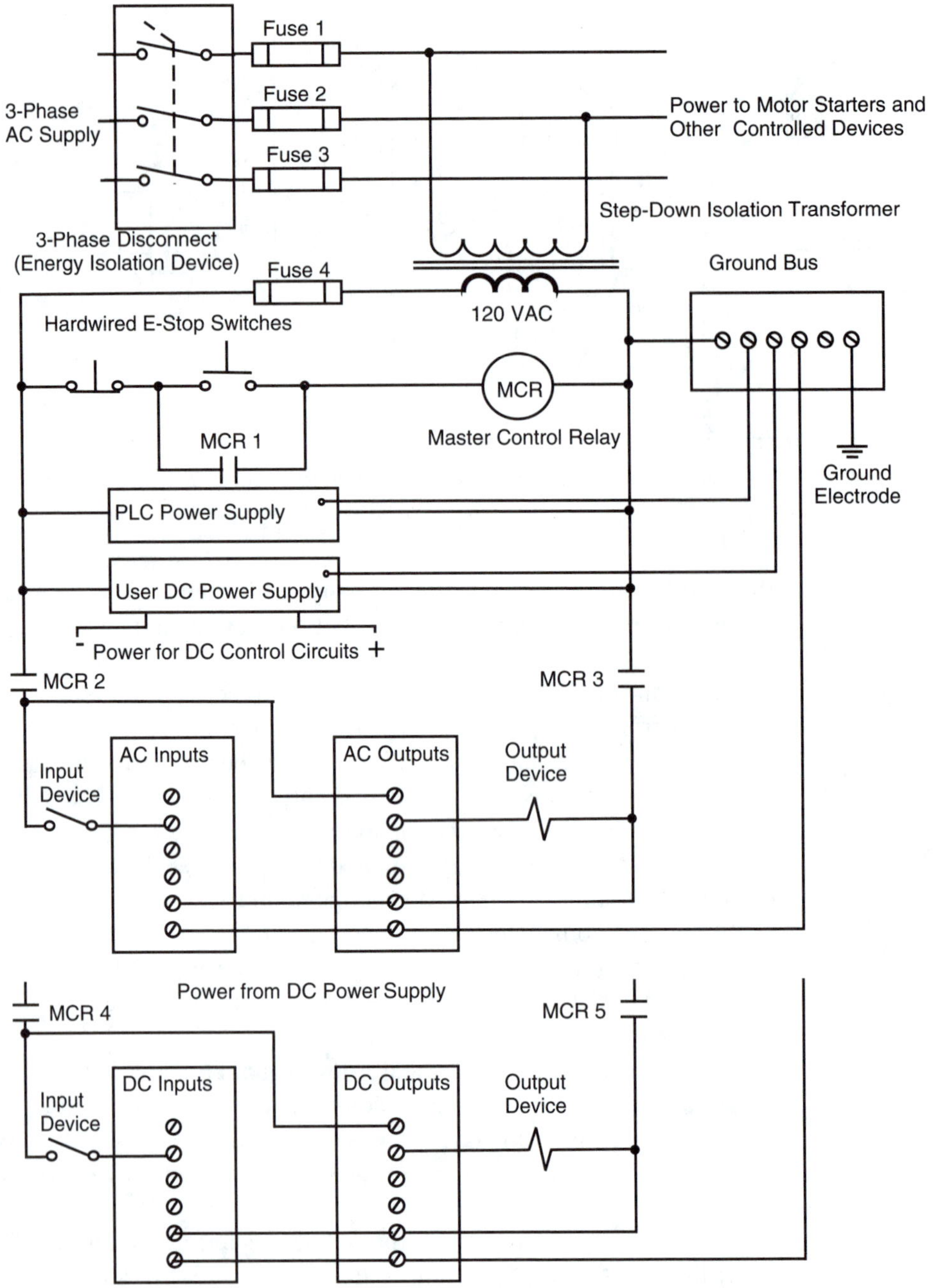

Figure 23–7 Control wiring diagram.

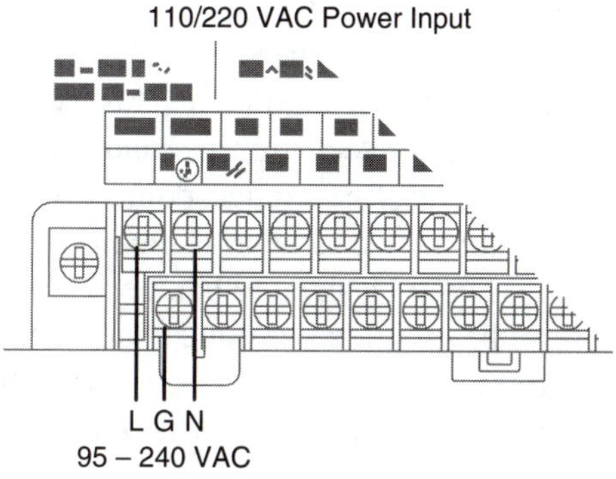

Figure 23–8 Diagram of PLC wiring. *(Courtesy AutomationDirect Inc.)*

Wiring Guidelines The following guidelines should be considered when wiring a system:

- Always use the shortest possible cable.
- Use a single length of cable between devices. Do not connect pieces of cable to make a longer cable. Avoid sharp bends in wiring.
- Avoid placing system and field wiring close to high-energy wiring.
- Physically separate field input wiring, output wiring, and other types of wiring.
- Separate DC and AC wiring when possible.
- A good ground must exist for all components in the system (0.1 ohm or less).
- If long return lines to the power supply are needed, do not use the same wire for input and output modules. Separate return lines will minimize the voltage drop on the return lines of the input connections.
- Use cable trays for wiring.

Grounding

Neutral versus Ground Grounding of electrical equipment is one of the least understood details about electrical systems. In addition to ensuring safety, grounding must also ensure that the electrical devices do not interfere with each other in a cell or plant environment. The terms *ground* and *neutral* tend to cause confusion. They are not the same.

A ground represents an electrical path that is designed to carry current when an insulation breakdown occurs in a system. For example, a technician who drops a tool in an electrical cabinet may cause a breakdown if the tool touches a voltage source and the cabinet or other metal. It should be noted that some current will always flow through the ground path. This can be caused by capacitive coupling or inductive coupling between current conductors and the ground path.

Neutral represents a reference point within an electrical distribution system. Wires that are connected to neutral should normally be non–current-carrying conductors. They should be sized to handle short-term faults that may occur.

Neutral can be grounded; ground is not neutral.

Proper grounding is essential for safety and proper operation of a system. Grounding also helps limit the effects of noise due to electromagnetic interference (EMI). Check the appropriate electrical codes and ordinances to ensure compliance with minimum wire sizes, color coding, and general safety practices.

Connect the PLC and components to the subpanel ground bus (Figure 23–9). Ground connections should run from the PLC chassis and the power supply for each PLC expansion unit to the ground bus. The connection should exhibit very low resistance. Connect the subpanel ground bus to a single-point ground, such as a copper bus bar to a good earth-ground reference. There must be low impedance between each device and the single-point termination; a rule of thumb is less than 0.1 ohm DC resistance. This can be accomplished by removing the anodized finish and using copper lugs and star washers.

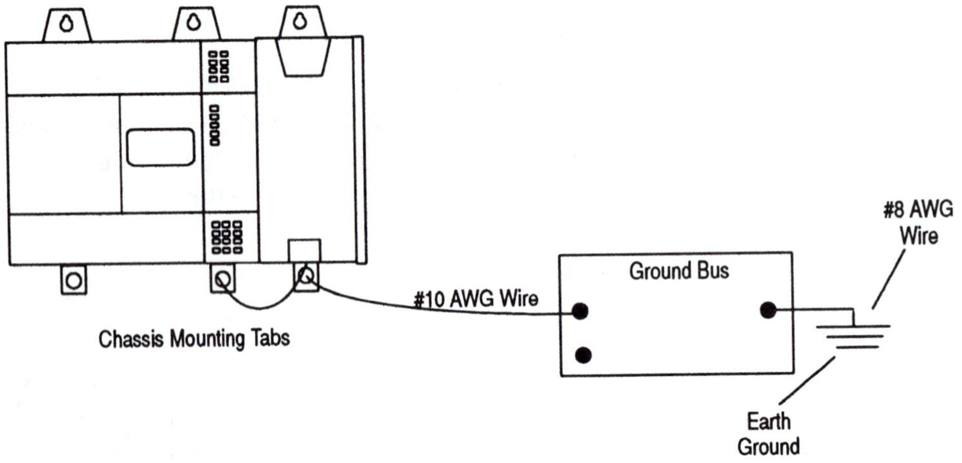

Figure 23–9 Grounding of an SLC-500 controller with a two-slot expansion chassis. *(Courtesy Rockwell Automation Inc.)*

PLC manufacturers will provide details in the hardware installation manual for their equipment. The National Electrical Code is an authoritative source for grounding requirements.

Grounding Guidelines Figure 23–10 shows examples of ground connections. Grounding braid and green wires should be terminated at both ends with copper eye lugs to provide good continuity. Lugs should be crimped and soldered. Copper no. 10 or 12 bolts work well for fasteners that are used to provide an electrical connection to the single-point ground. This applies to device mounting bolts and braid termination bolts for subpanel and user-supplied single-point grounds. Tapped holes rather than nuts and bolts should be used. Note that a minimum number of threads are also required for a solid connection.

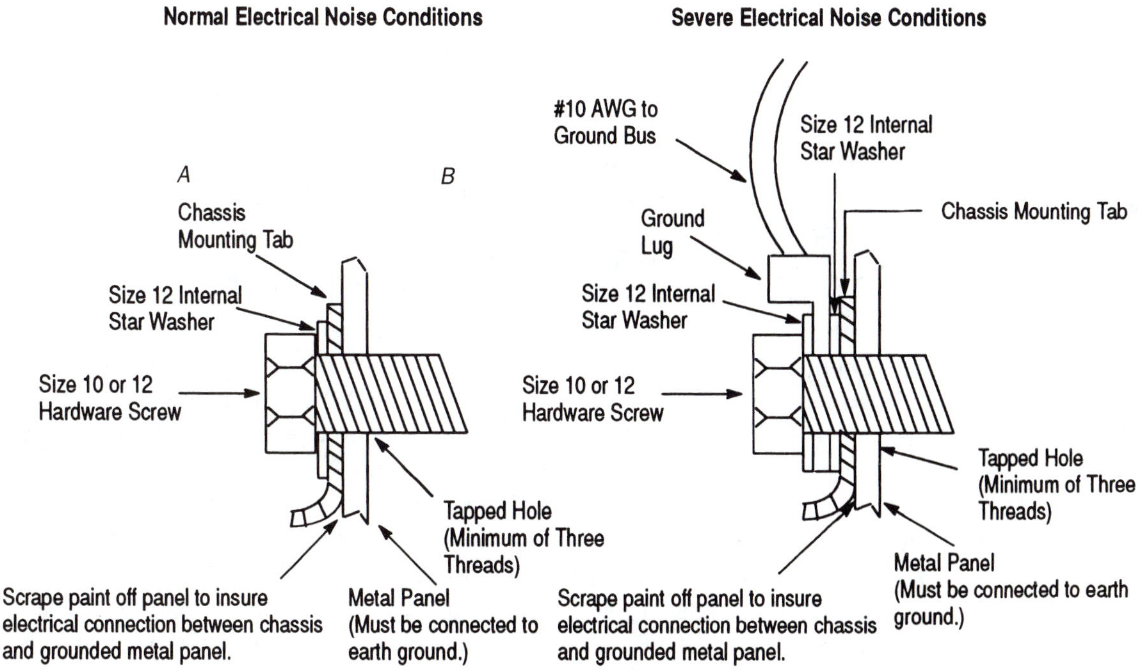

Figure 23–10 Ground connections. *(Courtesy Rockwell Automation Inc.)*

Paint, coatings, or corrosion must be removed from the areas of contact. Use external toothed lock washers (star washers). This practice should be used for all terminations: lug to subpanel, device to lug, device to subpanel, subpanel to conduit, and so on.

Check the appropriate electrical codes and ordinances to ensure compliance with minimum wire sizes, color coding, and general safety practices.

Handling Electrical Noise

Electrical noise is unwanted electrical interference that affects control equipment. The control devices in use today utilize microprocessors, which constantly fetch data and instructions from memory. Noise can cause the microprocessor to misinterpret an instruction or fetch bad data, which can cause minor problems or severe damage to equipment and people.

Noise is caused by a wide variety of manufacturing devices. Devices that switch high voltage and current are the primary sources of noise. Such devices include large motors and starters, welding equipment, and contactors that are used to turn devices on and off. Noise is not continuous; thus, it can be very difficult to find intermittent noise sources.

Noise can be created by power-line disturbances, transmitted noise, or ground loops. Power-line disturbances are generally caused by devices that have coils. When they are switched off, they create a line disturbance. Power-line disturbances can normally be overcome through the use of line filters. Surge suppressors such as MOVs or an RC network across the coil can limit the noise. Coil-type devices include relays, contactors, starters, clutches/brakes, and solenoids.

Transmitted noise is caused by devices that create radio frequency noise, in high-current applications. Welding causes transmitted noise. When contacts that carry high current open, they generate transmitted noise. Application wiring carrying signals can often be disrupted by this type of noise. Imagine wiring carrying sensor information to a control device. In severe cases, false signals can be generated on the signal wiring. This problem can often be overcome by using twisted-pair shielded wiring and connecting the shield to ground.

Transmitted noise can also “leak” into control cabinets. The holes that are put into cabinets for switches and wiring allow transmitted noise to enter the cabinet. The effect can be reduced by properly grounding the cabinet.

Ground loops can also cause noise, and they are often difficult to find because they produce intermittent noise. They generally occur when multiple grounds exist. The farther the grounds are apart, the more likely the problem. A potential problem can exist between the power supply earth and the remote earth, which can create unpredictable results, especially in communications.

Proper installation technique can avoid problems with noise. The two main ways to deal with noise are suppression and isolation.

Noise Suppression Suppression attempts to deal with the device that is generating the noise. A very high-voltage spike is caused when the current to an inductive load is turned off.

Inductive devices include relays, solenoids, motor starters, and motors. Suppression is even more important if the inductive device is in series or parallel with a hard contact such as a push button or switch.

This high voltage can cause trouble for the device PLC. Lack of surge suppression can contribute to processor faults and intermittent problems. Noise can also corrupt RAM and may cause intermittent problems with I/O modules. Note that many of these problems are hard to troubleshoot because of their sporadic nature. Excessive noise also can significantly reduce the life of relay contacts. Some PLC modules include protection circuitry to protect against inductive spikes. A suppression network can be installed to limit the voltage spikes. Surge suppression circuits connect directly across the load device to help reduce the arcing of output contacts.

Figure 23–11 shows how AC and DC loads can be protected against surges. Noise suppression is also called snubbing, which can be used to suppress the arcing of mechanical contacts caused by turning inductive loads off (Figure 23–12). Surge suppression should be used on all coils.

An RC or a varistor circuit can be used across an inductive load to suppress noise (1000 ohm, 0.2 microfarad). These components must be sized appropriately to meet the characteristic of the inductive output device that is being used. Check the installation manual for the PLC that you are using for proper noise suppression. A diode is sufficient for DC load devices. A 1N4004 can be used in most applications. Surge suppressors also can be used.

Noise Isolation Besides suppression, the other way to deal with noise is isolation. The device or devices that cause trouble are physically separated from the control system. The enclosure also helps separate the control system from noise. In many cases field wiring must be placed in very noisy environments to allow sensors to monitor the process. This presents a problem especially when low voltages are used. Shielded twisted-pair wiring should be used for the control wiring in these cases. The shielding should be grounded only at one end. The shield should be grounded at the single-point ground.

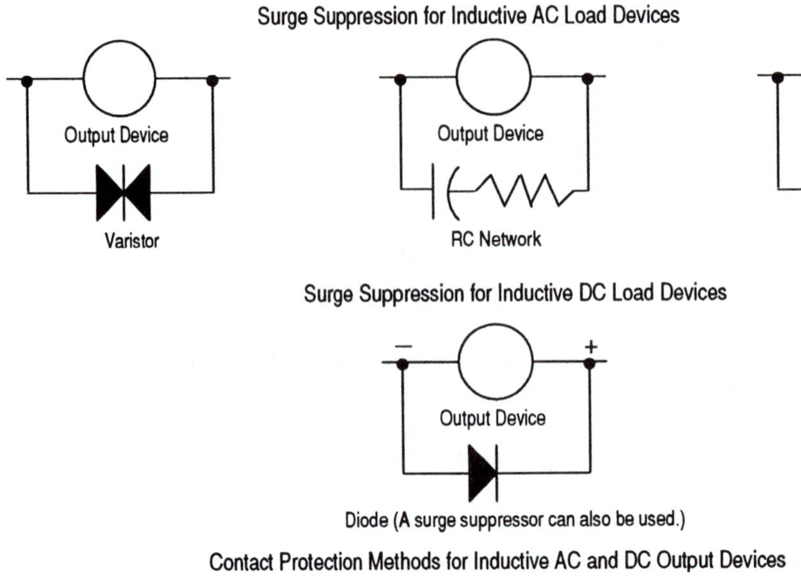

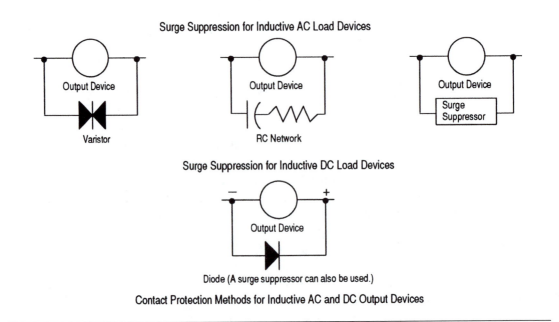

Figure 23–11 Surge suppression methods for AC and DC loads. *(Courtesy Rockwell Automation Inc.)*

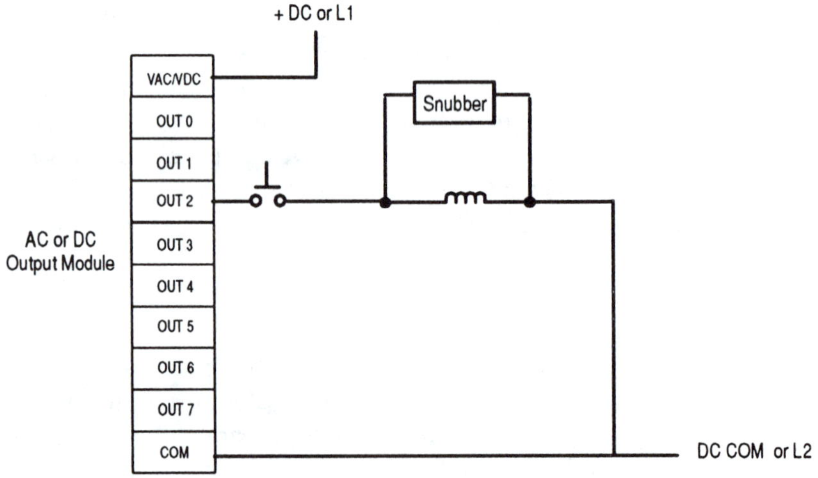

Figure 23–12 Example of snubbing. *(Courtesy Rockwell Automation Inc.)*

You can reduce the effects of noise by using one of the following methods:

Properly mount the PLC within a suitable enclosure.

Properly route all wiring.

Properly ground all equipment.

Install suppression to noise-generating devices.

Industrial Controller Maintenance

Industrial controllers are designed to be reliable devices. They are used in automated systems that are costly to operate. Downtime is costly in industry. The technician will be expected to keep the systems in operation and downtime to a bare minimum. Although industrial controllers have been designed to be low maintenance devices, certain tactics will help to reduce downtime.

It is important to keep the controllers clean. Industrial controllers are normally mounted in enclosures, which are typically air-cooled. A fan mounted in the wall of the enclosure circulates fresh air through the cabinet to cool the components. This enclosure can lead to the accumulation of dust, dirt, and other contaminants, which can cause short-circuits or intermittent problems in electronic equipment. The fans for such enclosures must have adequate filters that are cleaned regularly. A preventive maintenance schedule will include a check of the enclosure to make sure the inside is free of contamination. This check should also include an inspection for loose wires or termination screws that could cause problems later. Vibration can loosen screws and cause intermittent or permanent problems. Modules should also be checked to make sure they are securely seated in the backplane, especially in high-vibration environments.

Many control devices have battery backup. Long-life lithium batteries are a good choice—they can have lifetimes of two or more years; however, batteries will fail eventually. They will probably fail when least convenient. Their failure can cause unnecessary system downtime, which is often measured in hundreds or thousands of dollars. For this reason, regular battery replacement should be part of the preventative maintenance schedule. Replacing the batteries once a year is a cheap investment to avoid costly downtime and potentially larger problems.

Companies must maintain an inventory of spare parts to minimize downtime. With the high cost of downtime, there is no reason to spend additional time repairing boards or other components. The technician must be able to find and correct the problem in a minimum amount of time, which means quick fault isolation and component replacement. It is wise to keep approximately one spare for each of the ten most-used devices, including each type of input and output module, CPU, and other communication and special-purpose modules. Spare parts must also be maintained for sensors, drives, and other crucial devices. In many cases, modules may be returned to the manufacturer for repair. Having spare parts ensures that production can resume while the defective one is returned for repair. A reasonable selection of spare parts can drastically reduce downtime, firefighting, and frustration.

PLC Troubleshooting

The first consideration in troubleshooting and maintaining systems is safety. When you encounter a problem, remember that less than one-third of all system failures will be due to the PLC. Approximately 50 percent of the failures are due to input and output devices.

Several years ago, a technician was killed when he isolated the problem to a defective sensor. He bypassed the sensor and the system restarted with him in it. He was killed by the system he had fixed. You must always be aware of the possible outcomes of changes you make.

Troubleshooting is actually a straightforward process in automated systems. The first step is to think. This may seem rather basic, but many people jump to improper, premature conclusions and waste time finding problems. The first step is to examine the problem logically. Think the problem through using common sense first. This will point to the most logical cause. Troubleshooting is much like the game 20 Questions. Every question should help isolate the problem. In fact, every question should eliminate about half of the potential causes. Remember, a well-planned job is half done.

Think logically.

Ask yourself questions to isolate the problem.

Test your theory.

Next use the resources you have available to check your theory. Often the error-checking that is present on the PLC modules is sufficient. The LEDs on PLC CPUs and modules can provide immediate feedback on what is happening. Many PLCs have LEDs to indicate blown fuses and other problems. Check these indicators first.

The usual problem is that an output is not turning on when it should. There are several possible causes. The output device could be defective. The PLC output that turns it on could be defective. One of the inputs that allow, or cause, the output to turn on could be defective. The sensor or the PLC input that it is attached to could be defective. The ladder logic could even be faulty. It is possible that a ladder performs perfectly the vast majority of the time but fails under certain conditions. Again the module I/O LEDs provide the best source of answers (see Figure 23–13). If the PLC module output LED is on for that output, the problem is probably not the inputs to the PLC. The device is defective, the wiring is defective, or the PLC output is defective.

Output Condition	Output LED	Status in Ladder	Probable Problem
On	On	True	None
Off	Off	True	Bad fuse or bad output module
Off	Off	False	None
Off	On	True	Wiring to output device or bad output device

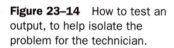

Figure 23–13 Isolating a PLC problem by comparing the states of inputs/outputs, indicators, and the ladder status.

Figure 23–14 How to test an output, to help isolate the problem for the technician.

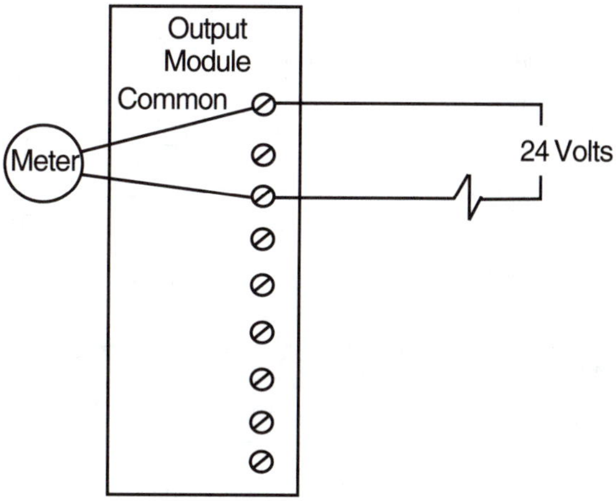

The next step is to isolate the problem further. A multimeter is invaluable at this point. If the PLC output is off, a meter reading should show the full voltage with which the device is turned on. If the output is used to supply 115 V to a motor starter, the meter should show the full 115 volts between the output terminal and common on the PLC module (Figure 23–14).

If the PLC output is on, the meter should read 0 volts because the output acts as a switch. If we measure the voltage across a switch, we should read 0 volts. If the switch is open, we should read the full voltage. If we read no voltage in either case, the wiring and power supply should be checked. If the wiring and power supply are good, the device is the problem. Depending on the device, there may be fuses or overload protection present.

Now pretend that the output LED is not turning on, so we must check the input side. If the input LED is on, we can assume that the sensor, or other input device, is operational (Figures 23–15 and 23–16). Next we must see if the PLC CPU really sees the input as true. At this point a monitor, such as a handheld programmer or a computer, is often used. The ladder is then monitored under operation. Note that many PLCs allow the outputs to be disabled for troubleshooting, which is a safe

Actual Input Condition	Actual Module LED Status	Ladder Status		Probable Problem
Off	Off	False –| |–	True –|/|–	None
Off	On	True –| |–	False –|/|–	Short in the input device or wiring or a bad input module
On	Off	False –| |–	True –|/|–	Wiring/power to I/O module or I/O module
On	On	False –| |–	True –|/|–	I/O module
On	On	True –| |–	False –|/|–	None

Figure 23–15 Chart for troubleshooting inputs.

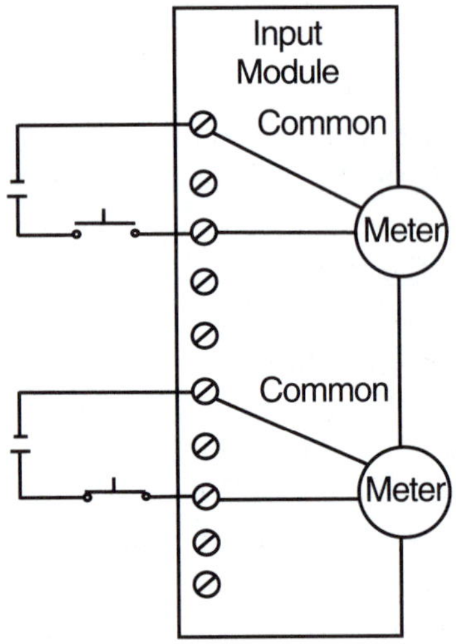

Figure 23–16 How to isolate input problems.

procedure. Check the ladder to see if the contact is closing. If it is seeing the input as false, the problem could be a defective PLC module input. Figures 23–17 and 23–18 show troubleshooting for input and output modules.

Potential Ladder Diagram Problems

Some PLC manufacturers allow output coils to be used more than once in ladders. This means that multiple conditions can control the same output coil. Sometimes technicians feel sure the output module is defective because when they monitor the ladder, the output coil is on but the actual output LED is off. In this case, the

Input LED State	Real-World State of Device	Condition	Cause	Action
Off	Off	Program operates as if the input is on.	Input is forced on in the program.	Check the forced I/O on the processor and remove forces, verify wiring. Try another input or replace the module.
			Input circuit is damaged.	Verify wiring. Try another input or replace the module.
	Off	Input device will not turn on.	The input device is damaged.	Verify device operation or replace if defective.
On	On	Input device will not turn off.	Device is damaged or shorted.	Verify device operation or replace if defective.
		Program operates as if the input device is off.	Input is forced off in program.	Check the forced I/O on the processor and remove forces.
			Damaged input circuit.	Verify the wiring. Try another input or replace the module.
	Off	Program operates as though the input is on and/or the input circuit will not turn off.	Damaged input circuit.	Verify the wiring. Try another input or replace the module.
			Input device is damaged or shorted.	Verify device operation or replace if defective.
			The leakage current of the input device exceeds the input circuit specification.	Use load resistor to bleed off current.

Figure 23–17 Input troubleshooting.

Output LED State	Real-World State of Device	Condition	Cause	Action
Off	On	Output device will not turn off but the program indicates it is off.	Incorrect wiring.	Check the wiring. Disconnect the device and test.
			Output device is damaged or shorted.	Verify device; replace if necessary.
			Output circuit is defective.	Check the wiring. Try another output circuit; replace module if possible.
	Off	Program shows that the output circuit is on or the output circuit will not turn on.	Damaged output circuit.	Use the force function to turn the output on. If the output turns on, there is a programming problem, If not, there is an output circuit problem. Try a different output circuit and replace the module if necessary.
			Programming problem.	Check for duplicate output addresses. If subroutines are used, outputs are in their last state when not executing subroutines. Use the force function to force the output on. If the output does not turn on, the output circuit is damaged. Try a different output circuit and replace the module if necessary. If the output does force on, check for a programming problem.
			The output is forced off in the program.	Check the processor forced I/O or force LED and remove forces.
On	On	The program shows the output circuit is off or the output circuit will not turn off.	Programming problem.	Check for duplicate output addresses. If subroutines are being used, outputs are in their last state when not executing subroutines. Use the force function to force the output off. If the output does not turn off, the output circuit is damaged. Try a different output circuit and replace the module if necessary. If the output does force off, check for a logic or programming problem.
			Damaged output circuit.	Use the force function to turn the output off. If the output turns off, there is a programming problem. If not, there is an output circuit problem. Try a different output circuit and replace the module if necessary.
			Output is forced on in the program.	Check the processor forced I/O or force LED and remove forces.
	Off	Output device will not turn on but the program indicates it is on.	Incorrect wiring or an open circuit.	Check the wiring and the common connections.
			Defective output circuit.	Try a different output circuit and replace the module if necessary.
			Low or no voltage across the load.	Measure the source voltage.
			Output device incompatibility.	Check the specifications.

Figure 23–18 Output troubleshooting *(Concept courtesy of Rockwell Automation Inc.)*

technician inadvertently programmed the same output coil twice with different input logic. The rung being monitored was true, but one farther down in the ladder was false, so the PLC kept the output off.

The other potential problem with ladder diagramming is that the problem can be intermittent. Timing is crucial in ladder diagramming. Devices are often exchanging signals to signify that an action has occurred. For example, a robot finishes its cycle and sends a digital signal to the PLC to let it know. The programmer of the robot must be aware of the timing considerations. If the robot programmer turns it on for one step, the output may be on for only a few milliseconds. The PLC may be able to "catch" the signal every time, or it may occasionally miss it because of the length of the scan time of the ladder diagram.

This can be a tough problem to find because it happens only occasionally. The better way to program is to handshake instead of rely on the PLC to see a periodic input. The programmer should have the robot output stay on until it is acknowledged by the PLC. The robot turns on the output and waits for an input from the PLC to ensure that the PLC saw the output from the robot.

SUMMARY

The installation of a control system must be carefully planned. People and devices must be protected. Hardwired switching should be provided to drop all power to the system. Lockouts should be provided to ensure safety during maintenance. Proper fusing to ensure protection of individual devices is a must. The cabinet must meet the needs of the application environment. Control and power wiring should be separated to reduce noise. Proper grounding procedures must be followed to ensure safety.

Troubleshooting is done in a logical way. Think the problem through. Ask questions that will help isolate the potential problems. Above all, apply safe work habits while working on systems.

QUESTIONS

1. What is a NEMA enclosure?
2. Why should enclosures be used?
3. Describe how an enclosure is chosen.
4. What is a fusible disconnect?
5. What is a contactor?
6. What is the purpose of an isolation transformer?
7. What is the major cause of failure in systems?
8. Describe a logical process for troubleshooting.
9. Describe proper grounding techniques.

10. Draw a block diagram of a typical control cabinet.
11. Describe at least three precautions that should be taken to help reduce the problem of noise in a control system.
12. A technician has been asked to troubleshoot a system. The output device is not turning on for some reason. The output LED is working as it should. The technician turns the output on and places a meter over the PLC output, which reads 115 V. The device is not running. What is the most likely problem? How might it be fixed?
13. A technician has been asked to troubleshoot a system. The PLC does not seem to be receiving an input because the output it controls is not turning on. The technician notices that the input LED is not turning on but that the output indicator LED on the sensor seems to be working. A meter placed across the input with the sensor on reads 24 V, but the input LED is off. What is the most likely problem?
14. A technician is asked to troubleshoot a system. An input seems to be defective. The technician notices that the input LED is never on. A meter placed across the PLC input reads 0 volt. The technician removes and tests the sensor. (The LED on the sensor comes on when the sensor is activated.) The sensor is fine. What is the most likely problem?
15. A technician is asked to troubleshoot a system. An output device is not working, but the technician notices that the output LED seems to be working. A meter placed over the PLC output with the output on reads 0 volt. Describe what the technician should do to find the problem.
16. What items should be included in system documentation?

appendix A

Input and Output Device Symbols

Input Device Symbols

	Normally Open		Normally Open - Held Closed
	Normally Closed		Normally Closed - Held Open

Liquid Level Switches

	Closes on Rise		Opens on Rise

Temperature Switches

	Closes on Rise		Opens on Rise

Pressure or Vacuum Switches

	Closes on Increase		Opens on Increase

Flow Switches

	Closes on Increase		Closes on Decrease

Time Delay Contacts

	Normally Open Time Delay Opening		Normally Closed Time Delay Closing
	Normally Open Time Delay Closing		Normally Closed Time Delay Opening

Output Device Symbols

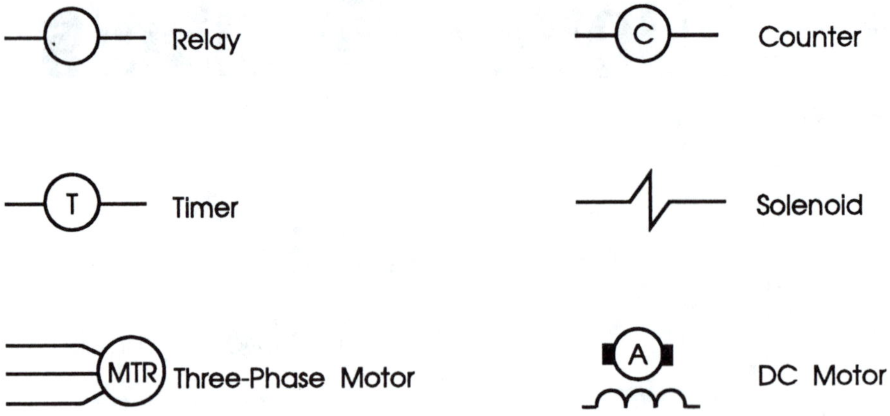

appendix B

Control Logic Addressing and Project Organization

ControlLogix Addressing

ControlLogix memory and addressing may be confusing at first for those who are already familiar with PLC addressing. The major difference is that there are no predefined data tables in the ControlLogix processor. There are, however, predetermined types of data. With ControlLogix, you define tags instead of addresses. This is certainly the future of programming. Tags can be created before or after you develop the logic. Question marks (?) can be entered in place of the tag while programming and then the tag can be defined later. Once you are familiar with tag addressing, you may like it more than traditional PLC memory addressing.

Tags

The naming of tags follows IEC 61131-3 rules. Tagnames must begin with an alphabetic character (A_Z or a-z) or an underscore. They may be up to forty characters long. Tagnames are not case sensitive. They cannot have consecutive or trailing underscores.

Tag Types

There are three types of tags: base, alias, and consumed. A base tag is the most common. A base tag defines the memory location where a data element is stored. An alias tag references a memory location that has been defined by another tag. An alias tag can reference a base tag or another alias tag. In other words, you can have more than one name for a specific tag. You can have a base tag and then other names to reference it. A consumed tag references data that comes from another controller.

The following information must be specified to define a tag:

Name	The tagname.
Description	This is optional. Descriptions can be up to 120 characters long. They are stored in the offline project file but are not downloaded to the controller.
Tag Type	The user chooses base, alias, or consumed.
Data Type	The user selects the type from a list of predefined types or any user-defined types. If the tag will be an array, the user must specify the number of elements in up to three dimensions.
Scope	The user selects the scope in which the tag is valid. The user can make the tag scope the whole controller or any of the existing program scopes. Note that a controller can have multiple programs.
Display Style	This defines the display type when the user is monitoring the tag in the programming software. The user chooses from the available types.
Produce This Tag	The user selects whether or not to make the tag available to other controllers. The user must also specify the number of other controllers that can consume the tag.

Data Types

ControlLogix has two predefined data types: basic and structure types. The basic types conform to the IEC61131-3 standard.

Basic Types Basic types are also called atomic data types. They are listed below:

BOOL	1-bit Boolean
SINT	1-byte integer
INT	2-byte integer

DINT	4-byte integer
REAL	4-byte floating point

Structure Types

AXIS	Control structure for an axis (controller-only tag)
CONTROL	Control structure for arrays
COUNTER	Control structure for counters
MOTION_INSTRUCTION	Control structure for motion instructions
PID	Control structure for a PID instruction
TIMER	Control structure for a timer
MESSAGE	Control structure for a message instruction (controller-only tag)
MOTION GROUP	Control structure for a motion group (controller-only tag)

Base Tags

All data in the controller is stored in a minimum of 4 bytes (32 bits). The types that can be used to define base tags are BOOL, SINT, INT, DINT, and REAL.

A BOOL (Boolean) base tag is stored in 4 bytes. A BASE tag's bit is stored in bit 0 of the 32 bits reserved for the tag. Bits 1–31 are not used.

A SINT (single integer) base tag consists of 8 bits. Its value is stored in bits 0–7 (the low byte) of the 32 bits reserved for the tag. Bits 8–31 are not used.

An INT base tag consists of 16 bits and uses bits 0–15 (the lower 2 bytes) of the 32 bits reserved for the tag. Bits 16–31 are not used.

DINT (double integer) and REAL base tags use all 32 bits (4 bytes) of the memory reserved for them.

Structures

There are three types of structures: predefined, module defined, and user defined. Structures are used for groups of related data. Structures can contain multiple data types. An address book is like a structure. Many friends can be listed in an address book. For each friend there are several pieces and types of information: a phone number, street address, e-mail address, name, maybe even things like birthdays, and so on. Some structures are predefined to make it easy for the programmer. Some were listed earlier.

Controller module-defined structures are created automatically when you configure I/O modules for the system. These structures contain information such as data, status, and fault information.

The user can define (create) structures that group different types of data into a single named entity, much like an address book groups similar data. User-defined structures contain one or more data definitions called members. The data type determines the amount of memory that is allocated for each. The data type for each member can be a basic type, a user-defined structure, a predefined structure, a single dimension array, a single-dimension array of a predefined structure, or a single dimension of a user-defined structure.

Creating a User-Defined Structure

To create a structure, the user first enters a name and may enter a description (optional). The following parameters must be entered for each member:

Name	Name the member.
Data Type	The correct type is selected from the list.
Style	Select the desired style of the member from the list.
Description	Describe the member. This is optional.

Figure B–1 shows an example of a structure. This user-created structure will hold a bit (BOOL), a single integer (SINT), and a real (REAL) number. The figure also shows how they are stored in memory.

User-Defined Structure
Name: DATA

Member	Data Type
Inp_1	BOOL
Temp	SINT
Pressure	REAL

Memory Allocation for the Data

	Bits 24–31	Bits 36–23	Bits 5–15	Bits 0–7
Bit_0	Unused	Unused	Unused	Bit 0 used
Temp	Unused	Unused	Unused	First 8 bits
Pressure	Alsl 32 bit			

Figure B–1 Structure example.

Member Addressing

Members within a structure are addressed by using the tagname and a period, followed by the member name. For example, tag_name.member_name. If a structure is within a structure, the tagname of the highest level is followed by the substructure tagname, followed by the member name. For example: tag_name.structure_name.member_name.

Arrays

A user can utilize arrays to group sets of data, of the same type, in a contiguous block of controller memory. Note that arrays consist of one type of data; structures can hold multiple types. For example, an address book holds many types of data, so it is a structure. A table of temperatures can be an array because it holds one type of data. Each tag within the array is one element of the array. The element can be a basic type or a structure type. Element numbering in arrays starts with 0.

Figure B–2 shows an example of a one-dimensional array. This array was created to hold five temperatures. Each element has an address. Temp[0] is the first address of the array and holds the first temperature. Temp[4] is the fifth element of the array and holds the fifth temperature. Note that the array was created with the name Temp. The array Temp was declared to be of type SINT (single integer) and has five elements. A particular element is addressed by using the array name followed by the number of the particular element in brackets, for example, Temp[2]. A variable can also be used to identify the particular element, for example, Temp[var1]. Whatever number is in var1 specifies the particular element of the

Array-Temp	
Data Type Sint[5]	
	Temp[0]
	Temp[1]
	Temp[2]
	Temp[3]
	Temp[4]

Figure B–2 A one-dimensional array to hold five temperature values.

Array-Machine_One			
Data Type DINT[5,2]			
Pressure[0,0]			Temp[0,1]
Pressure[1,0]			Temp[1,1]
Pressure[2,0]			Temp[2,1]
Pressure[3,0]			Temp[3,1]
Pressure[4,0]			Temp[4,1]

Figure B–3 A two-dimensional array with ten rows and two columns.

Temp array. Math expressions can also be used in array addressing, for example, Temp[count – 1]. This specifies the element that equals the value of count minus 1. If the present value of count is equal to 4, element 3 would be specified. Remember that this would be the fourth element because the first element is 0.

You can also specify the address for an element to the bit level. For example, Temp[1].2 specifies the third bit (bit 2) of Temp element 1 (the second element). You may also use variable addressing for bits, for example, Temp[1].count + 2.

If you are using variable addressing, you must be sure that the address is within the array boundaries. If it is outside the array, a major fault will occur.

Figure B–3 shows an example of a two-dimensional array. The first column is used to store pressure values and the second column is used to hold temperature values. Note that both have to be the same type. Note that the array was named Machine_One and the data type for the array was specified to be DINT[5,2]. This means that the array was created to have ten rows and two columns. The array will hold double integers (DINT).

Three-dimensional arrays can also be created.

ControlLogix Project Organization

To use a ControlLogix controller, create a project file. The project file contains the programming and configuration information. The following list shows the information that is entered by the user when creating a project:

- The chassis size/type
- The slot number of the processor

- A description (optional)
- A file path
- A project name (this is also the file name)

  The file name is also assigned to the controller name.
  You may change the controller name if you wish.
  If you save the file under a different name, the controller name does not change.
  A file name can consist of up to forty characters. It may be letters, numbers, and underscores. The file name cannot have consecutive or trailing underscores.

There are three major components in a project: tasks, programs, and routines. Tasks are used to configure the controller's execution. Programs are used to group logic and data. Routines contain the executable code. A project may have multiple tasks. Only one task executes at a time. Other tasks may interrupt the controller and execute. When the interrupting task is done, executing the original task is resumed. Only one task can operate at a time. Only one program within a task can operate at one time.

Tasks

Tasks schedule and prioritize programs or a group of programs that are assigned to that task. There are two types of tasks: periodic and continuous. The Logix5550 controller can support thirty-two tasks. There can be only up to one continuous task. If there is a continuous task, there can be only thirty-one periodic tasks (total tasks = 32). Tasks are assigned priority numbers to determine which task has priority if multiple tasks are called. A continuous task always has the lowest priority; 1 is the highest priority and 15 is the lowest. A higher-priority task interrupts a lower-priority task.

Periodic tasks are triggered at a repetitive time period. The interruption can be set from between 1 ms to 2000 milliseconds. The default time is 10 milliseconds. Periodic tasks that have the same priority execute on a time-slicing scheme at 1-millisecond intervals.

Programs

A task may have up to thirty-two programs assigned to it. Each program can have its own routines and tags. If a task is triggered, all of the programs associated with the task will execute in the order in which they are grouped. A program can be assigned to only one task in the controller.

Routines

A routine is a set of logic instructions that is written in a single programming language. Remember that IEC 61131-3 specifies different languages, and ControlLogix can utilize some of them. The main program in an SLC 500 is similar to a routine. It consists of one language (ladder logic). Logic in the ladder logic (jump to subroutines [JSRs]) can be used to call other subroutines (routines).

Glossary

absolute pressure: The pressure of a gas or liquid measured in relation to a vacuum (zero pressure).

accumulated value: Applies to the use of timers and counters. The present count or time.

accuracy: The deviation between the actual position and the theoretical position.

acidity: A measure of the hydrogen ion content of a solution.

AC input module: A module that converts a real-world AC input signal to the logic level required by the PLC processor.

AC output module: Module that converts the processor logic level to an AC output signal to control a real-world device.

action: Refers to the action of a controller. Defines what is done to regulate the final control element to effect control. Types include on-off, proportional, integral, and derivative.

actuator: Output device normally connected to an output module. An example is an air valve and cylinder.

address: Number used to specify a storage location in memory.

alkalinity: A measure of the hydroxyl ion content of a solution.

ambient temperature: Temperature that naturally exists in the environment. For example, the ambient temperature of a PLC in a cabinet near a steel furnace is very high.

analog: Signal with a smooth range of possible values. For example, a temperature that can vary between 60 and 300 degrees is analog in nature.

ANSI: American National Standards Institute.

ASCII: American Standard Code for Information Interchange. A coding system used to represent letters and characters. Seven-bit ASCII can represent 128 different combinations. Eight-bit ASCII (extended ASCII) can represent 256 different combinations.

asynchronous communications: Method of communications that uses a series of bits to send data between devices. There is a start bit, data bits (7 or 8), a parity bit (odd, even none, mark, or space), and stop bits (1, 1.5, or 2). One character is transmitted at a time. RS-232 is the most common.

atmospheric pressure: The pressure exerted on a body by the air. It is equal to 14.7 pounds per square inch at sea level.

backplane: Bus in the back of a PLC chassis. It is a printed circuit board with sockets that accept various modules.

batch process: A process in which a given amount of material is processed in one operation to produce what is required.

baud rate: Speed of serial communications. The number of bits per second transmitted. For example, RS-232 is normally used with a baud rate of 9600, or about 9600 bits per second. It takes about 10 bits in serial to send an ASCII character, so a baud rate of 9600 transmits about 960 characters per second.

bellows: A pressure-sensing element consisting of a metal cylinder that is closed at one end. The difference in pressure between the inside and outside of the cylinder causes the cylinder to expand or contract its length.

beta ratio: Diameter of orifice divided by the internal diameter of the pipe.

BEUG:BITBUS European Users Group A nonprofit organization devoted to spreading the BITBUS technology and organizing a basic platform where people using BITBUS can share application experiences.

binary: Base two number system. Binary is a system in which 1s and 0s are used to represent numbers.

binary-coded decimal (BCD): A number system. Each decimal number is represented by four binary bits. For example, the decimal number 967 is represented by 1001 0110 0111 in BCD.

bit: Binary digit. The smallest element of binary data. A bit will be either a 0 or a 1.

BITBUS: BITBUS was created by Intel in 1983. It is one of the most widely used fieldbuses. It was promoted as a standard in 1990 by a special committee of the IEEE (standard IEEE-1118 1990).

boolean: Logic system that uses operators such as AND, OR, NOR, and NAND. This system is utilized by PLCs, although it is usually made invisible by the programming software for the ease of the programmer.

bounce: The erratic make and break of electrical contacts (an undesirable effect).

branch: Parallel logic path in a ladder diagram.

byte: Eight bits or two nibbles. A nibble is 4 bits.

calibration: A procedure used to determine, correct, or check the absolute values corresponding to the graduations on a measuring instrument.

cascade: Programming technique used to extend the range of timers and counters.

celsius: The centigrade temperature scale on which the freezing point of water is 0 degrees and the boiling point is 100 degrees.

CENELEC: European Committee for Electrotechnical Standardization. It develops standards for dimensional and operating characteristics of control components.

central processing unit (CPU): Microprocessor portion of the PLC. It is the portion of the PLC that handles the logic.

closed loop: A closed-loop system has feedback and can automatically correct for errors.

CMOS: Complementary metal-oxide semiconductor. Integrated circuits that consume very little power and also have good noise immunity.

cold junction: The point at which a pair of thermocouple wires is held at a fixed temperature. It is also called the reference junction.

color mark sensor: Sensor designed to differentiate between two different colors. They differentiate on the basis of contrast between the two colors.

compare instruction: PLC instruction used to test numerical values for equal, greater than, or less than relationships.

complement: The inverse of a digital signal.

contact: Symbol used in programming PLCs. Used to represent inputs. There are normally open and normally closed contacts. Contacts are also the conductors in electrical devices such as starters.

contactor: Special-purpose relay used to control large electrical current.

CSA: Canadian Standards Organization. Develops standards, tests products, and provides certification for a wide variety of products.

current sinking: Refers to an output device (typically an NPN transistor) that allows current flow from the load through the output to ground.

current sourcing: Output device (typically a PNP transistor) that allows current flow from the output through the load and then to ground.

cyclic redundancy check (CRC): A calculated value based on the content of a communication frame. It is inserted in the frame to enable a check of data accuracy after receiving the frame across a network. BITBUS uses the standard SDLC CRC.

dark-on: Refers to a photosensor's output. If the sensor output is on when no object is sensed, it is called a dark-on sensor.

data highway: A communications network that allows devices such as PLCs to communicate. They are normally proprietary, which means that only like devices of the same brand can communicate over the highway. Allen Bradley calls their PLC communication network Data Highway.

data table: A consecutive group of user references (data) of the same size that can be accessed with table read/write functions.

debugging: Process of finding problems (bugs) in any system.

derivative gain: The derivative in a PID system acts based on the rate of change in the error. It has a damping effect on the proportional gain.

DeviceNet: An industrial bus for field level devices. It is based on the controller area network (CAN) chip.

diagnostics: Devices normally have software routines that aid in identifying and finding problems in the device. They identify fault conditions in a system.

digital output: An output that can have two states: on or off. Also called discrete outputs.

distributed processing: The concept of distributed processing allows individual discrete devices to control their area and still communicate to the others via a network. The distributed control takes the processing load off the host system.

documentation: Descriptive paperwork that explains a system or program. It describes the system so that the technician can understand, install, troubleshoot, maintain, or change the system.

downtime: The time a system is not available for production or operation. Downtime can be caused by breakdowns in systems.

EEPROM: Electrically erasable programmable read only memory.

energize: Instruction that causes a bit to be a 1. This turns an output on.

examine-off: Contact used in ladder logic. It is a normally closed contact. The contact is true (or closed) if the real-world input associated with it is off.

examine-on: Contact used in ladder logic programming. Called a normally open contact. This type of contact is true (or closed) if the real-world input associated with it is on.

expansion rack: A rack added to a PLC system when the application requires more modules than the main rack can contain. A remote rack is sometimes used to permit I/O to be located remotely from the main rack.

false: Disabled logic state (off).

fault: Failure in a system that prevents normal operation of a system.

firmware: A series of instructions contained in read-only memory (ROM) that are used for the operating system functions. Some manufacturers offer upgrades for PLCs. This is often done by replacing a ROM chip. Thus, the combination of software and hardware lead to it being called firmware.

flowchart: Used to make program design easier.

force: Refers to changing the state of actual I/O by changing the bit status in the PLC. In other words, a person can force an output on by changing the bit associated with the real-world output to a 1. Forcing is normally used to troubleshoot a system.

frame: Packet of bits that will be transmitted across a network. A frame contains a header, user data, and an end of frame. The frame must contain all the necessary information to enable the sender and receiver(s) of the communication to decode the user's data and to ensure that this data is right.

full duplex: Communication scheme where data flows in both directions simultaneously.

ground: Direct connection between equipment (chassis) and earth ground.

half duplex: Communication scheme where data flows in both directions but in only one direction at a time.

hard contacts: Physical switch connections.

hard copy: Printed copy of computer information.

HDLC: High-level data link control. Standard protocol of communication oriented in message transmission (frames). The user's data field in an HDLC-frame can be of a free number of bits. The SLDC is a subset of the HDLC that defines the whole protocol in more detail and is byte-oriented.

hexadecimal: Numbering system that utilizes base 16.

host computer: One to which devices communicate. The host may download or upload programs, or the host might be used to program the device. An example is a PLC connected to a microcomputer. The host (microcomputer) "controls" the PLC by sending programs, variables, and commands. The PLC controls the actual process but at the direction and to the specifications of the host.

hysteresis: A dead band that is purposely introduced to eliminate false reads in the case of a sensor. In an encoder, hysteresis is introduced in the electronics to prevent ambiguities if the system happens to dither on a transition.

IEC: International Electrotechnical Commission. Develops and distributes recommended safety and performance standards.

IEEE: Institute of Electrical and Electronic Engineers.

image table: Area used to store the status of input and output bits.

incremental: Typically refers to encoders. Encoders provide logic states of 0 and 1 for each successive cycle of resolution.

instruction set: Instructions available to program the PLC.

integral gain: The integral gain in a PID system corrects for small errors over time. The proportional gain cannot correct for small errors. The integral gain eliminates offset and permanent error.

intelligent I/O: PLC modules that have a microprocessor built in. An example is a module that controls closed-loop positioning.

interfacing: Connection of a PLC to external devices.

I/O: Input/output. Used to speak about the number of inputs and outputs needed for a system, or the number of inputs and outputs that a particular programmable logic controller can handle.

IP rating: Rating system established by the IEC that defines the protection offered by electrical enclosures. It is similar to the NEMA rating system.

isolation: Used to segregate real-world inputs and outputs from the central processing unit. Isolation ensures that even with a major problem with real-world inputs or outputs (such as a short), the CPU is protected. This isolation is normally provided by optical isolation.

K: Abbreviation for the number 1000. In computer language, it is equal to 2^{10}, or 1024.

keying: Technique to ensure that modules are not put in the wrong slots of a PLC. The user sets up the system with modules in the desired slots. The user then keys the slots to ensure that only a module of the correct type can be physically installed.

ladder diagram: Programmable controller language that uses contacts and coils to define a control sequence.

LAN: *See* local area network.

latch: An instruction used in ladder diagram programming to represent an element that retains its state during controlled toggle and power outage.

leakage current: Small amount of current that flows through load-powered sensors. The small current is necessary for the operation of the sensor. The small amount of current flow is normally not sensed by the PLC input. If the leakage is too great, a bleeder resistor must be used to avoid false inputs at the PLC.

LED: Light-emitting diode. A solid-state semiconductor that emits red, green, or yellow light or invisible infrared radiation.

light-on sensor: Refers to a photosensor's output. If the output is on when an object is sensed, the sensor is a light-on sensor.

linear output: Analog output.

line driver: A differential output driver intended for use with a differential receiver. Usually used where long lines and high frequency are required and noise may be a problem.

line-powered sensor: Normally, three-wire sensors, although four-wire sensors also exist. The line-powered sensor is powered from the power supply. A separate wire (the third) is used for the output line.

load: Any device that current flows through and produces a voltage drop.

load-powered sensor: Has two wires. A small leakage current flows through the sensor even when the output is off. The current is required to operate the sensor electronics.

load resistor: A resistor connected in parallel with a high-impedance load to enable the output circuit to output enough current to ensure proper operation.

local area network (LAN): A system of hardware and software designed to allow a group of intelligent devices to communicate within a fairly close proximity.

lockout: The placement of a lockout device on an energy-isolating device, in accordance with an established procedure, to ensure that the energy-isolating device and the equipment being controlled cannot be operated until the lockout device is removed.

lockout device: A device that utilizes a positive means such as a lock, either key or combination type, to hold an energy-isolating device in the safe position and prevent the energizing of a machine or equipment.

LSB: Least significant bit.

machine language: Control program reduced to binary form.

MAP: Manufacturing automation protocol. Standard developed to make industrial devices communicate more easily. Based on a seven-layer model of communications.

master: On a network, the device that controls communications traffic. The master of a network usually polls every slave to check if it has something to transmit. In a master-slave configuration, only the active master can place a message on the bus. The slave can reply only if it receives a frame from the master that contains a logical token that explicitly enables the slave to reply.

master control relay (MCR): Hardwired relay that can be de-energized by any hardwired series-connected switch. Used to de-energize all devices. If one emergency switch is hit, it must cause the master control relay to drop power to all devices. A master control relay is also available in most PLCs. The master control relay in the PLC is not sufficient to meet safety requirements.

memory map: Drawing showing the areas, sizes, and uses of memory in a particular PLC.

microsecond: A microsecond is one millionth (0.000001) of a second.

millisecond: A millisecond is one thousandth (.001) of a second.

mnemonic codes: Symbols designated to represent a specific set of instructions for use in a control program. An abbreviation given to an instruction, usually made by combining the initial letters or parts of words.

MSB: Most significant bit.

NEMA: National Electrical Manufacturers Association. Develops standards that define a product, process, or procedure. The standards consider construction, dimensions, tolerances, safety, operating characteristics, electrical rating, and so on. They are probably best known for their rating system for electrical cabinets.

network: System that is connected to devices or computers for communications purposes.

node: Point on the network that allows access.

noise: Unwanted electrical interference in a programmable controller or network. It can be caused by motors, coils, high voltages, welders, and so on. It can disrupt communications and control.

nonretentive coil: A coil that will turn off on removal of applied power to the CPU.

nonretentive timer: Timer that loses the time if the input enable signal is lost.

nonvolatile memory: Memory in a controller that does not require power to retain its contents.

NOR: The logic gate that results in 0 unless both inputs are 0.

NOT: The logic gate that results in the complement of the input.

octal: Number system based on the number 8 and utilizing numbers 0 through 7.

off-delay timer: A type of timer that is on immediately when it receives its input enable. It turns off after it reaches its preset time.

off-line programming: Programming that is done while not attached to the actual device. For example, a PLC program can be written for a PLC without being attached. The program can then be downloaded to the PLC.

on-delay timer: Timer that does not turn on until its time has reached the preset time value.

one-shot contact: Contact that is on for only one scan when activated.

open loop: Refers to a system that has no feedback and no autocorrection.

operating system: The fundamental software for a system that defines how it will store and transmit information.

optical isolation: Technique used in I/O module design that provides logic separation from field levels.

OR: Logic gate that results in 1 unless both inputs are 0.

parallel communication: A method of communications where data is transferred on several wires simultaneously.

parity: Bit used to help check for data integrity during a data communication.

peer-to-peer: Communication that occurs between similar devices. For example, two PLCs communicating are peer-to-peer. A PLC communicating to a computer is device-to-host.

PID (proportional, integral, derivative) control: Control algorithm used to control processes such as temperature, mixture, position, and velocity closely. The proportional portion takes care of the magnitude of the error. The integral takes care of small errors over time. The derivative compensates for the rate of error change.

PLC: Programmable logic controller.

PPR (pulses per revolution): Refers to the number of pulses an encoder produces in one revolution.

programmable controller: A special-purpose computer programmed in ladder logic. It is also designed so that devices can be interfaced with it easily.

proportional gain: Looks at the magnitude of error and attempts to correct a system.

pulse modulated: Turning a light source on and off at a high frequency. In sensors, the sending unit pulse modulates the light source. The receiver responds only to that frequency. Helps make photo sensors immune to ambient lighting.

quadrature: Two output channels out of phase with each other by 90 degrees.

rack: PLC chassis. Modules are installed in the rack to meet the user's need.

radio frequency (RF): Communications technology with a transmitter/receiver and tags. The transmitter/receiver can read or write to the tags. Active and passive tags are available. Active tags are battery powered. Passive tags are powered from the RF emitted from the transmitter. Active tags have a much wider range of communication. Either tag can have several K of memory.

RAM: Random access memory. Normally considered user memory.

register: Storage area typically used to store bit states or values of items such as timers and counters.

repeatability: The ability to repeat movements or readings. A robot's reliability is reflected in its ability to return to a position accurately time after time. Repeatability is unrelated to resolution and is usually three to ten times better than accuracy.

resolution: A measure of how closely a device can measure or divide a quantity. For example, in an encoder, resolution is defined as counts per turn. For an analog-to-digital card, it is the number of bits of resolution. For example, for a 12-bit card, the resolution is 4096.

retentive coil: A coil that remains in its last state even though power was removed.

retentive timer: Timer that retains the present count even if the input enable signal is lost. When the input enable is active again, the timer begins to count again from where it left off.

retroreflective: Photosensor that sends out a light that is reflected from a reflector back to the receiver (the receiver and emitter are in the same housing). When an object passes through, it breaks the beam.

RF: *See* radio frequency.

ROM: Read-only memory. Operating system memory. ROM is nonvolatile. It is not lost when the power is turned off.

RS-232: Common serial communications standard. This standard specifies the purpose of each of twenty-five pins. It does not specify connectors or which pins must be used.

RS-422 and RS-423: Standards for two types of serial communication. RS-422 is a balanced serial mode. The transmit and receive lines have their own common instead of sharing one like RS-232. Balanced mode is more noise immune and thus allows for higher data transmission rates and longer transmission distances. RS-423 uses the unbalanced mode. Its speeds and transmission distances are much greater than RS-232 but less than RS-422.

RS-449: Electrical standard for RS-422/RS-423. It is a more complete standard than the RS-232. It also specifies the connectors to be used.

RS-485: Similar to the RS-422 standard. Receivers have additional sensitivity, which allows for longer distances and more communication drops. Includes some extra protection for receiver circuits.

rung: Group of contacts that control one or more outputs. In a ladder diagram, it is the horizontal lines on the diagram.

scan time: Amount of time it takes a programmable controller to evaluate a ladder diagram. The PLC continuously scans the ladder diagram. The time it takes to evaluate it once is the scan time. It is typically in the low-millisecond range.

SDLC: Serial data link control, a subset of the HDLC used in a large number of communications systems like Ethernet, ISDN, BITBUS, and others. This protocol defines the structure of the frames and the values of several specific fields in these frames.

sensitivity: Refers to a device's ability to discriminate between levels. In a sensor it relates to the finest difference it can detect. In an analog module for a PLC, it is the smallest change it can detect.

sensor: Device used to detect change. Normally it is a digital device. The outputs of sensors change state when they detect the correct change. Sensors can be analog or digital. They can also be purchased with normally closed or normally open outputs.

sequencer: Instruction type used to program a sequential operation.

serial communication: Sending of data 1 bit at a time. The data is represented by a coding system such as ASCII.

slave: On a master-slave configured network, there is usually one master and several slaves. The slaves are nodes of the network that can transmit information to the master only when they are polled (called) from it. The rest of the time, a slave doesn't transmit anything.

speech modules: Used by a PLC to output spoken messages to operators. The sound is typically digitized human speech stored in the module's memory. The PLC requests the message number to play it.

tagout: The placement of a tagout device on an energy-isolating device, in accordance with an established procedure, to indicate that the energy-isolating device and the equipment being controlled may not be operated until the tagout device is removed.

tagout device: A prominent warning device, such as a tag and a means of attachment, that can be fastened securely to an energy-isolating device in accordance with an established procedure to indicate that the energy-isolating device and the equipment being controlled may not be operated until the tagout device is removed.

thermocouple: A thermocouple is a sensing transducer. It changes a temperature to a current. The current can then be measured and converted to a binary equivalent that the PLC can understand.

thumbwheel: Device used by an operator to enter a number between 0 and 9. Thumbwheels are combined to enter larger numbers. Thumbwheels typically output BCD numbers to a device.

timer: Instruction used to accumulate time until a certain value is achieved. The timer then changes its output state.

TOP: Technical and office protocol. Communication standard developed by Boeing. Based on the contention access method. The MAP standard is meant for the factory floor, and TOP is meant for the office and technical areas.

transducer: A device that changes one form of energy to another. For example, a tachometer changes a mechanical rotation into an electrical signal.

transitional contact: Contact that changes state for one scan when activated.

true: The enabling logic state. Generally associated with a "one" or "high" state.

UL: Underwriters Laboratory. Organization that operates laboratories to investigate the safety of systems.

user memory: Memory used to store user information. The user's program, timer/counter values, input/output status, and so on, are all stored in user memory.

volatile memory: Memory that is lost when power is lost.

watchdog timer: Timer that can be used for safety. For example, if an event or sequence must occur within a certain amount of time, a watchdog timer can be set to shut down the system if the time is exceeded.

word: Length of data in bits that a microprocessor can handle. For example, a word for a 16-bit computer would be 16 bits long, or 2 bytes. A 32-bit computer has a 32-bit word.

Index